T0413690

Traditional Forest-Related Knowledge

WORLD FORESTS

World Forests Description

As forests stay high on the global political agenda, and forest-related industries diversify, cutting edge research into the issues facing forests has become more and more transdisciplinary. With this is mind, Springer's World Forests series has been established to provide a key forum for research-based syntheses of globally relevant issues on the interrelations between forests, society and the environment.

The series is intended for a wide range of readers including national and international entities concerned with forest, environmental and related policy issues; advanced students and researchers; business professionals, non-governmental organizations and the environmental and economic media.

Volumes published in the series will include both multidisciplinary studies with a broad range of coverage, as well as more focused in-depth analyses of a particular issue in the forest and related sectors. Themes range from globalization processes and international policies to comparative analyses of regions and countries.

For further volumes:
http://www.springer.com/series/6679

John A. Parrotta • Ronald L. Trosper
Editors

Traditional Forest-Related Knowledge

Sustaining Communities, Ecosystems
and Biocultural Diversity

Editors
Dr. John A. Parrotta
U.S. Forest Service
Research and Development Branch
1601 North Kent Street
4th floor RP-C
Arlington, VA 22209
USA
jparrotta@fs.fed.us

Dr. Ronald L. Trosper
American Indian Studies
The University of Arizona
1103 E. Second Street
Tucson, AZ 85721
USA
rltrosper@email.arizona.edu

ISSN 1566-0427   e-ISSN 1566-0427
ISBN 978-94-007-2143-2   e-ISBN 978-94-007-2144-9
DOI 10.1007/978-94-007-2144-9
Springer Dordrecht Heidelberg London New York

Library of Congress Control Number: 2011937448

Springer is part of Springer Science+Business Media (www.springer.com)

*To those who came before and taught us
what they had learned.*

Preface

In this era of rapid population growth and ever-increasing consumption of natural resources, concern grows for the fate of the world's forests and their capacity to provide multiple goods and environmental services. Forest management worldwide is evolving from its earlier emphasis on wood production towards a broader consideration of multiple economic, environmental, cultural and social objectives. New approaches such as "sustainable forest management" and the "ecosystem approach" seek to achieve a balance between society's increasing demands for forest products and benefits and the preservation of forest health and biological diversity. This balance is increasingly recognized as critical to the survival of forests and to the prosperity of forest-dependent communities. Such approaches, however, are not new. Long before the birth of forest science and 'scientific' forest management, traditional societies – indigenous and local communities – have managed forests and associated ecosystems in ways that sustained their livelihoods and cultures without jeopardizing the capacity of forest ecosystems to provide for future generations.

Forest managers, policy-makers, and the scientific community have too long ignored, denigrated, and even suppressed the knowledge and experiences of traditional societies. Holders and users of traditional knowledge face an uphill battle in most parts of the world to protect their lands, their practices and institutions, and their cultural identities in the face of injustice and an array of political, economic, social, cultural, and environmental pressures. Growing awareness of the importance of broader environmental, social and cultural values of forests supports increased recognition of traditional forest-related knowledge. These 'alternative' knowledge systems complement formal forest science and have a vital role to play in our quest for sustainability at local, regional, and global levels. Fulfilling this role will require a better understanding of, and respect for, traditional forest-related knowledge, practices and innovations – as well as support for the cultural values and social institutions of indigenous and local communities.

Current efforts to bridge the significant gaps that exist between the forest science community and the holders and users of traditional forest knowledge need to expand. Recognizing this need, the International Union of Forest Research Organizations

(IUFRO) established a Task Force on Traditional Forest Knowledge in 2005. Its objectives were to foster a broader understanding of traditional forest knowledge within the forest science community and to critically evaluate the opportunities and limitations for enhanced collaboration among these two broad communities and decision-makers. Between 2006 and 2009, the IUFRO Task Force organized a series of conferences in Europe, Asia, Africa and North America in partnership with a variety of collaborating institutions and organizations. Several publications resulting from these meetings have helped to raise the profile of traditional forest-related knowledge within the global forest science community.[1]

We hope that this book, the collective effort of an international, multi-disciplinary group of scientists, serves to improve our understanding of traditional forest-related knowledge, its history and its relationships with formal forest science. We believe that this effort will help both to inform and to promote greater attention to, and consideration of, the knowledge possessed by traditional communities. We further hope that it will encourage increased collaborative research aimed at the preservation, development and application of traditional knowledge to enhance culturally, ecologically, and economically sustainable forest ecosystem management.

U.S. Forest Service, Research and Development   John A. Parrotta
University of British Columbia, Faculty of Forestry   Ronald L. Trosper

---

[1] For more information see: http://www.iufro.org/science/task-forces/former-task-forces/traditional-forest-knowledge/

Acknowledgements

The preparation of this book would not have been possible without the help of many individuals and organizations, whose assistance we gratefully acknowledge.

We would like to thank those organizations whose financial and logistical support enabled the IUFRO Task Force on Traditional Forest Knowledge to organize indispensible international conferences in Italy, the United States, China, Ghana, and Kyrgyzstan. These organizations include: IUFRO's Special Programme for Developing Countries; the Korea Forest Research Institute; the U.S. Forest Service; the University of British Columbia, Faculty of Forestry; the Austrian Federal Ministry of Agriculture, Forestry and Water Management; the Italian Academy of Forestry Science; the University of Florence; the Ministry of Agricultural, Food and Forestry Policies (Italy); the College of Menominee Nation (USA); the Sustainable Forest Management Network (Canada); the Chinese Academy of Forestry; the Food and Agriculture Organization of the United Nations (FAO); Seoul National University; the Asia-Pacific Association of Forestry Research Organizations (APAFRI); United Nations University; the Council for Scientific and Industrial Research of Ghana; the Swedish International Development Cooperation Agency (Sida); the Ministry of Foreign Affairs of Finland; the Netherlands Federal Ministry of Foreign Affairs; the Global Forest Coalition; and the NGO BIOM (Kyrgyzstan).

We gratefully acknowledge the generous contributions of The Christensen Fund, whose support to IUFRO (via grant no. 2008–2255987) was indispensible, providing essential funding for two authors' workshops held in Vienna during the preparation of this book.

We would also like to express our appreciation to the members of the Task Force who are chapter authors, as well as those who provided critical support and guidance during the early stages of our work, particularly Rob Doudrick and Cheryl Kitchener.

We thank the following individuals who served as chapter reviewers and contributed greatly to the relevance and quality of this volume: Tania Ammour, Brian Belcher, Harold Brookfield, Jeff Campbell, Oracio Ciancio, David Cohen, John Dargavel, Ronnie de Camino, Craig Elevitch, Jesus García Latorre, Adam Gerrand, Peter Kanowski, Keith Kirby, Su See Lee Florencia Montagnini, Gary Morishima, Colin Nicholas, Oliver Rackham, Hugh Raffles, Nitin Rai, Gleb Raygorodetsky,

Enrique Salmon, Robin Sears, Peggy Smith, Marjorie Sullivan, Peter Szabo, Sonia Tamez, Viktor Teplyakov, Nancy Turner, and Ludmila Zhirina.

Special thanks are due to Erik Lindquist and LarsGunnar Marklund of FAO's Global Forest Resources Assessment Programme (FAO Forestry Department, Rome) for their generous assistance in the preparation of the forest cover maps included in each of the regional chapters based on the FAO's 2000 Global Forest Resources Assessment.

We are deeply indebted to Mary A. Carr from the U.S. Forest Service for her outstanding help in the final editing of all chapters, and to Margaret Deignan and Takeesha Moerland-Torpey at Springer for their cheerful support, and patience, during the preparation of this book.

Contents

Contributors

Mauro Agnoletti Dipartimento di Scienze e Teconolgie Ambientali Forestali, Università di Firenze, Facoltà di Agraria, Florence, Italy, mauro.agnoletti@unifi.it

Per Angelstam Faculty of Forest Sciences, School for Forest Engineers, Swedish University of Agricultural Sciences, Skinnskatteberg, Sweden, per.angelstam@smsk.slu.se

Santiago Barros Instituto Forestal, Santiago de Chile, Chile, sbarros@infor.cl

Sebastián Bessonart Directorate of Forestry, Secretariat of Environment and Sustainable Development, Buenos Aires, Argentina, sbessonart@ambiente.gob.ar

Vladimir Bocharnikov Pacific Institute of Geography, Russian Academy of Science, Vladivostok, Russia, vbocharnikov@mail.ru

János Bölöni Institute of Ecology and Botany, Hungarian Academy of Sciences, Vacratot, Hungary, jboloni@botanika.hu

Dominic Byarugaba Department of Biology, Faculty of Science, Mbarara University of Science and Technology, Mbarara, Uganda, dbbyarugaba@yahoo.com

Leni D. Camacho University of the Philippines, Los Baños, College Laguna, Philippines, camachold@yahoo.com.ph

U.M. Chandrashekara Kerala Forest Research Institute, Nilambur Campus, Chandrakunnu, Malappuram, Kerala, India, umchandra@rediffmail.com

N. Chhetri International Centre for Integrated Mountain Development, Kathmandu, Nepal, chettrin@rediffmail.com

Fred Clark U.S. Forest Service, Washington, DC, USA, fclark@fs.fed.us

Edwin A. Combalicer University of the Philippines, Los Baños, College Laguna, Philippines, eacombalicer@yahoo.com

Sakuma Daisuke Osaka Museum of Natural History, Osaka, Japan, sakuma@ mus-nh.city.osaka.jp

Ilya Domashov "BIOM" Ecological Movement, Bishkek, Kyrgyzstan, idomashov@gmail.com

David Eastburn Landmark Communications, Chapman, ACT, Australia, Eastburn. landmark@gmail.com

Marine Elbakidze Faculty of Geography, Lviv University of Ivan Franko, Lviv, Ukraine, marine.elbakidze@smsk.slu.se

Seçil Yurdakul Erol Faculty of Forestry, Istanbul University, Istanbul, Turkey, sechily@hotmail.com

Suzanne A. Feary Conservation Management, Vincentia, NSW, Australia, suefeary@hotkey.net.au

Mónica Gabay Directorate of Forestry, Secretariat of Environment and Sustainable Development, Buenos Aires, Argentina, monagabay@yahoo.com

Christian Gamborg Danish Centre for Forest, Landscape and Planning, University of Copenhagen, Copenhagen, Denmark, chg@life.ku.dk

Jesús García Latorre Federal Ministry of Agriculture, Forestry, Environment and Water Management, Vienna, Austria, jesus.garcia-latorre@lebensministerium.at

Juan García Latorre Association for Landscape Research in Arid Zones, Almería, Spain, jglatorre2@gmail.com

Patrica Gerez-Fernandez Instituto de Biotechnologia y Ecología Aplicada, Universidad Veracruzana, Xalapa, Veracruz, Mexico

H.K. Gupta Forest Survey of India North Zone, Shimla, India, hemantgup@ gmail.com

Susanna Hecht Department of Urban Planning, School of Public Affairs, UCLA, Los Angeles, CA, USA, sbhecht@ucla.edu

Kate Holl Forest Research, Northern Research Station, Bush, Midlothian, UK, Kate.Holl@snh.gov.uk

Masahiro Ichikawa Faculty of Agriculture, Kochi University, Kochi, Japan, ichikawam@kochi-u.ac.jp

Elisabeth Johann Austrian Forest Association, Vienna, Austria, elisabet.johann@ aon.at

Olesya Kaspruk Ukrainian National Forestry University, Lviv, Ukraine, ok@nepcon.net

Jean Kennedy Department of Archaeology and Natural History, School of Culture, History and Language, The Australian National University, Canberra, Australia, Jean.Kennedy@anu.edu.au

Kim Kiweon Department of Forest Resources, Kookmin University, Seoul, Republic of Korea, kwkim@kookmin.ac.kr

Jürgen Kusmin State Forest Management Centre, Tallinn, Estonia, jurgen@velise.ee

Frank Lake U.S. Forest Service, Arcata, CA, USA, flake@fs.fed.us

Andrey Laletin Friends of the Siberian Forests, Krasnoyarsk, Russia, laletin3@gmail.com

Liang Luohui Institute of Sustainability and Peace, United Nations University, Tokyo, Japan, liang@hq.unu.edu

Lim Hin Fui Forest Research Institute Malaysia, Kepong, Selangor, Malaysia, limhf@frim.gov.my

Liu Jinlong School of Agricultural Economics and Rural Development, Renmin University of China, Beijing, China, liujinlong@ruc.edu.cn

William Armand Mala Department of Plant Biology, Yaounde I University, Yaoundé, Cameroon, williammala@yahoo.fr

Deborah McGregor University of Toronto, Toronto, ON, Canada, d.mcgregor@utoronto.ca

Zsolt Molnár Institute of Ecology and Botany, Hungarian Academy of Sciences, Vacratot, Hungary, molnar@botanika.hu

Doris Mutta Kenya Forestry Research Institute, Nairobi, Kenya, doris_mutta@yahoo.com

Alfred Oteng-Yeboah Department of Botany, University of Ghana, Legon, Ghana, alfred.otengyeboah@gmail.com

Christine Padoch Center for International Forest Research (CIFOR), Bogor, Indonesia

New York Botanical Garden (NYBG), Bronx, NY, USA, c.padoch@cgiar.org; cpadoch@nybg.org

John A. Parrotta U.S. Forest Service, Research and Development, Arlington, VA, USA, jparrotta@fs.fed.us

Reg Parsons Canadian Forest Service, Atlantic Forestry Centre, Natural Resources Canada, Fredericton, Canada, Reg.Parsons@NRCan-RNCan.gc.ca

S. Patnaik UNESCO Regional Office, New Delhi, India, s.patnaik@unesco.org

Charles M. Peters The New York Botanical Garden, Bronx, NY, USA, cpeters@nybg.org

Miguel Pinedo-Vasquez Center for Environmental Research and Conservation (CERC), Columbia University, New York, NY, USA,

Center for International Forest Research (CIFOR), Bogor, Indonesia, map57@columbia.edu

Silvia Purata People and Plants International, Xalapa, Veracruz, Mexico

Rajindra K. Puri School of Anthropology and Conservation, University of Kent at Canterbury, Canterbury, UK, R.K.Puri@kent.ac.uk

P.S. Ramakrishnan School of Environmental Sciences, Jawaharlal Nehru University, New Delhi, India, psr@mail.jnu.ac.ac.in

K.S. Rao Department of Botany, University of Delhi, Delhi, India, srkottapalli@yahoo.com

Xavier Rochel Département de Géographie, Université Nancy 2, Nancy, France, Xavier.Rochel@univ-nancy2.fr

Ian D. Rotherham Sheffield Hallam University, Sheffield, UK, I.D.Rotherham@shu.ac.uk

Teresa Ryan University of British Columbia, Vancouver, BC, Canada, ryantl@shaw.ca

Nalish Sam Fenner School of Environment and Society, The Australian National University, Canberra, Australia, Nalish.Sam@anu.edu.au

Peter Sandøe Danish Centre for Bioethics and Risk Assessment, University of Copenhagen, Copenhagen, Denmark, pes@life.ku.dk

Eirini Saratsi Department of Politics, Centre for Rural Policy Research, University of Exeter, Exeter, UK, E.Saratsi@exeter.ac.uk

K.G. Saxena School of Environmental Sciences, Jawaharlal Nehru University, New Delhi, India, kgsaxena@mail.jnu.ac.in

Hovik Sayadyan Armenian State Agrarian University, Yerevan, Republic of Armenia, hovik_s@yahoo.com

E. Sharma International Centre for Integrated Mountain Development, Kathmandu, Nepal, icimod@icimod.org.np

Shin Joon-Hwan Korea Forest Research Institute, Seoul, Republic of Korea, kecology@forest.go.kr

Emil Shukurov "Aleyne" Ecological Movement of Kyrgyzstan, Bishkek, Kyrgyzstan, shukurovemil@mail.ru

Savinder Kaur Kapar Singh Malaysian Environmental Consultants, Sdn. Bhd, Ampang Kuala Lumpur, Malaysia, savinder.gill@gmail.com

Mike Smith Scottish Natural Heritage, Edinburgh, UK, Mike.Smith@forestry.gsi.gov.uk

Igor Solovyi Ukrainian National Forestry University, Lviv, Ukraine, soloviy@yahoo.co.uk

Lembitu Tarang Estonian Society of Foresters, Haapsalu, Estonia, letarang@hot.ee

Alan Thomson Canadian Forest Service, Victoria, BC, Canada, ajthomson@shaw.ca

Ronald L. Trosper American Indian Studies, The University of Arizona, Tucson, AZ, USA, rltrosper@email.arizona.edu

Tengiz Urushadze Ivane Javakhishvili Tbilisi State University, Tbilisi, Georgia, t_urushadze@yahoo.com

Mark van Benthem Probos Foundation, Wageningen, The Netherlands, mark.vanbenthem@probos.net

Jim van Laar Forest and Nature Conservation Policy Group, Wageningen University, Wageningen, The Netherlands, Jim.vanlaar@wur.nl

Alan E. Watson Aldo Leopold Wilderness Research Institute, Missoula, MT, USA, awatson@fs.fed.us

Stephen Wyatt Université de Moncton, Campus d'Edmundston, Edmundston, NB, Canada, swyatt@umce.ca

Youn Yeo-Chang Department of Forest Sciences, Seoul National University, Seoul, Republic of Korea, youn@snu.ac.kr

Yuan Juanwen Guizhou College of Finance and Economy, Guiyang, China, yuanjuanwen@yahoo.com

List of Boxes

List of Figures

List of Tables

Chapter 1
Introduction: The Growing Importance of Traditional Forest-Related Knowledge

Ronald L. Trosper and John A. Parrotta

Abstract The knowledge, innovations, and practices of local and indigenous communities have supported their forest-based livelihoods for countless generations. The role of traditional knowledge—and the bio-cultural diversity it sustains—is increasingly recognized as important by decision makers, conservation and development organizations, and the scientific community. However, there has long existed a lack of understanding of, and an uneasy relationship between, the beliefs and practices of traditional communities and those of formal forest science. This mutual incomprehension has a number of unfortunate consequences, both for human societies and our planet's forests and woodlands, which play out both on solid ground in many parts of the world as well as in international policy arenas. In this chapter, we define traditional forest-related knowledge, and explore the relationships between traditional knowledge systems and scientific approaches. We follow with an overview of the scope and central questions to be addressed in subsequent chapters of the book, and then provide an overview of international and intergovernmental policy processes that affect traditional knowledge and its practitioners. Finally, we introduce some of the major international programmes and research initiatives that focus on traditional forest-related knowledge and its applications for sustaining livelihoods in local and indigenous communities in a world struggling to deal with environmental, cultural, social, and economic change.

Keywords Biocultural diversity • Forest policy • Forest management • Indigenous peoples • Knowledge systems • Sustainability • Traditional communities • Traditional knowledge • United Nations

R.L. Trosper
Faculty of Forestry, University of British Columbia, Vancouver, BC, Canada
e-mail: rltrosper@email.arizona.edu

J.A. Parrotta (✉)
U.S. Forest Service, Research and Development, Arlington, VA, USA
e-mail: jparrotta@fs.fed.us

J.A. Parrotta and R.L. Trosper (eds.), *Traditional Forest-Related Knowledge: Sustaining Communities, Ecosystems and Biocultural Diversity*, World Forests 12, DOI 10.1007/978-94-007-2144-9_1,
© Springer Science+Business Media B.V. (outside the USA) 2012

1.1 Introduction

After being dominated for centuries by the singular purpose to create and maintain the flow of wood fibre from forests, forest management throughout the world has moved in major ways to incorporate other values. This change in the 'paradigm' of forestry is widely recognized (Mery et al. 2005) and was mostly driven by pressures from outside of forestry—particularly the increased political power of environmental non-governmental organizations. In the United States, for example, the passage of laws protecting endangered species, combined with a strong judicial system, created a change in orientation for forest management towards species protection. This change occurred at the national level, however, and local values regarding forests are not necessarily fully devoted to preservation of species. The use of forests for other purposes, such as generation of ecosystem services or wildlife, is also considered essential at local levels. Among the ecosystem services of significance are clean water, recreation, and provision of livelihood resources such as firewood, food, and medicine.

Clearly, the concerns at national and international levels for preservation of biodiversity and sequestration of carbon, while contributing to the shifting forest management paradigm, do not necessarily include the concerns of local communities. In some regions, large corporations concerned with timber values and large environmental non-governmental organizations, have agreed among themselves to divide up landscapes between protected areas and areas of timber production, leaving out the other interests that exist in such landscapes. Indigenous peoples and other local forest users have objected (Davidson-Hunt et al. 2010).

These objections by local people raise the vital issue of how the interests and knowledge of all people can be incorporated in forest management. As Fortmann (2008) suggests, both 'conventional science' and 'civil science' are needed, and the question that applies to both is: How can we learn what we need to know and understand in order to create, sustain, and enhance healthy ecosystems and human communities? The question then arises of how to add such knowledge to the knowledge pool in a mutually beneficial way—that is, in a manner allowing the respectful and appropriate use of traditional forest-related knowledge, and discouraging the extraction of such knowledge without benefitting those who developed it?

Traditional knowledge, innovations, and practices have long sustained the livelihoods, culture, identities, and the forest and agricultural resources of local and indigenous communities throughout the world. Traditional forest-related knowledge (TFRK) is of particular importance to indigenous communities, peoples, and nations, whom Martínez Cobo (1986/7) defined as:

> … those which, having a historical continuity with pre-invasion and pre-colonial societies that developed on their territories, consider themselves distinct from other sectors of the societies now prevailing on those territories, or parts of them [who] form at present non-dominant sectors of society and are determined to preserve, develop and transmit to future generations their ancestral territories, and their ethnic identity, as the basis of their continued existence as peoples, in accordance with their own cultural patterns, social institutions and legal system....

Indigenous peoples comprise approximately 5% of the world's population (between 250 and 300 million people), representing up to 5,000 different cultures on all continents and throughout the Pacific islands. The Asia-Pacific region is home to the largest number of indigenous peoples, approximately 60–70% of the world's total (Galloway-McLean 2010). These communities account for most of the world's cultural diversity (Gray 1991). Indigenous communities manage an estimated 11% of the world's forest lands and customarily own, occupy, or use, 22% of the earth's total land surface; they protect and manage an estimated 80% of the planet's biodiversity. These communities reside in or adjacent to approximately 85% of the world's protected areas (Galloway-McLean 2010).

Yet the survival of indigenous peoples and the cultural diversity they represent continues to be threatened today, as it has for centuries in most parts of the world. Land dispossession, large-scale development projects, and efforts by governments and the dominant societies they represent to 'assimilate' indigenous peoples are major challenges for these communities and their cultures. Of the some 7,000 languages today, it is estimated that more than 4,000 are spoken by indigenous peoples. Language specialists predict that up to 90% of the world's languages are likely to become extinct or threatened with extinction by the end of the century (UN 2009).

The link between cultural diversity and biological diversity has been increasingly recognized in recent years by the international scientific and policy communities. In 1988 during the First International Congress of Ethnobiology held in Belém, Brazil, scientists and representatives of indigenous and local communities from around the world discussed a common strategy to halt the rapid loss of the world's biological and cultural diversity. These discussions considered the unique ways in which indigenous and traditional peoples perceive, use, and manage their natural resources and how programmes can be developed to guarantee the preservation and strengthening of indigenous communities and their traditional knowledge. The Declaration of Belém, an output of this meeting, outlined the responsibilities of scientists and environmentalists in addressing the needs of local communities and acknowledged the central role of indigenous peoples in all aspects of global planning. In the intervening years the concept of 'bio-cultural diversity' (the total variety exhibited by the world's natural and cultural systems) has emerged as an increasingly important concept through the recent work of a number of scholars (Posey 1999; Maffi 2001, 2005; Harmon 2002; Moore et al. 2002; Sutherland 2003; Loh and Harmon 2005).

Among indigenous communities, traditional knowledge is embedded and expressed in their languages, cultural values, rituals, folklore, land-use practices, and community-level decision-making processes. It is inextricably linked to indigenous peoples' identity, their experiences with the natural environment, and their territorial and cultural rights. In these communities, this knowledge is usually collectively owned and is transmitted orally from generation to generation. Passing this knowledge on to future generations is therefore considered to be very important to sustain their knowledge as well as their cultures and identities (Collings 2009). Developed from long experience and experimentation within local and indigenous communities, traditional forest-related knowledge has historically been dynamic, adapting to changing environmental, social, economic, and political conditions.

The evolution of this knowledge in the face of these changes has enabled traditional communities to manage forest resources to provide tangible (foods, medicines, wood and other non-timber forest products, water, and fertile soils) and intangible (spiritual, social, and psychological health) benefits for present and future generations.

There is a growing appreciation of the value and importance of traditional forest-related knowledge, and of traditional knowledge more generally, not only to local and indigenous communities, but also to broader metropolitan, increasingly globalized, societies. Widely used products such as plant-based medicines and cosmetics, agricultural and non-wood forest products, and handicrafts are derived from traditional knowledge. Traditional forest-related knowledge, innovations, and practices can contribute to sustainable development in several ways. Most indigenous and local communities live in areas containing the vast majority of the world's forest (and agricultural) genetic resources, including most of the world's terrestrial so-called biodiversity hotspots. The traditional knowledge and techniques used to sustainably manage and use these genetic resources and ecosystems can provide useful insights and models for biodiversity conservation practices and policies.

The combined influences of economic and cultural globalization, land-use change, climate change, and increased climate variability have significant implications for the world's forests and their biological diversity (UN 2009). These trends present major challenges, and perhaps opportunities, for the long-term conservation and management of forests for their social, cultural, environmental, and economic values. In the face of these challenges, we maintain that both formal science and traditional ecological knowledge have important, complementary, roles to play in the development of viable, locally adapted forest management approaches and practices.

1.2   Objectives and Scope of This Book

This book is intended to provide readers with an overview of the history, current status, and trends in the development and application of traditional forest-related knowledge (TFRK) by local and indigenous communities worldwide. It will consider the relationship between traditional beliefs and practices and formal forest science, and the often uneasy relationship among these different knowledge systems. It also highlights efforts in many parts of the world to conserve and promote traditional forest management practices to balance environmental, economic, and social objectives of forest management in light of recent trends towards devolution of forest management authority in many parts of the world.

The central issues that this book considers are:

- The relevance and roles of traditional forest-related knowledge for the maintenance of cultural values, livelihood security, and sustainable management of forest ecosystems and cultural landscapes;
- The historical context and current trends regarding relationships between traditional and formal scientific knowledge and practices with respect to forest resource management;

- Experiences related to resolution of conflicts between formal science-based approaches to forest resource management and conservation, and traditional forest-related knowledge and practices, as well as lessons learned from experiences in different parts of the world on ways to foster mutually beneficial collaboration and exchange of knowledge between traditional and formal knowledge systems and their practitioners; and
- Approaches for including both traditional knowledge and formal forest science in natural resource education, research, resource assessments, and forest management activities, based on studies of how this has been done successfully, or how and why attempts have failed.

In this introductory chapter, we define traditional knowledge, explore the relationships of traditional knowledge to different scientific approaches, survey international and intergovernmental policy processes that affect traditional knowledge and its practitioners, and introduce some of the major international programmes and research initiatives that focus on traditional forest-related knowledge and its applications for sustaining livelihoods in local and indigenous communities.

In the following chapters, regional overviews—covering Africa (Chap. 2), South America (Chaps. 3 and 4), North America (Chap. 5), Europe (Chap. 6), Russia and neighbouring countries (Chap. 7), Northeast Asia (Chap. 8), South Asia and the Himalayas (Chap. 9), Southeast Asia (Chap. 10), and the Western Pacific (Chap. 11)—aim to summarize and evaluate the history and current status of TFRK and its relationship to formal forest science and 'modern' forest management. Each of these chapters highlights regional trends and key issues pertaining to the preservation, development, and application of TFRK in forest research, education, and management.

Chapters 12 and 13 examine the historical, current, and potential future impacts of two of the major challenges faced by indigenous and local communities in their efforts to preserve, develop, and use traditional forest-related knowledge—economic and cultural globalization and climate change. Discussion focuses not only on the threats posed by economic, social, and environmental change to the survival and development of traditional forest-related knowledge and practices, but also on the opportunities that these changes may present. Examples include the promotion of traditional forest (and related agricultural) management systems for climate change mitigation and adaptation.

Chapter 14 analyzes ethical issues and best practices for the scientific study of traditional forest-related knowledge and exchange of information between holders and users of traditional forest-related knowledge and forest scientists.

Finally, Chap. 15 summarizes the characteristics of traditional knowledge that both distinguish it from other knowledge and also make it vulnerable to disappearance in the modern world. Contributions of traditional forest-related knowledge to forestry science are discussed. The chapter suggests how to promote collaborative activities by forest scientists and professional organizations working with local and indigenous communities seeking to promote bio-cultural diversity and cultural, economic, and environmental sustainability in forest resource management.

1.3 Traditional Forest-Related Knowledge—Definition and Scope

The terms of reference for the task force on traditional forest-related knowledge incorporated the following definition of traditional knowledge, based upon the widely used definition of traditional ecological knowledge provided by Fikret Berkes, which also was adopted by the UN Forum of Forests:

> *Traditional forest-related knowledge*: 'a cumulative body of knowledge, practice and belief, handed down through generations by cultural transmission and evolving by adaptive processes, about the relationship between living beings (including humans) with one another and with their forest environment' (UN 2004, adapted from Berkes et al. 2000).

This may be considered as a forest ecosystem-focused variant of the more general term *traditional knowledge* as defined by the Convention on Biodiversity's Article 8(j):

> … the knowledge, innovations and practices of indigenous and local communities around the world. Developed from experience gained over the centuries and adapted to the local culture and environment, traditional knowledge is transmitted orally from generation to generation. It tends to be collectively owned and takes the form of stories, songs, folklore, proverbs, cultural values, beliefs, rituals, community laws, local language, and agricultural practices, including the development of plant species and animal breeds. Traditional knowledge is mainly of a practical nature, particularly in such fields as agriculture, fisheries, health, horticulture, and forestry.

Applied to forest environments, the scope of TFRK is broadly similar to that described by other commonly used terms, as they apply to forest environments. For example, indigenous knowledge (IK), traditional environmental (or ecological) knowledge (TEK), and local ecological knowledge, all generally refer to the long-standing traditions and practices of certain regional, indigenous, or local communities, and encompass the wisdom, knowledge, and teachings of these communities. While the richest body of traditional forest-related knowledge resides within the world's indigenous communities, peoples, and nations, there exists in non-indigenous communities throughout the world as well a wealth of traditional knowledge related to trees, forests, and managed forested cultural landscapes.

In substituting 'forest-related' for 'ecological,' the task force recognized that forests are ecosystems and that knowledge related to forests would be knowledge of ecosystems. Although the terminology has been widely used, each of the terms in 'traditional forest-related knowledge' needs to be carefully considered.

The word *traditional*, while it accurately refers to the idea of a cumulative body of knowledge, practice and belief,' could be interpreted to say that the knowledge is 'old' or 'unchanging.' The definition emphasizes that traditional knowledge does change and is 'evolving by adaptive processes.' Thus, although traditional, the knowledge can also be modern; the emphasis of the word is on local or indigenous knowledge, as opposed to universal knowledge, or knowledge from conventional science. One could use 'indigenous,' but not all local knowledge is indigenous in the sense of known by people that have been colonized by outsiders; rather, most traditional knowledge is held by people who have become 'indigenous to a place;' they may not, however, be the original inhabitants.

We use the term *related* to recognize that forests produce more than wood fibre, and that the knowledge applies to all ecosystem processes that relate to forested lands. Many agroforestry and fishery issues depend upon forest-related knowledge. The word *related* has the additional advantage of suggesting other matters, such as issues having to do with human relationships to the forest. This aspect is referred to in the definition given above, which includes relationships among humans as part of traditional forest knowledge, in addition to including relationships between humans and non-humans.

The third word, *knowledge*, might seem at first to be less of an issue, but the breadth of the meaning of the word needs to be considered. Often, knowledge refers only to matters about the nature of the world and how it works, rather than including ideas related to ethics and morality. In considering traditional knowledge, one must adopt a broad concept, as will become clear in the following sections that compare and contrast traditional ecological knowledge to conventional science. Traditional knowledge holders often stress the importance of respecting nature, of being grateful for the generosity of non-humans in allowing themselves to be harvested and used by humans.

The correct words to use for *science* also need attention. Many times science is described as 'Western science,' 'state science,' or 'reductionist science.' Each of these adjectives contains its own issues. While science may have originated in Western Europe, it utilized knowledge from Asia and from Africa, and also may well have included ideas from the Americas. Forest science may have originated in a situation in which foresters worked for an emerging German state, but many others have contributed to forest science in subsequent years. Perhaps science can attribute some of its great success to applying reductionist techniques, particularly through laboratory and field experiments.

1.4 Relationship Between Traditional and Scientific Knowledge Systems

This section considers two aspects about the relationship between traditional forest-related knowledge and science: scientific credibility, and the actual differences and similarities between the two kinds of knowledge systems. The issue of the credibility of science creates a diversion from considering the more fundamental aspects of the relationship. Science and traditional knowledge have different origins and different relationships to their societies. However, differences between traditional knowledge and science have been narrowing in recent years because of differentiation among the different sciences, particularly the ecological sciences. The second part of this section examines the extent to which differing ecological sciences make assumptions similar to those made by traditional knowledge holders.

1.4.1   The Credibility Issue

The definition of traditional forest-related knowledge may imply that it is useful to distinguish traditional knowledge from conventional scientific knowledge. While drawing such distinctions is valuable for explaining and identifying traditional knowledge, there is a disadvantage related to recent concerns on the part of those who conduct science to protect the 'credibility' of science. Philosophers and social scientists have asked questions such as, 'what kind of knowledge is science?' and 'how do scientists do their work?' In exploring the answers, those engaged in science studies have appeared to question the objectivity, rationality, and truth of standard science. Defenders of science have responded so vehemently that the ensuing discussion has been labelled the 'science wars.' Sociologist Thomas Gieryn (1999) has described the science wars as examples of 'boundary work': the utilization of a series of strategies to define what is and is not science, with the goal of protecting the credibility of science.

The stakes in these 'wars' have been high. In the United States, the National Science Foundation (NSF) allocates scarce research funds. Fields outside of science cannot be supported by the NSF, which has taken a broad view that includes many social sciences within 'science.' As cited by Gieryn, defenders of science identify themselves as expert and credible because their knowledge is tested by nature. "Science is nature, and therefore the very opposite of culture," wrote Michael Holmquest (quoted by Gieryn 1999, p. 342). Scientific knowledge is said to be better and stronger than knowledge based on discourse, mere talk, because science is based on credible procedures and tests that are judged by the response of nature to identify truth.

Certainly, traditional knowledge holders can make the same claim; as explained by Nadasdy (2003), the test of a hunter's knowledge of moose is his ability to find a moose. Consequently, defenders of traditional ecological knowledge might be tempted to make the same claim as Holmquest: having successfully lived for years on a particular landscape, an elder recognized by his community as truly expert could also speak authoritatively on behalf of nature. That this may be possible could create a rivalry between the scientist and the elder, and the exploration of the rivalry would involve listing the many different ontological, epistemological, and ethical differences between their ways of knowing, with a goal of determining which methods were more credible. Going down this road, the scientist may dismiss the elder's knowledge as 'anecdotal,' while the elder will dismiss the scientist's knowledge as 'not applicable here.'

If a credibility contest proceeds through such comparisons, everyone loses in terms of advancing knowledge, although one or the other may gain with respect to other aspects, such as job security or research support. As the International Council for Science[1] points out, because 'pseudo science' attempts to pose as science and attack its credibility, pseudo-science depends on science for its existence; on the other

---

[1] Formerly the International Council of Scientific Unions (ICSU).

hand, traditional knowledge originates independently of science and competition with science is not necessary for traditional knowledge (ICSU 2002, p. 12).

If each respects the other, a trained scientist and a knowledgeable elder can have very productive discussions, as reported by botanist Nancy Turner (2005), ecologists P.S. Ramakrishnan (1992) and Fikret Berkes (2008), and fisheries biologist Robert Johannes (1981), to name a few scholars who have succeeded in establishing good working relationships with holders of traditional knowledge. The two types of experts can inform each other because their systems of knowledge rely on different sources and methods, while addressing similar problems. Traditional knowledge is typically based upon data observed over many, many years, while scientific knowledge usually relies on recent data collected under carefully controlled conditions, as in field experiments. Chapter 14 deals with ethical issues that are involved when scientists, usually connected to powerful institutions, interact with traditional knowledge holders, who usually are not connected to the powerful institutions.

The issue of credibility connects to the issue of epistemic injustice, a topic recently explored by Miranda Fricker (2007). Because traditional knowledge holders are often from groups who do not receive respect from the powerful groups, their testimony regarding events in the world are often discounted. Similarly, their contributions to the existing body of knowledge are not given credit. An example of this is the system of classification of species. The work of Rumphius originated with a classification system originally used by Malay women in Asia, while that of Linnaeus utilized Sami classifications (ICSU 2002, p. 13). Fortmann provides other examples of classification systems originating outside of Europe (Fortmann 2008). Another example is the use of vitamin C as a cure for scurvy: the Huron demonstrated the use of a tonic made from bark to Jacques Cartier in 1535; 200 years later, James Lind, reading of the experience, applied the cure to the British navy (Weatherford 1988, pp. 182–183). Weatherford provides other examples, some of which were credited to Indians—such as the word quinine, which originated in Quechua and remains the name of a drug used to cure malaria.

While contemporary recognition of the contributions of traditional forest-related knowledge can reverse the effects of epistemic injustice, the issue of the credibility of science in comparison to other knowledge systems remains an important barrier to joint learning by scientists and others. The International Council for Science has helped reverse this historical injustice by documenting many contributions of traditional knowledge to science (ICSU 2002, pp. 13–15).

1.4.2   Relationships Between Traditional Knowledge and the Sciences

We should set aside the credibility issue and instead examine methods of learning together, as Fortmann (2008) recommends. Table 1.1 highlights the relationship between many of the tenets of traditional knowledge and the characteristic assumptions of various ecological sciences. Perhaps traditional forest-related knowledge is

Table 1.1 Classification of components of traditional knowledge in relation to the sciences

Level	1	2	3	4	5	6	7
Label	Reductionist science	Ecology	Social-ecological systems	Resilience theory	Sustainability, ecological economics	Actor-network-theory	Sacred traditional knowledge
Connectedness: 'all things are connected'		X	X	X	X	X	X
Humans are part of the system: no society/nature division			X	X	X	X	X
The history of a place matters			X	X	X	X	X
Expect change and emergence at multiple scales; use humility in application of current knowledge				X	X	X	X
Knowledge is transmitted orally through narratives, stories				X	X	X	X
Training needs to focus on development of the capacity to learn and recognize new situations				X	X	X	X
Stewardship of the land is paramount					X	X	X
Generosity, sharing and equity among humans are important					X	X	X
Everything is there for a reason, and deserves respect and the right to live its way (limit the marketability of things; value growing stock for itself)					X	X	X
Practical experience on land is the main source of knowledge						X	X

Level	1	2	3	4	5	6	7
Label	Reductionist science	Ecology	Social-ecological systems	Resilience theory	Sustainability, ecological economics	Actor-network-theory	Sacred traditional knowledge
Doubts exist about the generality of knowledge						X	X
The fact/value distinction is unimportant						X	X
Knowledge, language, and identity are local.						X	X
Reciprocity governs human-prey relationships; humans must give thanks and reciprocate for the gifts given by plants and animals							X
Non-humans have consciousness							X
Humans permanently belong to their place							X

becoming more acceptable because of increasing overlap between the approaches of traditional knowledge holders and the approaches of new branches of science, particularly the ecological sciences. The first column of the table provides some examples of ideas used in TFRK; the remainder of the table shows which of the ideas have become used in the various ecological sciences. The ideas from traditional knowledge are organized in a manner that allows identification of seven levels of increasing consistency between traditional knowledge and different ecological sciences, with the seventh level being 'sacred traditional knowledge,' which has ideas that no science as yet has adopted. The list of ideas in the first column summarizes the literature on traditional knowledge, with an emphasis on the Americas. Since the main purpose of this section is to explore the relationships, the ideas in traditional knowledge are not explored in detail here. Many of these are dealt with in the chapters that follow.

1.4.2.1 Reductionist Science

The second column, Level 1 in the comparison, indicates that none of the statements of the first column can be said to be acceptable to that part of science called 'reductionist.' By using the experimental method to control for conditions not under study, scientists have become knowledgeable about microscopic or small components of systems. The success of reductionist science has of course contributed to its credibility, as those components are readily observed with the equipment scientists have developed. Traditional knowledge, on the other hand, is based on observations of the operation of whole systems, meaning that they describe relationships among the entities of an ecosystem, rather than the operations of each entity by itself. For instance, scientists understand why Pandora moth populations exhibit cycles in their numbers, due to the interaction between the moths and a small pathogen, but how the forest responds to Pandora moth outbreaks can't be understood as easily without long-term data. When asked about the effects of moth abundance on forests, Paiute elders reported that sick trees would die and healthy ones would survive. This was consistent with the idea that other disease organisms in trees, such as mistletoe, would suffer from the deaths of their hosts due to moth outbreaks, which would reduce mistletoe infections (Blake and Wagner 1987).

1.4.2.2 Ecology

The importance of connections among components of an ecosystem distinguishes ecology from other biological sciences. Level 2 of the comparison of traditional knowledge to the sciences credits ecology with recognition of the idea that all things are connected. Textbooks such as those of Ricklefs and Miller (2000) and Kimmins (2004) stress such connections, recognizing that interactions among all parts of a system need to be understood, rather than focus upon just the parts. This recognition of 'holism' as an important approach has been controversial within scientific traditions, and the development of the disciplines has itself created a path-dependence in the structure of science. According to Paul Nadasdy, a Kluane hunter felt that

'The government could not effectively manage wildlife . . . because (it) has forestry experts, water experts, and mining experts. . . . sheep biologists, moose biologists, wolf biologists, and bear biologists; and none of these people know anything outside of their own specialty' (Nadasdy 2003, p. 123). While the hunter may be too harsh in the assertion that specialists know nothing outside of their specialty, that many are not interested in exploring system behaviour is accurate. Some wildlife biologists focus on habitat relationships; others, however, do not. That specialists in particular species should know about forest succession is an obvious statement of the need for interdisciplinary work, and it is no longer that controversial in the field of forestry.

1.4.2.3 Social Ecological Systems

Although ecology emphasizes connections, it differs from traditional knowledge in defining nature as separate from humanity; most ecosystems are described solely in terms of biotic and abiotic conditions, with biotic conditions omitting humans. Level 3 is the new field of social ecological systems, which emphasizes connections, and in addition recognizes the importance of history in understanding the particular configuration present at any one time in a system (Berkes et al. 2003).

The importance of the history of a place, and of human actions in that place, is another potential area of agreement among some conventional scientists and traditional knowledge holders. The concept of 'path dependence' captures this concern for the importance of history, a path, in understanding outcomes for complex systems.

This concern for history overlaps with the understanding of evolutionary principles of selection. Ray Pierotti (2010) argues that a 1911 quotation from Okute, a Lakota elder, is a recognition that organisms such as buffalo and people change in their characteristics in response to their specific history. After noting that, in his observations, no animals of the same species are exactly alike, and that variation also exists among plants, Okute observed:

> An animal depends upon the natural conditions around it. If the buffalo were here today, I think they would be very different from the buffalo of the old days because all the natural conditions have changed. They would not find the same food, nor the same surroundings…. We see the same change in our ponies…. It is the same with the Indians (Pierotti 2010, p. 78).

This view of the dynamics of systems, and the importance of individual variation within systems, is consistent with evolution theory and with the ideas of socio-ecological systems, which include humans.

1.4.2.4 Resilience Theory

Resilience theory, Level 4 in Table 1.1, focuses on the concept of ecological resilience originally defined by C.S. Holling (Holling 1986; Gunderson and Holling 2002). Many of the same scholars who have stressed the importance of including humans in the system models are also very interested in the concept of resilience, which has been emphasized by the Resilience Alliance, a loose network of scholars who founded the journal Ecology and Society (originally Conservation Ecology). These scholars hypothesize that ecological resilience has to be understood at different scales. Resilience results from successful social learning, which requires knowledge of the history of a system. Thus they have recognized the importance of oral narratives in providing such knowledge to humans in a system. They recognize the unpredictability of most systems, thus adopting humility about human understanding, a characteristic also emphasized by traditional knowledge holders. Given the stress on unpredictability, they see that the capacity to learn is a necessary component of achieving resilience (Berkes and Folke 1998; Folke et al. 2002).

A consequence of stressing unpredictability is a change in what is monitored—scientists create long lists with criteria and indicators, while traditional people use a few key signs. Berkes (2008) is particularly impressed by aboriginal systems of monitoring complex systems. This attention to signals that show results for the entire system is another consequence of the differing metaphors that underlie conventional science and traditional knowledge: attention to the detailed parts compared to attention to the relationships of the entities in the system.

The emphasis on unpredictability also is consistent with an insistence that training should focus on development of the ability to recognize new situations. Many have emphasized that training by traditional knowledge holders emphasizes teaching youth to recognize the signs of the land, and to let them make mistakes in implementation in order to learn how to evaluate what they see (Turner 2005).

1.4.2.5 Sustainability and Ecological Economics

The field of ecological economics, whose origin predates the establishment of the Resilience Network, uses many of the ideas present in the study of resilience but adds other concerns that are not so evident in the resilience literature. Prominent among the ideas of ecological economics is sustainability, and the concomitant idea that stewardship of the land is a paramount concern (Costanza et al. 1997). This stress on stewardship and sustainability marks the move to Level 5 of the table.

The concern for sustainability, broadly defined, is widespread in the literature on traditional knowledge. Attention is now given to the practices of the Menominee Tribe, for instance, as a model of the operation of sustainability goals in a forest environment. The Menominee like to say that they are borrowing their forest from their grandchildren (Pecore 1992; Davis 2000). Turner (2005) has also praised the concern of aboriginal peoples on the west coast of North America for their concern that 'the Earth's Blanket' be protected.

Another key characteristic of traditional systems is their stress on reciprocity and the importance of sharing among humans. This is evident on the Pacific Northwest coast of North America, where leaders are obligated to show their competence through feasting other leaders and their people (Daly 2005; Wa and Uukw 1992; Mills 1994, 2005; Trosper 2009; Turner et al. 2000; Umeek 2004; Walens 1981). This stress on equity and concern about unequal distributions of income is typical of writing in ecological economics.

Another characteristic of ecological economics is a vigorous debate about the use of market prices to value everything. While economists advocate the pricing of ecosystem services as a way to bring nature into economics, many ecological economists are far from sure (Norgaard 2010). Geoffrey Heal (1998), in considering the economics of sustainability, values the stock of natural resources for itself, without assigning a price, and derives what he and others call a 'green golden rule' by using a social welfare function that values the stock for itself rather than for its flow of services. This approach is what the Menominee Tribe applies in its forest management (Trosper 2007). Indigenous peoples of Mexico also resist full response to market prices, preserving a portion of their production for their own consumption, outside the market (Toledo et al. 2007).

Ecological economists, however, are fully in support of the search for generalizations that apply everywhere. In this they differ from those such as Bruno Latour, who seriously doubt such generalizations, which brings us to Level 6.

1.4.2.6   Latour's Actor-Network-Theory

Having carefully observed the way that scientists go about their work, sociologist Bruno Latour has challenged many of the epistemological and ontological assumptions that scientist claim to describe what they do. Among the ontological assumptions he has challenged is that of the existence of a single entity, nature, which is constant everywhere while cultures vary. He argues that no distinction should be made between nature and human culture (a division also rejected most clearly by those who advocate social ecological systems as the unit of analysis), arguing that what exists is 'multinaturalism': many different combinations of humans and nature (Latour 1993, 2004). As a consequence, he resists generalizations and makes that resistance a focus of his book on actor network theory (Latour 2005). He concludes by arguing that the best work a student can undertake is merely excellent description. This rejection of generalizations is quite evident among traditional knowledge holders, who do not believe that they can tell others what to do with their land using their own knowledge. An example is the strong view of Pikangikum elders to not make decisions for neighbours' land (O'Flaherty et al. 2008; Miller et al. 2010).

Another major similarity between Latour and traditional knowledge holders is the lack of attention and concern given to keeping facts separate from values. Some Menominee scholars, for instance, reject the term 'traditional ecological knowledge,' preferring 'indigenous wisdom,' because the term knowledge would seem to exclude ethical concerns about humanity's duty to care for the non-human world (Fowler 2005). Ethnobiologist Nancy Turner advocates traditional knowledge because of its explicit attention to the importance of taking care of the earth (Turner 2005). Latour centres his book, *Politics of Nature*, on a rejection of the standard fact/value distinction (Latour 2004). The International Science Council (ICSU 2002, p. 10) has also noted that the separation of fact from value leads scientists to misunderstand the depth of traditional knowledge.

A consequence of both the rejection of generalities and the fact/value distinction is an emphasis on the close connections among knowledge, language, and identity; all of these are assumed to be uniquely local characteristics. Part of the distrust of generalizations on the part of traditional knowledge holders is the desire that local applicability has to be demonstrated, an empirical concern that elevates practical knowledge above all other types.

1.4.2.7   Sacred Traditional Knowledge

Level 7 of the table asserts that sacred traditional knowledge has three ideas that are not shared with any of the ecological sciences: that reciprocity should govern human-prey relationships, that non-humans have consciousness, and that humans are permanently connected to their places. Although the sciences may have difficulty accepting these ideas, ecologist Fikret Berkes (2008) is willing to include the sacred in his consideration of traditional ecological knowledge. While other individual scientists may also be willing to accept the sacred, the organizations of science find sacred ideas much more difficult to include.

Another consequence of the rejection of the fact/value and nature/human distinction is that many traditional knowledge holders are very clear that they have standards of correct behaviour towards non-humans, particularly those species that are prey or sources of food and medicine. These entities deserve thanks for their contributions to humans. They also deserve respect and the right to live without undue interference from humans (O'Flaherty et al. 2008). Humans have a duty to act with reciprocity towards their prey, giving thanks for the willingness of these beings to give up their lives for human consumption. The Cree are particularly clear on this point (Tanner 1979; Feit 1992). Nonhumans have consciousness, and care must be taken as a result; even glaciers are recognized as conscious beings (Cruikshank 2005). This view of the consciousness of non-humans is seen as an extension of 'sacred' concepts in a way quite different from evident in other knowledge systems, and for this reason traditional knowledge is on occasion attacked for failing to be objective by bringing religious ideas into its purview.

Another aspect of sacred traditional knowledge is its emphasis on the strong connection between some humans and the land that they regard as sacred. In fact, the connection is so strong that a person cannot sever his bond with the land; humans are permanently connected to their place (Wa and Uukw 1992; Basso 1996; Cajete 1999; Burton 2002; Nadasdy 2003). To accept this link is so contrary to modern concepts of economic organization that no scholars are willing to go so far as to assert such unchangeable connections. In modern terms, humans are said to have to be able to move to places where they can earn good wages.

This brief survey of relationships between various ecological sciences and traditional knowledge reveals that many overlaps exist. Some difficulties also exist; for instance, few scientists are probably willing to give up the desire to find generalizations, or to accept the idea that people are permanently attached to their homelands. But these particular ideas may also be acceptable to some of those who possess traditional knowledge, which is probably more diverse than the ecological sciences. While traditional forest-related knowledge has become more relevant to the ecological sciences, it has also attracted considerable attention in international arenas. The following two sections summarize recent developments in policy processes such as those of the United Nations, in intergovernmental agreements, and among international nongovernmental organizations.

1.5   Traditional Forest-Related Knowledge in Intergovernmental Policy Processes

The protection, development, and utilization of traditional forest-related knowledge raise important questions about land tenure, ownership of genetic resources, intellectual property rights, and equitable sharing of benefits arising from the use of this knowledge. How these complex legal and ethical issues are resolved at local, national, and international levels has significant implications for cultural and biological diversity, and social justice.

Traditional forest-related knowledge and the constellation of issues surrounding it are increasingly recognized and addressed by the international community through a variety of legal instruments and policy processes. Those that focus on indigenous peoples' rights, including those related to traditional forest-related knowledge, include among others: the 2007 Declaration on the Rights of Indigenous Peoples, the UN Permanent Forum on Indigenous Issues and the Declaration on the Rights of Indigenous Peoples (UNPFII), the UN Working Group on Indigenous Populations, and the International Labour Organization's 1989 Convention 169 on Indigenous and Tribal Peoples.

Policy commitments developed at the international level that urge states to support, protect, and encourage the use of traditional knowledge and customary forest management and use are found in a number of instruments. Notable among these are the agreements negotiated during the 1992 United Nations Conference on Environment and Development (UNCED) in Rio de Janeiro in 1992 (the Earth Summit). These include the Rio Declaration on Environment and Development, Agenda 21 (Programme of Action on Sustainable Development), the Convention on Biological Diversity (CBD), the UN Convention to Combat Desertification (UNCCD), the UN Framework Convention on Climate Change (UNFCCC), and the non-legally binding Forest Principles. Traditional knowledge and its importance in education, science, and culture have received considerable attention by the United Nations Educational, Scientific and Cultural Organization (UNESCO).

With respect to access, benefit sharing, and intellectual property protection issues related to TFRK, the World Intellectual Property Organization (WIPO), the World Trade Organization (WTO), the United Nations Conference on Trade and Development (UNCTAD), and the International Treaty on Plant Genetic Resources for Food and Agriculture (ITPGRFA) also have important roles to play (UNU-IAS 2005).

Traditional knowledge, including forest-related knowledge pertaining to the therapeutic value of the thousands of plant species used in traditional systems of medicine, plays a very important role in meeting the health care needs of the majority of people throughout the world. Recognizing this, the World Health Organization (WHO) adopted a Traditional Medicine Strategy in 2002. Among the Strategy's objectives are to support relevant aspects of traditional medicine within national health care strategies and policies; promote the rational use of traditional medicine; and promote the safety, efficacy, and quality of traditional medicine practices (WHO 2002). WHO also supports a network of collaborating centres for promotion of traditional medicine, supports documentation of information on traditionally used medicinal plant species, and has worked with other organizations to develop guidelines for the sustainable collection and cultivation of medicinal plant resources (WHO/IUCN/WWF 1993; WHO 2003).

The commitments made by countries at the 1992 Earth Summit were reaffirmed in the Millennium Development Goals arising from the 2000 UN Millennium Declaration[2] (particularly Goal No. 7 on Environmental Sustainability), the Proposals for Action arising from the work of the Intergovernmental Panel on Forests and the Intergovernmental Forum on Forests from 1995 to 2000, and the ongoing work of the Rio Conventions and the United Nations Forum on Forests (UNFF, 2000–present). Further details regarding UNCED and the Rio Conventions, the United Nations Forum on Forests, UNESCO, and policies for and organizations dealing with international property rights issues related to TFRK may be found in the Appendix to this chapter.

Despite the significant progress made in the development of international forestry policy focused on sustainable forest management over the past 20 years, there is widespread concern that progress on the ground has been poor (Collings 2009). The lack of effective coordination among the different government agencies within countries, as well as among the various, largely autonomous, international organizations and processes dealing with TFRK issues from different perspectives, is a persistent obstacle (Rosendal 2001). A review of the implementation of key international commitments on traditional forest-related knowledge in a number of countries in the Americas, Africa, and Asia was conducted in 2004 by the International Alliance of Indigenous and Tribal Peoples of the Tropical Forest (Newing et al. 2005). The study concluded that beyond the production of numerous country-level planning documents and reports to the CBD and UNFF—such as national forest plans (NFPs), national forest action programmes (NFAPs), and national biodiversity strategy and action plans (NBSAPs)—evidence of effective implementation of laws, policies, and related actions on the ground is relatively scarce in most countries (Newing 2005).

---

[2] http://www.un.org/millenniumgoals/bkgd.shtml

1.6 TFRK in Intergovernmental Programmes, International Scientific Organizations, and Non-governmental Organizations

The growing recognition of traditional forest-related knowledge and its importance for biodiversity conservation and sustainable development in recent decades has encouraged an increased focus on TFRK in a variety of intergovernmental and scientific organizations, as well as in international programmes of non-governmental organizations (NGOs). While an exhaustive survey of these is beyond the scope of this chapter, UNESCO's Man and the Biosphere (MAB) and its Local and Indigenous Knowledge Systems in a Global Society (LINKS) programme, as well as United Nations University–Institute of Advanced Studies' Traditional Knowledge Initiative are excellent examples. Further details on these are provided in the Appendix.

Research in a variety of biophysical social science disciplines related to traditional forest-related knowledge and practices, is currently undertaken by scientists from universities, research institutes, and other organizations worldwide. There is a long history of such research in social science disciplines such as anthropology, and keen interest among botanists and other forest scientists—particularly on traditional knowledge related to forest species of economic value, such as plants used in traditional medicine and other commercially valuable non-timber forest products. However, most research in this field (particularly within the biophysical sciences such as ecology, forestry, and agriculture) is relatively recent. Inter- or multi-disciplinary research, which one could argue is essential for a proper understanding of traditional forest-related knowledge, is even more recent, mostly dating from the 1970s.

Among the many international scientific organizations with research programmes relevant to traditional forest-related knowledge and its practitioners, the Center for International Forestry Research[3] (CIFOR, based in Bogor, Indonesia); the World Agroforestry Center[4] (ICRAF, based in Nairobi, Kenya); and the International Institute of Tropical Agriculture[5] (IITA, based in Ibadan, Nigeria) are notable. All are part of the network of 15 internationally funded research centres established by the Consultative Group on International Agricultural Research[6] (CGIAR, established in 1971), and each has strong, integrated, multi-disciplinary programmes oriented towards the practical challenges faced by local and indigenous communities, primarily those in tropical and subtropical regions of the world. While IITA's work is focussed on Africa, with research stations located in ten sub-Saharan countries, ICRAF and CIFOR's research programmes are global in scope (or at least pan-tropical), with major initiatives undertaken in many countries in Asia, Africa, and Latin America.

---

[3] http://www.cifor.cgiar.org/

[4] http://www.worldagroforestrycentre.org/

[5] http://www.iita.org/

[6] http://www.cgiar.org/centers/index.html

ICRAF's research programmes involve close engagement with local communities and aim towards the development more productive, diversified, integrated, and intensified trees and agroforestry systems to provide livelihood and environmental benefits to communities. One of its better-known programmes—the Alternatives to Slash and Burn (ASB) Partnership for the Tropical Forest Margins (established in 1994)—combines local knowledge, policy perspectives, and science to understand the trade-offs associated with different land uses and the roles of markets, regulation, property rights, and rewards (ASB 2011).

CIFOR conducts research and policy analysis in collaboration with a broad array of partners—research organizations, local communities, NGOs, and others. Through this work, CIFOR aims address the needs and perspectives of people who depend on forests for their livelihoods, as well as improve the management of tropical forests. Much of CIFOR's research is multidisciplinary, and some is long-term work, such as its well-known global programmes on forests and livelihoods, and on adaptive collaborative forest management (Colfer 2005). These and other programmes often involve participatory research with local and indigenous communities. At present, CIFOR's research programmes include a strong emphasis on issues and research questions relevant to climate change mitigation and adaptation and its relation to tropical forests and forest-dependent communities and their livelihoods.

A large number of local grass-roots, national, regional, and international non-governmental organizations work on issues relevant to the preservation, development, and promotion of traditional forest-related knowledge, as well as the protection of the rights and interests of local and indigenous communities whose traditional knowledge, practices, and cultures are under threat. Many international networks, alliances, and individual NGOs promote indigenous and local communities' interests in global policy processes and programmes. Others support projects and other activities related to conservation and development of TFRK, including efforts to build greater understanding and appreciation among the scientific community, decision makers, and the people in traditional communities; or to develop approaches for integrating traditional and formal scientific forest-related knowledge to improve forest management for the benefit of both traditional communities and the broader suite of 'stakeholders' in sustainable forest management.

Two such organizations with somewhat different orientations, but both contributing in important ways towards harmonizing interests among forest stakeholders, are the International Model Forest Network and the Forest Peoples Programme. The International Model Forest Network[7] is a 'global community of practice' that includes local communities as well as other forest stakeholders. Model forests are voluntary, broad-based initiatives linking forestry, research, agriculture, mining, recreation, and other values and interests within a given landscape. The Network currently includes landscapes representative of most of the major forest ecosystems of the world whose management combines the social, cultural, and economic needs of local communities with the objective of long-term sustainability of these landscapes.

---

[7]http://www.imfn.net/

The Forest Peoples Programme[8] (FPP), a UK-based NGO actively engaged in international forest policy forums, works with indigenous and local communities South America, Central Africa, South and South East Asia, and Central Siberia, providing support for these communities to secure their rights, and strengthen their own organisations to negotiate with governments and companies as to how economic development and conservation is best achieved on their traditional lands. Among their activities are projects that promote community-based sustainable forest management based on traditional forest-related knowledge, practices, and related traditional governance. The Forest Peoples Programme produces a variety of publications, including useful syntheses of case studies and policy analyses based on scientific literature and other information sources.

1.7 Conclusion

The growing interest in traditional forest-related knowledge among international organizations and in policy arenas is a major indication of its currency and relevance to conservation of biocultural diversity and sustainable use of forests worldwide. The increasingly recognized overlap between traditional knowledge concepts and those of various ecological sciences is another sign of TFRK's growing importance. The following chapters by an international team of authors explore the tremendous pool of knowledge that exists in indigenous and local communities throughout the world. These chapters elaborate the many different ways in which traditional forest-related knowledge is relevant to communities, their livelihoods, and the history of forest resource management. The chapters also document examples of both conflict and collaboration between traditional knowledge holders and formal forest science. We hope that this extensive treatment of the current importance of traditional forest-related knowledge will stimulate further research and support for TFRK throughout the world.

Appendix—Traditional Forest-Related Knowledge in Intergovernmental Policy Processes and Selected Intergovernmental Programs

The UN Permanent Forum on Indigenous Issues and the Declaration on the Rights of Indigenous Peoples

The Permanent Forum on Indigenous Issues (UNPFII[9]), established in 2000, is an advisory body to the UN Economic and Social Council, with a mandate to discuss indigenous issues related to economic and social development, culture,

---

[8]http://www.forestpeoples.org

[9] http://www.un.org/esa/socdev/unpfii/index.html

the environment, education, health, and human rights. The Permanent Forum provides expert advice and recommendations on indigenous issues to the Council, as well as to programmes, funds, and agencies of the United Nations, through the Council. The UNPFII also seeks to raise awareness and promotes the integration and coordination of activities related to indigenous issues within the UN system, and prepares and disseminates information on indigenous issues, including traditional knowledge (UNPFII 2005).

The Declaration on the Rights of Indigenous Peoples[10] was adopted by the General Assembly in 2007, the result of more than 20 years of work by indigenous peoples and the United Nations system. The Declaration is the UN's most comprehensive statement of the rights of indigenous peoples, giving prominence to collective rights to a degree unprecedented in international human rights law. Its adoption is a clear indication that the international community is committing itself to the protection of the individual and collective rights of indigenous peoples. With respect to traditional knowledge, Article 31 of Declaration states that: (1) 'Indigenous peoples have the right to maintain, control, protect and develop their cultural heritage, traditional knowledge and traditional cultural expressions, including human and genetic resources, seeds, medicines, knowledge of the properties of fauna and flora, oral traditions, literatures, designs, sports and traditional games and visual and performing arts. They also have the right to maintain, control, protect and develop their intellectual property over such cultural heritage, traditional knowledge, and traditional cultural expressions,' and, (2) 'In conjunction with indigenous peoples, States shall take effective measures to recognize and protect the exercise of these rights.'

Other articles of the Declaration relate to issues that are fundamental to the protection and development of traditional forest-related knowledge and practices, including the collective and individual right to lands, territories, and resources that they have traditionally owned or occupied; the rights to self-government by their own institutions and authorities within their lands and territories; the rights to the conservation and protection of the environment and the productive capacity of their lands and resources; and the right to determine priorities and strategies for their development. The Declaration also calls on states to prevent dispossession of indigenous peoples from land, territories, and resources; allow indigenous peoples to participate in decision-making in matters affecting their rights; and to protect their right to be secure in their means of subsistence and development (Collings 2009; Lyster 2010).

UNCED and the Rio Conventions

Agenda 21 and the Forest Principles

The Programme of Action on Sustainable Development (Agenda 21) adopted by UNCED in Rio de Janeiro in 1992 includes a number of recommendations, directed primarily to countries with support from regional and international organizations,

---

[10] http://www.un.org/esa/socdev/unpfii/en/declaration.html

concerning the relevance of traditional knowledge for implementation of sustainable development policies and programmes (UNU-IAS 2005). These traditional knowledge-related recommendations address a variety of sustainability issues, including, among others: human health; deforestation; desertification; agriculture; and agricultural, forest, marine, and freshwater resource management. Dealing specifically with the connections between traditional and formal scientific knowledge, recommendation 35.7 (h) (on science for sustainable development) urges countries, with support of international organizations, to:

> Develop methods to link the findings of the established sciences with the indigenous knowledge of different cultures. The methods should be tested using pilot studies. They should be developed at the local level and should concentrate on the links between the traditional knowledge of indigenous groups and corresponding, current "advanced science", with particular focus on disseminating and applying the results to environmental protection and sustainable development.[11]

The Forest Principles—whose objective is to contribute to the management, conservation, and sustainable development of forests and to provide for their multiple and complementary functions—encourages countries to pursue forest policies that recognize and support the identity, culture, and rights of indigenous and local communities who depend on forests for their livelihoods. They also stress the importance of recognizing, respecting, recording, and developing indigenous capacity and local forest knowledge and the equitable sharing of benefits derived from indigenous forest-related knowledge.[12]

Convention on Biological Diversity

The Convention on Biological Diversity (CBD), which entered into force in 1993, encourages states to:

> Protect and encourage customary use of biological resources in accordance with traditional cultural practices that are compatible with conservation or sustainable use requirements[13] (CBD Article 10c).

Likewise, CBD Article 8(j) emphasizes:

> the need to respect, preserve and maintain knowledge, innovations and practices of indigenous and local communities embodying traditional lifestyles for the conservation and sustainable use of biological diversity and [promotion of] their wider application with the approval and involvement of such knowledge, innovations and practices and encourage the equitable sharing of the benefits arising from the utilization of such knowledge, innovations and practices.

---

[11] http://www.un.org/esa/dsd/agenda21/res_agenda21_33.shtml

[12] http://www.un.org/documents/ga/conf151/aconf15126-3annex3.htm

[13] The CBD Secretariat notes that this implies that governments should ensure that national legislation and policy account for and recognize, among others, indigenous legal systems, corresponding systems of governance and administration, land and water rights, and control over sacred and cultural (CBD 1997).

Since 1994, CBD's Conference of the Parties decisions related to the Convention's thematic areas and cross-cutting issues have routinely emphasized the important role of traditional knowledge and practices towards the achievement of the Convention's three major objectives: conservation of biological diversity, sustainable use of the components of biological diversity, and the fair and equitable sharing of the benefits arising out of the utilization of genetic resources. Traditional forest-related knowledge also is also explicitly considered in CBD's expanded programme of work on forest biological diversity adopted in decision VI/22 of the Convention's 6th Conference of the Parties in 2002, and calls on parties, governments, international, and regional organizations and processes, civil society organizations, and other relevant bodies to:

Support activities of indigenous and local communities involving the use of traditional forest-related knowledge in biodiversity management;

Encourage the conservation and sustainable use of forest biological diversity by indigenous and local communities through their development of adaptive management practices, using as appropriate traditional forest-related knowledge; [and to]

Implement effective measures to recognize, respect, protect and maintain traditional forest-related knowledge and values in forest-related laws and forest planning tools, in accordance with Article 8(j) and related provisions of the Convention on Biological Diversity.[14]

The CBD's Working Group implementation of Article 8 (j) and related provisions, established in 1998, has sought to raise the profile of indigenous and local community issues throughout the Convention. It has developed and monitored the implementation of the work programme on Article 8(j) and related provisions (adopted by the Conference of the Parties in 2000), as well a plan of action for the retention of traditional knowledge, innovations and practices, including the development of voluntary Akwé: Kon[15] guidelines (SCBD 2004) for the conduct of cultural, environmental and social impact assessment regarding developments proposed to take place on, or likely to have an impact on, sacred sites and lands and waters traditionally occupied or used by indigenous and local communities (adopted in 2004). The Working Group has also reported on the status and trends of traditional knowledge from all regions of the world and the identification of processes at national and local levels that may threaten the maintenance, preservation, and application of traditional knowledge.

At its 10th meeting in October 2010, the CDB Conference of the Parties (COP-10) adopted the Nagoya Protocol on Access to Genetic Resources and the Fair and Equitable Sharing of Benefits Arising from their Utilization (SCBD 2011), which includes traditional knowledge associated with genetic resources that is held by indigenous and local communities. Article 9 of the Protocol calls on parties (governments) to not restrict customary use and exchange of genetic resources and associated traditional knowledge within and amongst indigenous and local communities in accordance with the objectives of the CBD, and to take into consi-

---

[14] CBP COP Decision VI/22: http://www.cbd.int/decisions/cop/?m=cop-06

[15] Akwé: Kon is a Mohawk term meaning 'everything in creation.'

deration, and support development of, indigenous and local communities' customary laws, community protocols, and procedures with respect to access to traditional knowledge associated with genetic resources and the fair and equitable sharing of benefits arising from utilization of such knowledge.

COP-10 also adopted a decision (COP/10/L.38) including an ethical code of conduct,[16] and inviting parties and other governments to make use of this code to develop their own models on ethical conduct for research, access to, and use of information concerning traditional knowledge for the conservation and sustainable use of biodiversity, and to undertake education, awareness-raising, and communication strategies on the code for incorporation into policies and processes governing interactions with indigenous and local communities.

Currently under discussion within the CBD is the development of a joint programme of work on biological and cultural diversity, proposed during the International Conference on Biological and Cultural Diversity: Diversity for Development—Development for Diversity,[17] held in Montreal in June 2010. The proposed work program, led by the Secretariat of the CBD and UNESCO (acting as the global focal point on issues related to cultural diversity), would aim to better coordinate closely related activities of the CBD and UNESCO; to broaden the knowledge base on linkages between biological and cultural diversity including, among others, the ways in which cultural diversity has shaped biodiversity in sacred natural sites, cultural landscapes, and traditional agricultural systems; and to raise awareness of these linkages through educational activities and related work with decision makers.

UN Convention to Combat Desertification

The United Nations Convention to Combat Desertification[18] (UNCCD) was adopted in 1994 and entered into force in 1996. The Convention contains a number of provisions related to the protection, development, and use of traditional and local knowledge, technologies, know-how, and practices to combat desertification and mitigate the effects of drought on agricultural and pastoral systems and traditional livelihoods in dryland ecosystems. Since 1997, the UNCCD has been involved in the inventory of traditional knowledge systems through the work of the two ad hoc panels—regional cooperation within the Convention's Thematic Programme Networks, and National Action Programmes. Based on the work of the ad hoc panels, the Convention's Committee on Science and Technology prepared a compilation of official documents and reports focusing on the use of traditional and local

---

[16]Tkarihwaié:ri Code of Ethical Conduct on the Respect for the Cultural and Intellectual Heritage of Indigenous and Local Communities Relevant to the Conservation and Sustainable Use of Biological Diversity. Tkarihwaié:ri is a Mohawk term meaning 'the proper way.' Available via www.cbd.int/doc/decisions/cop-10/cop-10-dec-42-en.doc. Cited 24 March 2011.

[17] http://www.cbd.int/doc/meetings/development/icbcd/official/icbcd-scbd-unesco-en.pdf

[18]http://www.unccd.int/main.php

technologies, knowledge, know-how, and practices relevant to efforts to combat desertification and mitigate the effects of drought (UNCCD 2005).

UN Framework Convention on Climate Change

The United Nations Framework Convention on Climate Change (UNFCCC) is the principal international forum for negotiations among governments on issues related to climate change, and measures to be taken for its mitigation. Until recently the inclusion of forests, and forest-dependent people, in development of policies and programmes for climate change mitigation under the Kyoto Protocol was constrained until 2007, when the UNFCCC's 13th Conference of the Parties adopted the so-called Bali Action Plan,[19] or 'Bali roadmap,' for a future international agreement on climate change. The Bali Action Plan's overall goal is to develop a 'shared vision for long-term cooperative action, including a long-term global goal for emission reductions, to achieve the ultimate objective of the Convention ....' The Bali Road Map contains detailed lists of issues to be considered under topics related to climate change mitigation and adaptation actions, among others. There are no specific references to indigenous people or traditional knowledge in these issues. The issues do, however, refer to the economic and social consequences of response measures.

Indigenous peoples' groups (as well as many other key stakeholders in climate change policy) have for many years been disappointed by the limited consideration given to their views, knowledge, and interests within UNFCCC negotiations, and space allowed within the UNFCCC process for their involvement. For example, while indigenous peoples' organizations have been acknowledged as a constituency in climate change negotiations within UNFCCC since 2001, they are still awaiting UNFCCC's approval of an Ad Hoc Working Group on Indigenous Peoples and Climate Change, which would allow them to actively participate in the meetings of the Conference of Parties in the same way they are able to under the Convention on Biological Diversity (Collings 2009, Galloway-McLean 2010). However, recent developments since adoption of the Bali Action Plan should create opportunities within the UNFCCC (as well as the Intergovernmental Panel on Climate Change) processes for indigenous views and traditional knowledge about climate change to be incorporated in the development of future policies and commitments.

United Nations Forum on Forests

The United Nations Forum on Forests (UNFF[20]) was established in 2000 as a subsidiary body of the Economic and Social Council of the UN together with the Collaborative Partnership on Forests (CPF), comprising forest-related UN agencies

---

[19] http://unfccc.int/resource/docs/2007/cop13/eng/06a01.pdf#page=3

[20] http://www.un.org/esa/forests/about.html

and international and regional organizations, institutions, and instruments. Its main objective is to promote the management, conservation, and sustainable development of all types of forests and to strengthen long-term political commitment to this end.

Prior to the establishment of the UNFF, between 1995 and 2000, the Intergovernmental Panel on Forests (IPF), and the Intergovernmental Forum on Forests (IFF), both under the auspices of the United Nations Commission on Sustainable Development, were the main intergovernmental forums for international forest policy development. IPF and IFF examined a wide range of forest-related topics over a 5-year period. The final reports of these processes included 270 non-legally binding proposals for action towards sustainable forest management. These proposals for action include numerous references to traditional forest-related knowledge (TFRK) related to: the use of TFRK for sustainable forest management; development of intellectual property rights for TFRK and promotion of equitable benefit-sharing; technology transfer and capacity-building; and promotion of participation of people who possess TFRK in the planning, development, and implementation of national forest policies and programmes.

Although the UNFF failed to adopt a decision on traditional forest-related knowledge during its 4th meeting in 2004, the Non-Legally Binding Instruments on All Types of Forests (NLBI) adopted by the UNFF in 2007 (United Nations 2007) commits member states to:

> Support the protection and use of traditional forest-related knowledge and practices in sustainable forest management with the approval and involvement of the holders of such knowledge, and promote fair and equitable sharing of benefits from their utilization, according to national legislation and relevant international agreements (NLBI, para. 6(f)).

The first inter-governmental instrument on sustainable forest management, the NLBI covers issues ranging from protection and use of traditional forest-related knowledge and practices in sustainable forest management, to the need for enhanced access to forest resources and relevant markets to support the livelihoods of forest-dependent indigenous communities living inside and outside forest areas.

Despite UNFF's recognition of the role that indigenous and local communities play in achieving sustainable forest management, indigenous peoples' organizations and civil society have generally been disappointed with the results achieved by the UNFF (Collings 2009). Many argue that the UNFF does not build on what were seen as the more open and progressive practices of the IPF/IFF and Commission on Sustainable Development (Forest Peoples Programme 2004). The NLBI has also been criticized for its failure to recognize, respect and support the implementation of customary rights of indigenous peoples who live in and depend on forests and for failing to comply with best practices in environment management (Forest Peoples Programme 2007).

United Nations Educational, Scientific and Cultural Organization (UNESCO)

Under the auspices of UNESCO, two conventions have been adopted that are relevant to the preservation and development of traditional forest-related knowledge: the Convention for the Safeguarding of Intangible Cultural Heritage (2003), and the Convention on the Protection and Promotion of the Diversity of Cultural Expressions (2005).

However, UNESCO has been criticized for exclusion of indigenous peoples in their drafting, and the inadequate acknowledgement that a large part of 'cultural heritage' and 'cultural expressions' that these conventions deal with is the heritage of indigenous peoples and indigenous cultures (Kipuri 2009).

TFRK and Intellectual Property Protection

Issues related to access, benefit sharing, and protection of intellectual property rights (IPRs) with respect to traditional knowledge are considered in a number of international agreements and in ongoing policy debates, largely within the UN system. These involve a wide array of international forums and intergovernmental organizations concerned with food and agriculture, conservation, health, human rights, indigenous issues, and development and trade (UNU-IAS 2005). Prominent among them, discussed above, are the UN Permanent Forum on Indigenous Issues, the Declaration on the Rights of Indigenous Peoples, the Convention on Biological Diversity, the UN Convention to Combat Desertification, and the UNFF's Non-Legally Binding Instruments on All Types of Forests. Traditional forest-related knowledge issues also figure prominently in the work of bodies and processes that deal more directly with intellectual property protection, such as the World Intellectual Property Organization (WIPO), established as a UN agency in 1967, and the World Trade Organization (WTO).

Although international human rights standards since 1948 have recognized the importance of protecting intellectual property (IP), these rights have not yet been extended to the holders of traditional knowledge. Requirements for IP protection under current IP regimes are widely considered to be inconsistent or incompatible with the nature of traditional knowledge (Hansen and Van Vleet 2007). The approaches taken towards traditional knowledge in these organizations and processes, each with their own mandate, are often very different, not very well-coordinated, and at times conflicting. For example, while the Convention on Biological Diversity is generally more open to approaches that recognize collective ownership of traditional knowledge, innovations, and cultural expressions that are more compatible with the philosophical and cultural standards of indigenous and local communities, the emphasis within WIPO and WTO is on promoting existing IPR regimes (i.e., copyright, trademark, and patent systems) that have evolved based on assumptions of individual 'ownership' of such knowledge and expressions.

The issue of patenting of life forms (species and varieties of plants, animals, and micro-organisms), relevant to the conservation and development of traditional forest-related knowledge, is a particularly contentious one in all forums dealing with intellectual property protection. It is of particular concern to farmers in traditional communities who, through many generations, have developed and applied traditional knowledge and practices to develop new crop varieties that have created and continue to enrich the world's agricultural biodiversity.

The World Trade Organization's Agreement on Trade-Related Aspects of Intellectual Property Rights (TRIPS), negotiated in the 1986–1994 Uruguay Round, introduced IPR rules into the multilateral trading system for the first time in its Article 27.3(b), which deals with patentability or non-patentability of plant and animal inventions, and the protection of plant varieties. The 2001 Doha Declaration broadened discussion of IPR issues by directing the TRIPS Council to review TRIPS Article 27.3(b), specifically the relationship between the TRIPS Agreement and the UN Convention on Biological Diversity with respect to the protection of traditional knowledge and folklore.

WIPO's Intergovernmental Committee on Intellectual Property and Genetic Resources, Traditional Knowledge and Folklore (IGC) undertakes negotiations with the objective of reaching agreement on principles for, and an international legal instrument (or instruments) intended to ensure, the effective protection of traditional knowledge (TK), traditional cultural expressions (TCEs), folklore, and genetic resources (WIPO 2010a, b). Standards under development by the IGC focus on protection against misappropriation of TK, and they attempt to complement other international instruments and processes dealing with other aspects of the preservation, safeguarding, and conservation of such knowledge (UNU-IAS 2005). At a recent meeting of the IGC's Second Intersessional Working Group (in February 2011), consolidated, streamlined text on the protection of traditional knowledge was produced (on issues related to a definition of traditional knowledge, beneficiaries of protection, and the scope of rights to be granted, and how they would be managed and enforced). The results of this meeting[21] were considered by many to be a step towards an international legal instrument to ensure the effective protection of traditional knowledge.

The United Nations Conference on Trade and Development (UNCTAD) also seeks to address the protection of traditional knowledge as part of its work, recognizing the need to harmonize its work in this area with relevant organizations such as WIPO, WHO, and CBD. In 2001, UNCTAD Trade and Development Board's Commission on Trade in Goods and Services and Commodities recommended that international efforts be made to promote capacity-building to implement protection regimes for traditional knowledge, promote fair and equitable sharing of benefits derived from this knowledge in favour of local and traditional communities, exchange information on national systems to protect traditional knowledge, and explore minimum standards for internationally recognized sui generis systems for traditional

---

[21] The agreed-upon text was scheduled to be considered at the next meeting of the IGC in May 2011.

knowledge protection. UNCTAD's mandate for work on traditional knowledge was reaffirmed at its 11th Conference in São Paolo in 2004 (UNU-IAS 2005).

Intellectual property rights and benefit-sharing issues relevant to traditional forest-related knowledge have been given significant attention by the Food and Agriculture Organization of the United Nations (FAO) and the Consultative Group on International Agricultural Research (CGIAR). FAO's International Treaty on Plant Genetic Resources for Food and Agriculture (PGRFA), which entered into force in 2004, focuses on the conservation and sustainable use of plant genetic resources for food and agriculture, and the fair and equitable sharing of benefits derived from their use, for sustainable agriculture and food security (FAO 2009). Closely related to this treaty, FAO's Global Plan of Action for the Conservation and Sustainable Use of Plant Genetic Resources for Food and Agriculture promote a number of specific actions by countries to realize the objectives of the treaty (UNU-IAS 2005).

Within the CGIAR, the Genetic Resources Policy Committee (GRPC) provides guidance to the CGIAR on policy issues surrounding genetic resources in line with the provisions of the PGRFA (SGRP 2010). These include research ethics guidelines that are relevant to the study of traditional knowledge and practices within CGIAR centres and the agriculture and forest science community more generally. These and other guidelines for researchers are discussed in Chap. 14 of this book.

UNESCO's Man and the Biosphere (MAB) and LINKS Programmes

Concerned with problems at the interface of scientific, environmental, societal, and development issues, UNESCO's Man and the Biosphere (MAB) Programme,[22] initiated in the early 1970s, aims to improve human livelihoods and safeguard natural ecosystems by promoting innovative approaches to economic development that are socially and culturally appropriate and environmentally sustainable. The programme supports research involving natural and social sciences, economics, and education. Its interdisciplinary research agenda and capacity-building activities target the ecological, social, and economic dimensions of biodiversity loss and the reduction of this loss through empowerment of local and indigenous peoples in various aspects of environmental management. The biosphere reserve concept, developed through the MAB Programme (Batisse 1982), has emerged as a widely used model for integrating conservation of biological and cultural diversity while promoting sustainable economic and social development of cultural landscapes based on traditional values, local community efforts, and conservation science, particularly in the 'buffer zones' of protected areas (Ramakrishnan et al. 2002).

---

[22] http://www.unesco.org/new/en/natural-sciences/environment/ecological-sciences/man-and-biosphere-programme/

In 2002, UNESCO launched the Local and Indigenous Knowledge Systems in a Global Society (LINKS) programme, a collaborative multidisciplinary effort (involving all programme sectors of UNESCO) focusing on local and indigenous knowledge. The programme aims at empowering local and indigenous peoples in various aspects of environmental management by advocating recognition and mobilization of their traditional knowledge. Its goals include exploration of synergies between traditional and scientific knowledge as a means to: enhance biological and cultural diversity; revitalize intergenerational transmission of traditional knowledge within local communities; identify customary rules and processes that govern knowledge access and control; improve dialogue amongst traditional knowledge holders, natural and social scientists, resource managers, and decision makers to enhance biodiversity conservation efforts; and promote active and equitable roles for local communities in natural resource governance (UNU-IAS 2005; Kipuri 2009).

United Nations University–Institute of Advanced Studies (UNU–IAS)

The UNU–IAS Traditional Knowledge Initiative[23] seeks to build greater understanding and facilitate awareness of traditional knowledge to inform action by indigenous peoples, local communities, and domestic and international policy makers. The Initiative supports research activities, policy studies, capacity development, and online learning and dissemination in several thematic areas. These include: traditional knowledge and climate change, with activities including the Indigenous Peoples Climate Change Assessment and Indigenous Perspectives on Climate Change; traditional knowledge and biological resources; and traditional knowledge and natural resources, with a special focus on forests and forest management, marine resources, and water management.

Information services provided by the Initiative include newsletters, reviews of thematic areas (such as the REDD site, available online at: http://www.unutki.org/redd/), focusing on indigenous peoples and Reducing Emissions from Deforestation and Degradation (REDD), and the Traditional Knowledge Bulletin (available online at: http://unu.edu/tk/). The Bulletin is a weekly review of traditional knowledge issues in the global media and postings on issues of relevance to traditional knowledge at a global level, including issues discussed at international meetings and fora, such as:

- World Intellectual Property Organization's Inter-governmental Committee on Intellectual Property, Genetic Resources, Traditional Knowledge and Folklore;
- World Trade Organization;
- United Nations Permanent Forum for Indigenous Peoples;
- United Nations Inter-Agency Support Group on Indigenous Issues;
- United Nations Educational, Scientific and Cultural Organization (UNESCO);

---

[23]http://www.unutki.org/

- Food and Agriculture Organization of the United Nations (FAO) (e.g., the International Treaty on Plant Genetic Resources);
- Convention on Biological Diversity (CBD);
- United Nations Conference on Trade and Development (UNCTAD);
- processes of the World Bank and regional development banks;
- initiatives such as the Millennium Ecosystem Assessment and the International Assessment of Agricultural Science and Technology; and
- work of regional organizations such as the Organization of American States.

The Initiative's TK and Higher Education project explores the integration of indigenous knowledge in higher education programmes through relevant topics and methodologies in an attempt to gain recognition and valuation for traditional knowledge in academic and scientific circles, an important step in finding points of convergence between Western scientific and traditional indigenous understanding. The International Policy Making project supports research and training to facilitate the participation and empowerment of traditional knowledge holders in relevant policy processes. Initial activities of the project have focussed on information and policy analysis for indigenous and local communities on emerging issues in traditional knowledge discussions in international fora.

References

Alternatives to Slash and Burn [ASB] (2011) ASB partnership for the tropical forest margins. Available via http://www.asb.cgiar.org/. Cited 25 Mar 2011

Basso KH (1996) Wisdom sits in places: landscape and language among the western Apache. University of New Mexico Press, Albuquerque

Batisse M (1982) The biosphere reserve: a tool for environmental conservation and management. Environ Conserv 9:101–111

Berkes F (2008) Sacred ecology, 2nd edn. Routledge, New York

Berkes F, Folke C (eds) (1998) Linking social and ecological systems: management practices and social mechanisms for building resilience. Cambridge University Press, Cambridge

Berkes F, Colding J, Folke C (2000) Rediscovery of traditional ecological knowledge as adaptive management. Ecol Appl 10:1251–1262

Berkes F, Colding J, Folke C (eds) (2003) Navigating social-ecological systems: building resilience for complexity and change. University of Cambridge Press, Cambridge

Blake EA, Wagner MR (1987) Collection and consumption of Pandora moth, *Coloradia pandora lindseyi* (Lepidoptera: Saturniidae), larvae by Owens Valley and Mono Lake Paiutes. Bull Entomol Soc Am 33(1):22–27

Burton L (2002) Worship and wilderness: culture, religion, and law in the management of public lands and resources. University of Wisconsin Press, Madison

Cajete G (1999) Native science: natural laws of interdependence. Clear Light Publishers, Santa Fe

Colfer CJP (ed) (2005) The complex forest: communities, uncertainty, and adaptive collaborative management. Resources for the Future, Washington, DC

Collings N (2009) Environment. In: United Nations The state of the world's indigenous peoples. Department of Economic and Social Affairs, Division for Social Policy and Development, Secretariat of the Permanent Forum on Indigenous Issues Report No. ST/ESA/328, pp 84–127

Convention on Biological Diversity [CBD] (1997) Traditional knowledge and biological diversity. Doc. UNEP/CBD/TKBD/1/2. Available via http://www.cbd.int/doc/meetings/tk/wstkbd-01/official/wstkbd-01-02-en.pdf. Cited 5 Mar 2011

Costanza R, Cumberland J, Daly H, Goodland R, Norgaard R (1997) An introduction to ecological economics. St. Lucie Press, Boca Raton

Cruikshank J (2005) Do glaciers listen? Local knowledge, colonial encounters, and social imagination. University of British Columbia Press, Vancouver

Daly R (2005) Our box was full: an ethnography for the Delgamuukw plaintiffs. University of British Columbia Press, Vancouver

Davidson-Hunt I, Smith P, Burlando C (2010) When a bill passes in the wilderness, does anybody hear? CEESP News 7 September. Available via http://www.iucn.org/about/union/commissions/ceesp/ceesp_news/?5968/When-a-Bill-Passes-in-the-Wilderness-Does-Anyone-Hear. Cited 6 Apr 2011

Davis T (2000) Sustaining the forest, the people, and the spirit. State University of New York Press, Albany

Feit HA (1992) Waswanipi Cree management of land and wildlife: Cree ethno-ecology revisited. In: Cox BA (ed) Native people, native lands: Canadian Indians, Inuit and Metis, vol 142, Carleton Library Series. Carleton University Press, Ottawa, pp 75–91

Folke C, Carpenter S, Elmqvist T, Gunderson L, Holling CS, Walker B, Bengtsson J, Berkes F, Colding J, Danell K et al (2002) Resilience and sustainable development: building adaptive capacity in a world of transformations. The Environmental Advisory Council to the Swedish Government, Stockholm

Food and Agriculture Organisation [FAO] (2009) International treaty on plant genetic resources for food and agriculture. FAO, Rome. Available via http://www.planttreaty.org/texts_en.htm. Cited 5 Mar 2011

Forest Peoples Programme (2004) Briefing on the United Nations Forum on Forests (UNFF) and Collaborative Partnership on Forests (CPF). Available via http://www.forestpeoples.org. Cited 5 Mar 2011

Forest Peoples Programme (2007) The UNFF fails indigenous peoples again. Briefing Note. Available via http://www.forestpeoples.org. Cited 5 Mar 2011

Fortmann L (2008) Introduction: doing science together. In: Fortmann L (ed) Participatory research in conservation and rural livelihoods: doing science together. Wiley, Hoboken, pp 1–17

Fowler V (2005) Introduction. In: YoungBear-Tibbetts H, Van Lopik W, Hall K (eds) Sharing indigenous wisdom: an international dialogue on sustainable development. College of Menominee Nation Press, Keshena

Fricker M (2007) Epistemic injustice: power and the ethics of knowing. Oxford University Press, Oxford

Galloway-McLean K (2010) Advance guard: climate change impacts, adaptation, mitigation and indigenous peoples—a compendium of case studies. United Nations University-Traditional Knowledge Initiative, Darwin. Available via http://www.unutki.org/news.php?doc_id=101&news_id=92. Cited 5 Mar 2011

Gieryn TF (1999) Cultural boundaries of science: credibility on the line. University of Chicago Press, Chicago

Gray A (1991) Between the spice of life and the melting pot: biodiversity conservation and its impact on indigenous peoples. IWGIA document no. 70, International Work Group for Indigenous Affairs, Copenhagen

Gunderson LH, Holling CS (eds) (2002) Panarchy: understanding transformations in human and natural systems. Island Press, Washington, DC

Hansen SA, Van Fleet JW (2007) Issues and options for traditional knowledge holders in protecting their intellectual property. In: Krattiger A, Mahoney RT, Nelsen L, Thomson JA, Bennett AB, Satyanarayana K, Graff GD, Fernandez C, Kowalski SP (eds) Intellectual property management in health and agricultural innovation: a handbook of best practices. MIHR/PIPRA, Oxford/Davis, pp 1523–1538. Available via awww.ipHandbook.org. Cited 5 Mar 2011

Harmon D (2002) In light of our differences: how diversity in nature and culture makes us human. Smithsonian Institution Press, Washington, DC

Heal GM (1998) Valuing the future: economic theory and sustainability. Economics for a Sustainable Earth Series. Columbia University Press, New York

Holling CS (1986) Resilience of ecosystems: local surprise and global change. In: Clark WC, Munn RE (eds) Sustainable development of the biosphere. Cambridge University Press, Cambridge, pp 292–317

International Council for Science [ICSU] (2002) Science, traditional knowledge and sustainable development, 4th edn, ICSU Series on Science for Sustainable Development, ICSU, Paris. Available via http://www.icsu.org/Gestion/img/ICSU_DOC_DOWNLOAD/65_DD_FILE_Vol4.pdf. Cited 11 Feb 2011

Johannes RE (1981) Words of the lagoon: fishing and marine lore in the Palau District of Micronesia. University of California Press, Berkeley

Kimmins JP (2004) Forest ecology: a foundation for sustainable forest management and environmental ethics in forestry, 3rd edn. Pearson Prentice Hall, Upper Saddle River

Kipuri N (2009) Culture. In: United Nations, The state of the world's indigenous peoples. Department of Economic and Social Affairs, Division for Social Policy and Development, Secretariat of the permanent forum on indigenous issues report no. ST/ESA/328, pp 52–81

Latour B (1993) We have never been modern (trans: Porter C). Harvard University Press, Cambridge

Latour B (2004) Politics of nature: how to bring the sciences into democracy. Harvard University Press, Cambridge

Latour B (2005) Reassembling the social: an introduction to actor-network-theory. Clarendon lectures in management studies. Oxford University Press, Oxford

Loh J, Harmon D (2005) A global index of biocultural diversity. Ecol Indic 5:231–241

Lyster R (2010) REDD+, transparency, participation and resource rights: the role of law. Legal studies research paper no. 10/56. The University of Sydney Law School, Sydney. Available via http://ssrn.com/abstract=1628387. Cited 5 Mar 2011

Maffi L (ed) (2001) On biocultural diversity: linking language, knowledge, and the environment. Smithsonian Institution Press, Washington, DC

Maffi L (2005) Linguistic, cultural, and biological diversity. Annu Rev Anthropol 29:599–617. doi:10.1146/annurev.anthro.34.081804.120437. Cited 5 March 2011

Martínez Cobo J (1986/7) Study of the problem of discrimination against indigenous populations. UN Doc. E/CN.4/Sub.2/1986/7 and Add. 1–4. Available via http://www.un.org/esa/socdev/unpfii/en/spdaip.html. Cited 5 Mar 2011

Mery G, Alfaro R, Kanninen M, Lovobikov M (eds) (2005) Forests in the global balance—changing paradigms, vol 17, IUFRO World Series. International Union of Forest Research Organizations [IUFRO], Helsinki

Miller AM, Davidson-Hunt IJ, Peters P (2010) Talking about fire: Pikangikum first nation elders guiding fire management. Can J For Res 40:2290–2301

Mills A (1994) Eagle down is our law: Witsuwit'en law, feasts, and land claims. UBC Press, Vancouver

Mills A (ed) (2005) 'Hang onto these words': Johnny David's Delgamuukw evidence. University of Toronto Press, Toronto

Moore JL, Manne L, Brooks T, Burgess ND, Davies R, Rahbek C, Williams P, Balmford A (2002) The distribution of cultural and biological diversity in Africa. Proc Royal Soc Lond B 269:1645–1653

Nadasdy P (2003) Hunters and bureaucrats: power, knowledge, and aboriginal-state relations in the southwest Yukon. University of British Columbia Press, Vancouver

Newing H (2005) A summary of case study findings on the implementation of international commitments on traditional forest related knowledge. In: Newing H, Pinker A, Leake H (eds) Our knowledge for our survival, vol 1. International Alliance of the Indigenous and Tribal Peoples of the Tropical Forest (IAITPTF) and Centre for International Forestry Research (CIFOR), Chiang Mai, pp 11–64

Newing H, Pinker A, Leake H (eds) (2005) Our knowledge for our survival, 2 vols. International Alliance of the Indigenous and Tribal Peoples of the Tropical Forest (IAITPTF) and Centre for International Forestry Research (CIFOR), Chiang Mai. Available via http://www.international-alliance.org/documents/overview-finaledit.pdf. Cited 5 Mar 2011

Norgaard RB (2010) Ecosystem services: from eye-opening metaphor to complexity blinder. Ecol Econ 69(6):1219–1227

O'Flaherty R, Davidson-Hunt I, Manseau M (2008) Indigenous knowledge and values in planning for sustainable forestry: Pikangikum first nation and the Whitefeather forest initiative. Ecol Soc 13(1):6–16

Pecore M (1992) Menominee sustained yield management: a successful land ethic in practice. J For 90:12–16

Pierotti R (2010) Indigenous knowledge, ecology and evolutionary biology. Routledge, London

Posey DA (ed) (1999) Cultural and spiritual values of biodiversity. Intermediate Technology Publications, UNEP, London

Ramakrishnan PS (1992) Shifting agriculture and sustainable development: an interdisciplinary study from north-eastern India, vol 10, Man and Biosphere Book Series. UNESCO/Parthenon Publishing, Paris/Caernforth

Ramakrishnan PS, Rai RK, Katwal RPS (2002) Traditional ecological knowledge for managing biosphere reserves in South and Central Asia. Oxford/IBH Publishing, New Delhi

Ricklefs RE, Miller GL (2000) Ecology, 4th edn. Freeman, New York

Rosendal GK (2001) Overlapping international regimes—the case of the Intergovernmental Forum on Forests (IFF) between climate change and biodiversity. Int Environ Agree Polit Law Econ 1:447–468

Secretariat of the Convention on Biological Diversity [SCBD] (2004) Akwé: Kon—voluntary guidelines for the conduct of cultural, environmental and social impact assessment regarding developments proposed to take place on, or which are likely to impact on, sacred sites and on lands and waters traditionally occupied or used by indigenous and local communities. CBD Guidelines Series, Montreal. Available via http://www.cbd.int/doc/publications/akwe-brochure-en.pdf. Cited 5 Mar 2011

Secretariat of the Convention on Biological Diversity [SCBD] (2011) Nagoya protocol on access to genetic resources and the fair and equitable sharing of benefits arising from their utilization to the convention on biological diversity: text and annex. Montreal. Available via http://www.cbd.int/abs/doc/protocol/nagoya-protocol-en.pdf. Cited 24 Mar 2011

Secretariat of the Permanent Forum on Indigenous Issues [UNPFII] (2005) Background note prepared by the Secretariat of the permanent forum for the international workshop on traditional knowledge, Panama City, 21–23 Sept 2005. Doc. PFII/2005/WS.TK. Available via http://www.un.org/esa/socdev/unpfii/documents/workshop_TK_background_note.pdf. Cited 5 Mar 2011

Sutherland WJ (2003) Parallel extinction risk and global distribution of languages and species. Nature 423:276–279

System-wide Genetic Resources Programme [SGRP] (2010) Booklet of CGIAR Centre policy instruments, guidelines and statements on genetic resources, biotechnology and intellectual property rights, version III. System-wide Genetic Resources Programme [SGRP] and the CGIAR Genetic Resources Policy Committee [GRPC]. Biodiversity International, Rome. Available via http://sgrp.cgiar.org/?q=publications. Cited 5 Mar 2011

Tanner A (1979) Bringing home the animals: religious ideology and mode of production of Mistassini Cree hunters. Hurst, London

Toledo VM, Ortiz-Espejel B, Moguel P, Ordoñez MDJ (2007) The multiple use of tropical forests by indigenous peoples in Mexico: a case of adaptive management. Conserv Ecol 7(3):9

Trosper RL (2007) Indigenous influence on forest management on the Menominee Indian Reservation. For Ecol Manag 249:134–139

Trosper RL (2009) Resilience, reciprocity and ecological economics: sustainability on the northwest coast. Routledge, New York

Turner NJ (2005) The earth's blanket: traditional teachings for sustainable living. Douglas and McIntyre, Vancouver

Turner NJ, Ignace MB, Ignace R (2000) Traditional ecological knowledge and wisdom of aboriginal peoples in British Columbia. Ecol Appl 10(5):1275–1287

Umeek ERA (2004) Tsawalk: a Nuu-chah-nulth worldview. University of British Columbia Press, Vancouver

United Nations [UN] (2004) Traditional forest-related knowledge: report of the Secretary-General, United Nations Forum on Forests 4th session. Document E/CN.18/2004/7. Available via http://daccess-dds-ny.un.org/doc/UNDOC/GEN/N04/261/74/PDF/N0426174.pdf?OpenElement. Cited 5 Mar 2011

United Nations [UN] (2007) United Nations Forum on Forests, report of the seventh session. Economic and Social Council, Official Records, Supplement No. 22, E/2007/42, E/CN.18/2007/8. Available via http://daccess-dds-ny.un.org/doc/UNDOC/GEN/N07/349/31/PDF/N0734931.pdf?OpenElement. Cited 5 Mar 2011

United Nations [UN] (2009) The state of the world's indigenous peoples. Department of Economic and Social Affairs, Division for Social Policy and Development, Secretariat of the permanent forum on indigenous issues report no. ST/ESA/328

United Nations Convention to Combat Desertification [UNCCD] (2005) Revitalizing traditional knowledge. A compilation of documents and reports from 1997 to 2003. Committee on Science and Technology, Bonn

United Nations University-Institute of Advanced Studies [UNU-IAS] (2005) Establishing a UNU initiative on traditional knowledge. Draft report, Traditional Knowledge Initiative, Darwin

UNPFII (2005) See Secretariat of the Permanent Forum on Indigenous Issues

Wa G, Uukw D (1992) The spirit in the land: statements of the Gitksan and Wet'suwet'en hereditary chiefs in the supreme court of British Columbia, 1987–1990. Reflections, Gabriola

Walens S (1981) Feasting with cannibals: an essay on Kwakiutl cosmology. Princeton University Press, Princeton

Weatherford JM (1988). Indian giving: how the Indians of the Americas transformed the world. Crown Publishers, New York

World Health Organization [WHO] (2002) WHO traditional medicine strategy 2002–2005. Publ. WHO/EDM/TRM/2002.1. WHO, Geneva. Available via http://www.who.int/medicines/publications/traditionalpolicy/en/index.html. Cited 5 Mar 2011

World Health Organization [WHO] (2003) WHO guidelines on good agricultural and collection practices (GACP) for medicinal plants. WHO, Geneva. Available via http://apps.who.int/medicinedocs/en/d/Js4928e/. Cited 5 Mar 2011

World Health Organization [WHO], International Union for Conservation of Nature and Natural Resources [IUCN], World Wide Fund for Nature [WWF] (1993) Guidelines on the conservation of medicinal plants. WHO, IUCN, WWF, Gland. Available via http://apps.who.int/medicinedocs/documents/s7150e/s7150e.pdf. Cited 5 Mar 2011

World Intellectual Property Organization [WIPO] (2010a) The protection of traditional knowledge: revised objectives and principles. Secretariat document prepared for the 17th session (December 2010) of the Intergovernmental Committee on Intellectual Property and Genetic Resources, Traditional Knowledge and Folklore. WIPO/GRTKF/IC/17/5. Geneva. Available via http://www.wipo.int/tk/en/igc/index.html. Cited 5 Mar 2011

World Intellectual Property Organization [WIPO] (2010b) The genetic resources: revised list of options and factual update. Secretariat document prepared for the 17th session (December 2010) of the Intergovernmental Committee on Intellectual Property and Genetic Resources, Traditional Knowledge and Folklore. WIPO/GRTKF/IC/17/6. Geneva. Available via http://www.wipo.int/tk/en/igc/index.html. Cited 5 Mar 2011

Chapter 2
Africa

Alfred Oteng-Yeboah, Doris Mutta, Dominic Byarugaba, and William Armand Mala

Abstract The rich body of traditional forest-related knowledge (TFRK) in Africa has been widely acknowledged as important for its contribution to current global efforts towards sustainable forest management and biodiversity conservation. While many rural communities in Africa continue to observe their age-old traditions in relation to forests to ensure the provision of their livelihoods, other communities have lost their traditions for many reasons, including their forced or voluntary cultural alienation from forests, reduced dependence on forests for rural livelihoods, and extensive urbanization. Nonetheless, many communities throughout Africa are still living in or near the continent's diverse range of forest ecosystems and continue to depend on these forests for their livelihoods. A documentation of how communities have successfully managed these forests to provide for their needs until the present day can serve many useful purposes, including for evidence-based sharing of experiences or case studies, research adoption and uptake, and knowledge transfer and training in forestry curricula. In this chapter, we provide a general background on traditional forest-related knowledge in Africa; its historical and present contributions to food security and rural livelihoods; the present

A. Oteng-Yeboah (✉)
Department of Botany, University of Ghana, Legon, Ghana
e-mail: alfred.otengyeboah@gmail.com

D. Mutta
Kenya Forestry Research Institute, Nairobi, Kenya
e-mail: doris_mutta@yahoo.com

D. Byarugaba
Department of Biology, Faculty of Science, Mbarara University of Science
and Technology, Mbarara, Uganda
e-mail: dbbyarugaba@yahoo.com

W.A. Mala
Department of Plant Biology, Yaounde I University, Yaoundé, Cameroon
e-mail: williammala@yahoo.fr

J.A. Parrotta and R.L. Trosper (eds.), *Traditional Forest-Related Knowledge:*   37
Sustaining Communities, Ecosystems and Biocultural Diversity,
World Forests 12, DOI 10.1007/978-94-007-2144-9_2,
© Springer Science+Business Media B.V. (outside the USA) 2012

challenges faced by the holders and users of this knowledge; and opportunities for its preservation, enhancement, and application to help solve pressing environmental, economic, and social challenges, including the conservation and sustainable use of forest biodiversity.

Keywords Agroforestry • Biodiversity • Food security • Forest ecosystems • Forest governance • Local communities • Livelihoods • Non-timber forest products • Shifting cultivation • Traditional knowledge

2.1   Introduction

In African countries that have been endowed with forests and savanna woodlands, local populations have depended through countless generations on the fauna and flora of these ecosystems for their survival. Through long experience, these communities have accumulated a significant body of knowledge on the sustainable conservation, harvesting, processing, and utilization of forest resources. The close associations between local people and the different forest landscapes in which they live illustrate the adaptations that local communities have undergone while reshaping their cultural identities. Stories of these adaptations underpin the history of traditional forest-related knowledge (TFRK) in Africa.

An examination traditional forest-related knowledge in the African continent underscores the dependence of people on forests for their livelihoods, food security, energy requirements, medicines, and several other environmental services including social, cultural, and spiritual benefits. This attachment of people to their forest environments has defined the cultural heritage of almost all the forest-dependent peoples of Africa, in which forests are the focus of myths, artefacts, and folklores. This knowledge has become a source of pride for people. When an African refers to his or her ancestry, there is always a mention of the intricate system of cultural heritage that is associated with the past history of the people's interactions with their environment and from which several of management and governance systems of natural resources have emerged.

Unfortunately, this body of knowledge is being lost because of alienation of communities from their lands and traditional livelihoods, as a result of rapid urbanization, particularly during the past half century, and the promotion of foreign technologies and (formal) science-based approaches in agriculture and forestry. Despite the erosion of traditional forest-related knowledge, much of this knowledge still exists, as do opportunities to properly identify, acknowledge, and document it; to combine it with modern science and technology; and for communities and others to use this knowledge more broadly in a variety of forests and savanna woodland management applications.

In this chapter we provide an overview of Africa's forest ecosystem and biocultural diversity; traditional perspectives on forests in relation to livelihoods, culture, biodiversity conservation and poverty reduction goals; the role of forest-related knowledge and practices for food security, sustainable forest management, and forest governance; policy issues and processes affecting the preservation and development of traditional knowledge; and forestry education and the role of forest science in the African context of traditional knowledge.

2.1.1 Africa's Forest Ecosystems and Forest-Dependent Communities

The overwhelming majority of Africa's forests and wooded savannas are found in the tropical sub-Saharan regions, including the Sahel, from Senegal eastwards through central Sudan and Eritrea, and in the Horn of Africa (eastern Ethiopia and Somalia), southwards through Angola, eastern Namibia, Zimbabwe, and eastern and south-coastal South Africa (Fig. 2.1). These forests represent approximately 16% of the world's total forest cover (FAO 2006). They include a variety of dry and moist tropical forest types, and less extensive temperate and Mediterranean forest types in the Atlas Mountains (of Morocco, Algeria, and Tunisia) and in southern Africa. Tropical dry forests and wooded savannas (parklands) comprise nearly two-thirds (63.0%) of the total, followed by tropical moist forests (31.6%), and temperate forests (5.4%) (UNEP-WCMC 2011).

Dry forest types predominate throughout sub-Saharan Africa, with the exception of the moist forest regions of the Congo Basin and parts of West Africa, the Ethiopian Highlands, Uganda and easternmost Democratic Republic of Congo (DRC). They occur in all countries from the Sahel to Namibia, Zimbabwe, Angola, and Madagascar. These forest formations include sclerophyllous dry forests, deciduous/semi-deciduous broadleaf forests, thorn forests, and sparse tree/parkland forests. Finally, temperate forests include mixed broadleaf and needle leaf, deciduous broadleaf, and sclerophyllous dry forests found mainly at higher elevations from Morocco to Tunisia, and in eastern and south-coastal South Africa, southern Botswana, and southern Angola (Iremonger et al. 1997; UNEP-WCMC 2011).

The major tropical moist forest types include:

- Mangrove forests, occurring along the coasts of West Africa from Senegal to the DR Congo, and along the east coast from Kenya to Mozambique;
- Freshwater swamp forests in the central Congo Basin along the Congo (or Zaïre) River;
- Lowland evergreen broadleaf rainforests occurring in relatively small areas of West Africa from Sierra Leone eastwards to Nigeria, and over very extensive areas of the Congo Basin region from Cameroon to DR Congo eastwards to the Mitumba Mountains of eastern DR Congo, as well as in the eastern highlands of Madagascar;
- Lower and upper montane forests, widely scattered throughout the sub-Saharan region but including more extensive areas in Ethiopia, Kenya, Angola, and Madagascar; and

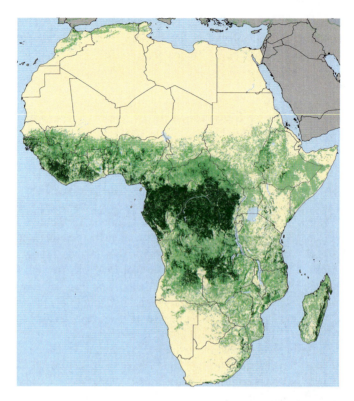

Fig. 2.1 Forest and woodland cover in Africa (Source: Adapted from FAO (2001)). Key: *Dark green* = closed forest; *light green* = open or fragmented forest; *pale green* = other wooded land; *yellow* = other land

- Semi-evergreen moist broadleaf forests in Gambia and Senegal as well as in Uganda, Rwanda, and Burundi.

Of these, which were estimated to cover 25,875 km^2 in 2000, lowland evergreen broadleaf rainforest comprised 65% (Table 2.1; Iremonger et al. 1997; UNEP-WCMC 2011).

According to data from the FAO Global Forest Resources Assessment, Africa's total forest area in 2005 was estimated to be 6.35 million km^2, a decline from the 1990 and 2000 estimates of 6.99 and 6.56 million ha, respectively (FAO 2006; Table 2.1).[1] Forest cover loss in all regions of Africa over the past 20 years, as

---

[1] Forest areas estimated by the UNEP-WCMC (Table 2.1) are somewhat higher—8.12 km^2 in 2000. Differences between FAO and UNEP-WCMC forest cover estimates may be attributed to the different definitions of "forest" used in their assessments, with more open woodlands being included the UNEP-WCMC estimates.

Table 2.1 Extent and protection status of African forests in 2000. Total land area = 29.6 million km^2, of which 27.6% is forested. Total protected area figures include IUCN categories 1a-VI

Forest type	Total forest (km^2)	Total protected (km^2)	% Protected
Temperate and boreal			
Mixed broadleaf/needleleaf forest	32,627	1,743	5.3
Deciduous broadleaf forest	140,727	4,523	3.2
Sclerophyllous dry forest	17,447	133	0.8
Sparse trees/parkland	253,001	4,753	1.9
Tropical moist			
Lowland evergreen broadleaf rainforest	1,685,244	127,171	7.5
Lower montane forest	33,232	384	1.2
Upper montane forest	156,341	25,145	16.1
Freshwater swamp forest	196,729	2,877	1.5
Semi-evergreen moist broadleaf forest	47,512	2,868	6.0
Mangrove	53,805	1,179	2.2
Disturbed natural forest	414,480	7,966	1.9
Tropical dry			
Deciduous/semi-deciduous broadleaf forest	2,243,590	271,187	12.1
Thorn forest	192,927	10,065	5.2
Sparse trees/parkland	2,714,396	299,404	11.0
Total	**8,182,058**	**759,396**	**9.3**

Source: Adapted from UNEP-WCMC (2011). Available via http://www.unep-wcmc.org/forest/africa_map.htm

shown in Table 2.2, are cause for concern, particularly given the high degree of dependence on forest resources by people throughout the continent. FAO's State of the World's Forests report (FAO 2009) includes an analysis of forest conditions and trends in Africa, and discussion of underlying causes and impacts of forest loss and degradation on biodiversity, socio-cultural conditions, and economic aspirations of the people. The socio-cultural impacts have been severe in many regions, ranging from loss of control over forest lands and erosion of traditional forest and agricultural management practices to customary law and governance institutions.

Throughout Africa, large populations of people live in local communities in or near forest landscapes or savanna woodlands. Irrespective of the forest types with which these communities are associated, a wide variety of well-established and documented land-use systems based on traditional knowledge have long sustained these communities, supporting their agricultural livelihoods and sustainable utilization of natural resources (Fig. 2.2). Communities have also developed elaborate social institutions, beliefs, and practices to ensure the maintenance of resilience of their socio-ecological systems. Some developed as part of very strong political systems and the enforcement of taboos and regulations to control social behaviour. Spiritual, cultural,

Table 2.2 Forest area in Africa: extent and change

Subregion	Area (1,000 km²)			Annual change 1,000 km²		Annual change rate (%)	
	1990	2000	2005	1990–2000	2000–2005	1990–2000	2000–2005
Central Africa	485	2,394	2,361	−9.1	−6.73	−0.37	−0.28
East Africa	890	810	771	−8.01	−7.71	−0.94	−0.97
Northern Africa	848	795	768	−5.26	−5.44	−0.64	−0.69
Southern Africa	1,884	1,176	1,711	−11.52	−11.54	−0.63	−0.66
West Africa	887	788	743	−9.85	−8.99	−1.17	−1.17
Total Africa	6,994	6,556	6,354	−43.75	−4.040	−0.64	−0.62
World	40,773	39,886	39,520	−88.68	−7.317	−0.22	−0.18

Source: Adapted from FAO (2006)

Fig. 2.2 Traditional integrated rural land use around rice paddies Antananarivo Madagascar (Photo courtesy of Coert Geldenhuys)

and religious educational methods have also been used to bridge intergenerational cultural divides, including folklore, traditional ritual ceremonies, and customary laws.

The history of the relationship between people and forests in Africa, from the Iron Age to the present, is replete with stories of how African societies have depended on forests and their rich biodiversity for medicine, food, crafts, construction, shelter, energy, water, and income. Many African cultures have also evolved from people's close spiritual associations with forests. For example a number of cultures—particularly in West, Central, and Eastern Africa—have traditional beliefs

Fig. 2.3 Mother and child collecting wild fruit from tropical moist forest, Baobato, Central African Republic. For people in communities within and adjacent to forests wild plant and animal species provide essential food and nutrition, particularly during 'hunger periods' in the agricultural cycle as well as medicine, fodder, fuel, construction materials, mulch, and non-farm income (Photo courtesy of Coert Geldenhuys)

tied to sacred groves (Dei 1993) and other special forest areas (Rodgers et al. 2002, Pandey n.d.). The preservation of sacred groves may be considered a functional conservation policy with rules obeyed by everyone in the society. Today such forests provide an important source of baseline information on past vegetation in otherwise altered forest landscapes (Gadgil et al. 1993, 1998).

After decades of over-emphasis on the monetary value of timber, there is growing appreciation for the immense importance of the spiritual, cultural, and socioeconomic roles of non-timber forest products (NTFPs) to local communities (Figs. 2.3 and 2.4). This broadened societal perspective on forest conservation for the multiple values that forests provide has contributed significantly to the recognition of their true market and livelihood potentials for forest-dependent communities (Ndoye and Tieguhong 2004, 2009; Belcher et al. 2005; Sunderlin et al. 2005). It has also stimulated renewed interest in community-based natural resources management (Nhira 1994; Murphree 1995, 1997; Amanor 2003; Berkes 2004; Kojwang 2004). From the perspective of societies that rely fundamentally on local natural resources for their subsistence and well-being, these may be seen as positive developments. However, the continued failure of states (and many conservation organizations) to recognize and respect traditional forest-related knowledge, practices, and customary land tenure in conservation management planning is a source of considerable tension (Joiris 1998; Chloé et al. 2009).

Fig. 2.4 Traditional rural houses, Port St Johns, Eastern Cape Province South Africa. The main species used in the house construction are *Millettia grandis, Ptaeroxylon obliquum, Brachylaena discolor, Englerophytum natalense, Duvernoia adhatodioides* and *Eucalyptus* sp. as poles, and *Buxus natalensis, Tricalysia capensis* and *Grewia occidentalis* as laths (Photo courtesy of Coert Geldenhuys)

2.1.2 Traditional Perspectives on African Forests in Relation to Livelihoods, Culture, and Biodiversity Conservation

Local conceptions of nature and local knowledge systems in traditional communities affect forest management and agricultural practices in terms of understanding and interpretation of states of nature where human activities will take place. For example, studies in southern Cameroon in Central Africa have illustrated how the concept of nature is deeply embedded in how people and communities design and experience their homes and public spaces (Mala 2009; Mala et al. 2009a). Forest management knowledge systems derived from the same conception of nature also combine space, time, and the supernatural. But these perceptions are bipolarized: one relationship (the personal relationship) looks towards the spiritual world, while the other (the forest management relationship) involves a combination of social systems and natural resources systems and their entities such as trees, animals, fishes, and water. Moreover, these studies show that the dynamics of local bio-ecological knowledge over time and on the ground depend on the seasons and their transitions. This confirms the existence of a cognitive background of local knowledge systems that support adaptive forest and natural resources management practices.

The relationship between traditional forest-related knowledge and conservation attitudes towards biodiversity outside protected area systems are significant. For example, Byarugaba and Nakakeeto (2009) examined the management of forest biodiversity by local communities in Igara County, Uganda, (on the fringe parts of the great Albertine Rift), where the majority of forest biodiversity resources are located outside of protected forest reserves and national parks. The study found that these resources are still abundant, not by fashion but by the choice of people in local communities who manage these resources using their traditional knowledge and agricultural and forest management to sustain their livelihoods and cultures.

Studies have shown that forest wildlife conservation efforts can greatly benefit from an understanding and support of local (traditional) institutions and their embedded knowledge systems. This was illustrated in the study of traditional knowledge, management practices, and socio-cultural importance of small mammals in villages around protected areas in the Pendjari Biosphere Reserve in northern Benin (Djagoun et al. 2009). Similarly, a recent project funded by the International Tropical Timber Organization (ITTO) project on Sustainable Community Management, Utilization and Conservation of Mangrove Ecosystems in Ghana, has explored the rich knowledge of mangrove species taxonomy, life histories, ecology, and utilization of individual mangrove species (Derkyi et al. 2009).

2.2   The Importance of Traditional Forest-Related Knowledge for Food Security, Poverty Reduction, and the Conservation and Sustainable Use of Biodiversity

There is a growing recognition that traditional forest-related knowledge, practices, and associated local institutions provide the foundation for participatory approaches to development that are both cost-effective and sustainable. This has been facilitated by a rapidly expanding scientific knowledge base generated by both biological and social scientists regarding the complexity and sophistication of many indigenous natural resource management systems. For example, shifting cultivation (also known as slash-and-burn) practices, discussed below, are widespread throughout Africa (Fig. 2.5). These traditional techniques have long been a subject of intensive

Fig. 2.5   Tropical moist forest clearance for shifting cultivation in Guietso, Gabon. These traditional agricultural practices, widespread throughout the continent, have ensured food security in rural areas for generations and have much to offer for biodiversity conservation (Photo courtesy of Coert Geldenhuys)

debate as to their past, present, and potential future contribution to food security, poverty reduction, and biodiversity conservation. A growing body of scientific literature is emerging that these practices indeed have much to offer in terms of their high returns to labour, as well as to biodiversity conservation (Brown and Schreckenberg 1998; Ngobo et al. 2004).

Traditional agricultural, forest and woodland management, and animal husbandry practices based on traditional knowledge have often developed over countless generations, and by necessity have required adaptation to uncertain environmental conditions such as periodic drought (Ajibade and Shokemi 2003; Nyong et al. 2007). In the context of climate change, which is likely to exacerbate the vulnerability of local communities throughout much of sub-Saharan Africa (Parry et al. 2007), traditional knowledge and practices represent an important element of the adaptive capacity of rural communities (Simel 2008; Galloway-McLean 2010; Osman-Elasha et al. 2009). For example, the monitoring of grazing pressure and the state of the pasture by herders of the Sahel enables them to make informed decisions about rotating or relocating herds (Niamir-Fuller 1998), an important adaptation to their unpredictable, low-rainfall environment (Coughenour et al. 1985). Such knowledge and practices are found in traditional communities throughout the forest-savanna transition zones in Africa (Fairhead and Leach 1996; Rodgers et al. 2002). For example, the Maasai pastoralists from Kenya and Tanzania are known for their abilities to monitor the onset of rains, to predict water availability in their rangelands, to assess both quality and quantity of grazing, to diagnose animal diseases, and to prescribe plant-based therapies for diseases. Their abilities to co-exist peacefully with wildlife bears testimony to the abundance of wildlife in their territory even before national parks and game reserves were established in those areas (Kenrick 2000).

2.2.1 Traditional Forest-Related Knowledge, Land Use Classification and Management

A broad array of traditional forest management system and practices are found in Africa, ranging from strictly protected forests (or species) to intensively managed systems. These include (Wiersum 1997):

- Protected natural forests such as sacred forests and sacred groves that are believed to be abodes of ancestral spirits, as well as ceremonial and rainmaking forests;
- Water protection forests, such as spring forests in Tanzania or riverine vegetation in Kenya;
- Clan or village forests and woodlands, managed forest belts, and strictly protected species (taboo trees);
- Enriched natural forests such as expanded forest islands and gallery forests (as in Guinea),
- Shifting cultivation fallows, often enriched with fruit and other valued tree species;

- Reconstructed (natural) forests such as fortification forests established for defence around human habitations, as found in some regions of the Sahel and in Guinea; and
- Mixed arboriculture systems including home gardens and smallholder plantations (Wiersum 1997).

Within communities that practice shifting cultivation, traditional knowledge is used to describe and characterize forest dynamics (i.e., categories of land use successions in traditional shifting agriculture) based on socio-ecological descriptors and associated natural resource management practices. The study conducted in southern Cameroon by Mala et al. (2009a) identified 11 socio-ecological indicators that are often used for such descriptions: (1) indications (or not) of human disturbance; (2) diversity, abundance, and size of animals; (3) perception of distance of land use from the villages; (4) indicators of human activity such as huts used for seasonal migrations; (5) presence of isolated old oil palm trees; (6) presence of *Irvingia* spp. and *Cola acuminata;* (7) presence, age, and abundance of oil palm trees *(Elaeis guineensi*s); (8) abundance of pioneer species such as *Musanga* spp and *Macaranga* spp; (9) abundance of rodents and other small mammals attracted by the farming activities; (10) abundance of *Chromolaena odorata* and Maranthaceae as indicators of the distance of agricultural fields from villages; and (11) presence of food crops such as *Musa* spp. and *Manihot esculenta* (cassava).

A variety of management practices are closely associated with the classification of the land use successions. These are (1) the commercial or domestic orientation of hunting or trapping; (2) the domestic or commercial orientation/use of wild fruits and medicinal plants (NTFPs) and their level of intensity of gathering; (3) the cash orientation given to crops such as *Cucumeropsis mannii* (melon seed grains) or *Musa spp.*; (4) the level of intensity in extracting *Raphia* and Palm tree sap (for making wine); and (5) the level of suitability of agricultural lands where indicators of fertility are abundant. These indicators are often combined for the valorisation of biophysical and socioeconomic functions of landscape, either by modification of landscape structure through shifting agriculture or though extractive forest management practices that contribute to both forest productivity and biodiversity conservation (Wiersum 2004). There is a relationship between the description of traditional forest systems and management practices that effectively regulate vegetation development, land and forest productivity, wildlife abundance, and management of other natural resource components (Fairhead and Leach 1994; Dounias 1996; Dounias and Hladik 1996; Diaw and Oyono 1998).

Traditional knowledge of forest and land use dynamics determines natural resource management within villages to sustain livelihoods and to enhance adaptive capacity of these socio-ecological systems (Diaw 1997; Vermeulen and Carrière 2001; Wiersum 2004). It reflects the broad perspective of traditional societies that combines their knowledge of the biology and ecology of plants and animals, and the symbiotic interactions among humans, plants, animals, and landscapes (Gillon 1992; Fairhead and Leach 1994; Altieri 2002; Oyono et al. 2003; Joshi et al. 2004). These findings reflect similar conclusions from studies of socioeconomic and cultural uses of and relationships to forests among Pygmy and Bantu peoples in southern Cameroon and Central Africa (Mviena 1970; Bahuchet 1996; Oyono 2002; Ohenjo et al. 2006).

2.2.2   Traditional Agricultural Practices: Shifting Cultivation and Agroforestry Systems

Shifting cultivation is both a land use management practice and a traditional farming technique. As a land use management practice, it helps to determine and regulate land rights within communities through the dynamics of land use successions, e.g., through alternating periods of cropping and abandonment of lands (fallows) that characterize traditional shifting cultivation systems (Diaw 1997; Diaw and Oyono 1998). Mixed food-crop farming, which preceded most of current land uses in the humid forest zone in Cameroon and Central Africa (Carrière 1999; Mala 2009), is another key land use with a vital socioeconomic role. It provides for household consumption needs and income within the forest margins while the land is not in fallow (Gockowski et al. 2004). Mixed food-crop agroforests (a non-permanent land use) are often converted into permanent human-modified land uses such as cocoa agroforest with permanent land rights.

Shifting cultivation management includes a variety of practices that affect forest regeneration and regrowth, while contributing to sustainable food production and production of desired forest products (Fig. 2.6). It typically involves periodic clearing of secondary forest patches; retaining useful plant and tree species (such as wild fruit, medicinal, timber, and nitrogen-fixing trees); and planting of food or cash crops to create agroforests (Mala 2009). Depending on the types of vegetation associations used, these practices modify the structure and composition of forest landscapes, increasing their value to farmers and providing for domestic needs and additional income for households (Dounias 1996; Gockowski et al. 2004; Sonwa 2004; Mala 2009; Wala et al. 2009).The associations of crops and trees species that are created are based on farmers' knowledge of complementary, competitive, and supplementary interactions among tree and crop species.

When agroforests are managed for cash crops such as cocoa (*Theobroma cacao*) and oil palm (*Eleias guineensis*), as they are in many parts of West Africa, remnant (non-crop) trees such as *Dacryodes edulis* (African pear), *Irvingia* spp. (wild or bush mango), *Ricinodendron heudelotii,* ('Njangsa,' valued for its edible seed kernels), are typically retained to provide food and income. The tree components of these agroforestry systems are also important source of carbon sequestration during the life of agroforests.

Knowledge about the life histories and ecology of tree species, and their interactions with crops, has a long history in traditional farming communities. Farmers' wealth of knowledge about tree-crop associations has been crucial for the development of agroforestry systems. As noted above, in traditional farming (predominantly shifting cultivation) in Africa, trees deliberately left on farms serve

Fig. 2.6 Shifting cultivation management in Miombo woodland, Sofala Province, Mozambique. (**a**) land preparation involving selective tree cutting and burning. The main canopy tree species are *Brachystegia spiciformis*, *Julbernardia globliflora* and *Albizia adianthifolia*. (**b**) Initial crop cultivation phase (Photos courtesy of Coert Geldenhuys)

different functions (Wiersum 1997; Carrière 1999; Owusu-Sekyere 2009; Mala 2009). Typically, such trees left are selected based on several characteristics, including their habit and canopy architecture, compatibility with agricultural crops, and traditional subsistence and commercial uses and values (e.g. for medicines, housing, food, fuelwood, ceremonial or religious purposes, etc.). The diversity of species in agroforestry is influenced by the traditional knowledge of their uses (Wiersum 2004). A study conducted in South-West Nigeria to understand fuelwood species selection in traditional agroforestry systems showed that 12 woody species are pre-dominantly used, and that food and medicinal values were more important criteria for selection than their other uses (Erakhrumen 2009).

Eastern Africa is a region that is highly susceptible to repeated cycles of drought, which appear to be worsening with climate change and thereby increasing the vulnerability of local communities to its effects. Traditional communities in this region have accumulated a rich body of traditional forest-related knowledge, handed down over generations, that has helped them develop and adapt their agricultural, forest, and rangeland management practices to drought and other environmental uncertainties. An example of this is found among the indigenous Tigrigna speaking people who live in the mountainous dryland landscapes of central Tigray in northern Ethiopia (Balehegn and Eniang 2009). Here, farmers have long cultivated indigenous (and some exotic) multipurpose fodder trees for diverse uses, using a number of criteria for selecting desirable species, including: feed quality, value of wood for timber or fencing, biomass productivity, soil and water conservation value, evergreenness, spiritual value, drought resistance, and absence of allelopathic effects on understory plants and agricultural crops.

While many local and introduced species are used in these communities, certain species were more commonly preferred by farmers based on their consumptive qualities, namely: *Ficus thonningii, Cordia africana, Eucalyptus camaldulensis,* and *Acacia etbaica*. To adapt to changing environmental conditions, farmers in these communities have adopted and intensified vegetative propagation of *F. thonningii* (by cuttings); they use an indigenous protocol for evaluating, propagating, and conserving the species, including its tending at different stages of growth. These evolving traditions have created a forested island (and wildlife habitat) in the midst of a highly denuded dryland landscape, which has helped to sustain the agricultural livelihoods of the communities in the face of increasing environmental uncertainty. The tree species preferred by farmers in these communities, given their varied ecological, economic, and social values, could be adopted in agro-silvopastoral systems for climate change adaptation and mitigation (carbon sequestration) in vulnerable regions.

As in other tropical regions, the use of fire is fundamental in traditional agriculture practices in many parts of Africa (ASB 1995, 2000; Oyono et al. 2003; Shaffer 2010). Within the forest zone of Ghana, for example, farmers consider farming without fire to be impractical, and regularly use fire to increase cultivable area for crops such as maize, clear competing weeds, release nutrients to enhance crop growth, prevent microfaunal pests from attacking crops, and manage shade. Fire is also used both in traditional hunting practices and for gathering honey. Local people in northern Ghana set fires under trees such as *Parkia biglobosa* (dawadawa) to promote their fruiting. These and other burning practices, and indicators for their use, have been developed over the years by local farmers. Major fire management activities used by local people in Ghana include the maintenance of fire belts to reduce the risk of fires spreading to homes, sacred groves, and cash crops. Current indigenous practices for fire usage and management in Ghana are comparable to scientific practices, although these practices appear to have certain weaknesses, such as burning at inappropriate times or repeated burning at short intervals, or poorly developed alternatives to the use of fire (Korem 1985; Amissah 2009). In Uganda, the Ik people of Karamoja use fire as a forest management tool in a number of ways, and have established guidelines for fire regimes, tree cutting, and grass harvesting, all of which demonstrate a wealth of traditional knowledge regarding forest ecosystems and their management (Rodgers et al. 2002).

2.2.3 Traditional Knowledge and Agricultural Biodiversity

Traditional farming systems are an extremely rich source of agro-biodiversity (Akundabweni and Chweya 1993; Thrupp 1997; Wiersum 2004). Their abundant biodiversity is the product of the diverse indigenous technologies that have been developed by traditional farming communities to suit different ecological and socio-economic needs. These systems and practices, which include rotation, the use of transitional crops, agroforestry, homestead gardens, and mixed crop farming, represent strategies to meet both predictable and unforeseen ecological and economic circumstances (Altieri and Anderson 1986; Mala et al. 2009a, b).

Based on centuries of experimentation, traditional farmers have evolved cropping patterns that suit their particular situations and have demonstrated their ability to manage intra-specific diversity of land races, which include native or 'wild' species and naturalized species (Wiersum 2004; Mala 2009). Crop varieties found in traditional farming systems tend to have a greater intra-varietal diversity than is found in modern systems, and in many African countries small farmers play a central role in the conservation of agricultural germplasm.

In their efforts to promote sustainable livelihoods, eradicate poverty, and contribute towards the achievement of the Millennium Development Goals (MDGs), international organizations and development agencies are giving increasing attention to both the natural capital of biodiversity as well as the human and social capital that is found in even the poorest rural communities. This human and social capital includes the knowledge, skills, and institutions that enable communities to make optimal use of their assets, which includes not only land and water resources but also agricultural biodiversity and that of associated forests and woodlands upon which communities depend for the many goods and services they provide (IPGRI 2005).

2.2.4 Traditional Knowledge and Non-Timber Forest Products

Non-timber forest products (NTFPs) are extremely important to forest-dependent and agricultural communities throughout Africa, and contribute significantly to domestic energy and to food- and health-security needs. These products include fuelwood and charcoal, the primary fuel source in rural areas (Figs. 2.7 and 2.8);

Fig. 2.7 Fuelwood collection. Forests provide the domestic energy needs for the overwhelming majority of people in rural areas throughout Africa. Its collection is a daily activity primarily for women and children. (**a**) Fuelwood harvests from coastal forests Port St Johns, Eastern Cape Province, South Africa. (**b**) Collectors of firewood and fruit returning home, Central African Republic (Photos courtesy of Coert Geldenhuys)

Fig. 2.8 Charcoal on sale along road through Miombo woodland between Ndola and Kitwe, Zambia. There is a high demand for charcoal derived from forests and woodlands in rapidly growing urban communities throughout Africa (Photo courtesy of Coert Geldenhuys)

wood used for tools, carving, and other household purposes; livestock fodder; gums, resins, fruits, nuts, tubers, mushrooms, spices, fish, bush meat, and other wild foods (Figs. 2.9 and 2.10); and plants and oils for pharmaceuticals and cosmetic products (Campbell 1996; FAO 1995, 1999). Honey hunting and traditional beekeeping practices are an important part of the subsistence economy of many communities throughout Africa (Fischer et al. 1993; Endalamaw and Wiersum 2009). Traditional forest-related knowledge throughout Africa also includes a wide variety of techniques for processing non-timber forest products such as seed oil, medicinal plant extracts, dyes, and tannins.

Fig. 2.9 Bushmeat collection for local consumption from tropical moist forest, Baobato, Central African Republic. Hunting of forest wildlife, and harvesting of fruits and other wild foods, provide a important 'safety net' for food security in forest-dependent communities during lean seasons in the agricultural cycle. The increasing urban demand for African wildlife for food (bushmeat) is leading to unsustainable harvesting of some species, with implication for both biodiversity conservation and traditional management and utilization practices (Photo courtesy of Coert Geldenhuys)

Fig. 2.10 Village market in southern Angola. The increasing urban demand for wild foods, medicinal plants and other non-timber forest products, while providing important non-farm income to people in rural areas, raises issues concerning forest biodiversity and its sustainable use (Photo courtesy of Coert Geldenhuys)

In some regions, NTFPs are increasingly sold in urban centres and also internationally to meet demands of the African diaspora in Europe and North America.
This increasing demand is leading to unsustainable harvesting of some species, with
implication for both biodiversity conservation and traditional management and utilization practices (Sunderland and Ndoye 2004; Tabuna 2007; Nasi et al. 2009).

For people in communities within and adjacent to forests and woodlands, forest plant and animal species provide essential food and nutrition, particularly
during 'hunger periods' in the agricultural cycle (Falconer 1990; Arnold and
Townson 1998; Tieguhong et al. 2009; Ndoye and Tieguhong 2009), as well as
medicine, fodder, fuel, thatch and construction materials, mulch, and non-farm
income. In eastern and northern Sudan, for example, Doum (*Hyphaene thebaica*)
forests provide a diversity of non-timber forest products of great importance in
the rural economy. These products include fuelwood and charcoal, fibre ('sa'af')
from the leaves of young trees used to make of ropes, baskets, and mats; and
edible nuts, the kernels of which produce 'vegetable ivory.' In addition, the
timber from mature trees provides a durable building material for house construction and posts. Handicrafts made from this species, predominantly by
women, provide an important source of household income (Abdel Magid 2001,
cited in Osman-Elasha et al. 2009).

Other NTFPs also provide a means of cultural expression, in particular as symbols of traditional festivals and celebrations. In Ghana, the first harvest of 'new'
yams (wild species of *Dioscorea*) in the traditional household and community is a
celebration. This is considered as the end of hunger and heralds the emergence of
plentiful supply of food during the year. Again in Ghana, the traditional capture of
live antelope from the wild by any of their traditional warrior groups (called Asafo
companies) is considered by the Efutu clan of coastal Ghana as a good omen for the
year, and calls for extensive celebrations and merrymaking during the season.

Local names of trees and other plant species are very often derived from their
functions, attributes, uses, other special characteristics, and the history of their discovery (Owusu-Sekyere 2009; Mala 2009). The use of wild plants and animals for
food and medicine from selected African countries are shown in Table 2.3.

2.2.4.1 Traditional Forest-Related Knowledge, Wild Foods, and Health Care

The contribution of non-timber forest products to rural incomes, health security,
livelihood security, and quality of life, is well-acknowledged in the literature in
Africa (Sunderland and Ndoye 2004; Masozera and Alavalapati 2004). Africa's
tropical and subtropical forests and woodlands are rich in multipurpose tree species
used on a daily basis by rural communities. A well-known example is tamarind
(*Tamarindus indica*), native to the dry savanna regions of Africa from Sudan,
Ethiopia, Kenya, and Tanzania westward to Senegal. Recent studies in northern
Benin (Fandohan et al. 2009) examined the important role that tamarind plays in
local communities' livelihoods, with special reference to its food value (from the

Table 2.3 Use of wild plants and animals for food and medicine by farming communities

Country/location	Importance of wild resources
Botswana	The agro-pastoral Tswana use 126 plant species and 100 animal species as sources of food
Ghana	16–20% of food supply from wild animals and plants
Kenya, Bungoma	100 species of wild plants collected; 47% of households collected plants from the wild and 49% maintained wild species within their farms to domesticate certain species
Kenya, Machakos	120 species of medicinal plants are used, plus many wild foods
Nigeria, near Oban National Park	150 species of wild food plants
South Africa, Natal/Kwa Zulu	400 indigenous medicinal plants are sold in the areas
Swaziland	200 species collected for food
Sub-Saharan Africa	60 wild species in desert, savannah and swamp lands utilized as food
Democratic Republic of Congo (DRC)	20 tonnes of chanterelle mushrooms collected and consumed by people of upper Shaba
Zimbabwe	20 wild vegetable, 42 wild fruits, 4 edible grasses and one wild finger millet; trees fruit in dry season provide 25% of poor people' diet

Sources: FAO (1995), Ndoye and Tieguhong (2004), Sunderlin et al. (2005), and Wiersum (1997)

fruit pulp and leaves) and its many medicinal uses (from its bark, leaves, and fruits). The values associated with multipurpose trees are influenced by the nature of the species, socio-cultural group affiliation, and gender. The gender differentiation was found to be very important in communities surrounding protected areas in Benin, where women had a preference for species with high commercial and nutritional values, while men preferred trees and other plant species that provide construction material and medicine (Vodouhê et al. 2009).

In addition to trees that yield food products, lianas are also intensively used and valorised. For example, a study of the use of 50 forest species by local communities in semi-deciduous forest areas in west-central Côte d'Ivoire revealed that trees represent 60%, the lianas 34%, and palms 6% of species used (N'Dri and Gnahoua 2009). The most commonly used species—*Irvingia robur, Beilschmiedia mannii, Ricinodendron heudelotii,* and *Myrianthus arboreus*—are maintained by small farmers in their cocoa and coffee plantations and in fallows. Generally, fruits are the most consumed plant parts, followed by leaves, barks, stalks, tubers, and stems. This study highlighted the nutritional values of the more commonly used forest species and their importance for food security within villages of the study region.

For communities living near protected areas, non-timber forest products offer opportunities for co-management. A recent study assessing NTFP utilization in villages surrounding the W National Park in northern Benin found that 172 species of trees play important roles in the livelihoods of these communities

(Bonou et al. 2009). These species are valued for economic (income), cultural, and non-monetary utilitarian purposes (food, firewood, and medicine) as well as being used in traditional industries. This study found that certain species were used by all households: *Anogeissus leiocarpus* and *Crossopteryx febrifuga* (for fuelwood); *Vitellaria paradoxa* (seed kernels, for food); *Parkia biglobosa* (fruits); and *Adansonia digitata* (leaves). In a study on the utilization and local knowledge of *Sclerocarya birrea* (Anacardiaceae) by the rural population in the same protected area in northern Benin, Gouwakinnou et al. (2009) found that this species is commonly managed in agroforestry settings with a majority of farmers each having at least one *S. birrea* tree on their farms. This species is highly valued, and all parts of the plant species (bark, fruits, kernel, leaves, root, and wood) are used traditionally. It is particularly valued for food and medicine, the bark alone being used to treat a variety of illnesses. Apart from harvest of individual trees for wood, and clearing for agricultural purposes, current utilization of this species does not appear to threaten its viability. *Sclerocarya birrea*, well-integrated into the livelihoods of the local Karimama communities in Benin, is also cultivated in shifting cultivation systems in other African countries (Carrière 1999; Mala 2009). *Pentadesma butyracea,* a multipurpose tree that provides fruit kernels that are processed to make butter for cooking and cosmetics, is also widely used in traditional medicine in Benin—its bark, roots, leaves, flowers, and fruits being used to treat a variety of ailments and conditions in local communities (Avocèvou et al. 2009).

Multiple uses of forest plants for medicinal purposes are common throughout Africa. Forest-dependent indigenous communities who have traditionally used and managed Africa's remaining natural forest areas demonstrate strong conservation ethics through their traditional practices and beliefs. Traditional practices are used in the conservation of medicinal and sacred plants, sacred forests, cultural forests, and plants at grave sites; these practices often involve taboos and maintaining secrecy regarding the location and uses of some medicinal plants (Derkyi and Derkyi 2009). The importance of traditional knowledge and practices for conservation of forest medicinal plants in rural communities has been studied with reference to Mozambique (Albano 2001) and Zimbabwe (Clarke 1994; Afolayan and Kambizi 2009).

In recent times, plants used in traditional health care in Africa have received increasing attention (Ampofo 1983; Mshana et al. 2001; Asase et al. 2005). Ethnobiological studies of traditional herbal medicine practices often provide shortcuts to drug discovery. The documentation of such research, in some cases without the prior informed consent or even acknowledgement of the traditional knowledge holders, raises issues related to the protection of intellectual property rights. Other 'knowledge hunters' clandestinely engage the services of unscrupulous wild plant collectors who destroy valued wild plant populations through overharvesting or otherwise unsustainable harvesting practices.

In the case of tree species whose bark is the source of traditional (or modern) drugs, the destructive practice of collection by tree debarking is of particular concern. For example, *Prunus africana,* whose bark is widely used in traditional medicine, is highly valued for treatment of prostate cancer. However, due to

Fig. 2.11 Second bark stripping of *Prunus africana* after recovery following complete bark removal up to the branches, Umzimkulu, South Africa. High market demand for plant drugs used in traditional African medicine has led to destructive or otherwise unsustainable harvesting of a number of important species (Photo courtesy of Coert Geldenhuys)

excessive exploitation by debarking,[2] the species is nearing local extinction in Cameroon and elsewhere within its native range (Hall et al. 2000). Recognizing this widespread problem, there have been successful efforts in South Africa to integrate traditional and scientific forest-related knowledge on *P. africana* and other tree species (e.g., *Ocotea bullata, Curtisia dentata,* and *Rapanea melanophloeos*) whose bark is also harvested for traditional medicine. A research and development programme reported by Geldenhuys (2004) yielded management guidelines for all aspects of the sustainable harvesting and processing of bark from these species, and has been integrated into the forest management plan for the Umzimkulu District in South Africa (Fig. 2.11).

2.3   Traditional and Modern Forest Governance Systems and Conflicts

2.3.1   Traditional Social Institutions and Governance Structures

Traditional forest management systems are typically supported by customary institutions based on principles of equity, reciprocity, and equilibrium in the management of the resource and the sharing of benefits (Mutta et al. 2009). Historically, traditional governance structures have created the environment within which people interacted with their local environment to generate knowledge, technologies, and innovations to fulfil their physical, spiritual, and socio-ecological needs (UNEP 1997; Johnson 1992a, b).

---

[2] *Prunus africana* is listed in Appendix 2 of CITES: Convention on International Trade in Endangered Species (http://www.cites.org/).

For example, the indigenous Mijikenda community from the coastal region of Kenya has a well-defined social structure that is closely associated with forest management, particularly of their sacred Kaya forests (Willis 1996). Through their traditional governance structure, the 'ngambi,' composed of Kaya elders respected by all in the community, these communities managed the Kayas sustainably for countless generations (Fig. 2.12). As primary custodians of the sacred forests, the elders of the ngambi were responsible for making and enforcing rules related to management, control, and access to biological resources and traditional knowledge within the forests. Access to the Kaya forests was subjected to traditional rules and rituals, which were organized by elders, under the advice of a seer who communicated with the spirit world to explain the immediate future (Munyi and Mutta 2007). This system played an important role in maintaining the integrity of the natural forests and conservation of the biodiversity therein. Over time, various processes have influenced the effectiveness of the traditional governance system to conserve the forests. These include: formal governance institutions, anthropogenic factors, modernization, and religion. Introduction of the central governance structure has been the most significant in eroding the influence of the community elders to enforce customary laws and principles that sustained the forests. The positive elements of both systems can be used to explore opportunities for integrating them for enhanced management of sacred forests.

The Baka, Efe, and Mbuti peoples of Central Africa—who spend long periods of time in the forest, hunting for meat, gathering plant foods, and collecting honey—have traditionally relied on social structures, customary laws, and rituals embedded in spiritual cosmology to establish close relationships within their communities and with the forests (Kenrick 2000). These communities have accumulated an intimate knowledge of the forest, including the life histories, behaviour, and ecology of forest fauna and flora. They also acquired knowledge of properties of thousands of plant species and their uses for food; medicines to heal wounds, cure fevers, and dull pain; and to make poisons used in hunting.

In the Horn of Africa (Ethiopia, Eritrea, and Somalia), as in many other forested regions in Africa, traditional forest resource management practices are being eroded or lost for several reasons. These include urbanization and increased market demand for timber and non-timber forest products; increased influence and dependence on cash economies in rural areas; and other socioeconomic, political, and cultural changes. These changes have resulted in the progressive loss of forests and highly valued species and have important consequences for biodiversity conservation (Cotton 1996). For forest biodiversity conservation programmes to succeed, sustainable traditional management practices, and associated local governance institutions, should be given very serious consideration (Chloé et al. 2009).

A study of local communities in the Jimma zone of Oromyia regional state in Ethiopia (Hundera 2007) identified traditional practices (such as shade coffee cultivation and apiculture) that contribute positively to the conservation of forest resources, and discussed ways in which these practices could be used to conserve remaining natural forests. The study highlighted several important findings. Local

Fig. 2.12 Documenting traditional biocultural knowledge of the Kaya Fungo sacred forest in Kenya's coastal region. Traditional governance systems can serve as models for the development in participatory forest management approaches. (**a**) Group discussion with Kaya Fungo elders. (**b**) Kaya elders showing a valuable indigenous food plant at the entrance of Kaya Fungo sacred forest (Photos courtesy of Florence Mwanziu and Doris Mutta)

communities in these communities are well aware of the important benefits they derive from the forest, especially among those who live within and have greater affinity to the forest. Further, traditional conservation of larger trees such as *Ficus vasta, Podocarpus falcatus, Ekebergia capensis,* and *Ficus sycomorus* in the area formerly hinged on religious beliefs and cultural attachments that are presently non-existent. However these tree species are still maintained because of their associations with the other trees species that are of benefit to the people, such as coffee shade trees (*Albiza gummifera, Milletia ferruginea,* and *Acacia abyssinica*) and trees used for apiculture. For these reasons, forests are conserved because of the benefits the community derives from the resources rather than the non-binding sanctions of the traditional management systems of social sanctions (Cotton 1996; Marena Research Team 2002).

Elsewhere in Ethiopia, Black et al. [n.d.] discuss the traditional beliefs, social institutions, and tenure arrangements that have historically helped to preserve the now rare remnant forests in the Amhara region. These authors concluded that community cohesiveness and homogeneity, backed by shared history and religion, helped to preserve forests, and that increased state control and open-access to forests led to the breakdown of traditional rules regulating their use and the degradation and loss of forest in the region.

2.3.2 The Conflict Between Modern Land Laws and Traditional Tenure Systems

Within the traditional African tenure systems, pastoralists practiced well-organized grazing strategies encouraged by their ability to move freely to obtain their livestock needs in ever-changing grazing environments. This was regulated by sociopolitical controls, cooperation among individual herding units, and technical knowledge and innovation (Niamir 1990). For example, the Zaghawa of Chad move their sheep and camels to Saharan pastures in separate parallel paths, leaving a portion of the land ungrazed for their return south (Tubiana and Tubiana 1977). The Fulani of West Africa, in particular in the northern Sierra Leone, practice 'shifting pasturage,' grazing one area for 2–3 years and then moving elsewhere, allowing the first area to 'rest' for 15–20 years (Allen 1965). The Sukuma (south of Lake Victoria) use a similar technique but with a 30–50 year rest period (Brandstrom et al.1979).

Many of these types of land tenure and land-use rotations are no longer easily practiced. This is because of political changes associated with new territorial land laws and instruments that define nations (such as those that separate countries within the East African community of Kenya, Uganda, and Tanzania), with attendant conflicts in 'ownership' of trans-boundary migratory wildlife populations (especially between Kenya and Tanzania regarding elephants), and strict border controls by new and emerging independent African countries since the early 1960s and 1970s. This has created several challenges for African pastoral communities (Lusigi 1984), especially communities whose traditional activities (and rangelands) are now divided between several countries.

The position and health of two indigenous communities in Africa, especially with regard to land rights, is precarious (Fortmann and Nhira 1992). The Pygmies of Central Africa and the San of southern Africa are no longer able to maintain traditional livelihoods and sustain traditional culture, knowledge, and institutions because of loss of land and other natural resources (Ohenjo et al. 2006). The forced removal of the San from the Central Kalahari Game Reserve in Botswana provides a stark example of prejudice and the exercise of majority power, and of the consequences to the social structure and health of the dispossessed. A similar situation is seen also in the creation of national parks that led to the exclusion of the Twa people from the forests in Uganda and many other parts of Central Africa, which has badly affected these people.

These cases highlight the challenges of recognition, marginalization, and discrimination faced by many traditional communities in Africa as well as the policy changes that are needed for integrating community livelihoods and conservation using traditional forest-related knowledge. There is no doubt that in the post-colonial period, attempts have been made to suppress ethnic differences in the interests of building national consciousness, and to portray indigenous peoples both as a danger to national survival and as an obstacle to development. As a result, the major determinants of the poor health status of indigenous peoples such as the Pygmies are dispossession from their lands, destruction of their culture, discrimination, and marginalization leading to poverty and the associated high prevalence of both communicable and non-communicable disease. In the case of the Pygmies, the situation has been exacerbated by widespread conflict (Ohenjo et al. 2006).

Indigenous people define themselves in terms of their relationships with their land, environment, and community, all of which they depend on for their physical and cultural survival. Land rights, consistently cited as a primary barrier for indigenous peoples in Africa, are described by the African Commission on Human and Peoples' Rights (ACHPR) as 'fundamental for the survival of Indigenous communities.' The Minority Rights Group[3] states that 'conflicts over land rights and access to land are the source of many violations of the rights of minorities and Indigenous peoples.' Defining their relationship to the land and participation in decisions on its use does not imply a wish to secede or maintain themselves in a primitive state, but a recognition of the right to choose a concept of and path to development within the modern nation state. Unless indigenous peoples are recognized and empowered to negotiate on equal terms with majority populations, they will not able to secure equitable settlements with more dominant groups. Local contexts vary but the importance of establishing equal rights and ending discrimination remains constant. The emergence of an indigenous peoples' movement in Africa, part of a wider international process, signals the determination of such peoples to secure recognition of both their existence and their contributions. These contributions include traditional forest-related knowledge and practices that have enabled these traditional societies to sustainably use and effectively conserve biodiversity, land, and water resources in often very harsh environmental conditions; and that this knowledge and these practices and innovations could be harnessed for the benefit of the wider society.

In seeking solutions, other African states could start by looking at the example provided by South Africa in recognizing claims to land, and the adoption of a memorandum creating the first indigenous peoples' policy process in Africa. The adoption of ACHPR report note above offers further encouragement, but much remains to be done. Achieving equity and justice in health will need wider changes to protect

---

[3] http://www.minorityrights.org/

the rights claimed in the Declaration on the Rights of Indigenous Peoples[4] adopted by the United Nations General Assembly in 2007. Certainly it requires new approaches to dealing with diversity, cultural difference and self-determination.

2.3.3   Forest Governance Conflicts and Their Resolution

As we trace the emergence of current forest governance systems in all African countries, the existence of conflicts in management practices and governance systems between traditional/local/indigenous and state/government cannot escape our scrutiny.

The history of forest management and governance in Africa can be considered in three periods (Rodgers 1993; Wily and Mbaya 2001), namely: the pre-colonial period when people were free to convert and use forest land; the colonial and post-colonial period during which people were progressively barred or removed from closed forests (which were increasingly controlled by the state); and the current period of changing conservation strategies including forms of community forest management. Out of this can be gleaned four forest policy categories: (1) traditional natural resources conservation with forests preserved primarily for their value as refuges (often having a religious significance), with development of customary rules, regulations and community sanctions related to their use (Rodgers et al. 2002); (2) colonial era policies which managed forests primarily for timber destined for the colonizing (European) country; (3) forestry for economic development in the export substitution era in the newly independent African states in the latter half of the twentieth century; and (4) forestry policies under conditions of economic liberalization and the structural adjustment era, with emphasis placed on farm forestry, agroforestry, and community-level management of natural forests (Kojwang 2004).

While forests are no longer being managed purely for timber production but rather for the provision of a variety of products for a much larger and mixed group of stakeholders, conflicts have emerged in relation to forest management practices and governance systems. On one side, the state assumes forest governance authority and oversight, including responsibility of issuing licenses for timber operations (in most African countries). On the other side are the mixed stakeholders, the majority of whom are forest communities whose immediate interest is in livelihood support from forest products. One major conflict, or trade-off, which occurs *within* forest communities faced with industrial logging, is between the immediate benefits in employment, income, roads, schools, and hospitals received from timber operations (i.e., logging concessions); and negative impacts of logging, i.e., destruction of their cultural heritage and decreased availability of the vast array of subsistence and commercial NTFPs such as for medicines, food, and energy that these forests formerly provided (Ndoye and Tieguhong 2004).

---

[4] http://www.un.org/esa/socdev/unpfii/en/declaration.html

Introduction of centralized governance structures has been the most significant factor contributing to the erosion of the influence of the community elders to enforce customary laws and principles that long sustained traditionally managed cultural forests. The forest policy structure in many African countries has been informed by the assumption that indigenous communities mismanage their natural environment. For example, the Kenyan forest policies and other rural development policies and strategies thus systematically increased central governance and diminished the role indigenous communities in the decision making process. Most of the Kaya forests of the Mijikenda, discussed above, have been converted to either forest reserves or national monuments managed under national statutes.

These changes in governance and authority have effectively revoked historic rights derived from community-based legal systems, opened up community rights to exploitation and use by persons considered outsiders by the community, replaced community based traditional leaders and authority systems by state appointed leaders, and invalidated community enforcement systems. The effect of these changes has been alienation of indigenous communities from their heritage and marginalization of their traditional governance systems, rendering them ineffective and secondary in status (Mumma 2004). One result of the exclusion of indigenous communities in decision-making, loss of traditional governance systems and customary laws, and alienation from their lands has been an increased insensitivity by the people towards forest conservation and increased encroachment (and degradation) of cultural forests (Mutta et al. 2009).

Despite the apparent (and real) conflicts that resulted from lack of understanding of, or interest in, traditional forest-related knowledge, practices, and institutions on the part of the colonial masters, management systems of traditional communities are emerging as good examples of sustainable forest management. Such systems, where they have maintained their integrity, are extremely valuable for the development in participatory approaches for forest management, as illustrated by cases in Uganda (Byarugaba and Nakakeeto 2009), in the Nzema East and Mfantseman districts of Ghana (Derkyi et al. 2009), and in Cameroon (Bele and Jum 2009). The management of sacred forests by the Tiriki in Kenya (Darr et al. 2009) provides a good example of how traditional perceptions and forest conservation management practices can be viewed through the lens of sustainable forest management to reveal the values of traditional knowledge for maintaining intangible values of forests in Africa.

Traditional approaches for managing forest resources and resolving conflicts related to forest resource access and use offer insights and perhaps helpful models that can be used elsewhere to deal with conflicts between traditional and non-traditional forest resource users. In the case of the southwestern Ethiopia forest region, forest beekeeping is an ancient form of forest exploitation that involves a variety of traditional management practices, tree tenure arrangements, and community decision-making processes related to the use of trees for hives on private and common lands (Endalamaw and Wiersum 2009). In their study of the Dimiko council forest in Cameroon, Mvondo and Oyono (2004) examined how community-based social negotiation tools can be used effectively for management of conflicts related to forest access. While such traditional approaches to forest conflict resolution have not been widely recognized until now by forest scientists and decision makers, they underscore the value of such approaches that are tailored to local situations using socially and culturally acceptable.

The understanding of stakeholder needs and priorities, and balancing trade-offs between near-term community livelihoods and the imperatives of conservation for the future generation, is a prerequisite for implementing sustainable natural resource management. Studies conducted in Burkina Faso to ensure that the needs of local people were taken into account in policies and decisions for sustainable use and management of plant resources showed that tree and other forest species can be classified according to informants' preferences, and that this information can be used to develop management strategies better meet local needs (Belem et al. 2009; Bognounou et al. 2009).

2.4   Challenges and Opportunities for the Study, Preservation, and Enhancement of Traditional Knowledge

It is widely assumed that anthropogenic activities inevitably have negative impacts on agro-ecosystem dynamics, i.e., productivity, regrowth, and regeneration (ASB 1995, 2000). This overly simplistic conventional perspective has contributed to the artificial segregation of 'agriculture' from 'forest' issues in Africa, as elsewhere. This has had serious implications for traditional forest-related knowledge and innovations that link forestry and agriculture activities in complex tropical environments, such as those that characterize much of the African continent (Colfer 2005; Mala 2009; Mala et al. 2009a, b). Additionally, attempts to address biodiversity conservation at different spatial scales, i.e. farms, the larger landscapes in which they are situated, and protected areas, have been disjointed. While most agricultural research to date has focused almost exclusively on the small plot (i.e., farm or household) level, forestry research has generally focused on larger spatial scales (protected areas, community forests, council forests, state forest plantations, etc.), without sufficient collaboration and communication between the two research communities. These conceptual and professional disjunctions have hampered effective study of agro-ecosystems and their management by the scientific community as well as development of useful innovations in traditional agriculture (particularly shifting cultivation) in Africa.

There is no single approach for forest landscape management that combines timber and non-timber harvesting with agricultural cropping needs to sustain livelihoods (household consumption needs, i.e., food security), commercial/industrial development (small-scale agroforestry and forest enterprises), and environmental and biodiversity conservation needs. Tropical forest landscapes are typically a complex mosaic of agricultural and non-agricultural land uses fulfilling a range number of biophysical and socioeconomic functions (Wiersum 2004; Mala et al. 2009b).

The contribution of non-timber forest products to livelihoods and incomes in forest-dependent communities is usually significant, and often much greater than that attainable from timber and fuelwood. In southern Cameroon, for example,

NTFPs may contribute up to 45% of farmers' incomes (van Dijk 1999; Ndoye and Tieguhong 2009). This underscores the importance of considering NTFPs and their management in the design of adaptive co-management strategies within forest land-scape mosaics ,where state-owned timber resources typically co-exist with crop-lands, which are communities' capital in non-permanent forest domains. The integration of social, cultural, economic, ecological, and legal aspects of timber and NTFP harvesting is a crucial step for better policy formulation and improved man-agement (Ndoye and Tieguhong 2004, 2009).

Conventional forest management systems differ in fundamental ways from tradi-tional approaches. Indigenous cultural belief systems, or environmental philosophies, are chiefly concerned with ensuring that resources are used in a way that preserves them for future generations. In contrast, formal governance systems are mainly concerned with ensuring that resources are used to maximize present or near-future benefits. Given the critical poverty alleviation, food security, and other immediate needs of people in most rural communities in Africa, we believe a critical balance of these two systems is needed (Mutta et al. 2009). Conventional natural resource plan-ners and managers could benefit from greater understanding and appreciation of the adaptive management practices derived from traditional forest-related knowledge systems, which typically include monitoring, responding to, and managing ecosys-tem processes and functions with special attention to ecological resilience (Murphree 1995, 1997; Mangani-Kamoto 1999; Berkes et al. 2000; Wiersum 2004; Mala 2009). Rethinking conventional forest management strategies means above all recognizing the key roles of indigenous people, their knowledge, and social organization, in the management and maintenance of biocultural resources. Recognizing these roles is the proper basis for greater integration of traditional approaches in forest management (Murphree 1995, 1997; Mutta et al. 2009).

Recognition that rural communities have been the custodians of traditional forest-related knowledge and biocultural resources, and are still the key to their conservation (Jackson 2004), calls for participatory approaches to forest manage-ment involving forest managers and local communities as equal partners. An exam-ple of new applications of traditional knowledge to sustain both rural livelihoods and biodiversity conservation is found in the development of village botanical gar-dens, which are making major contributions to plant conservation and human well-being worldwide (Waylen 2006). Village botanical gardens have recently been established in Benin as strategy to conserve the country's medicinal and threatened plants (Hamilton 2002; Akpona et al. 2009).

Participatory approaches to forest management are believed to offer the best opportunity for the conservation and sustainable use of natural resources. In such collaboration, forest managers and local communities should discuss and share the decision making on an equal footing, and traditional knowledge, customary laws, and principles should inform natural forest management plans, as for example in the protection of biodiversity (Munyanziza and Wiersum 1999; Mutta et al. 2009; Mala 2009; Chloé et al. 2009). For this to occur, greater priority needs to be given to strengthening and protecting existing customary law systems because of the impor-tant values inherent in those systems, which are critical to the maintenance of the cultures concerned and also to the maintenance and enhancement of biological diversity. The legitimization of traditional governance systems, reinstatement of historical ownership, community leadership and traditional structures that define authority will be valuable (Mutta et al. 2009).

2.4.1 Traditional Forest-Related Knowledge and Forestry Education

There are many tertiary institutions in African countries promoting forestry education. Their curricular development has generally matched prevailing government forest policies. In recent decades there has been a shift in emphasis from forestry for timber production (primarily for the benefit of colonizing countries) towards forestry that aims to contribute to the socioeconomic development of the people and the maintenance of biological diversity, with the participation of local communities and the wider public (Kojwang 2004). With the involvement of local communities in forest issues, including forest management, traditional forest-related knowledge has gradually become an important component in research and training in forest education. For example, knowledge about traditional sacred groves is now being appreciated, taught, and researched (Hartnett 2010). Social sciences research methodologies are being used to complement studies in silviculture involving farm forestry, agroforestry, and the community-level management of natural forests. One important need is a standard text on African traditional forest-related knowledge, which would bring together the varied experiences in the cultural and environmental interfaces in the African wilderness. This will be an invaluable contribution to forestry in the process of documentation of African TFRK.

In some communities in Africa, local conservation projects that use traditional forest-related knowledge are being encouraged. The traditional fire management practices of the Ik people in the Karamoja district in Uganda are being shared with their Karimpong, Tepeth, Dodoth, and Pokot neighbours. Such extension to other tribal groups who inhabit forests/miombo woodlands between Kenya and Tanzania and between Uganda and Tanzania is particularly valuable for efforts to reverse environmental degradation (Kessy 1998; Rodgers et al. 2002).

Through funding provided through the Global Environment Facility's Small Grants Program, several NGOs have developed links with communities in Ghana to transfer traditional forest-related knowledge to promote livelihood development (Ghana Ministry of Environment, Science and Technology 2009). Ghana's Forestry Commission is using the concept of community-based natural resources management to promote its 'community resource management associations' (CREMA) for the sustainable use of forest resources. Within these associations, special efforts are made by the members to formally and informally (as was done in the olden days) pass on information. Through these efforts, the people in these communities have acquired self discipline and are responsible for policing the resource areas.

Traditional forest-related knowledge is not a 'sacred black box' (Haruyama 2002). For its effective application to biodiversity conservation, it is necessary to clarify which aspects of traditional knowledge contribute to biodiversity conservation, how to apply this knowledge, and how to conserve the knowledge we are in danger of losing, possibly by developing appropriate incentives. However, since there are different kinds of traditional forest-related knowledge, that which should be conserved or created within communities should be in accordance to their respective situations. Some indigenous peoples rely on natural resources for their cash income, while others do not. Thus, the policy measures for preventing traditional forest-related knowledge loss and its application to biodiversity conservation should be based on the specific conditions of the indigenous peoples and traditional knowledge in question.

2.4.2 The Role of Forest Science in Relation to African Traditional Forest-Related Knowledge

In considering the potential for combining formal scientific and traditional forest-related knowledge to address natural resource management challenges in Africa, it is important to identify the ways and means of creating positive reinforcements between the two knowledge systems. These revolve around:

1. Rethinking/challenging conventional portrayals of forest agriculture (namely shifting cultivation) and of the concepts associated with it (i.e., deforestation, poverty, and degradation);
2. Selection of appropriate scales for agro-ecosystem analysis and intervention by taking into account the interactions between ecology, society, and economy in the real dynamics of agro-ecosystems;
3. Use of innovation and flexible processes that can allow for enhancement of social learning and adaptive capacity to cope with uncertainty and change; adaptive co-management already provides a framework to deal with such complexity of knowledge-practice-belief (Niamir 1990; Dei 1993; Berkes 1999); and
4. Recognition that knowledge constantly evolves and that understanding both similarities and differences between traditional and scientific knowledge systems to deal with the complexity of traditional agro-ecosystems in their heterogeneous contexts can yield positive outcomes.

According to Shackeroff and Campbell (2007), traditional ecological knowledge (TEK) research can be considered as an opportunity to revisit conservation commitments with special treatment of TEK in conservation-oriented research. The conservation interests, objectives, and sometimes advocacy activities of researchers pose perhaps the greatest challenge for successful TEK research. This is not to say that indigenous communities do not want, understand, or practice conservation, but their approaches to it may vary greatly from those of Western researchers and even of other indigenous groups (Berkes 1999; Martello 2004). Just as assumptions regarding what conservation should look like can undermine inter-disciplinary scientific collaboration, they also underlie and magnify the issues of politics and politicization, ethics, and situated knowledge in TEK research.

Shackeroff and Campbell (2007) suggest that challenges associated with TEK research may be overcome by adopting appropriate research protocols and engaging in participatory and inter-disciplinary research, which will also provide the opportunity to revisit conservation commitments and the normative assumptions embedded in these. At the very least, researchers need to make their conservation commitments explicit to traditional knowledge holders to satisfy their ethical obligations to human research subjects (Fortmann and Ballard 2009). But at a more fundamental level, open conversations between both researchers and traditional knowledge holders about conservation commitments can provide important opportunities for questioning, challenging, re-envisioning, and sometimes reinforcing such commitments, and, in fact, may end up with 'better' conservation. In other words, this is the kind of conservation that achieves biological and socioeconomic goals in a culturally appropriate manner, that recognizes and respects the traditional knowledge, and that is supported by both researchers and traditional knowledge holders with whom they engage.

2.5 Conclusion: The Way Forward for Traditional and Scientific Forest-Related Knowledge in Africa

Traditional knowledge and practices have sustained the livelihoods, cultures, and the forest and agricultural resources of local and indigenous communities throughout Africa for millennia. This knowledge is tightly interwoven with traditional religious beliefs, customs, folklore, land-use practices, and community-level decision-making processes. Historically, these traditional knowledge systems have been dynamic, responding to changing environmental, social, economic, and political conditions to ensure that forest resources continue to provide tangible benefits (food, medicines, wood and other non-timber forest products, water, and fertile soils) as well as intangible benefits (spiritual, social, and psychological health) for present and future generations.

Despite their importance and contributions to sustainable rural livelihoods, traditional forest-related knowledge and practices are under pressure in most African countries (as elsewhere in the world) for a number of reasons. These include imbalanced power relations between state forest management authorities and local and indigenous communities whose traditional governance systems and customary laws are often at odds with those of the state (Cotton 1996; Mutta et al. 2009); government policies and regulations within and outside of the forest sector restricting access and traditional use of forest resources; a general erosion of traditional culture and of traditional land and forest management knowledge and practices; and declining interest in traditional wisdom, knowledge, and lifestyles among younger generations.

The negative implications of this loss of traditional forest-related knowledge on livelihoods, cultural and biological diversity, and the capacity of forested landscapes to provide environmental goods and services remain poorly understood, largely unappreciated, and undervalued by policy makers and the general public in most countries. It is important that such knowledge be documented to illustrate, reflect, and authenticate the historical continuum of diversity in African cultures that has survived through time, as well as the variability of traditional knowledge and practices that have been developed within, and contributed to the sustainable management of, Africa's diversity of forest ecosystems and forested landscapes.

In the emerging global knowledge economy, a country's ability to build and mobilize knowledge capital is equally essential for sustainable development as the availability of physical and financial capital. The basic component of any country's knowledge system is its indigenous knowledge, encompassing the skills, experiences, and insights of people, applied to maintain or improve their livelihood. According to World Bank (2005), eight natural resource management projects in Africa aiming to contribute to the Millennium Development Goals (MDGs) were implemented based on local knowledge, including both tangible knowledge (e.g., of traditional medicinal plants and extractive resources) and intangible knowledge (e.g., mediation and consensus-building strategies for natural resources management and shared use of environmental resources). Such projects illustrate how traditional and modern knowledge systems can be leveraged to address significant problems, in this case reducing maternal mortality, treating HIV/AIDS, increasing food security, and conserving biodiversity.

While the importance of traditional forest-related knowledge is already widely acknowledged, major issues remain to be resolved regarding how this knowledge should be codified and used. This would include means for its effective integration with modern science to further the objectives of sustainable forest management. This integration could be improved and become productive by: (1) selection of the appropriate scale of analysis and intervention based on an adaptive co-management framework; (2) the adoption of flexible approaches for integrating the linkages between knowledge, decision-making, and innovations; (3) the initiation of a science-policy-practices dialogue on issues relevant for the integration of both agriculture and forest management; and (4) a better definition of units of forest landscape mosaics management and research relevant to improvement of forest and agricultural policies to reflect the integration of timber, non-timber products, and crops. The achievement of such complex goals requires the development of technologies, and appropriate institutional frameworks and socio-organizational settings.

Community-level social and political processes reflecting different interests in forests require more attention in policy analysis. National policies, laws, and regulation can be effective only if they are meaningful to and accepted by local and indigenous communities. Forests issues have become global concerns subject to political efforts to foster greater cooperation in their management. Policy research needs to give greater attention to international forest policy developments and their possible impacts at the national and local levels (Schmithüsen 2003).

In view of the above, it can be inferred that there is more to do in African forest management and governance in order to achieve the desired and finer products of effective synergies between traditional forest-related knowledge and modern forestry science.

References

Abdel Magid TD (2001) Forests biodiversity: its impact on nonwood forest products. Book prepared on the occasion of the centennial anniversary of Sudan's Forests Service. National Forests Corporation, Khartoum

Afolayan AJ, Kambizi L (2009) The impact of indigenous knowledge system on the conservation of forests medicinal plants in Guruve, Zimbabwe. In: Parrotta JA, Oteng-Yeboah A, Cobbinah J (eds) Traditional forest-related knowledge and sustainable forest management in Africa, vol 23, IUFRO World Series. International Union of Forest Research Organizations [IUFRO], Vienna, p. 112

Ajibade LT, Shokemi OO (2003) Indigenous approaches to weather forecasting in Asa LGA, Kwara State, Nigeria. Indilinga African Journal of Indigenous Knowledge Systems 2:37–44

Akpona HA, Sogbohossou E, Sinsin B, Houngnihin RA, Akpona JDT, Akouehou G (2009) Botanical gardens as a tool for preserving plant diversity, threatened relic forest and indigenous knowledge on traditional medicine in Benin. In: Parrotta JA, Oteng-Yeboah A, Cobbinah J (eds) Traditional forest-related knowledge and sustainable forest management in Africa, vol 23, IUFRO World Series. International Union of Forest Research Organizations [IUFRO], Vienna, pp 5–13

Akundabweni LSM, Chweya JA (1993) Field survey report on indigenous African food crops and useful plants, food preparation from them and home gardens. Situation report submitted to the United Nations University Program Kenya/ Natural Resources in Africa (UNU/INRA) Legon, Ghana

Albano G (2001) Indigenous management practices and conservation of *Dalbergia melanoxylon* (Guill. & Perr.) in Mozambique. Thesis, Wageningen University, Department Environmental Sciences, Wagingen

Allen W (1965) The African husbandman. Oliver and Boyd, London

Alternatives to Slash and Burn Programme [ASB] (1995) Alternatives to slash and burn phase i report: forest margins benchmark of Cameroon. Mimeograph, IITA Humid Forest Ecoregional Centre, Yaoundé

Alternatives to Slash and Burn Programme [ASB] (2000) Alternatives to slash and burn phase ii report: forest margins benchmark of Cameroon. Mimeograph, IITA Humid Forest Ecoregional Centre, Yaoundé

Altieri AM (2002) Agroecology: the science of natural resource management for poor farmers in marginal environments. Agric Ecosyst Environ 93:1–24

Altieri MA, Anderson MK (1986) An ecological basis for the development of alternative agricultural systems for small farmers in the Third World. Am J Altern Agric 1:30–38

Amanor KS (2003) Natural and cultural assets and participatory forest management in West Africa. Conference paper series no. 8. In: International conference on natural assets, Tagaytay, 6–11 Jan 2003

Amissah L (2009) Indigenous fire management practices in Ghana. In: Parrotta JA, Oteng-Yeboah A, Cobbinah J (eds) Traditional forest-related knowledge and sustainable forest management in Africa, vol 23, IUFRO World Series. International Union of Forest Research Organizations [IUFRO], Vienna, pp 131–135

Ampofo O (1983) First aid in plant medicine. Waterville, Accra

Arnold M, Townson I (1998) Assessing the potential of forest product activities to contribute to rural incomes in Africa. ODI natural resource perspectives no. 37.Overseas Development Institute, London

Asase A, Oteng-Yeboah AA, Odamtten GT, Simmonds MSJ (2005) Etnobotanical study of some Ghanaian anti-malarial plants. J Ethnopharmacol 99:273–279

ASB (1995, 2000) *See* Alternatives to Slash and Burn Programme

Avocèvou C, Sinsin B, Oumorou M, Dossou G, Donkpègan A (2009) Ethnobotany of *Pentadesma butyracea* in Benin: a quantitative approach. In: Parrotta JA, Oteng-Yeboah A, Cobbinah J (eds) Traditional forest-related knowledge and sustainable forest management in Africa. IUFRO world series vol 23. International Union of Forest Research Organizations [IUFRO], Vienna, pp 154–164

Bahuchet S (1996) La mer et la forêt: Ethnoécologie des populations forestières et des pêcheurs du Sud-Cameroun. In: Froment A, De Garine I, Binam Bikoï C, Loung J-F (eds) Anthropologie alimentaire et développement en Afrique intertropicale: du biologique au social, Actes du colloque tenue à Yaoundé (1993), ORSTOM, Paris

Balehegn M, Eniang EA (2009) Assessing indigenous knowledge for evaluation, propagation and conservation of indigenous multipurpose fodder trees towards enhancing climate change: adaptation in Northern Ethiopia. In: Parrotta JA, Oteng-Yeboah A, Cobbinah J (eds) Traditional forest-related knowledge and sustainable forest management in Africa. IUFRO world series vol 23, vol 23. International Union of Forest Research Organizations [IUFRO], Vienna, pp 39–46

Belcher B, Ruíz-Pérez M, Achdiawan R (2005) Global patterns and trends in the use and management of commercial NTFPs: implications for livelihoods and conservation. World Development 33(9):1435–1452

Bele MY, Jum C (2009) Local people and forest management in Cameroon. In: Parrotta JA, Oteng-Yeboah A, Cobbinah J (eds) Traditional forest-related knowledge and sustainable forest management in Africa. IUFRO world series vol 23. International Union of Forest Research Organizations [IUFRO], Vienna, pp. 136

Belem B, Olsen CS, Guinko S, Theilade I, Bellefontaine R, Lykke AM, Boussim JI, Diallo A (2009) Identification participative des arbres hors forêt préférés par les populations locales dans la Province du Sanmatenga au Burkina Faso. In: Parrotta JA, Oteng-Yeboah A, Cobbinah J (eds) Traditional forest-related knowledge and sustainable forest management in Africa. IUFRO world series vol 23. International Union of Forest Research Organizations [IUFRO], Vienna, pp 172–180

Berkes F (1999) Sacred ecology. Traditional ecological knowledge and resource management. Taylor & Francis, Philadelphia

Berkes F (2004) Rethinking community-based conservation. Conserv Biol 18(3):621–630

Berkes F, Colding J, Folke C (2000) Rediscovery of traditional ecological knowledge as adaptive management. Ecol Appl 10:1251–1262

Black R, Harrison E, Matakala P, Serra A, Watson E, Admassie Y, Pankhurst A, Yibabie T, Schafer J, Ribeiro A (undated) Characteristics of 'Traditional Forest Management'. Briefing ET10. Forum for Social Studies (Ethiopia), Brighton, Centro de Experimentação Florestal (Mozambique), and University of Sussex (UK). Available via http://www.geog.sussex.ac.uk/research/development/marena/pdf/ethiopia/Eth10.pdf. Cited 13 Apr 2011

Bognounou F, Savadogo P, Thiombiano A, Oden PC, Boussim IJ, Guinko S (2009) Ethnobotany and utility evaluation of five Combretaceae species among four ethnic groups in western Burkina Faso. In: Parrotta JA, Oteng-Yeboah A, Cobbinah J (eds) Traditional forest-related knowledge and sustainable forest management in Africa, vol 23, IUFRO World Series. International Union of Forest Research Organizations [IUFRO], Vienna, pp 181–189

Bonou A, Adégbidi A, Sinsin B (2009) Endogenous knowledge on non-timber forest products in northern Benin. In: Parrotta JA, Oteng-Yeboah A, Cobbinah J (eds) Traditional forest-related knowledge and sustainable forest management in Africa. IUFRO world series vol 23. International Union of Forest Research Organizations [IUFRO], Vienna, pp. 48

Brandstrom P, Hultin J, Lindstrom J (1979) Aspects of agro-pastoralism in East Africa. Research report no.51, Scandinavian Institute for African Studies, Uppsala

Brown D, Schreckenberg K (1998) Shifting cultivators as agents of deforestation: assessing the evidence. ODI natural resource perspectives no. 29. Overseas Development Institute, London

Byarugaba D, Nakakeeto RF (2009) Indigenous knowledge and conservation attitudes on biodiversitry outside protected area systems in Uganda: Igara County scenario. In: Parrotta JA,

Oteng-Yeboah A, Cobbinah J (eds) Traditional forest-related knowledge and sustainable forest management in Africa. IUFRO world series vol 23, vol 23. International Union of Forest Research Organizations [IUFRO], Vienna, pp. 111

Campbell B (ed) (1996) The Miombo in transition: woodlands and welfare in Africa. Center for International Forestry Research (CIFOR), Bogor

Carrière S (1999) Les orphelins de la forêt: influence de l'agriculture itinérante sur brûlis de Ntumu et des pratiques agricoles associées sur les dynamiques forestière du sud Cameroun. These. Université de Montpellier, Montpellier

Chloé NM, Sibelet N, Dulcire M, Rafalimaro M, Danthu P, Carrière SM (2009) Taking into account local practices and indigenous knowledge in an emergency conservation context in Madagascar. Biodivers Conserv 18(10):2759–2777

Clarke J (1994) Building on indigenous natural resource management: forestry practices in Zimbabwe's communal lands. Forestry Commission, Harare

Colfer CJP (ed) (2005) The complex forest: communities, uncertainty, and adaptive collaborative management. Resources for the Future, Washington, DC

Cotton CM (1996) Ethnobotany, principles and applications. Wiley, Chichester

Coughenour MB, Ellis JE, Swift DM, Coppock DL, Galvin K, McCabe JT, Hart TC (1985) Energy extraction and use in a nomadic pastoral ecosystem. Science 230:619–625

Darr B, Pretzsch J, Depzinsky T, Darr B, Pretzsch J, Depzinsky T (2009) Traditional forest perception and its relevance to forest conservation among the Tiriki in Kenya. In: Parrotta JA, Oteng-Yeboah A, Cobbinah J (eds) Traditional forest-related knowledge and sustainable forest management in Africa. IUFRO world series vol 23. International Union of Forest Research Organizations [IUFRO], Vienna, pp 113–120

Dei GJS (1993) Indigenous African knowledge systems: local traditions of sustainable forestry. Singap J Trop Geogr 14:28–41

Derkyi NSA, Derkyi MAA (2009) Traditional knowledge on conservation of seed oil, medicinal and dyes and tannin plants. In: Parrotta JA, Oteng-Yeboah A, Cobbinah J (eds) Traditional forest-related knowledge and sustainable forest management in Africa. IUFRO world series vol 23. International Union of Forest Research Organizations [IUFRO], Vienna, pp 148–153

Derkyi MAA, Boateng K, Owusu B (2009) Local communities' traditional knowledge of the mangrove forest ecosystem in Nzema East and Mfantseman Districts in Ghana. In: Parrotta JA, Oteng-Yeboah A, Cobbinah J (eds) Traditional forest-related knowledge and sustainable forest management in Africa. IUFRO world series vol 23. International Union of Forest Research Organizations [IUFRO], Vienna, pp 137–147

Diaw MC (1997) Nda Bot and Ayong shifting cultivation, land use and property rights in Southern Cameroon. Rural Development Forestry Network paper 21. Overseas Development Institute, London

Diaw MC, Oyono PR (1998) Dynamiques et représentations des espaces forestiers au Sud Cameroun: pour une relecture sociale des paysages. Arbres Forêts et Communautés Rurales 15(16):36–43

Djagoun CAMS, Kindomihou V, Sinsin B (2009) Diversity and ethnozoological study of small mammals in villages of the Pendjari Biosphere Reserve in northern Benin. In: Parrotta JA, Oteng-Yeboah A, Cobbinah J (eds) Traditional forest-related knowledge and sustainable forest management in Africa. IUFRO world series vol 23. International Union of Forest Research Organizations [IUFRO], Vienna, pp 191–198

Dounias E (1996) Recrus forestiers post-agricoles: perceptions et usages chez les Mvae du Sud-Cameroun. Journal d'Agriculture Tropicale et de Botanique Appliquée 38:153–178

Dounias E, Hladik M (1996) Les agroforêts Mvae et Yassa au Cameroun littoral: functions socio-culturelle, structure and composition floristique. In: Hladik CM, Hladik A, Pagezy H, Linares OF, Koppert GJA, Froment A (eds) L'alimentation en forêt tropicale; interactions sociocul-turelles et perspectives de développement. Editions UNESCO, Paris, pp 1103–1126

Endalamaw TB, Wiersum F (2009) Traditional access and forest management arrangements for beekeeping: the case of Southwest Ethiopia forest region. In: Parrotta JA, Oteng-Yeboah A, Cobbinah J (eds) Traditional forest-related knowledge and sustainable forest management in

Africa. IUFRO world series vol 23. International Union of Forest Research Organizations [IUFRO], Vienna, pp 165–171

Erakhrumen A (2009) Some other uses of accepted agroforestry fuelwood species based on traditional knowledge in selected rural communities of Oyo State, Southwest Nigeria. In: Parrotta JA, Oteng-Yeboah A, Cobbinah J (eds) Traditional forest-related knowledge and sustainable forest management in Africa. IUFRO world series vol 23. International Union of Forest Research Organizations [IUFRO], Vienna, pp 85–92

Fairhead J, Leach M (1994) Représentations culturelles africaines et gestion de l'environnement. Politique Africaine 53:11–24

Fairhead J, Leach M (1996) Misreading the African landscape: society and ecology in a forest-savanna mosaic. Cambridge University Press, Cambridge

Falconer J (1990) The major significance of minor forest products: the local use and value of forest in the West African humid forest zone. Community Forestry Note no. 6. Food and Agriculture Organization of the United Nations [FAO], Rome

Fandohan AB, Assogbadjo AE, Sinsin B (2009) Endogenous knowledge on tamarind (*Tamarindus indica* L.) in northern Benin. In: Parrotta JA, Oteng-Yeboah A, Cobbinah J (eds) Traditional forest-related knowledge and sustainable forest management in Africa. IUFRO world series vol 23. International Union of Forest Research Organizations [IUFRO], Vienna, pp 57–62

Fischer FU, Packham J, Amadi RM (1993) NTFPs—three views from Africa. Rural Development Forestry Network paper 15c. Overseas Development Institute, London. Available via http://www.mekonginfo.org/mrc/rdf-odi/english/papers/rdfn/15c-i.pdf. Cited 5 Apr 2011

Food and Agriculture Organization [FAO] (1995) Non-wood forest products in nutrition. In: Report of the expert consultation on non-wood forest products, Yogyakarta, Indonesia. Non-Wood Forest Products 3, Rome

Food and Agriculture Organization [FAO] (1999) The role of wood energy in Africa: wood energy today for tomorrow. Regional studies. Working paper FOPW 99/3, Rome

Food and Agriculture Organization [FAO] (2001) Global forest resource assessment, 2000. FAO Forestry paper 140, Rome

Food and Agriculture Organization [FAO] (2006) Global forest resources assessment 2005—progress towards sustainable forest management. FAO Forestry paper no. 147, Rome. Available via www.fao.org/docrep/008/a0400e/a0400e00.htm. Cited 28 Feb 2011

Food and Agriculture Organization [FAO] (2009) State of the world's forests 2009. FAO, Rome

Fortmann L, Ballard H (2009) Sciences, knowledges, and the practice of forestry. Eur J For Res. doi:10.1007/s10342-009-0334-y

Fortmann L, Nhira, C (1992) Local management of trees and woodland resources in Zimbabwe: a tenurial niche approach. O.F.I. occasional paper no. 43. Oxford Forestry Institute, Oxford

Gadgil M, Berkes F, Folke C (1993) Indigenous knowledge for biodiversity conservation. Ambio 22:151–156

Gadgil M, Hemam NS, Reddy BM (1998) People, refugia and resilience. In: Berke F, Folke C (eds) Linking social and ecological systems: management practices and social mechanisms for building resilience. Cambridge University Press, Cambridge, pp 30–47

Galloway-McLean K (2010) Advance guard: climate change impacts, adaptation, mitigation and indigenous peoples—a compendium of case studies. United Nations University-Traditional Knowledge Initiative, Darwin. Available via http://www.unutki.org/news.php?doc_id=101&news_id=92. Cited 20 Mar 2011

Geldenhuys CJ (2004) Bark harvesting for traditional medicine: from illegal resource degradation to participatory management. Scand J For Res 19(suppl 4):103–115

Ghana Ministry of Environment, Science and Technology (2009) Fourth national report to the Convention on Biological Diversity. Available via http://www.cbd.int/doc/world/gh/gh-nr-04-en.pdf. Cited 13 Apr 2011

Gillon Y (1992) Empreinte humaine et facteurs du milieu dans l'histoire écologique de l'Afrique tropicale. Afrique Contemporaine 161:30–41

Gockowski J, Tonye J, Baker D, Legg C, Weise S, Ndoumbé M, Tiki-Manga T, Fouaguegué A (2004) Characterization and diagnosis of farming systems in the ASB forest margins: benchmark of Southern Cameroon. International Institute for Tropical Agriculture, Ibadan

Gouwakinnou GN, Kindomihou V, Sinsin B (2009) Utilisation and local knowledge on *Sclerocarya birrea* (Anacardiaceae) by the rural population around the W National Park in Karimama District (Benin). In: Parrotta JA, Oteng-Yeboah A, Cobbinah J (eds) Traditional forest-related knowledge and sustainable forest management in Africa. IUFRO world series vol 23. International Union of Forest Research Organizations [IUFRO], Vienna, pp 49–56

Hall JB, O'Brien EM, Sinclair FL (eds) (2000) *Prunus africana*: a monograph. Publication No.18, School of Agriculture and Forestry Science. University of Wales, Bangor

Hamilton A (2002) Medicinal plants and conservation: issues and approaches. Worldwide Fund for Nature (WWF), International Plants Conservation Unit, Godalming

Hartnett DC (2010) Into Africa: promoting international ecological research and training in the developing world. Bull Ecol Soc Am 91(2):202–206

Haruyama T (2002) Traditional ecological knowledge: from the sacred black box to the policy of local biodiversity conservation. [Ritsumeikan University] Policy Science 10(1):85–96

Hundera K (2007) Traditional forest management practices in Jimma Zone, South West Ethiopia. Ethiopia J Educ Sci 2(2):1–10. Available via http://www.ajol.info/index.php/ejesc/article/view/41982. Cited 4 Apr 2011

International Plant Genetic Resources Institute [IPGRI] (2005) Diversity for well-being: making the most of agricultural biodiversity. Bioversity, Rome

Iremonger S, Ravilious C, Quinton T (1997) A statistical analysis of global forest conservation. In: Iremonger S, Ravilious C, Quinton T (eds) A global overview of forest conservation. Center for International Forestry Research (CIFOR) and World Conservation Monitoring Unit (WCMC), Cambridge

Jackson D (2004) Implementation of international commitments on traditional forest-related knowledge: indigenous peoples' experiences in Central Africa. Forest People Programme, Moreton-in-March (UK). Available via http://archive.forestpeoples.org/documents/africa/tfrk_expert_mtg_oct04_eng.pdf. Cited 5 Apr 2004

Johnson M (ed) (1992a) Lore: capturing traditional environmental knowledge. International Development Research Centre, Dene Cultural Institute, Ottawa

Johnson M (1992b) Dene Traditional knowledge. Northern Perspectives 20(1):3–4

Joiris D (1998) Indigenous knowledge and anthropological constraints in the context of conservation programs in central Africa: resource use in the trinational Sangha River Region of Equatorial Africa: histories, knowledge forms and institutions. Bulletin series 102. Yale School of Forestry and Environmental Studies, New Haven, pp 130–140

Joshi L, Shrestha P, Moss C, Sinclair FL (2004) Locally derived knowledge of soil fertility and its emerging role in integrated natural resource management. In: Van Noordwijk M, Cadisch G, Ong CK (eds) Belowground interactions in tropical agroecosystems: concepts and models with multiple plant components. CABI, London

Kenrick J (2000) The forest people of Africa in the 21st century. Present predicament of hunter-gatherers and former hunter-gatherers of the Central African rainforests. Indigenous Affairs 2/00:10–24

Kessy JF (1998) Biodiversity management and local interests; alternatives for the East Usambara Forest Reserve, Tanzania. Tropical Resource Management paper no. 18, Wageningen Agricultural University, Wageningen

Kojwang HO (2004) Forest science and forest policy development: the challenges of Southern Africa. Scand J For Res 19(suppl 4):116–122

Korem A (1985) Bushfire and agricultural development in Ghana. Ghana Publishing Corporation, Accra

Lusigi WJ (1984) The integrated project on arid lands (IPAL), Kenya. In: Joss PJ, Lynch PW, Williams OB (eds) Rangelands: a resource under siege. Proceeding of 2nd international Rangeland congress, Adelaide, pp 338–349

Mala WA (2009) Knowledge systems and adaptive collaborative management of natural resources in southern Cameroon: decision analysis of agrobiodiversity for forest-agriculture innovations. Dissertation, Stellenbosch University, Faculty of AgriSciences, Stellenbosch

Mala WA, Geldenhuys CJ, Prabhu R, Mala WA, Geldenhuys CJ, Prabhu R (2009a) Local conceptualization of nature, forest knowledge systems and adaptive management in southern Cameroon. In: Parrotta JA, Oteng-Yeboah A, Cobbinah J (eds) Traditional forest-related knowledge and sustainable forest management in Africa, vol 23, IUFRO World Series. International Union of Forest Research Organizations [IUFRO], Vienna, pp 102–110

Mala WA, Geldenhuys CJ, Prabhu R (2009b) A predictive model of local agricultural biodiversity knowledge management in southern Cameroon. Biodiversity 9(1/2):96–101

Mangani-Kamoto JF (1999) Indigenous silvicultural practices of Miombo woodlands in Malawi: a case in five villages close to Chimaliro Forest Reserve. Thesis, Wageningen Agricultural University, Department of Forestry, Tropical forestry, Wageningen

Marena Research Team (2002) Characteristics of 'traditional' forest management. Policy Briefing #10. University of Sussex, East Sussex. Available via http://www.geog.sussex.ac.uk/research/development/marena/#briefings. Cited 28 Feb 2011

Martello ML (2004) Expert advice and desertification policy: past experience and current challenges. Global Environmental Politics 4(3):85–106

Masozera MK, Alavalapati JRR (2004) Forest dependency and its implications for protected areas management: a case study from the Nyungwe Forest Reserve, Rwanda. Scand J For Res 19(suppl 4):85–92

Mshana RN, Abbiw DK, Addae-Mensah I, Adjanouhoun E, Ahyi MRA, Ekpere JA, Enow-Rock EG, Gbile ZO, Noamesi GK, Odei MA, Odunlami H, Oteng-Yeboah AA, Sarpong K, Sofowora A, Tackie AN (2001) Traditional medicine and pharmacopoeia: contribution to the revision of ethnobotanical and floristic studies in Ghana. Science and Technology Press, Council for Scientific and Industrial Research, Accra

Mumma A (2004) Community based legal systems and the management of world heritage sites. In: de Merode E, Smeets R, Westrik C (eds) Linking universal and local values: managing a sustainable future for world heritage. World Heritage series 13. UNESCO World Heritage Centre, Paris. Available via http://whc.unesco.org/documents/publi_wh_papers_13_en.pdf. Cited 5 Apr 2011

Munyanziza E, Wiersum KF (1999) Indigenous knowledge of Miombo trees in Morogoro, Tanzania. Indigenous Knowledge and Development Monitor 7(2):10–13

Munyi P, Mutta D (2007) Protection of community rights over traditional knowledge: implications for customary laws and practices in Kenya. Science Press, Nairobi

Murphree MW (1995) Optimal principles and pragmatic strategies: creating an enabling politico-legal environment from community-based natural resources management (CBNRM). In: Steiner E, Rihoy A (eds) The commons without the tragedy? strategies for community-based natural resources management in Southern Africa. USAID-Regional NRMP, SADC Wildlife Technical Coordination Unit, Malawi

Murphree MW (1997) Congruent objectives, competing interest and strategic compromise: concept and process in the evolution of Zimbabwe's CAMPFIRE programme. Paper presented to the conference on representing communities: histories and politics of community-based resources management, Helen

Mutta D, Chagala-Odera E, Wairungu S, Nassoro S (2009) Traditional knowledge systems for management of Kaya forests in Coast region of Kenya. In: Parrotta JA, Oteng-Yeboah A, Cobbinah J (eds) Traditional forest-related knowledge and sustainable forest management in Africa, vol 23, IUFRO World Series. International Union of Forest Research Organizations [IUFRO], Vienna, pp 122–130

Mviena P (1970) Univers culturel et religieux du peuple beti. Librairie Saint Paul, Yaoundé

Mvondo SA, Oyono PR (2004) An assessment of social negotiation as a tool for local management: a case study of the Dimako council forest, Cameroon. Scand J For Res 19(suppl 4):78–84

N'Dri MTK, Gnahoua GM (2009) Arbres et lianes spontanées alimentaires de la zone de forêt semi décidue (Centre-Ouest de la Côte d'Ivoire): flore des espèces rencontrées, organes consommés, valeurs alimentaires. In: Parrotta JA, Oteng-Yeboah A, Cobbinah J (eds) Traditional forest-related knowledge and sustainable forest management in Africa, vol 23, IUFRO World Series. International Union of Forest Research Organizations [IUFRO], Vienna, pp 73–84

Nasi R, Brown D, Wilkie D, Bennett E, Tutin C, Van Tol G, Chistophersen T (2009) Conservation and use of wildlife-based resources: the bushmeat crisis. CBD technical series No. 33. Secretariat of the Convention on Biological Diversity and Center for International Forestry Research (CIFOR), Montreal

Ndoye O, Tieguhong JC (2004) Forest resources and rural livelihoods: the conflict between timber and non-timber forest products in the Congo Basin. Scand J For Res 19(suppl 4):36–44

Ndoye O, Tieguhong JC (2009) NTFPs and services for sustainable livelihoods in Central Africa. In: Diaw MC, Aseh T, Prabhu R (eds) Search of common grounds: adaptative collaboration management in Cameroon. Center for International Forestry Research (CIFOR), Bogor, pp 353–378

Ngobo MP, Weise SF, McDonald MA (2004) Revisiting the performance of natural fallows in central Africa. Scand J For Res 19(suppl 4):14–24

Nhira C (1994) Efforts at reorienting forest policy in Zimbabwe: a pilot resource-sharing scheme. In: Roundtable discussions on forest policies in East and southern Africa, Addis Ababa, 24–26 Oct 1994

Niamir M (1990) Traditional woodland management techniques of African pastoralists. Unasylva 41(160). Available via http://www.fao.org/DOCREP/T7750E/t7750e08.htm. Cited 5 Apr 2011

Niamir-Fuller M (1998) The resilience of pastoral herding in Sahelian Africa. In: Berkes F, Folke C (eds) Linking social and ecological systems: management practices and social mechanisms for building resilience. Cambridge University Press, Cambridge, pp 250–284

Nyong A, Adesina F, Osman EB (2007) The value of indigenous knowledge in climate change mitigation and adaptation strategies in the African Sahel. Mitig Adapt Strateg Glob Chang 12:787–797

Ohenjo N, Willis R, Jackson D, Nettleton C, Good K, Mugarura B (2006) Health of indigenous people in Africa. Lancet 367(9526):1937–1946. Available via http://www.thelancet.com. Cited 5 Apr 2011

Osman-Elasha B, Parrotta J, Adger N, Brockhaus M, Colfer CJP, Sohngen B, Dafalla T, Joyce LA, Nkem J, Robledo C (2009) Future socio-economic impacts and vulnerabilities. In: Seppälä R, Buck A, Katila P (eds) Adaptation of forests and people to climate change—a global assessment report, vol 22, IUFRO World Series. International Union of Forest Research Organizations [IUFRO], Helsinki, pp 101–122

Owusu-Sekyere E (2009) Traditional knowledge on tree characteristics and use for agroforestry in Ghana. In: Parrotta JA, Oteng-Yeboah A, Cobbinah J (eds) Traditional forest-related knowledge and sustainable forest management in Africa, vol 23, IUFRO World Series. International Union of Forest Research Organizations [IUFRO], Vienna, pp. 47

Oyono PR (2002) Usages culturels de la forêt au Sud-Cameroun: Rudiments d'ecologie sociale et matériau pour la gestion du pluralisme. Africa LVII 3:334–355

Oyono PR, Mala AW, Tonyé J (2003) Rigidity versus adaptation: contribution to the debate on agricultural viability and forest sustainability in Southern Cameroon. Cult Agric 25(2):32–40

Pandey DN (Not dated) Ethnoforestry. The Overstory #76, The Overstory Agroforestry Ejournal. Available via http://www.agroforestry.net/overstory/overstory76.html. Cited 1 Mar 2011

Parry ML, Canziani OF, Palutikof JP, van der Linden PJ, Hanson CE (eds) (2007) Contribution of working group ii to the fourth assessment report of the Intergovernmental Panel on Climate Change. Cambridge University Press, Cambridge

Rodgers WA (1993) The conservation of the forest resources of eastern Africa: past influences, present practices and future needs. In: Lovett J, Wasser S (eds) Biogeography and ecology of the rain forests of eastern Africa. Cambridge University Press, Cambridge, pp 283–331

Rodgers WA, Nabanyumya R, Mupada E, Persha L (2002) Community conservation of closed forest biodiversity in East Africa: can it work? Unasylva 209(53):41–47

Schmithüsen F (2003) Understanding cross-sector policy impacts—policy and legal aspects. FAO Forestry paper 142:5–44. Food and Agriculture Organization of the United Nations, Rome, pp 5–44. Available via http://www.fao.org/DOCREP/006/Y4653E/Y4653E00.HTM. Cited 13 Apr 2011

Shackeroff JM, Campbell LM (2007) Traditional ecological knowledge in conservation research: problems and prospects for their constructive engagement. Conserv Soc 5(3):343–360

Shaffer LJ (2010) Indigenous fire use to manage savanna landscapes in Southern Mozambique. Fire Ecology 6(2):43–59

Simel JO (2008) The threat posed by climate change to pastoralists in Africa. Indigenous Affairs–Climate Change and Indigenous Affairs 1-2/08:34–43. Available via http://www.iwgia.org/. Cited 25 Mar 2011

Sonwa JD (2004) Biomass management and diversification within cocoa agroforests in the humid forest zone of southern Cameroon. Dissertation, Universitat Bonn, Bonn

Sunderland T, Ndoye O (eds) (2004) Forest products, livelihood and conservation: cases studies of non-timber forests products systems, vol 2, Africa. Center for International Forestry Research [CIFOR], Bogor

Sunderlin W, Angelsen A, Belcher B, Burgers P, Nasi R, Santoso L, Wunder S (2005) Livelihoods, forests, and conservation in developing countries: an overview. World Development 33(9):1383–1402

Tabuna J (2007) Commerce régional et international des produits forestiers non ligneux alimentaires et des produits agricoles traditionnels en Afrique Centrale. Food and Agriculture Organization of the United Nations [FAO], Rome

Thrupp LA (1997) Linking biodiversity and agriculture: challenges and opportunities for sustainable food security. World Resources Institute, Washington

Tieguhong JC, Ndoye O, Vantomme P, Grouwels S, Zwolinski J, Masuch J (2009) Coping with crisis in Central Africa: enhanced role for non-wood forest products. Unasylva 233:49–54

Tubiana M-J, Tubiana J (1977) The Zaghawa from an ecological perspective. Balkema, Rotterdam

United Nations Environment Programme [UNEP] (1997) Final document of the second international indigenous forum on biodiversity. UNEP/CBD/TKBD/1/3/ Annex 1. Available via http://www.biodiv.org. Cited 5 Apr 2011

United Nations Environment Programme [UNEP], World Conservation Monitoring Center [WCMC] (2011) UNEP-WCMC forests programme. Available at: http://www.unep-wcmc.org/forest/africa_map.htm. Cited 10 Mar 2011

Van Dijk JFW (1999) An assessment of non-wood forest products resources for the development of sustainable commercial extraction. In: Sunderland TCH, Clark LE, Vantomme P (eds) The NWFP of Central Africa: current research issues and prospects for conservation and development. Food and Agriculture Organisation of the United Nations [FAO], Rome

Vermeulen C, Carrière S (2001) Stratégies de gestion des ressources naturelles fondées sur les maîtrises foncières coutumières. In: Delvingt W (ed) La forêt des hommes: terroirs villageois en forêt tropicale africaine. Les presses agronomiques de Gembloux, Gembloux, pp 109–141

Vodouhê GF, Coulibaly O, Sinsin B (2009) Estimating local values of vegetable non-timber forest products to Pendjari Biosphere Reserve dwellers in Benin. In: Parrotta JA, Oteng-Yeboah A, Cobbinah J (eds) Traditional forest-related knowledge and sustainable forest management in Africa, vol 23, IUFRO World Series. International Union of Forest Research Organizations [IUFRO], Vienna, pp 63–72

Wala K, Guelly AK, Batawila K, Dourma M, Sinsin B, Akpagana K (2009) Traditional agroforestry systems in Togo: variability according to latitude and local communities. In: Parrotta JA, Oteng-Yeboah A, Cobbinah J (eds) Traditional forest-related knowledge and sustainable forest management in Africa, vol 23, IUFRO World Series. International Union of Forest Research Organizations [IUFRO], Vienna, pp 21–27

Waylen K (2006) Botanic Gardens: using biodiversity to improve human well-being. Botanic Gardens Conservation International, Richmond

Wiersum KF (1997) Indigenous exploitation and management of tropical forest resources: an evolutionary continuum in forest-people interactions. Agric Ecosyst Environ 63:1–16

Wiersum KF (2004) Forest gardens as an 'intermediate' land-use system in the nature-culture continuum: characteristics and future potential. Agrofor Syst 61:123–134

Willis J (1996) The Northern Kayas of the Mijikenda: a gazetteer, and an historical reassessment. Azania 31:75–98

Wily LA, Mbaya S (2001) Land, people and forests in eastern and southern Africa at the beginning of the 21st century: the impact of land relations on the role of communities in forest future. World Conservation Union, Eastern Africa Regional Office, Nairobi

World Bank (2005) The World Bank Group indigenous knowledge program. Available via www.worldbank.org/afr/ik/achieve.htm. Cited 1 Apr 2011

Chapter 3
Latin America—Argentina, Bolivia, and Chile

Mónica Gabay, Santiago Barros, and Sebastián Bessonart

Abstract Argentina, Chile, and Bolivia are home to a wealth of biodiversity and a wide variety of landscapes, which have shaped over many centuries a very rich cultural tradition among the numerous indigenous peoples present in their territory. This chapter deals with the traditional knowledge generated and preserved by these indigenous peoples, their encounter with Western science, and the resulting knowledge hybridisation processes. Key issues such as multipurpose forest management, non-timber forest products, co-management of protected areas, certification, and key elements related to indigenous forest management are presented here as examples of the encounter of these sometimes conflicting paradigms about forests and nature. A brief discussion on cross-cutting issues such as ethical considerations and equitable benefit sharing is included, stressing the need of further advancements along the line of the full recognition of the worth of indigenous traditional knowledge as well as the enforcement of indigenous communities' property rights.

Keywords Biodiversity • Co-management • Cultural diversity • Certification • Forest history • Forest management • Indigenous peoples • Knowledge hybridisation • Non-timber forest products • Traditional knowledge

M. Gabay (✉) • S. Bessonart
Directorate of Forestry, Secretariat of Environment and Sustainable Development,
Buenos Aires, Argentina
e-mail: monagabay@yahoo.com; sbessonart@ambiente.gob.ar

S. Barros
Instituto Forestal, Santiago de Chile, Chile
e-mail: sbarros@infor.cl

J.A. Parrotta and R.L. Trosper (eds.), *Traditional Forest-Related Knowledge:*
Sustaining Communities, Ecosystems and Biocultural Diversity,
World Forests 12, DOI 10.1007/978-94-007-2144-9_3,
© Springer Science+Business Media B.V. (outside the USA) 2012

3.1　Introduction

This chapter examines the major issues related to traditional forest-related knowledge in Argentina, Chile, and Bolivia, and explores the potential for interactions and integration of current forest management practices and procedures with ancestral tribal wisdom. Since these countries are very diverse in both ecological and cultural terms, it is not intended to present detailed descriptions here, but rather to focus on meaningful cases that contribute to a better understanding of this rich cultural capital and how it is developing.

The status of indigenous peoples and, hence, of their traditional knowledge has evolved in the three countries. Before the Spanish conquest, there was a wide cultural diversity with indigenous groups thriving in a dialectic relationship with their landscape. These peoples developed very rich cosmogonies in line with their holistic approach to nature. However, the European newcomers endeavoured to homogenise these cultures after their own. Traditional knowledge was delegitimized and made invisible. In spite of this new lowly status, oral tradition preserved at least part of the ancient heritage. At present, indigenous communities' movements reclaim their rights, both to their identities and to their ancestral land. This dynamic is parallel to Western science's interest in traditional knowledge. Integrating both types of knowledge offers an opportunity to improve current forest management, while it also poses some challenges.

Section 3.2 deals with the ecological diversity of these countries, and their indigenous peoples—their beliefs, customs, and uses. It also provides a conceptual framework regarding the encounter of cultures that occurred with the European conquest and the cultural hybridisation that followed. Section 3.3 discusses key opportunities and best practices to combine traditional and modern knowledge in forest management, and the main elements related to indigenous forest management. Section 3.4 presents cross-cutting issues that should be considered when dealing with traditional knowledge. Section 3.5 contains the conclusions and final remarks.

3.2　Indigenous People in Argentina, Bolivia, and Chile: Their Ecological Context and Traditional Forest-Related Knowledge

This section presents a brief overview of the ecological context in which the indigenous peoples of Argentina, Bolivia, and Chile developed their traditional knowledge on forest management, and their ancestral culture. It is worth noting that these indigenous peoples are not a homogeneous category but are most diverse.

3.2.1　Ecological Context and Diversity

Argentina, Bolivia, and Chile are located in South America. The area comprises a wide variety of landscapes and climates that account for a rich biodiversity in a

Fig. 3.1 Forest and
woodland cover in southern
South America (Source:
Adapted from FAO (2001)).
Key: *Dark green* closed
forest, *light green* open or
fragmented forest, *pale green*
other wooded land, *yellow*
other land

wealth of forest regions (Fig. 3.1). This diversity has shaped very distinct cultures.
The indigenous peoples of this region have developed different strategies in order
to make the most of the available resources, which can partly be explained by the
local conditions as well as external economic and political factors. Therefore, in
order to analyse their traditional knowledge, it is necessary to present a brief description of the ecological context in which this knowledge has been developed.

3.2.1.1 Argentina

The Argentine Republic covers a total area of 2,766,890 km², spanning 3,694 km,
between 22 and 55°s. This vast territory determines a wide climatic diversity, from

subtropical in the north to cold in the south, with most of the territory having a temperate climate.

About 31,443,873 ha (UMSEF 2005), some 12% of the territory, are covered with native forests, and about 1,115,655 ha (SAGPyA 2002) with plantation forest. About 1,133,000 ha of the native forests are protected as national parks, while some 1,253 000 ha have been included under UNESCO's Man and Biosphere (MAB) reserve scheme (FAO 2005). During the past century there has been a decline in native forest cover, mainly due to the expansion of the agricultural frontier and the incidence of forest fires—mostly of anthropogenic origin.

There are six distinctive forest regions: Parque Chaqueño; Selva Tucumano Boliviana; Selva Misionera; Bosque Andino Patagónico; Monte; and Espinal. There is a wide diversity of ecosystems, ranging from exuberant subtropical rainforests to semi-arid and shrubby steppes, with elevations ranging from 200 m to 2,500 m above sea level.

3.2.1.2 Bolivia

Bolivia has a total area of 1,098,580 km^2, of which about 48% or some 59,963,200 ha are covered with native forests, located mostly in the lowlands (BOLFOR II 2009; Van Dam 2007). Bolivian native forests cover 48% of the total area, with about 53 million ha, and are located mostly in the eastern region of the country (CADEFOR 2009). There are about 30,000 ha of plantation forests. The National System of Protected Areas (Sistema Nacional de Áreas Protegidas [SNAP]) comprises 40 legally created areas, but less than 50% of these have clear boundaries, and consolidated area and management. The core of the system consists in 21 active protected areas that account for 17% of the Bolivian territory (Muñoz 2004).

The deforestation process in Bolivia has intensified during the past decade, particularly in the lowlands. For example, in the Department of Santa Cruz an estimate shows that for the period from 1993 to 2000 about 1,424,033 ha, or some 200,000 ha per year, have been subject to deforestation (Camacho et al. 2001). The total national deforested area for the years 2004 and 2005 was 275,128 ha and 281,283 ha, respectively (Wachholtz et al. 2006).

The main reason for this situation has been the promotion of extensive crops, like soybean, with incentives provided by the World Bank through the 'Lowlands Project' (Van Dam 2007; Camacho et al. 2001), and cattle breeding (Camacho et al. 2001). Other reasons that could help explain the accelerated deforestation are the uncertainty of land tenure rights, migrations from the highlands, illegal logging, problems with local economies, and poor control from the government.

Ibisch and Mérida (2003) describe the following forest regions: (a) in the lowlands, Bosques del Sudoeste de la Amazonía, Bosque Seco Chiquitano, and Gran Chaco; (b) in the eastern Andean slopes and inter-Andean valleys, Yungas (Bolivia–Peru), Bosque Tucumano–Boliviano, and Chaco Serrano. The wide range of landscapes goes from Amazon rainforests to Chaco dry forests, with elevations varying from 100 m to 4,200 m above sea level.

3.2.1.3   Chile

The continental portion of Chile is located between 18 and 55°s, and spans between the Andean mountains and the Pacific Ocean, over a total area of about 756,950 km². There are 15,637,232.5 ha of forests in Chile, of which 13,430,602.8 ha are native forests—with some 3.9 million hectares under some protection scheme—and 2,119,004.5 ha are forest plantations (CONAF 2009). More than 80% of the native forest is concentrated in the southern area of the country.

Native forests comprise a wide variety of species, from the xeromorphic and meso-morphic formations of the arid and semiarid regions of the north and centre of Chile, to the temperate and cold forests of the southern regions where the *Nothofagus* genus is dominant, accompanied by other species, including very long-lived species such as the *Araucaria araucana* and the *Fitzroya cupressoides*. Chilean forest types are defined by the Forest Law as follows: Lenga, Alerce, Araucaria, Esclerófilo, Roble Hualo, Siempreverde, Palma Chilena, Roble–Raulí–Tepa, Roble–Raulí–Coihue, Coihue de Magallanes, Ciprés de la Cordillera, Ciprés de las Guaitecas (Donoso 1981).

3.2.2   Cultural Diversity and Traditional Forest-Related Knowledge

Before the arrival of the European conquerors, there were many different ethnic groups with a great diversity of cultures. The indigenous peoples of Argentina, Chile, and Bolivia had developed a profound knowledge about the forest and its resources. The forest and the people were inseparable parts of a cosmic unity. The forest supported their lives by providing food, shelter, medicines, energy, fibre, and building materials, along with high spiritual value. Indigenous communities in the three countries emphasise the importance of non-timber goods and services. Traditional knowledge focuses on the use of flora (seeds, fruits, flowers, tubercles, and fungi) for products and services such as food, medicine, and basket weaving. Woody plants are also used for energy, lumber, and constructions.

Many of the indigenous peoples living in the lowlands were nomadic, whereas Andean groups were mostly sedentary. Nomadic groups were mainly hunter-gatherers whereas sedentary ones developed such crops as potatoes, corn, yucca, peppers, beans, kiwicha (amaranth), cañahua, and quinoa. A common rule present in the three countries' indigenous populations was the respect for the guardian spirits of the forest and its animals. Hence, the waste of resources was taboo.

3.2.2.1   Argentina

The indigenous peoples account for 3–5% of the total Argentine population and up to 17–25% in some provinces; many of them still live in rural communities (IWGIA

Table 3.1 Indigenous population per ethnic group in Argentina

Ethnic group	Province	Households
Mapuche	Chubut, Neuquén, Rio Negro and Tierra del Fuego	76,606
Kolla	Jujuy and Salta	53,019
Toba	Chaco, Formosa and Santa Fe	47,591
Wichí	Chaco, Formosa and Salta	36,135
Ava Guarani; Guarani; Tupi Guarani	Jujuy and Salta	29,703
Ava Guarani; Guarani; Tupi Guarani	Buenos Aires and the 24 administrative districts of Greater Buenos Aires	20,340
Toba	Buenos Aires and the 24 administrative districts of Greater Buenos Aires	14,456
Diaguita Calchaquí	Jujuy, Salta and Tucumán	13,773
Huarpe	Mendoza, San Juan and San Luis	12,704
Total		383,132

Source: International Work Group for Indigenous Affairs–IWGIA, available via http://www.iwgia.org/sw3184.asp (Cited 10 May 2009)

2009). During the past 50 years, there has been a significant urbanisation process, which has also affected the indigenous communities. Thus, many people have moved to big cities, particularly the national and provincial capitals (IWGIA 2009). Table 3.1 presents the indigenous population per ethnic group.

The main ethnic groups, classified according to their geographic location, are (IWGIA 2009):

- **Northeast region**: Charrúa, Lule, Mbya-Guaraní, Mocoví, Pilagá, Toba, Tonocoté, Vilela, and Wichí.

- **Northwest region**: Atacama, Avá-Guaraní, Chané, Chorote, Chulupí, Diaguita-Calchaquí, Kolla, Ocloya, Omaguaca, Tapiete, Toba, Tupí-Guaraní, and Wichí.

- **South region**: Mapuche, Ona, Tehuelche, and Yamana.

- **Central region**: Atacama, Avá-Guaraní, Diaguita-Calchaquí, Huarpe, Kolla, Mapuche, Rankulche, Toba, Tupí-Guaraní, and Comechingon.

The territorial distribution of the indigenous peoples is shown in Fig. 3.2.

Many of these peoples' social reproduction[1] was—and still is—associated with the rivers and the forests, for they were hunter-gatherers. The indigenous groups were rather small and were led by a chief. He was appointed as such because of his capacity to balance intra- and inter-group relationships, and his skills to find resources for the community. His power lay in his good judgement and his ability to settle conflicts through dialogue with other groups' chiefs (Carrasco 2000).

---

[1] The notion of social reproduction is used here as a social phenomenon related to families and the actions aimed at the transmission of material and symbolic capital from one generation to the next one.

INDIGENOUS PEOPLES OF ARGENTINA

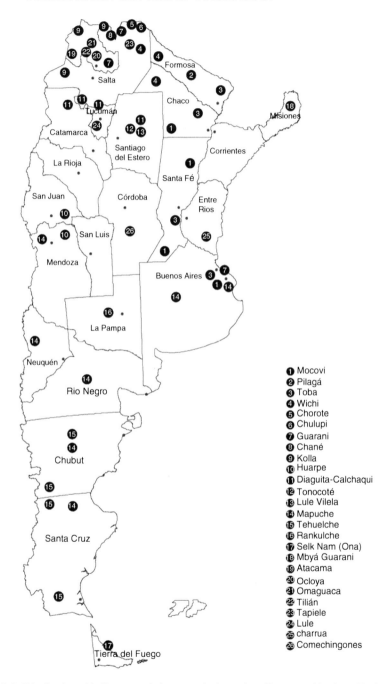

❶ Mocovi
❷ Pilagá
❸ Toba
❹ Wichi
❺ Chorote
❻ Chulupi
❼ Guarani
❽ Chané
❾ Kolla
❿ Huarpe
⓫ Diaguita-Calchaqui
⓬ Tonocoté
⓭ Lule Vilela
⓮ Mapuche
⓯ Tehuelche
⓰ Rankulche
⓱ Selk Nam (Ona)
⓲ Mbyá Guarani
⓳ Atacama
⓴ Ocloya
㉑ Omaguaca
㉒ Tilián
㉓ Tapiele
㉔ Lule
㉕ charrua
㉖ Comechingones

Fig. 3.2 Distribution of indigenous ethnic groups in Argentina. Courtesy of Instituto Nacional de Asuntos Indígenas (INAI 2011)

Box 3.1 Indigenous Peoples in the Semi-Arid Chaco

The semi-arid Chaco region peoples were located in the Upper Bermejo River and Tucumán plains and, towards the south, along the Salado and Dulce Rivers. They grew crops in floodplains and were also nomadic hunters, fishers, and gatherers. The population density was low, each extended family using some 40,000 ha. According to a Jesuit priest, at the time of the Spanish conquest 'their way of life is not living together in large crowds because in two days they would exhaust the hunting, fishing and fruits, which are the only repositories they have for their subsistence.'

Indigenous technologies have caused some impacts on the ecosystem, since they used weapons and fire for hunting, communicating news, and calling for meetings. This produced large forest fires and created grasslands in burnt areas, thus resulting in the typical semi-arid Chaco landscape mosaic. Gathering activities contributed to the dissemination of some tree species, such as the *Prosopis* sp. However, anthropogenic pressure on ecosystems was not high because the indigenous economy was not based upon capital accumulation and maximisation of utilities, but rather on family group consumption.

Indigenous knowledge about certain animal species' natural history did not involve management practices. Notions such as ecologism are related to the advanced capitalist culture, concerned about its own environmental disasters, rather than the hunter-gatherer cultures. In spite of claims that indigenous peoples managed wildlife (based on the association of hunters and the spirits that 'own' the animals), this is not correct because the spirits' true mandate was to not waste the product of the hunt, rather than to impose limits on the number of animals that could be hunted.

The calendar was, and still is, strongly linked to natural cycles. Winter–May to August–is the dry season, when fish are available in rivers, lakes, and wetlands. August, September, and October are rigorous for fish run-out; also, due to the lack of rain there are no flowers–hence, no honey or fruits. During the rainy season, from November to April, indigenous families grow pumpkins, corn, tubers, watermelons, and melons. They also harvest fruits from the forest, such as the algarrobo (*Prosopis alba, P. nigra*), chañar (*Geofrea decorticans*), mistol (*Zyzipus mistol*), sacha pera (*Acanthosyris falcata*), bola verde (*Capparis speciosa*), tala negra (*Celtis spinosa / palida*), and forest beans (*Capparis retusa*), as well as wild honey. All year round they hunt small mammals and birds. In the old days, when the rich times for fishing and gathering arrived, the different groups got together to collectively perform this tasks. These were also the times for making political alliances through marriages.

(continued)

Box 3.1 (continued)

In the old days there were no white men. The earth did not have kings or president; there were no kingdoms, only the indigenous people governed. We did not have identification cards, nor registries or census, but we had a shared idea. In those times we did not suffer anything, we had everything to live on. Our custom was to use the animals and fruits of the forest. Men looked for fish, women looked for chaguar and burnt it, and so we ate the fish with chaguar... Nobody disturbed us then. People moved from one place to another, we camped, when we grew tired of eating honey, we went to the river to eat fish, and this was on November. We grilled the fish so that it would last longer, we also made charque. October was the time of the suri, suri grease was mixed with chaguar, and we did the same with fish grease. We used clothes made of chaguar and bear skin shoes to go hunting.[2]

The indigenous people had extensive knowledge about the uses of non-timber forest products. This traditional knowledge has been partially preserved through oral transmission. People like the Wichí have a wide diversity of herbal remedies they gather from the forest, such as the hukw (*Bulmesia sarmientoi*), for kidney and rheumatic conditions; the Chilayhi lhok (*Croton bomplandianus*), for toothaches; the Hi'nulcha (*Schaefferia argentinensis*), smoked with tobacco to attract the opposite sex; the Tsunhakajlhech'e (*Pithecoctenium cynanchoides*), for kidney and stomach conditions; the Ahot ch'uté (*Pycnoporus sanguineous*), used for abortions and as a contraceptive; the Fwit'hitas (*Cyclolepis genisboides*), for kidney, stomach, and cholesterol problems; the Mawu kafway (*Senna morongii*), for skin diseases; Tsehekw (*Capparis tweediana*), Fwijten (*Acanthosyris falcaba*), and Tulu kawufwja (*Schinus fasciculata*) for respiratory conditions; Hal'o kae's—a kind of lichen—for mouth infections; Tekw (*Cyriapodium punchatum (L)*), for the kidney; Intian (*Ruellia hygophila Mart.*), for internal haemorrhages; Mankuen (*Plantago major*), as an antiseptic for pimples, and for respiratory conditions; Ases ka hal'oy (*Heimia salicifolia*), for respiratory problems and as flocculants for clarifying water; Ch'anhu khos (*Armenia tomentosa*), as a coagulant for post child-birth haemorrhages; Note lhok (*Commelia erecta*), for conjunctivitis; and Fwina (*Typha*) for the flu, among many others.

Notes: *Chaguar*: *Bromelia hieronymi* and *Deimacanthon urbanianum*, were used to extract fibre for a wide range of textiles, from clothing to fishing nets, ropes, and bags. *Charque*: a traditional way of preserving meat by means of drying it. *Suri*: an American ostrich.

Source: Barbarán and Saravia Toledo (2000); Spagarino (2008); Carrasco (2000).

---

[2] Wichí elder testimony, cited in Spagarino (2008).

Fig. 3.3 Wichí woman collecting chaguar (Photo courtesy of Programa Nacional de Bosques Modelo–SAyDS)

Some 300,000/500,000–1,000,000 individuals pertaining to 20 ethnic groups lived in Argentina at the time of the Spanish conquest in the sixteenth century (Golluscio 2008). The conquerors reduced the indigenous peoples to a lowly status, depriving them of their freedom and their ancestral lands. The old religions were banished and the Catholic religion imposed on the peoples. However, many groups adopted the new religion, incorporating elements of the old religion. Ancestral knowledge has been preserved by means of oral tradition, although some practices are probably lost.

Argentina exhibits a low percentage of indigenous population when compared with other Latin American countries. From the late 1850s until the first decades of the twentieth century the dominant model was that of a monocultural, monolingual, homogeneous country, populated with European immigrants. A series of military campaigns was launched beginning in the 1860s to bring 'order and progress' to the indigenous territories. This fact, together with poor labour conditions in rural areas, epidemics, and low social status accounted for a sharp decline in the indigenous population (Golluscio 2008).

The 1994 constitutional reform recognises the 'ethnic and cultural pre-existence' of the Argentine indigenous peoples, and guarantees the respect for their identity and their right to an intercultural and bilingual education. The National Constitution also acknowledges communities' legal entity and the community possession and ownership of their traditional territories, which they cannot dispose of in any way. The indigenous peoples' participation in natural resources management is also recognised (cf. art. 75, §17). The country has also adopted the International Labour Organisation (ITO) Convention 169 in 1992. Box 3.1 and Fig. 3.3 present an example of indigenous uses of the forest in Chaco.

Table 3.2   Indigenous population in Bolivia

Department	Indigenous population		
	Total	Urban area	Rural area
Chuquisaca	345,010	114,889	230,121
La Paz	1,402,184	709,445	692,739
Cochabamba	999,963	446,960	553,003
Oruro	238,829	106,269	132,560
Potosí	572,592	134,518	438,074
Tarija	69,936	42,633	27,303
Santa Cruz	447,955	276,559	171,396
Beni	50,630	23,174	27,456
Pando	6,039	2,895	3,144
TOTAL	4,133,138	1,857,342	2,275,796

Source: INE/VA/UNFPA (2003)

3.2.2.2   Bolivia

According to the 2001 National Census, 62% of the population over 15 years of age self-identified as indigenous, of which Quechua and Aymara ethnic groups account for 30.7% and 25.2% respectively. These are located in the Andean areas, the western valleys and urban areas. In the eastern part of the country (the Chaco and Amazon, or two-thirds of the country), 17% of the population is indigenous. These areas are the richest in terms of cultural diversity, with 32 peoples, the largest populations being the Chiquitanos (2.2%), Guaraníes (1.6%), and Mojeños (0.9%) (IWGIA 2008). Table 3.2 shows the indigenous population of Bolivia, classified by department.

There is no consensus about the exact number of indigenous peoples in Bolivia. One possible reason could be the fact that there are ethnic groups that belong to one indigenous people, but nevertheless claim to be one people in their own right, because of their history and particular organisational structures (CEPAL 2005). In spite of this, it is possible to identify the following ethnic groups, classified according to the territory they occupy, as follows (IWGIA 2009):

- **Bolivian Andes**: The groups present in the Andean region are Aymara, Chipaya, Kallawaya, Quechua, and Uru. These groups account for a vast proportion of the population in the region. Migrations have caused the Quechua language to disseminate throughout the country.

- **Bolivian Lowlands**: These groups live in the plains and the humid forests of the Amazon River Basin, as well as in the dry Chaco forests and the Plata River Basin. The indigenous peoples of the East, Chaco, and Amazon live both in rural and urban areas, and include the following groups: Guaraníes, Chiquitanos, Moxeños, Guarayos, Movimas, Chimanes, Itonamas And Tacanas, Reyesano, Yuracare,

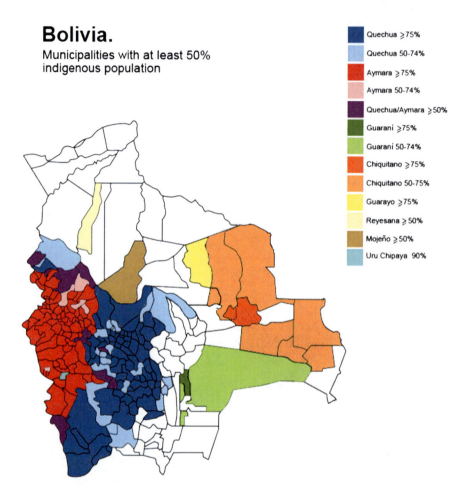

Fig. 3.4 Distribution of the major indigenous ethnic groups' in Bolivia. Only municipalities with >50% of its population of indigenous ethnicity are considered. (Adapted from Albó 2008)

Joaquinianos And Weenhayek, Cavineños, Mosetén, Loretano, Ayoreos, Cayubaba, Chácobo, Baure, Canichana And Esse-Ejja, Sirionó, Yaminahuas, Machineri, Yuki, Moré, Araona, Tapiete, Guarasugwe, Huaracaje, Pacahuara, Maropa, and Leco.

The location of the main indigenous groups is presented in Fig. 3.4.

There is archaeological evidence of ancient cultures present in Bolivia. The Wankarani culture inhabited the Andean Puna from 1,200 BC to 200 AD. The Chiripa culture was located in the Titicaca Lake around the fourteenth century BC until the beginning of the Christian era. The first period of the Tiahuanaco culture starts about 1,200 BC, and by the seventh century AD the imperial period started. Ceramic and textiles found throughout the conquered territory supports the hypothesis of the political expansion of the Tiahuanaco Empire during the eighth century AD (IWGIA 2009).

Although the Tiahuanaco emperors had a powerful rule, their decadence was marked by the twelfth century AD, and migrant Aymara groups occupied their territory. The disappearance of the Tiahuanaco Empire caused the fragmentation of the Aymara ethnic groups present in the territory. These Aymara groups shared the lake area with the Urus. Around the mid-fifteenth century, the Inca emperor Viracocha made the first incursions in the area, which was finally conquered by his son, Pachacutec. These indigenous groups were known as the Kolla (IWGIA 2009).

The Charca confederation was located in the southern area. It comprised the Carangas and the Quillacas, around Lake Poopo, and the Charcas in the north of Potosi and Cochabamba. The Cara-Caras and the Chichas were also members of this confederation. The Charcas and the Kolla shared the Aymara language. The Charca confederation was also conquered by the Inca emperor Tupac Inca Yupanqui (IWGIA 2009).

The diversity of indigenous peoples and cultures present in Bolivia was also translated into the systems of religious beliefs. In the Andean part, both the Tiahuanaco and the Inca empires had elaborate cosmologies that combined an advanced astronomical knowledge with a complex system of sacred beings and natural forces.

The religious systems of the lowlands peoples, like the Chiquitano, reflected the deep relationship between humans and nature, which permeated most aspects of everyday life. The Chiquitano consider that all beings have a soul; therefore, felling a tree or hunting an animal is a kind of murder. For them the 'jichis' are the 'masters of nature' and lords of the living beings; there are jichis for the water and fish, the animals, the forest, the plants, the stones, the mountains and hills, the plains, and the winds. The jichis make sure that people do not abuse natural resources and use only what they need. Before hunting or cutting down trees, it is mandatory to ask permission from the jichi, and to thank them afterwards. The Chiquitano have to observe rules and taboos in order to avoid being subject to punishments that include being unable to hunt, getting sick, suffering misfortunes, or having a poor harvest. Shamans are well-versed in the art of medicine—both to cure and to kill—and can also conjure up the forces of nature (the jichis) to produce rain, good hunting, and other services for their communities. They can also use these skills to attack their enemies (Centurión and Kraljevic 1996).

In the Chimane universe, society mirrors the natural world and they are related to each other. Duik and Mitsha—also called Jen when considered as a unique entity—are the mythical brothers responsible for the creation of the cosmos. They endowed men with weapons, fire, and food, among many useful goods. Respect for the natural world and equilibrium in the use of natural resources are fundamental in order to prevent evil. For these people the 'faratazik' are the master guardians of the world. Spirits are generally benevolent, unless humans fail to respect the norms. Shamans enforce cultural and religious norms and perform the main ceremonies (Wilbert 1994). Box 3.2 presents an example of the role of trees in the Chimane cosmovision.

After the Spanish conquest, the colonialist domination imposed a complex system of classification and categorisation that structured identities, values, and attitudes

---

Box 3.2 The Trees in the Chimane Universe

The Chimane people used the bark of the corocho or ashabá tree (*Ficus bopiana rusby*) to make dresses. In the ancient times, the Chimane had nothing to wear, but Duik transformed selfish women who would not share their belongings into ashabá. This tree grew big and until present it casts spells.

Dresses can also be made from the bark of the bibosi tree (*Ficus glabrata H.B.K*). Bibosi was a bad man, who was married to Corocho. Duik transformed this bad man into the bibosi tree.

Bibosi and Corocho were kukuitzí (old people).

In order to remove the bark, it is necessary to speak to the tree and ask for its skin, so that it does not get angry. This avoids making the bibosi angry, and prevents the Corocho from casting spells and producing diseases.

On the bark drawings of tigers and other figures are made using urucú (*Bixa orellana*) and bis.

All the trees have owners, because all the trees were people and have been transformed for some reason. When a tree is cut down, the owner moves into another one. But if it already has an owner, both get together in the same tree. When they get angry with people, they cast spells against them.

Trees get angry when they are cut down, so it is necessary to ask the tree not to kill the person who is going to cut it. That way they do not send faratazík (master guardians) to the people.

Source: Riester (1993).

---

related to the ethnicity. This system assigned virtues to the dominating sector and negative attributes to the dominated peoples. It had a strong impact on the indigenous peoples' identity, and caused in many cases the denial of their indigenous ethnicity in order to avoid being stigmatised (CEPAL 2005).

The 2008 Bolivian Constitution recognises the pre-existence of the indigenous peoples and considers them as nationalities whose traditions, cultures, and norms must be respected. Their languages are acknowledged as official languages in addition to Spanish. Specific provisions guarantee a number of rights for the indigenous peoples, including: their cultural identity and its registration in citizenship documents; free determination and territory; recognition that their institutions are part of the state; recognition of the collective property of their territory; protection of their traditional knowledge; respect for their sacred sites; community justice; the right to be consulted about any initiative dealing with non-renewable natural resources located in their territories as well as the participation in the distribution of the benefits those activities produce; the respect and protection of peoples on the verge of extinction; and the right of peoples in a state of voluntary isolation to remain so (IWGIA 2008).

3.2.2.3 Chile

According to the 2002 census, 4.6% of the population identified themselves as members of one of the eight indigenous peoples officially recognised by the Chilean Indigenous Law. The Mapuche account for 87.3% of the total indigenous population. The proportion of rural population is higher in indigenous than in non-indigenous population. Many indigenous people migrate to Santiago de Chile (Instituto Nacional de Estadística INE 2005). Table 3.3 presents the indigenous population in Chile.

There is little information about indigenous population in pre-hispanic times. The Araucano people were the largest ethnic group, with three distinctive sub-groups: the Picunche in the north, the Mapuche in the centre, and the Huilliche in the south. Besides the lowland Araucanos, there were the mountain people, the Pehuenche (Aagesen 2004).

The Mapuche used to clear land for their crops, but they also venerated forests, which provided food, shelter, fuel, medicine, and construction materials for their houses and enclosures. The Picunches used an agricultural practice called 'roza', consisting in felling and burning trees. They then used the ashes as a fertiliser in which they sowed corn, beans, pumpkin, potatoes, peppers, and quinoa. According to Aagesen (2004) the Pehuenche were nomadic and foraging people. They hunted the guanaco (*Lama guanicoe*) and ñandú (*Rhea americana*), and gathered piñones from the *Araucaria araucana*, which were central to their diet.

These peoples were also hunter-gatherers, both activities being of crucial importance for their economies. Gathering was not a lowly task, but one that involved a deep knowledge of the native flora, its uses, and taxonomy. Moreover, the Mapuche year was divided into 10 lunar periods closely linked to the native flora cycles (Aldunate and Villagrán 1992).

Box 3.3 presents a brief description of Mapuche traditional uses of forests.

The Chilean Constitution does not recognise the existence of the indigenous peoples, and the ITO Convention 169 has not been ratified by the Government. Among the reasons for this situation is the fact that right-wing parties' congressmen consider the expression 'indigenous peoples' as a threat to the unitary state,

Table 3.3 Indigenous population in Chile

Ethnic group		Total
1.	Alacalufe (Kawashkar)	2,622
2.	Atacameño	21,015
3.	Aimara	48,501
4.	Colla	3,198
5.	Mapuche	604,349
6.	Quechua	6,175
7.	Rapa Nui	4,647
8.	Yámana (Yagán)	1,685
Total		692,192

Source: INE (2005)

Box 3.3 Lelfunmapu: Mapuche Traditional Uses of the Forests

In the old days, the central valley of Chile was densely populated by the Mapuches, who settled by the rivers in a rather disperse pattern. The *Nothofagus* sp. deciduous forests and the evergreen forests in the south, provided a wealth of products for the gatherers, such as vegetables and fruits from trees and bushes such as the 'peumo' (*Cryptocaya alba*); the 'boldo' (*Peumus boldus*); the 'keule' (*Gomortega queule*); the 'gevuin' (*Gevuina avellana*); a variety of 'michay' (*Berberis damuinii, B. serata, B. dentata*); the 'litre' (*Lithraea caustica*); the 'pitra' (*Myrceugenia planzpes*); the 'murta' (*Ugni molinae*); the 'mulul' (*Ribes glandulosum*); and the 'luma' (*Amomyrtus luma*).

The *Drimys winteri* was the sacred tree of the Mapuche people. It was used as a medicine, skin tanner, tincture, insecticide, and repellent.

The lower forest stratum was also rich and varied, and was an important source of food. The lianas and epiphytes were sought for their fruits, like the 'copihue' (*Lapageria rosea*), the 'coguil' (*Lardizabala biternata*), the 'poe' (*Fascicularia bicolor*), the 'quilineja' (*Luzuriaga radicans*), the 'khelgen' (*Fragaria chiloensis*), and many other berries were eaten either fresh or dried, and were used to prepare fermented beverages called mudai.

Other popular species were a kind of celery called 'panul' (*Apium panul*), the 'panke' or 'nalka' (*Gunnera tinctoria*), and the 'chupón' (*Greigea sphacelata*). The 'madi' (*Madia sativa*) was used to extract high quality oil. Tubers or edible roots were collected from plants such as the 'lahue' or 'lam' (*Sisyrinchium* sp.), and more than 30 species of wild potatoes. The Mapuches used many species of ferns, such as the 'afipe' (*Lophosoria quadripinnata*), which they used to dry and grind in order to obtain a powder used to prepare thick soups. They also collected a wide variety of mushrooms, including more than 10 species of the *Cyttaria* genus, associated with the *Nothofagus* sp. forest.

The forest was a crucial source of medicine for the Mapuche people. Many medicinal plants were associated with sacred categories, and the word 'lawen' or 'lahuen' (medicine) was added to their names in order to express their healing properties. They were often used as herbal infusions, or applied on wounded areas. Ferns had many uses, like the 'llushu lawen' (*Hymenophyllum dentatum*) to heal newborns' belly buttons; the 'llanca lawen' (*Lycopodium paniculatum*) for tumours and ulcers; and the 'lafquen lawen' (*Euphorbia portulacoides*) or 'water remedy.' Other important medicinal plants are the 'cachan lawen' (*Eythraea chilensis*), which has many therapeutic applications; the Chusquea culeou, used as contraceptive and to ease menopause and hormonal problems; *Nothofagus pumilio* for headaches, muscle pains, and as a cure for fever; *Maytenus boaria*, for skin diseases and fever; *Cryptocarya alba*, to protect the liver and cure alcoholism; *Araucaria araucana* fruits, for back pains and ulcers; *Embothrium coccineum*, to heal wounds; *Fuchsia*

(continued)

Box 3.3 (continued)

magellanica, as a diuretic; *Quillaja saponaria,* for skin diseases and as soap and shampoo; and *Peumus boldus*, for liver problems.

The Mapuches also used plants for magic purposes. For example, the 'huentru lawen' (*Ophioglossum uulgatum*) was used by women in order to bear male children. The 'machi' or shamans used the 'huilel lawen' (*Hypolepis rugosula*) to predict evils caused by the 'huekufu' or demons. The 'huedahue' (*Gleichenia litoralis*) was good for preparing love potions or causing break ups. A most feared plant was the 'latue' (*Latua pubiflora.*), which could cause death by poisoning and had hallucinogenic effects when used in small doses.

Sources: Aldunate and Villagrán (1992), Catalán et al. (2006).

and a potential danger of future separatist attempts, as well as the risk that indigenous land could pose to property rights (Aylwin n/d). Moreover, the concession of benefits for indigenous people is regarded to be contrary to the constitutional principle of equality of all citizens. The indigenous organisations and the Government managed to pass an 'Indigenous Law' in 1993, as previously mentioned. This law recognises the existence of indigenous population, and establishes the obligation of the state and society to respect, protect and promote their development, and that of their cultures and families. It also recognises the main ethnic groups, namely Mapuche, Aymara, Rapa Nui, Atacameño, Aymara, Quechua, Colla, and Yámana (Aylwin n/d).

3.2.3 The Hybridisation Process Triggered After the Spanish Conquest

The Spanish conquerors brought with them new crops and cattle, together with European farming practices. They did not implement forest management practices, but rather used the forest as a source of timber, and forest land was considered to be 'unproductive'. Large areas of forests were cleared to give way to agriculture and ranching. Indigenous groups were reduced to servitude, their right to their ancestral land was denied, and their culture and religion doomed as inferior and 'heretic' (Golluscio 2008; Quijano 2005; Paz et al. 2001). Indigenous peoples assimilated Hispanic agriculture (Torrejón and Cisternas 2002; Torrejón et al. 2004).

From a sociological perspective however, the process triggered by this encounter is more complex than the mere incorporation of new technologies and practices. It is a process of hybridisation. Within the framework of postcolonial theory, Papastergiadis (2000, cited in de Mojica 2000, p. 90.) conceives hybridisation in two levels: as the constant process of differentiation and exchange between the centre and the periphery and different peripheries[3]; and as a metaphor of the identity that is being produced through these conjunctions.

García Canclini (2000a, b) defines hybridisation as 'socio-cultural processes in which discrete structures or practises, which exist in a separate way, are combined to generate new structures, objects and practises.' García Canclini points out that the so-called discrete structures have been a result of former hybridisations, therefore they cannot be considered to be pure sources. This transit from discrete to hybrid, and to new discrete forms, can be described as 'hybridisation cycles' (Stross 1999, cited in García Canclini 2000b), according to which, in history, we pass from more heterogeneous forms to others more homogeneous, and then to others relatively more heterogeneous, without any form being 'pure' or fully homogeneous. Many of these intercultural mixes are most productive and innovative (García Canclini 2000b).

Latin American countries are a result of sedimentation, juxtaposition, and inter-crossing of indigenous traditions and Catholic colonial Hispanism, and modern political, educational, and communicational actions (García Canclini 1990). Bermúdez (2002) considers this as an invitation to part from the sacralised vision of patrimony, and from the notions of 'authentic' that tend to associate identities with belonging to a territory and the collection of objects from the past. According to her, García Canclini focuses his analysis in the identities' transformation processes derived from the stakeholders' new cultural interactions and the power relationships built within this context.

The interactions of Spanish, Portuguese, English, and French colonisers with American indigenous peoples, and the African slaves, made the 'mestizaje'[4] a foundational process of the New World societies (García Canclini 2000b). The notion of mestizaje points to both biological and cultural aspects, indicating the mixture of European habits, beliefs, and ways of thinking with the American indigenous peoples.

An analysis of the whole process would hint at a pendular trend in the valuation of traditional knowledge. During the pre-Hispanic time, such knowledge was highly appreciated, the elders and shamans being the most respected figures within indigenous society. The conquest marked a rupture of indigenous communities' mechanisms of social reproduction, hence giving way to a period of denial and delegitimisation of traditional knowledge as lowly and evil. From the indigenous standpoint, ancestral and European knowledge hybridised, thus resulting in new agricultural activities. The recently emerging trend is to rescue and protect traditional knowledge.

---

[3] Proposed by Raúl Prebisch, the structuralist notions of 'centre' and 'periphery' refer to the relations between an 'advanced' metropolis and less developed countries. It was first applied to trade and power structures, and expanded into social and cultural relations.

[4] The French equivalent is 'metissage', a verb that in its literal meaning denotes the combination of races.

3.3 The Interactions Between Science-Based Forestry (Scientific Forestry) and Traditional Forest-Related Knowledge and Management Practices

Hybridisation reflects the diversity of the interests involved as well as the recognition of the variety of expertises available both within certified science and local communities of practice (Vessuri 2004). There is an ongoing inter-scientific dialogue between modern Western science and indigenous peoples, which is in the process of theoretical construction of its fundamentals and identifying the potential for mutual complementation, but also for confrontations. In order to carry on this inter-scientific dialogue, it is necessary to establish an inter-cultural dialogue that values local traditional knowledge (Delgado 2008).

This section offers an overview of forest management experiences with indigenous communities based on best practices, which combine traditional knowledge and 'Western'—i.e. Eurocentric—science. The implementation of such initiatives involves a strong capacity for intercultural dialogue, patience, and tolerance, plus the acknowledgement of the value of the traditional knowledge. However, professionals and practitioners trained in Western forestry tend to disregard traditional knowledge as a mere superstition or as a result of ignorance about 'scientific matters.' As a result, traditional knowledge is made invisible and not always reckoned with.

This section aims at pointing out key opportunities for integrating traditional and modern knowledge in forest management. Drawing from the experiences carried out in Argentina, Chile, and Bolivia, it is possible to propose the following areas of work as promising fields for rescuing and incorporating traditional knowledge into management practice:

- Multipurpose forest management;
- Non-timber forest products;
- Indigenous people's co-management of protected areas;
- Community forest inventories;
- Forest certification;
- Multicultural education; and
- Community based forest management.

The implementation strategy for these initiatives involved different degrees of public participation with a more decentralised and people-oriented approach. The incorporation of elements of traditional knowledge into project design has not always been as strong as desirable.

3.3.1 Multipurpose Forest Management

There is a growing trend in modern forest management of indigenous community land to promote the sustainable multipurpose management of community forests.

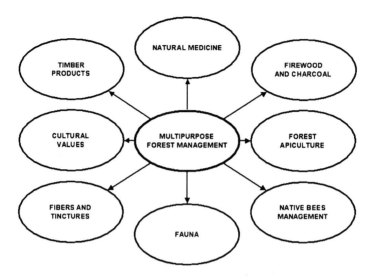

Fig. 3.5 Multipurpose forest management (Source: Spagarino 2008)

This is certainly in recognition of the traditional uses of the forest for food, shelter, medicine and tools to the indigenous peoples since the remote past. It also combines these ancestral uses with modern forestry science techniques and silvicultural practices, aimed at obtaining higher benefits for the communities.

The objectives of a forest management project should be set with broad-based community participation and respond to people's needs, beliefs, and demands. The management practices and tools should be consistent with local governance structures and the scale at which they will be implemented. Figure 3.5 shows the activities involved in multipurpose forest management.

Civil society organisations play an important role in capacity building in order to enable indigenous communities to manage their forests in a sustainable fashion. There are many ongoing experiences in Latin America. An interesting case is the initiative carried out by the Asociación para la Cultura y el Desarrollo (APCD) in support of the Wichí Lote 27-Laka Honhat community in the Province of Formosa, Argentina.[5]

The Provincial Government had given the Wichí community the legal ownership of 2,917.32 ha of land as an enlargement of their community land. The APCD provided support for community discussions on the future use of this land, between 2003 and 2004. Many people did not know the plot of land, particularly the youngsters. At the time, a forest company expressed its interest in the extraction of palo santo (*Bulnesia sarmientoi*). They decided to visit the plot to learn more about the available resources. The community was also analysing the possibility of building a new road, because the existing one got flooded during the rainy season, but was concerned about the impacts this would have on the forest.

---

[5] Based on Spagarino (2008).

As consequence of these analyses, the idea of sustainable forest management emerged and the Wichí community followed the advice of APCD's forest engineers to carry out a forest inventory, assess the natural resources, and produce a management proposal with the following objectives:

- Create opportunities for discussion and action related to the future of the plot.
- Explore options for production diversification for the community families.
- Set criteria for multiple uses with low impact on the forest, using the resources available in the territory. Merge sustainable forest use into the community's economic system.
- Contribute to job creation through the development of small local industries.
- Enable the replication of these systems in other similar communities within the area.

The project was carried out by APCD with the support of the Faculty of Agrarian and Forest Science of the National University of La Plata (UNLP, in Spanish). The first activity involved a series of workshops devoted to teach community members the basics about community forest inventory and management. After this, 23 community members with support from 3 members of APCD and the UNLP (which provided 7 student internships to cooperate with the tasks) carried out an inventory. During the activity, the community developed a stronger relationship with the land and the forest, and there was a knowledge exchange between them and the students. The Wichí elder men and the women provided extensive knowledge about species identification and description of their uses. The young members of the community therefore had a good chance to learn from them, and the concept of community forest management was strengthened. The combination of modern tools and techniques with traditional knowledge produced robust outcomes that would have not been possible otherwise.

Wichí communities use the forest intensively in their daily lives and have a profound knowledge about the uses of the forest flora and fauna. The inventory identified 74 vegetation species, of which 23 species have timber and food uses, 21 species are medicinal, and 14 species can be used as fodder for domestic cattle and game. Of these species, 9 from each group were useful to build houses, to make utensils and tools, and as tinctures; while 2 species were described as toxic.

The APCD developed a sustainable forest multipurpose management proposal for the plot of land, building upon the results of the inventory, with the following components:

- A multipurpose management proposal for the tall, well-developed forest

  - Forest management plan
  - Loro hablador (*Amazona aestiva*) management
  - Native bees (*Meliponas*) breeding[6]

---

[6]The most interesting species are pini or llana (*Sacaptotrigona jujuyensis*); kolo'pum or rubito (*Tetragonisca angustula*), and nakwu or moro moro (*Melipona favosa* or*bignyi*).

- Traditional Hapu'k apiculture
- Management and gathering of textile chaguar
- Use of medicinal plants and revalue of traditional medicine

• A management proposal for the low forest

- Forest management plan
- Vinal charcoal production.

This experience allowed the community to learn sustainable management practices that enabled them to improve their livelihoods. The Wichí elder's knowledge was passed to the young generations and shared with the practitioners, who systematised the information, thus contributing to rescue it. As mentioned above, implementing this kind of project requires time and patience from both parts, so that trust can be created. It is also necessary to learn how to communicate effectively in spite of cultural differences.

3.3.2 Non-timber Forest Products

As previously mentioned, traditional knowledge about non-timber forest products has been transmitted mostly orally from generation to generation. At the moment, Western foresters and scientists are in a learning process about the uses and potential of non-timber forest products, and so are transnational companies. This process involves the retrieval and systematisation of traditional indigenous knowledge and the registration of the biodiversity heritage.

Besides its spiritual significance and importance for indigenous peoples' livelihoods, there are economic implications that have been addressed by the governments through international environmental negotiations and agreements within the framework of the United Nations Convention on Biological Diversity. The main issue involves intellectual property rights on non-timber forest products with potential applications in the pharmaceutical industry.

There are a number of experiences involving non-timber forest products in Argentina, Bolivia, and Chile, mainly dealing with ornamental plants (e.g. ferns and orchids), medicinal plants, forest fruits, mushrooms, and wild honey. For illustrative purposes, the 'pewén' management practices of the Chilean Pehuenche people have been selected, as the *Araucaria araucana* fruit is an emblematic non-timber forest product for indigenous peoples in the south of Argentina and Chile (based on Aagesen 2004; Herrmann 2005, 2006).

During the past 35 years, the ancestral territory of the Mapuche people and, by extension, of the Pehuenches, has been drastically reduced. Large extensions of public and indigenous land were assigned to private landowners, and there were economic incentives for industrial plantations with exotic species such as the *Pinus radiata* and *Eucalyptus* spp. As a result, the indigenous communities have been relocated into marginal areas that are poorly productive and lack accessibility (Herrmann 2005). This has generated bitter confrontations between the

Mapuche—usually backed by national and international NGOs—on one side, and the government and forest industry on the other.

Another cause for conflicts is the fact that about 48% of the araucaria forest, much of it corresponding to ancestral indigenous territory, is currently under some protection category within the national system of protected areas administered by the Corporación Nacional Forestal (CONAF), which has established restrictions on the communities' access to the araucaria forest and the piñoneo activities. Indigenous knowledge and biodiversity management practices were rejected by the Chilean Government management and conservation policies (Herrmann 2005; Aagesen 2004).

The Pehuenche are strongly connected with the araucaria tree, so deeply that their name literally means 'people of the araucaria tree.' Its seeds, called piñones or ngülliw, are one of their main sources of food. It is a sacred tree, with a deep symbolic and spiritual meaning. It was created by Ngünemapun in order to feed his sons. The Pehuenche believe the araucaria male tree (wentrupewen) and the female tree (domopewen) get married and form extended families, called lobpewen. They reproduce through the interconnection of their roots, according to the will of Pewenucha and Pewenkuzé, spirits that live in the forest.

Just like many other indigenous peoples, the Pehuenche believe that each being possesses a spirit, called ngen, that owns it. It is therefore necessary to ask for the ngen's permission in order to use nature's goods, or else the trespasser could suffer some punishment. This also applies to the piñones harvest. The Pehuenche celebrate a ritual called Ngillatun and make offerings around an araucaria tree, to ask for the ngen's permission to collect the piñones and ensure a good harvest (Herrmann 2005, 2006; Aagesen 2004).

The piñoneo, or collection of piñones, takes place between February and April, during the late summer and early fall. During the late austral summer and early fall, the Pehuenche families establish summer camps and collect the piñones from the ground. For piñones collection purposes, each family has its own area. The piñones are eaten in different preparations, and are used to prepare a fermented beverage (Aagesen 2004; Herrmann 2005, 2006). The Pehuenche identified a 3-year cycle in the intensity of piñón production, in which a rich year, associated with hard and long winters, is followed by 2 years of scarcity (Herrmann 2005, 2006).

Combining their ancestral knowledge with modern forestry, in 2001 the Pehuenche communities have begun to produce araucaria seedlings in tree nurseries. The families also planted piñones between willow trees, so that the seeds would not feel lonely and had the chance to talk to the other young trees. In fact, the willows provide adequate conditions for the araucarias, by casting shade and protecting the seedlings from the wind. The willows' branches will afterwards be pruned and used for basketry, thus enabling the young araucarias to have enough light.

The Pehuenche have also developed araucaria agroforestry systems with co-cultures like corn and quinoa. Their knowledge of the areas and trees that produce high quality or quantity of piñones contributes to ensure that the trees they produce in these nurseries come from high quality seeds (Herrmann 2005, 2006). It is interesting to note that indigenous people are reforesting logged areas on slopes with native trees from their own nurseries.

The Pehuenche could make meaningful contributions to ex-situ and in-situ biodiversity conservation. Their knowledge on araucaria management has been ignored by the Chilean government; they had no voice in conservation policy design. There is also an opportunity for interactions between traditional forest management and Western forest science, involving traditional forest management, tree nursery management, weeds and pest control, and soil conservation techniques.

3.3.3   *Indigenous People's Co-Management of Protected Areas*

Latin America is leading an innovative shift from the traditional protected natural areas management into participatory co-management, involving local communities in the decision-making process. Many of the region's protected areas include traditional indigenous territories. This led to many conflicts between the government and the communities, whose ancestral activities in those areas were either limited or suppressed.[7]

The following case study refers to the evolution of the dominant traditional paradigm guiding natural protected areas management in Argentina, from social exclusion to social inclusion and participation of local communities. It deals with co-management of the Lanín National Park and the indigenous territories within it with participation of Mapuche communities (based on Carpinetti 2007, and 2006).

The National Parks Administration (NPA), a pioneering Argentine organisation, was established in the 1930s, following the American paradigm of people-free protected areas. It imposed severe restrictions on land use, which usually led to expropriation of land and the relocation of communities. Created in 1937, Lanín National Park comprises 420,000 ha representative of the Temperate Valdivian Forest eco-region (Bosque Andino Patagónico forest region). At that time, large private landowners had their properties declared national reservations, allowing for some productive activities, such as tourism. However, the indigenous communities' ancestral rights were not officially recognised. Those who were allowed to stay were assigned personal, non-transferrable precarious occupation and fodder permits (PPOP, in Spanish) for which they had to pay an annuity (Carpinetti 2006). By the late 1980s and 1990s indigenous peoples in Argentina were empowered by

---

[7]There have been formal requests for the creation of a new category of protected areas, the 'Indigenous Conservation Territory,' in order to acknowledge the fact that indigenous ancestral land is involved and thus the communities have to be integrated into a participatory protected areas governance body. Seven Mapuche communities from the Province of Neuquén, Argentina, made a similar proposal in 2000—'Indigenous Protected Territory'—for a disputed area of 110,000 ha known as Pulmarí (Nahuel 2000). Support for the new Indigenous Conservation Territory category has been expressed in the Bariloche Declaration approved by the Second Latin American Congress of National Parks and Other Protected Areas (Bariloche, Argentina, September 30–October 6, 2007). The conclusions of this Congress were presented at a IUCN-supported side event organised by the Latin American Indigenous Forum on Protected Areas, during the CBD 2nd Ad Hoc Open-ended Working Group Meeting (Rome, Italy, February 11–15, 2008). This new category would comply with international agreements that protect indigenous peoples' rights, such as ILO Convention 169, the UN Convention on Biological Diversity (CBD), and the UN Declaration on the Rights of Indigenous Peoples (UNDRIP).

new laws, and their rights were fully recognised by the 1994 constitutional reform. The NPA decided to harmonise communities' interests with the conservation goals for federal protected areas.

There are between 2,500 and 3,000 Mapuche within the Lanín National Park, which occupy some 24,000 ha distributed into seven communities. Over the past 15 years, some of the indigenous communities within the Lanin National Park received property titles within in the national reserve. The main activities are seasonal nomadic cattle breeding, forestry, quality textile handicrafts, and, more recently, an incipient tourism management. About 60% of the families live below the poverty line (Carpinetti 2007).

After political changes leading to the NPA centralisation, the Neuquén Mapuche Confederation (NMC)[8] occupied Lanin National Park's administration premises to enforce their demands. Following a NMC initiative, in May 2000 the NPA organised a workshop to discuss legislation, territory, and natural resources management. This workshop led to the creation of a joint management committee, and implied mutual recognition of both parties' rights.

The Intercultural Management Committee seeks to distribute responsibilities and competences, with a clear definition of public authority, and guidelines for natural resources use and management based on indigenous traditional knowledge. It is responsible for the delimitation and zonification of Mapuche territory for land transfer, the design of management plans, updating the NPA regulations when necessary, and fundraising for projects. Decisions are consensus-based and are implemented by local management committees.

Lanín National Park's Preliminary Management Plan contemplates the following general objectives for the national park and the national reserve areas:

- The conservation of representative samples of norpatagonic Andean ecosystems and their biodiversity;
- The protection of upper river basins, ensuring flow regulation;
- The preservation of the landscape and natural assets related to tourism potential;
- The conservation of cultural assets;
- The offering of tourist and recreational areas and facilities, in touch with nature;
- The contribution to regional development through the preservation of valuable resources for tourism, biodiversity, and the production of goods and services;
- The promotion of research on natural, cultural, and social aspects of the area; and
- The advancement of public knowledge about the protected resources, as well as an understanding of relevant natural and cultural processes.

---

[8]Neuquén Mapuche Confederation (Confederación Mapuche de Neuquén) is the organisation representing the Mapuche communities of the Province of Neuquén.

There are still some tensions, mainly associated with conservation decisions pertaining to the authorised use of the park's resources by the communities. According to the Mapuche, people's health, wealth, and quality of life are united with ecosystem diversity, productivity, and quality of which they are a part. Their economic future is deeply linked to their safe access to land, forests, animals, water, and the subsoil in a territory large enough to carry out their activities and strengthen their identity. From a cultural perspective, the territory is defined as 'the space in which Mapuche culture develops, that comprises as a whole (waj mapu), the history of their social, cultural, philosophical, and economical relationships, as well as their evolution.'[9]

The actual implementation of this notion is conflictive, for conservation measures have an impact on people's activities and livelihoods. Tension eased after an agreement was reached upon the following management principles:

- Systemic approach: people and ecosystems are considered one single system.
- Questioning attitude: action and reflection are part of a continuous feedback cycle.
- User-driven: people and their organisations must be responsible for managing the different activities.

Participatory governance structures, as in protected areas co-management, provide an enabling environment for land-use conflicts management. In this case, traditional knowledge combines with non-traditional organisational management to foster dialogues leading to the co-creation of effective solutions that allow for an improvement of communities' livelihoods without endangering the protected areas in the long run.

3.3.4 Community Forest Inventories

In order to advance community forest management, after the objectives and intensity of management are agreed upon, it is necessary to carry out an inventory so that a management plan can be developed. Since inventories are an alien practice for indigenous peoples, it is important to design sensitisation workshops and capacity-building activities with the indigenous community in order to empower them, enable them to own the whole process, and to define the criteria for forest management, including forest use and spatial organisation. The case presented here, based on Bessonart et al. (2009), is an example of these preliminary activities, which are crucial to ensure success in the implementation of community forest management plans.

The community of Carapari is located in the northern corner of the Province of Salta, 8 km away from the international border Argentina has with Bolivia. The

---

[9]Definition adopted at the 2000 workshop mentioned above.

population has Guaraní ancestry, and is composed of 120 families (672 people), distributed in four smaller villages: Madrejones, Yerba Buena, Playa Ancha, and Caraparí. The area is part of the Lower Mountain Floor of the 'Selva Tucumano Boliviana'.

The utilisation of forest inventories as a tool for forest management is not a common practice in northern Salta. Local producers, associated for the use of native forest resources, are not familiar with the advantages of using planning tools for woodlot management. The implementation of forest inventories is therefore an innovative practice in this region.

The following experience was carried out in two phases. The objective of the first stage was to 'get in touch' with the community. The concerns of this Guaraní community about the conservation of their forest and the upcoming formulation of a community carpentry project induced an approach between professional staff from the AER Tartagal-INTA[10] and the Chief, Hilario Vera. Caraparí's carpenter was old and the Chief wished the youngsters to learn his skills so that this knowledge would be preserved within the community. A major concern was to avoid the negative experience of a neighbouring Wichí community that had no experience in forest management and clear-cut their community forest. The Chief expressed his community's interest in managing the community forest, as well as in learning about the legal framework and procedures for the extraction, transportation, and marketing of native forest timber.

The second stage involved agreements about the importance of forest inventories in order to improve woodlot management. Seeking to build a common vision and consensus to launch the activities, the technical staff socialised the current situation of the community forest. They discussed the advantages of having a management plan and explained the general administrative aspects associated with forest resources. They also showed the community the results of a forest inventory and forest management practices undertaken in the area. An agreement was reached about the need to have basic information and to assess the community forest. To this end, the community agreed to carry out a forest inventory with the objective of 'obtaining information that allows interpreting the current status of the forest in order to prepare a forest management plan.'

As a result, the community learned that the quantification of forest resources using forest inventories and the preparation of management plans improve the value of the available resources and establish a viable alternative for community sustainable development.[11] The procedures involved, as well as the resulting documentation, are expected to be useful in order to prepare projects and action proposals for organisations related to sustainable native forest management. Community participation and the techniques applied to carry out this experience have legitimised the outcomes.

---

[10] Tartagal Rural Extension Agency of the National Institute of Agricultural Technology (INTA). The agency deals with agroforesty extension, adaptive experimentation, and the promotion of forest plantations and native forest management.

[11] Besides the technical support from INTA, Car/aparí also received support from an oil company for their carpentry project, and from the Catholic Church for production and marketing.

Finally, during the experience, many of the younger members of the community had a chance to learn more about their forest from their elders and the technical staff, together with community leaders, parents, and family. Moreover, the experience enabled a space for mutual learning and exchange of knowledge, and contributed to strengthen community ties that will allow them to face new challenges.

3.3.5 Forest Certification, Community Forestry, and Local Development

Forest certification schemes are linked to market conditions but include social components aimed at ensuring the sustainability of forest management from the labour and community perspectives. However, the relationship between the forest industry and neighbouring communities is not always peaceful. This case study shows the complex relationships emerging from forestry concessions, and the conflicts between large certified forestry companies and indigenous communities in southern Bolivia (based on Van Dam 2007). Notably, this country has the largest area of certified native tropical forests, with 2,209,083.83 ha under sustainable forest management.[12]

The population of Guarayos is 31,577 and, according to the map of poverty published by the National Institute of Statistics (INE) in 2001, 90.6% was below the poverty line, compared to 58.6% for Bolivia, and 38% for the Department of Santa Cruz. In spite of receiving land during the Agrarian Reform in the 1960s, the units were not sufficient for the subsistence of the families, and their property rights did not include the right to use the forest resources. The situation further deteriorated in the 1970s with the forestry business from Santa Cruz taking over vast expanses of forests, the arrival of new farmers, and Quechua and Aymara indigenous groups from the highlands (CEADES 2003).

The Guarayo people are 'guaranitic' hunter-gatherers and small farmers—growing corn, yucca, bananas, and rice—who settled in the area around the fifteenth century AD. Only a few Guarayos own cattle (CEADES 2003). There is an organisation, COPNAG (Centre of Guarayo Native Peoples' Organisations), which acts as the legal representative of the six Guarayo villages and owns the TCO.[13] Also present in the area is another organisation called Federación Campesina, which promotes the occupation of land.

Forestry is an old and significant sector; during the year 83% of the population conducts activities directly or indirectly linked with it. The main forest industries are sawmills and carpentry. At present, there are 1,611,973.1 ha of forestland

---

[12] Source: Cámara Forestal de Bolivia, http://www.cfb.org.bo/CFBInicio/ (Last access: 05/15/09).

[13] TCO: Tierras Comunitarias de Origen (original community land), was created by the INRA land reform act in 1996. The COPNAG demanded 2,205,369 ha of land as their TCO, of which they were granted 1,350,000 ha. By 2002, they were awarded 900,000 ha, and another 450,000 ha were under assessment.

suitable for sustainable management. Some 683,206 ha were under forest management plans, of which 395,534 ha were granted to 11 private companies under forestry concessions, and 287,672 were authorised to 8 OFCs[14] run by indigenous communities in the Guarayo TCO (BOLFOR II/CADEFOR 2008).

The approval of a new forestry law in 1996[15] produced major changes in the number of forestry concessions, which went down from 30 to 10, totalling 516,604 ha. Nine of these 10 concessions affected land claimed by the Guarayos as a TCO. One of the concessionaires asked to reduce the area of the concession, while another two concessionaires gave up their rights. Three of the concessions were granted to two large certified firms, La Chonta and CIMAL / IMR Guarayos, with a total area of 281,750 ha. These companies had little or no ties with the local economy, although they did relate to the neighbouring indigenous communities for the following reasons, according to Van Dam (2007): (1) the need to keep good relationships, after tensions aroused by the overlap between the Guarayo TCO and forest concessions; (2) the threat posed by migrant farmers both to the companies and the TCO and indigenous communities; (3) the potential for indigenous communities managing their forest resources to become commercial partners in the future; and (4) the need to demonstrate compliance with forest management certification criteria. This was quite a sensitive issue, for both companies took long to get certified because of the land tenure conflicts with the Guarayo communities.

There were six small sawmills in Ascensión de Guarayos, with insufficient equipment and little working capital. In spite of that, they were major local employers, and produced on demand for clients in cities such as Santa Cruz, La Paz, Sucre, Oruro, and Potosi. Many had lost their former forestry concessions, because they could not comply with the new legal requirements posed by the 1997 law. They had therefore devised ways to provide themselves with timber from informal sources, although they could have bought it from community forests (Van Dam 2007).

The owners of these small sawmills were high-ranking within local society and had strong ties with politicians. They were upset by the changes, and considered that the forestry superintendent should have supported them. Market crises affected their interests, and the fact that indigenous communities received support from NGOs and projects for sustainable forest management in their original community land (TCOs in Spanish) was perceived as unfair competition. Local carpentry businesses, numbering 20–25, produced windows, doors, and furniture for the local market, mainly by using timber from informal sources (Van Dam 2007).

The Centre of Guarayos Native Peoples' Organisations (COPNAG) had received support in order to launch a community forest management initiative in its TCO, involving 159,235 ha out of the 900,000 ha it owned (UPIP 2006). Each of the six

---

[14] OFC: Community Forestry Organisation (Organización Forestal Comunitaria).

[15] The new forestry regime opened the access to forest resources for indigenous communities that had previously been marginalised from forestry rights (BOLFOR II / CADEFOR 2008).

communities involved had an officially approved management plan, and was expected to produce 25,000–30,000 m^3 per year. Two-thirds of the harvested timber came from ochoó, a species of little commercial value because of the excess of offer. The areas the community managed were former forestry concessions that ended in 1996; logging companies had kept the most profitable areas (Van Dam 2007). The Guarayo conceived forest management as a means of occupying the territory, thus avoiding invasions. Besides, logging activities were complementary of the Guarayo traditional agricultural activities. Timber extraction occurred from April to October—the dry months—whereas agriculture was practised during the rainy season from April to October.

Van Dam (2007) points out that certified companies favoured Guarayo community forest management, because it helped to pacify the area after a serious conflict over TCO and forestry concessions. In contrast with the small sawmills, these companies did not regard the Guarayo as competitors, but as potential timber providers. Moreover, the fact that the Guarayo were actively managing their community forests increased the availability of local skilled workers.

On the other hand, Van Dam (2007) argues that the Guarayo received only small benefits from the certified companies, such as small donations of lumber and support for road maintenance. They were not aware of the requirements involved in certification. Guarayo regional development was thus not likely to be fostered by the forest concessions or forest certification, but rather from the positive impact that community forest management in the TCO had on the income of a large number of indigenous families.

As mentioned in previous sections, access to and control of their ancestral territory is essential for indigenous communities, because they depend on this access to be able to apply their traditional forest management practices. Conventional forest management instruments, such as forest certification, do not necessarily improve indigenous communities' livelihoods.

3.3.6   Forestry Education in a Multi-cultural High School[16]

The Department of San Martín, Province of Salta, has a large number of indigenous communities from the Tupí Guarani, Wichí, Chané, Chorote, Toba, Chulupí and Tapiete ethnic groups (Bazán et al. 2003). Most of them live in small villages and hamlets along National Route 34, within the rainforest in the 'Selva Tucumano Boliviana' forest region. There are many small rural schools near those villages and hamlets, usually named in Guaraní language after the nearby communities, such as the Yacuy indigenous community school, San Miguel Arcángel of Tuyunti, Che Sundaro of Tartagal, Sub-teniente Rodolfo Berdina of Tranquitas, and Manuel Dorrego of Tobatirenda, among others.

---

[16]Case study based on AER-INTA Tartagal's reports, and conversations with the headmaster of the high-school.

This case study refers to the E.E.T. (technical education school) N° 5130 of San José de Yacuy hamlet, which was established in 1992 within the framework of the Program for the 'Expansion and Improvement of the Agricultural Technical Education' (E.M.E.T.A., in Spanish). It first functioned at the same building as the elementary school, and moved to its own premises in 1994. Most of the students are Guaraní, and there are also some non-indigenous students as well.

The school seeks to provide the students with a holistic education by combining theory and practice based upon individual and group activities, in order to promote a better understanding of their surrounding environment. The integration of the school and the community is fostered as a way of strengthening students' responsibility and a positive attitude towards agricultural production (Saldaño 2009). Forestry education and production are considered key areas for the region, since the sector shows a strong development potential and timber is scarce.

Until 2000, there were general guidelines for the incorporation of this subject into the syllabus. There was a module on tree nurseries in the course on vegetation production, and a nursery was built with the cooperation of an oil company that has operations in the area. In the year 2000, the study program was enhanced in order to include a course on forestry production, aimed at advanced high-school students. The subjects include phytogeography and forest classification, silviculture and forestry production, history of the native forest uses in Argentina, study of forest masses, forestry, and elements of taxonomy.

The teachers established linkages with local public and private organisations to enrich the educational process, and the integration of the students into the technical and productive sectors. Many local companies provided support for the school's activities. An interesting example of a public-private partnership established in this context is the AER-INTA Tartagal[17] initiative for the creation of a forestry extension plan. A component for supporting the school was implemented within the framework of the AER-INTA Tartagal's cooperation agreement with an oil company, which provided funding to carry out the program.

The team of extension agents carries out technical capacity building activities at the school's arboretum in two modules, dealing with theoretical concepts and practical silvicultural activities in forest plantations. The teaching materials are prepared by AER-INTA Tartagal's professionals and include a written guide and support materials. The students are taught how to use instruments to measure tree attributes, and report the resulting data. They also make calculations such as the determination of tree volume, using formulas such as Huber, Smalian, and Newton, selected to suit students' interests. Figure 3.6 shows Guarani students involved in these activities.

This ongoing experience shows the potential of technical education of indigenous youth for promoting their integration into local economies in forest areas. Furthermore, the new skills acquired at school contribute to the improvement of the sustainability of indigenous communities' livelihoods and reduces youth migrations

---

[17] Tartagal Rural Extension Agency of the National Institute of Agricultural Technology (INTA).

Fig. 3.6 Guarani students engaged in forestry activities: (**a**) planting seedlings and (**b**) carrying out tree pruning treatments. Photos: Sebastián Bessonart

to large cities. This experience highlights the potential for incorporating local knowledge in educational programs, taking advantage of the traditions of indigenous communities. In return, the members of these communities might be much more interested in participating in educational processes.

3.3.7   Key Elements Referred to Indigenous Forest Management

The indigenous notion of the forest as the 'big house' is deeply related to its multiple uses as provider of materials, food, shelter, spiritual values, and medicine. As part of the political reform begun in 1996, Bolivia has implemented what the land reform act calls Tierras Comunitarias de Origen (TCOs, original community land—see previous case in Sect. 3.3.5). It is therefore important to consider that, when related to indigenous communities, community forestry deals with the multiple goods and services provided by forests that are usually under a collective property scheme.

Martínez (2002) proposes a simplified model in order to analyse the processes associated with the relationships between man and nature, as follows:

Process 1: Nature as a global order, with living beings as an undifferentiated part of it.

Process 2: The first men, in state of nature, are dependent upon it. In this stage men start the process of differentiation by creating and developing labour instruments and generating precarious livelihoods.

Process 3: The differentiation process between man and nature is enhanced. Some men keep in balance with nature, while others do not, thereby altering nature's reproductive cycles.

Process 4: Part of nature is altered and modified by human interactions. Altered nature has negative repercussions on itself and the populations. Some men keep reciprocity with parts of preserved nature.

Process 5: There is a need to reconstruct new dynamic equilibriums between men and nature.

After the arrival of the European conquerors, the indigenous people in Bolivia merged their culture with the foreign values and culture. The resulting social identity is also marked by the fact that their TCOs were occupied and used by others. Conflicts and competition for the use of natural resources were the results of these processes that led to the re-creation of a new indigenous social identity.

Martínez (2002) refers to the relationship between community property and resources management at the family or individual level. The management of resources within the collective property does not necessarily imply community labour. The latter is determined by the amount of labour that has to be carried out and is usually associated with external factors. Subsistence labours are individual for each family.

TCOs not only do not necessarily involve community labour, but also contribute to reaffirm 'individual property' tenure (chacos) connected to the family's uses. In a way, TCOs combine collective areas for traditional activities—like hunting, fishing, and gathering—with individualised chacos for use as farmland. As for community labour and community based forest management, it is mostly a result of external requirements by foreign development agencies and the government, more than being the consequence of internal needs of the indigenous community (Martínez 2002).

The promotion of community-based forestry has been mostly focused in the highlands and valleys, but it is possible to anticipate progress in this direction. In order to successfully implement community-based forestry initiatives in the lowlands, thus improving the sustainability of livelihoods and conserving the biodiversity, it will be necessary to combine this practice with the concept of the multiple goods and services of the forest, and the traditional knowledge of the indigenous communities.

There are certain key factors that should be considered when harmonising traditional knowledge and Western science, including:

- These initiatives are multi-cultural and require an open-minded approach. The Eurocentric prejudice should be overcome.
- Forest landscapes should be considered as a whole, thus management should be holistic. Unlike much of Western forestry science, indigenous knowledge is multi-dimensional.
- Both indigenous participants and Western practitioners should seek to enable mutual learning processes.
- Conventional scientific tools for forest management, such as forest inventories and certification systems, should be adapted to the objectives agreed upon by indigenous communities.
- New models of participatory governance should be explored in order to manage land use conflicts and produce synergies between Western and traditional knowledge for ecosystem co-management.
- Education of indigenous youth is a valuable means for fostering their integration into local economies in forest areas while improving communities' livelihoods.

It should be noted that the multiple uses of forests involve multiple dimensions that require an interdisciplinary approach if community-based forest management is to induce the sustainable development of the indigenous people. The challenge is to view the socio-economic–ecological system as a whole, and develop interdisciplinary teams and adequate planning and management tools that allow for the harmonic development of such a complex system.

3.4   Trends and Challenges for Traditional Forest-Related Knowledge Research

Vessuri (2004) contends that the prevalent narrative in Latin American science has been the adoption of European knowledge and techniques by 'cultural activists' and businessmen, seeking economic profits from natural resources exploitation, and political colonialist domination. Traditional indigenous knowledge was thus relegated to the peripheries and dismissed as illegitimate or inferior (Quijano

2005). This dominant scientific discourse is deeply rooted in the Renaissance, and the European rationalist paradigms that contend there is such thing as 'objective science,' and that a reality external to the observer can be described without any subjective 'contamination.' Although this positivist conception of science has been under much dispute for more than 30 years in Europe and is currently a minority position, it is still the dominant paradigm of researchers in South America, particularly those related to natural resources.

There are two cross-cutting issues involved in promoting the integration of indigenous communities' traditional knowledge into forest management: ethic considerations and the equitable distribution of benefits.

3.4.1   Ethical Considerations

All societies have social norms that structure them and organise relationships among their members. In many indigenous societies—and also in Western ones—knowledge is power. This means that researchers should take all necessary precautions to make sure they do not trespass social norms by socialising information that is not meant to be public.

Another important consideration is the existence of taboos. Researchers should be respectful of taboos and avoid hurting the community by, albeit inadvertently, infringing upon social norms. In order to do so, it is important to adopt adequate research protocols (Shackeroff and Campbell 2007, Ruiz Muller 2006).

3.4.2   Equitable Distribution of Benefits

As indigenous organisations grow stronger, there is an increased awareness about the need to enforce communities' intellectual property rights in order to prevent abuses. In spite of the provisions of the CBD, there are many cases of biopiracy in Latin America (Herrera Vásquez and Rodríguez Yunta 2004). The challenge is, therefore, to ensure that ethnobotanic research conducted with indigenous communities is done according to protocols that ensure the protection of indigenous peoples' traditional knowledge, and enable them to share the resulting benefits. This extends to all traditional knowledge, including forest management practices.

Latin America is a pioneering region in the field of protection of traditional knowledge, and has advocated for local and indigenous community rights in international arenas (Ruiz Muller 2006). There is still a long way to go to achieve reasonable levels of compliance with international regulations in this matter. The national outlook for Argentina and Bolivia is quite promising, for they have enacted legal regulations that empower indigenous communities and protect their rights. Chile has made some progress, but the adoption of ILO Convention 169 is still pending.

3.5 Conclusions

The encounter of traditional knowledge and Western science in Latin America has not been peaceful. More than 500 years have passed and there is still a strong deficiency concerning the recognition of traditional knowledge's value, from both the ethical as well as the epistemological perspectives. However, a most fruitful intercultural dialogue has developed over the past decade, which in time might lead to the joint construction of a new scientific paradigm involving both traditional knowledge and Western science as equally valuable and complementary for a better understanding of nature and its processes.

The experiences sampled in this chapter show that it is possible to integrate traditional knowledge and Western science in practice, and achieve good results. There is a promising outlook for further experimentation and integration in field work. This will require the intervention of interdisciplinary teams that consider not only the forestry aspects, but also the social and cultural issues involved.

Since these experiences are developed in a multicultural environment, they require an open-minded approach. Mutual learning is part of the process, thus Eurocentric prejudices should be left behind. It must be noted that unlike much of Western science, indigenous knowledge is multi-dimensional and is in line with a holistic perspective of forest management that considers forests as a whole. Education of indigenous youth is important in order to promote their integration into local economies so that they do not seek to migrate to urban centres.

The new models of participatory governance offer a good potential for conflict management as well as for creating synergies among stakeholders and their knowledge, be it Western or traditional. They provide a useful framework for consensus-building about forest management objectives. These objectives will provide the basis for management plans and implementation of tools such as forest inventories and forest certification. It is also important to take note of cross-cutting issues such as ethic considerations and equitable distribution of benefits.

Finally, any future action involving indigenous peoples and their traditional knowledge should make a substantial contribution to the improvement and the sustainability of their livelihoods. It is both a moral mandate and a legal obligation that would bring about a well-deserved compensation for centuries of inequality and exploitation.

References

Aagesen D (2004) Burning monkey-puzzle: native fire ecology and forest management in northern Patagonia. Agric Hum Values 21:129–138

Albó X (2008) Movimientos y poder indígena en Bolivia, Ecuador y Perú. Centro de Investigación y Promoción del Campesinado (CIPCA), La Paz, p 294

Aldunate C, Villagrán C (1992) Recolectores de los bosques templados del Cono Sur americano. In: Wilhelm de Mösbach E (ed.) Botánica indígena de Chile. Editorial Andrés Bello, Santiago de Chile

Aylwin J [N/d] Pueblos indígenas de Chile: antecedentes históricos y situación actual. Serie Documentos N° 1. Instituto de Estudios Indígenas. Universidad de la Frontera. Available via http://www.xs4all.nl/~rehue/art/ayl1c.html.Cited 10 Feb 2009

Barbarán FR, Saravia Toledo CJ (2000) Caza de subsistencia en la Provincia de Salta: su importancia en la economía de aborígenes y criollos del chaco semiárido. In: Bertonatti C, Corcuera J (eds.) Situación ambiental Argentina 2000. Fundación Vida Silvestre Argentina, Buenos Aires, pp 212–225

Bazán MC, Quiroga A, Jalil ML (2003) Opaete reve ya tape ipia ropi (Todos juntos vamos por el camino nuevo). Paper presented at the 1er. Encuentro Nacional de Educación e Identidades. Los Pueblos Originarios y la Escuela, Universidad Nacional de Luján, Luján, 26–27 Sept 2003. Available via http://www.ctera.org.ar/iipmv/encuentro.htm. Cited 10 Aug 2009

Bermúdez E (2002) Procesos de globalización e identidades: entre espantos, demonios y espejismos—rupturas y conjuros para lo 'propio' y lo 'ajeno.'. In: Mato D (ed.) Estudios y otras prácticas intelectuales latinoamericanas en cultura y poder. CLACSO, Caracas, pp 79–88

Bessonart S, Galarza M, Hernández H, Baldi B (2009) Experiencia en la implementación de un inventario forestal en la comunidad de Caraparí, Dpto. San Martín, Provincia de Salta. Revista Panorama Agropecuario, June 2009, pp. 26–29

BOLFOR II (2009) Dossier forestal. La Paz

BOLFOR II / CADEFOR (2008) Análisis de la cadena de la madera en la Provincia 'Guarayos.' Santa Cruz de la Sierra, Editorial El País

Camacho O, Cordero W, Martínez I, Rojas D (2001) Deforestación en el Departamento de Santa Cruz 1993–2000. Boletín BOLFOR 24:3–5

Carpinetti B (2006) Derechos indígenas en el Parque Nacional Lanín—…de la expulsión al comanejo. Editorial APN, Buenos Aires

Carpinetti B (2007) Estudio de caso—Una experiencia intercultural de co-manejo entre el Estado y las Comunidades Mapuches en el Parque Nacional Lanín, Argentina. Santiago de Chile, FAO. Available via http://www.rlc.fao.org/redes/parques. Cited 20 May 2009

Carrasco M (2000) Cazadores—recolectores en el Gran Chaco: una mirada. Asuntos Indígenas 2:64–70

Catalán R, Wilken P, Kandzior A, Tecklin D, Burschel H (eds) (2006) Bosques y comunidades del sur de Chile. Editorial Universitaria, Santiago de Chile

Centro Amazónico de Desarrollo Forestal (CADEFOR) (2009) Recursos forestales. http://www.cadefor.org/es/sectfor/recfor.php. Cited 20 May 2009

Centurión TR, Kraljevic IJ (eds) (1996) Las plantas útiles de Lomerío. BOLFOR, Santa Cruz de la Sierra

Colectivo de Estudios Aplicados al Desarrollo Social (CEADES), editors (ed.) (2003) Cultura democrática en municipios indígenas: Urubichá y Gutiérrez. CEADES/DIAKONIA, Santa Cruz de la Sierra

Comisión Económica para América Latina y el Caribe (CEPAL) (2005) Los pueblos indígenas de Bolivia: diagnóstico sociodemográfico a partir del censo del 2001. CEPAL, Santiago de Chile

Corporación Nacional Forestal (CONAF) (2009) Superficie nacional de bosques. http://www.conaf.cl. Cited 15 May 2009

de Mojica S (2000) Cartografías culturales en debate: culturas híbridas—no simultaneidad—modernidad periférica. In: de Mojica S (ed.) Culturas híbridas—no simultaneidad—modernidad periférica. Mapas culturales para la América Latina. Wissenschaftlicher Verlag Berlin, Berlin, pp 7–40

Delgado F (2008) Diálogo de saberes para el desarrollo endógeno sostenible en un mundo globalizado. Revista de Agricultura 58(38). Available via http://www.agruco.org. Cited 29 Jan 2009

Donoso C (1981) Tipos forestales de los Bosques nativos de Chile. Documento de Trabajo N°. 38. Investigación y Desarrollo Forestal (CONAF, PNUD-FAO). FAO, Santiago de Chile

Food and Agriculture Organization [FAO] (2001) Global forest resource assessment, 2000, FAO Forestry Paper 140. FAO, Rome

Food and Agriculture Organization [FAO] (2005). Evaluación de los recursos forestales mundiales 2005. Argentina—informe nacional, Rome

García CN (1990) Culturas híbridas:estrategias para entrar y salir de la modernidad. Grijalbo, México City

García Canclini N (2000a) Noticias recientes sobre la hibridación. Paper presented at the II meeting of the work group 'Culture and Social Transformations in Times of Globalisation' of the Consejo Latinoamericano de Ciencias Sociales (CLACSO), Caracas, 9–11 Nov 2000

García Canclini N (2000b) La globalización ¿productora de culturas híbridas?. In: Actas del III Congreso Latinoamericano de la Asociación Internacional para el Estudio de la Música Popular, Bogota, 23–27 Aug 2000. Available via http://www.hist.puc.cl/historia/iaspmla.html. Cited 20 May 2009

Golluscio L (2008) Los pueblos indígenas que viven en Argentina: actualización del año 2002. Secretaría Agricultura, Ganadería, Pesca y Alimentos, Buenos Aires

Herrera Vásquez S, Rodríguez YE (2004) Etnoconocimiento en Latinoamérica: apropiación de recursos genéticos y bioética. Acta Bioethica 10(2):181–190

Herrmann TM (2005) Knowledge, values, uses and management of the *Araucaria araucana* forest by the indigenous Mapuche Pewenche people: a basis for collaborative natural resource management in southern Chile. Nat Resour Forum 29(2):120–134

Herrmann TM (2006) Indigenous knowledge and management of *Araucaria araucana* forest in the Chilean Andes: implications for native forest conservation. Biol Conserv 15:647–662

Ibisch PL, Mérida G (eds.) (2003) Biodiversidad: la riqueza de Bolivia: estado de conocimiento y conservación. Editorial Fundación Amigos de la Naturaleza, Santa Cruz de la Sierra

Instituto Nacional de Asuntos Indígenas (INAI) (2011). Pueblos originarios. Available via http://www.desarrollosocial.gob.ar/Uploads/i1/Institucional/3.MapaDePueblosOriginarios.pdf. Cited 15 Feb 2011

Instituto Nacional de Estadística (INE) (2005) Estadísticas sociales de los pueblos indígenas en Chile - Censo 2002. INE, Santiago de Chile

Instituto Nacional de Estadística (INE)/VA/UNFPA (2003) Bolivia: características sociodemográficas de la población indígena. INE, La Paz

International Work Group on Indigenous Affairs [IWGIA] (2008) El mundo indígena 2008. Copenhagen

International Work Group on Indigenous Affairs [IWGIA] (2009) Los pueblos indígenas de Argentina hoy. Available via http://www.iwgia.org. Cited 10 Apr 2009

Martínez J (2002) Entendiendo la historia de los pueblos indígenas para promover la forestería comunitaria como una alternativa de desarrollo socio—económico local en las tierras comunitarias de origen (TCOs). Universidad Autónoma Gabriel René Moreno, Santa Cruz de la Sierra

Muñoz A (2004) Gestión sostenible de la diversidad biológica en Bolivia: factor desaprovechado en el Desarrollo Nacional. Ambiente Ecológico 88. Available via http://www.um.es/gtiweb/adrico/medioambiente/diversidad%20bolivia.htm. Cited 20 May 2009

Papastergiadis N (2000) The turbulence of migration. Polity Press/ Blackwell, Cambridge

Paz S, Chiqueno M, Cutamurajay J, Prado C (2001) Árboles y alimentos en comunidades indígenas. CERES, Cochabamba

Quijano A (2005) El 'movimiento indígena' y las cuestiones pendientes en América Latina. Revista Tareas 119:31–62

Riester J (1993) Universo mítico de los Chimane. Hisbol, La Paz

Ruiz MM (2006) La protección jurídica de los conocimientos tradicionales: algunos avances políticos y normativos en América Latina. UICN/BMZ/SPDA, Lima

Saldaño CA (2009) Escuela de Educación Técnica Nº 5130 EMETA II—Yacuy. Salta, Directorate General of Technical-Professional Education, Regional Supervision, Department of San Martín. Available via http://supervisionregional.blogspot.com. Cited 12 Oct 2009

Secretariat of Agriculture, Farming, Fisheries and Food (SAGPyA) (2002) Primer inventario de plantaciones forestales y actualización 2002. Buenos Aires

Shackeroff J, Campbell L (2007) Traditional ecological knowledge in conservation research: problems and prospects for their constructive engagement. Conserv Soc 5(3):343–360

Spagarino C (2008) Ampliación de tierras de la comunidad Wichí Lote 27—relevamiento de recursos naturales y propuesta de manejo. APCD, Las Lomitas

Stross B (1999) The hybrid metaphor: from biology to culture. J Am Folklore Theorizing the Hybrid 112(445):254–267

Torrejón F, Cisternas M (2002) Alteraciones del paisaje ecológico araucano por la asimilación mapuche de la agroganadería hispano-mediterránea (siglos XVI y XVII). Rev Chil Hist Nat 75:729–736

Torrejón F, Cisternas M, Araneda A (2004) Efectos ambientales de la colonización Española desde el río Maullín al archipiélago de Chiloé, sur de Chile. Rev Chil Hist Nat 77:661–677

Unidad de Monitereo del Sistema de Evaluación Forestal (UMSEF) (2005) Primer inventario nacional de Bosques nativos. Secretariat of Environment and Sustainable Development, Buenos Aires

Unidad de Planificación, Inversión y Programación (UPIP) (2006) Información área: identidad y pueblos indígenas—para el ajuste del Plan Departamental de Desarrollo Económico y Social (PDDES) 2006–2020. Santa Cruz de la Sierra: Secretaría General y de Planificación

Van Dam C (2007) Certificación forestal y desarrollo local: el caso de Guarayos. Rev Theomai 16:16–34

Vessuri H (2004) La hibridización del conocimiento: la tecnociencia y los conocimientos locales a la búsqueda del desarrollo sustentable. Convergencia 11(35):171–191

Wachholtz R, Artola JL, Camargo R, Yucra D (2006) Avance de la deforestación mecanizada en Bolivia. Superintendencia Forestal, Santa Cruz de la Sierra

Wilbert J (ed.) (1994) Encyclopedia of world cultures: volume VIII, South America. G.K. Hall, Boston

Chapter 4
Amazonia

Miguel Pinedo-Vasquez, Susanna Hecht, and Christine Padoch

Abstract Greater Amazonia—the Amazon Basin, which stretches from the Andes to the Atlantic—is roughly the size of the continental United States. It contains the largest planetary extension of humid forests as well as a complex array of more open forest formations, savanna ecosystems, and agricultural mosaics. About 40,000 plant species are found there. Historically, Amazonia was viewed as a place where ecosystems had been minimally affected by human activity, but modern archaeological discoveries ranging from anthropogenic soils, large scale earthworks, and historical ecological studies are changing this view; the region is now viewed as one of the main civilizational hearths of Latin America, on a par with the Inca, Maya, and Aztec cultures. Recent ethnographic studies of indigenous, traditional, and diasporic populations are also recasting our understanding of the extent and forms of ecosystem management from soil, succession, cultivar, and forest manipulations. These are reviewed in this chapter, and point to the complex managed forests produced today and in the past. What is clear is that there are suites of management techniques that provide income and resilience and that protect and enhance diversity while maintaining biomass through successional processes at the landscape level.

M. Pinedo-Vasquez (✉)
Center for Environmental Research and Conservation (CERC), Columbia University, New York, NY, USA

Center for International Forest Research (CIFOR), Bogor, Indonesia
e-mail: map57@columbia.edu

S. Hecht
Department of Urban Planning, School of Public Affairs, UCLA, Los Angeles, CA, USA
e-mail: sbhecht@ucla.edu

C. Padoch
Center for International Forest Research (CIFOR), Bogor, Indonesia
and New York Botanical Garden (NYBG), Bronx, NY, USA
e-mail: c.padoch@cgiar.org; cpadoch@nybg.org

J.A. Parrotta and R.L. Trosper (eds.), *Traditional Forest-Related Knowledge: Sustaining Communities, Ecosystems and Biocultural Diversity,* World Forests 12, DOI 10.1007/978-94-007-2144-9_4, © Springer Science+Business Media B.V. (outside the USA) 2012

This knowledge and practice certainly merit greater attention for the longer term, especially given the pivotal role of tropical forests in climate systems.

Keywords Amazonia • Ecosystem management • Environmental history • Ethnobotany • Forest ecosystem management • Non-timber forest products • Terra preta • Traditional agriculture • Traditional knowledge.

4.1   Amazonia: A Brief Introduction to Complex Ecosystems

Amazonia includes the largest tract of tropical forest on the planet. The basin in its entirety covers somewhat more than 7 million km². Much of the region—perhaps 5.5 million km²—is covered by lowland moist forests, with the rest divided among a great diversity of forests, savannas, and grasslands (including seasonally flooded grasslands), as well as considerable expanses of fragmented and converted forests, and a variety of agricultural mosaics (Fig. 4.1). The various forests and other land covers found in Amazonia have been put into classes and/or mapped by numerous scientists, using various tools, and for different purposes (e.g., Malleux 1983; Moran 1993; Tuomisto et al. 1995; Whitmore et al. 2001; Saatchi et al. 2007; GEOAmazonia 2009). These estimates certainly do not agree on how many significantly different forest types Amazonia may contain. For instance, botanist Prance and colleagues (1987) distinguished a total of 20 forest and non-forest habitat types for the Brazilian Amazon. Saatchi and his colleagues, who classified and mapped Amazonia to more accurately measure carbon stocks, used a combination of remote sensing and ground-truthing plots to come up with 16 distinct forest and wetland types (Saatchi et al. 2007). Tuomisto et al. (1995), however, argued cogently that far more than 100 types should be recognized in the Peruvian Amazon alone.

Amazonian peoples and societies, of course, also classify their environments, including their forests, in a variety of ways. The indigenous peoples whose classificatory systems have been studied include the Ka'apor (Balée 1994), Kuikuru (Carneiro 1978, 1983; Heckenberger 2005), and Kayapo (Posey 1996) of Brazil; and the Matses (Fleck and Harder 2000), Maijuna (Gilmore 2005), and Matsiguenka (Shepard et al. 2001) of the Peruvian Amazon. Halme and Bodmer (2007) have looked at forest classification by ribereño farmers and hunters in Peru, and Frechione et al. (1989) studied environmental zonation by Brazilian caboclos. Many of these studies found that rural Amazonian communities, whether indigenous or not, use complex classification systems that often include multiple criteria, including local geomorphology, physiognomy, soils, disturbance, and indicator plants (especially palms) and animal species to classify their environments; local communities often distinguish dozens of forest types over small scales (Anderson and Posey 1989; Elisabetsky and Posey 1989; Ellen et al. 2000; Posey and Balée 1989; Posey and Balick 2006; Posey and Plenderleith 2002; Rival 2002; Posey 1996). These systems are often not strictly hierarchical, but feature different and overlapping subsystems (Gilmore 2005). Many scholars have found that locally developed systems of forest

Fig. 4.1 Forest and woodland cover in the northern South America (Source: Adapted from FAO (2001)). Key: *Dark green* closed forest, *light green* open or fragmented forest, *pale green* other wooded land, *yellow* other land

and habitat classification could and should be used more widely by scientists, conservation managers and others (Posey and Balick 2006; WinklerPrins and Barrera-Bassols 2004). Halme and Bodmer (2007), for instance, suggest,

'for wildlife management and monitoring it is essential to know how habitat influences the density of wildlife populations. Since there is a strong agreement between the TEK (traditional ecological knowledge, in this case, forest classification) and the scientific classifications…TEK habitat analyses could be used as a valuable tool in the development and implementation of collaborative management plans. For instance, managers of the Mamirauá Sustainable Development Reserve in the Brazilian Amazon have used the "moradia system" a traditional knowledge for zoning riparian forest for the conservation and sustainable use of forests resources (Ayres 1996). The use of TEK nomenclature could increase the participation of the community members and facilitate their collaboration with conservation professionals.'

The great vegetational and other diversity that is encoded in these classifications reflects Amazonia's rarely appreciated wide variations in soils and drainage characteristics, temperature and rainfall patterns, differences in altitude, and other environmental variables, as well as the great number of ways that people have changed and continue to manage and manipulate vegetation.

Culturally, perhaps the most important division of Amazon terrestrial areas is the categorization of lands and their respective forests into 'terra firme' and 'várzea,' or dry lands /uplands and floodplains; and 'campo' and 'floresta,' or grasslands and forests. The várzea floodplains constitute a mere 2–3% of Amazonian territory, but have long had an economic and social importance that belies their limited size. Várzea refers technically only to areas flooded by the basin's great whitewater rivers including the Amazon, Solimões, Ucayali, and Marañon. The várzea is further subdivided into the tidal or estuarine floodplain, where twice-daily tides inundate significant areas of land near the mouth of the Amazon, and the middle and upper floodplains that lie above the tide's reach and where annual floods reach heights of 10–12 m and submerge broad swaths of land and forest along the rivers for 3 months or more. Large territories are also washed by black-water rivers; the greatest of these is the Rio Negro that drains parts of Venezuela and Brazil. The lands flooded by such blackwaters are known as 'igapó' and have very different forests and sets of terrestrial and aquatic species from those of the whitewater várzeas (although in some areas on the white water tributaries, igapo refers to perennially flooded forests). Several important Amazonian rivers are classified as clearwaters; these include Brazil's Tapajos and Xingu that flow off the Brazilian Shield. Finally, savannas, grasslands, or scrubland occur in some areas with white sand soils and particular drainage conditions, as well as in areas with less total rainfall and stronger dry seasons; these are also an outcome of extensive human interactions with these systems, creating forest and grassland mosaics. Cultures that form part of the Gê linguistic group were masters of the Brazilian savannas in pre-Colombian times and influenced the structure of these environments profoundly (Anderson and Posey 1989; De Toledo and Bush 2008b; Furley et al. 1992; Hecht 2009; Mistry and Berardi 2006).

Many of the Amazon's terrestrial and aquatic habitats are characterized by high species diversity. Some estimates place the plant diversity of the entire Amazon Basin at about 40,000 plant species. Studies in the terra firme forests of the western Amazon, where the broad flat plain begins to rise into the Andes, have been characterized as the most diverse on earth, with up to about 300 plant species (over 10 cm dbh) per hectare. But as noted above, the Amazon is far from uniform, and its forests differ substantially in diversity, structure, and other essential characteristics. Most floodplain forests, for instance, have been found to have lower tree diversity than their terra firme counterparts, but the diversity of the particular ecosystems and biomes that make up the várzea can be very high; local societies recognize that fact with a rich set of terms for various features of the várzea (Pinedo-Vasquez et al. 2002; Frechione et al. 1989). Amazonia famously also harbours remarkable numbers of animal species; it sustains the world's richest diversity of birds, fresh-water fishes, and butterflies.

4.2 Amazonians: History and Diversity

It is estimated that the total population of Amazonia is now more than 30 million people, although as the exact limits of the basin are disputed; so, too, are population totals (see for instance, GEOAmazonia 2009). A dispute over what might have been the population of the region before the invasion of Europeans (i.e., in 1491), however, has been going on for a long time, and is not apt to be resolved easily (e.g., Denevan 1964; Mann 2005). Estimates have ranged from 1 to 11 million. However, there is increasing agreement among most scholars that much of the Amazon was far more heavily settled in pre-Columbian times than early studies done by many prominent scholars had suggested (Meggers 1954, 1971). Many now believe that prior to the disastrous die-off of Amazonian populations that accompanied the entrance of Europeans, the peoples of Amazonia numbered in the millions, with large settlements located along major rivers and on the bluffs about the várzea floodplain as well as populations on the uplands (Church 1912; Erickson and Balée 2006; Heckenberger et al. 1999, 2008; McKey et al. 2001). It is now widely believed that some Amazonian societies not only were organized in complex ways politically and economically, but that they had the capacity to significantly alter their home environments including forests, and in fact might be usefully thought of as engineering societies (Fig. 4.2). A number of recent discoveries by archaeologists in widely separated and seemingly

Fig. 4.2 Geoglyph. At this writing more than 225 geoglyphs, that is, large-scale, geometric earthworks, have been found in the Purus watershed, including the Acre valley. These sites are taken to be the remnants of palisaded settlements in some cases, and others may have been ceremonial sites. Their most striking feature is their geometric precision (Photo: Susanna Hecht)

unlikely areas of Amazonia have helped make the case for the dense populations of the region and significant alteration of Amazonian environments. This is a large literature and expanding daily, but general overviews include Balée and Erickson (2006), Hecht (2003), Hill (1988), Hill and Santos-Granero (2002), Lehmann et al. (2006), Macia (2008), McEwan (2001), Roosevelt (1991), Woods et al. (2009).

The overwhelmingly important role of carefully planned and controlled fires in the management of soils, especially in the creation of some of the Amazon Basin's most fertile soils is only now being explored in depth. 'Terra preta' ('black earth' in Portuguese)—also known as 'Indian dark earths' or 'terra preta do índio' ('terra mulata' is a lighter-coloured, brownish version)—refers to dark, highly fertile anthropogenic soils that occur in several areas of the Amazon Basin, on every soil type. Terra preta owes its name to its very high charcoal content, and indeed appears to have been made by adding mixtures of charcoal, fish and animal bones, and manure to what may have otherwise been relatively infertile Amazonian soil over many years; thus they are distinctively the outcome of human intervention. The key element in their creation seems to be very low temperatures at burning, which creates a pyrolytic charcoal that is very stable over long periods of time and binds soil nutrients far more tightly than the typical result of 'hot fires' usually associated with deforestation. Terra preta seems to have two types of origins, one basically formed on urban residues (hence immense amount of potsherds), and the other, terra mulata, more a result of the slow cool burns on continuously managed agricultural sites.

Some of these soils have been dated to between 450 CE and 950 CE and are believed to have been created by low temperature burning of fields known as 'slash-and-burn'. Some contemporary Amazonian agriculturists, including the Kayapo who live in the state of Para in Brazil, continue to use complex sets of practices that include: specific ways of laying out agricultural fields, a range of burning technologies that produce 'cool' burns, the practice of scattering cooking residues and other garbage, and the practice of adding palm mulch and other nutrient-rich materials to soils (Hecht 2003; Hecht and Cockburn 1989; Lehmann et al. 2006; Liang et al. 2008; Peterson et al. 2001). These practices when used together have been described as potentially creating terra preta and terra mulata-like enriched soils (Hecht 2003). Terra preta soils have been identified mainly in Amazonian Brazil, although some sites have also been found in the Guaviare in Colombia, Peru, and Ecuador (Eden et al. 1984; Jackson 1999; Lima et al. 2002; Martínez 1998) and elsewhere in the Orinoco drainage. Terra preta has also been found in Central America including Panama and Costa Rica (WinklerPrins A. and W. Woods, personal communication, 2009). Originally believed to be confined only to small areas of several hectares apiece, recently terra preta sites have been found to cover areas of hundreds of hectares, again challenging preconceptions of the 'pristine' nature of much of the forested Amazon. These techniques are now part of a general effort to improve soil properties for small-scale producers (Woods et al. 2009). Overall, some 4% of Amazonia might display dark-earth characteristics (Fig. 4.3).

In the Beni region of Bolivia, geographers, archaeologists, and others have uncovered thousands of kilometres of raised fields standing above the seasonally flooded grasslands that are known as the Llanos de Moxos. Characterized by an

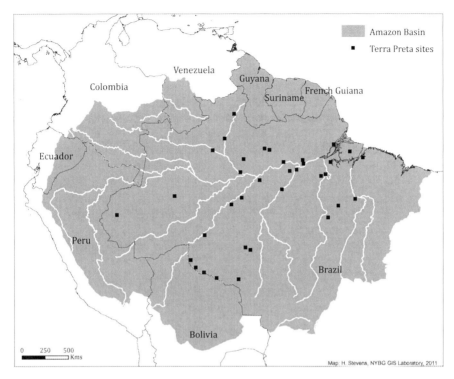

Fig. 4.3 Terra preta sites. These areas of anthropogenic soils are widespread throughout the Amazon, and reflect a unique adaptation to poor soil through the enrichment via cool burns and to create nutrient-rich, stable biochar. Terra preta and Terra mulata provided the basis for the development of large upland populations and may do so in the future (Map courtesy of H. Stevens, New York Botanical Garden, GIS Laboratory)

alternation between seasonal flooding with very intense rainfall and seasonal strong droughts accompanied by intense heat, and poor soil and drainage, the Llanos de Moxos would appear to be an unlikely region to harbour large numbers of agricultural peoples (e.g., Denevan 1970; Erickson 2006; Erickson and Balée 2006). Yet these well-developed raised bed earth work, coupled with mounds, and an array of large-scale drainage of other forms of engineering provide the production base for very large populations, perhaps even into the hundreds of thousands, who might have lived there (Erickson 2006). These ancient residents, apparently using both any existing natural differences in elevation and building thousands of artificial mounds or 'geoglyphs,' created a complex, built landscape that allowed large populations to subsist in the region on both agricultural and extracted, especially aquatic, resources. The most visible key feature is a system of planting surfaces elevated above the seasonally flooded and poorly drained flat grasslands. Research has demonstrated that management practices succeeded not only in elevating crops above the flood, but also in significantly improving the soils on the mounds, in controlling or improving drainage, and in enhancing conditions for fisheries and other provisioning activities. Archaeologist Clark Erickson and others have also suggested that the area may have been an extensive human-constructed landscape that was managed largely to increase fish production through an ingenious con-trolled flood-recession system that captured fish in hundreds of depressions that became ponds as floodwaters ebbed. Erickson has identified a large number of earthen fish weirs distributed across the Llanos de Moxos landscape (e.g., Erickson 2006). Similar structures have been found elsewhere as in the Xingu River (Heckenberger et al. 2008) and possibly in French Guyana.

There is also evidence that fire was an important tool used in the Beni and the Cerrado. The recurrent burning of the Llanos is assumed to have decidedly influenced the distribution of plant species, with a large assemblage of fire-resistant species favoured in the region. This active fire history is also documented in Kayapo and Xavante practices (Barbosa and Fearnside 2005; Coimbra-Junior 2004; De Toledo and Bush 2007, 2008a; Harris 1980; Hecht 2009; Mistry 2001; Santos et al. 1997).

Recent research on the southern edges of Amazonia, in the Brazilian State of Mato Grosso, has added important new information that considerably expands this story and alters yet again some of the most fervently held 'truths' about pre-Columbian Amazonia, its population and settlements, and the capacity of particular parts of the basin to support large numbers of humans and their developed societies. Along the upper Xingu River, archaeologist Michael Heckenberger has recently excavated sites of what he calls 'garden cities,' extending the concept of Ebenezer Howard's ideas for urban sustainability. (Heckenberger et al. 2008). Linked by extensive net-works of roads and canals, these settlements appear to indicate levels of population, political organization, and environmental management that were previously ascribed only to a few várzea sites or like the Llanos de Moxos, or sites to the western Amazon, such as the Chachapoya culture along the foothills of the Andes. Heckenberger identified several clusters of linked urbanized sites of varying sizes. Basing his calculations partly on present-day settlement and resource use patterns, he estimates a total population in the region that may have numbered about 50,000, distributed among numerous sites in an area the size of Belgium. Heckenberger writes that, 'within each cluster, we estimate 100–150 ha or more of settlement space and a population in the mid-thousands (≥2,500) distributed between walled towns, which are estimated to have 800–1,000 or more persons and smaller non-walled villages (250±100 persons)' (Heckenberger et al. 2008, page1216). The settlements were connected by multiple roads and canals, and surrounded by mosaic landscapes featuring woodlands—including mature forests, plots devoted to intensive agriculture, and even 'large compost plots' (Heckenberger et al. 2008, page1217). Not only are these finds surprising because of their location in an upland region long thought to be hostile to large-scale settlement, but also because they suggest that a very distinct type of urbanized and yet dispersed society existed in Amazonian forests. Much larger populations are recorded in many Amazonian oral histories and ritual activities (Hill and Santos-Granero 2002; La Torre-Cuadros 2008; Vidal 2000; Whitehead 2003).

These important examples are some of the most recent and most prominent addi-tions to a broad and increasingly convincing literature that suggests that numerous areas of the Amazon may have been densely populated and heavily managed before the European disruption of Amazonian societies. They are exciting additions to a set of evidence that appears to contradict assumptions that settlement and livelihood options in Amazonian forests, except for very limited zones, were severely limited by environments inimical to human settlement and subsistence, especially by extremely infertile soils. The corollary of this assumption was that most of Amazonia was able—at best—to support only small and unstable local populations that had little impact on forests and other environmental features. Both archaeological and historical data also point to the considerable trade and migration that occurred not only within the basin but also between Amazonia and the Andean realm. These new data powerfully question that early and still popularly held view that almost the whole of Amazonia was, on the eve of the European conquest, an uninhabited, ahistorical, pristine, and 'Edenic' forest landscape.

These complex Amazonian civilizations experienced the loss of much of their knowledge as their societies were destroyed by disease, slaving, and dislocations. However, a great deal of knowledge still resides in local societies and is encoded and reproduced in the landscapes and knowledge systems of today's inhabitants of Amazonia.

4.3 Local Knowledge and Use of Forest Plants

Researchers have estimated that at least 2,000 species of Amazonian plants have been given a particular use by humans, including as food, fibres, medicines, fuels, or for any number of other utilitarian, aesthetic, or ritual purposes (Duke and Vásquez 1994; Bennett 1992). In the Peruvian Amazon, for example, more than 1,250 species have been recorded as being used for an economic purpose by at least one local group (Pinedo-Vasquez et al. 1993), with at least 138 species of agricultural and tree crops among them (Clement 1998). Considerable work has been done on the uses of forest plants and plant lore of specific Amazonian communities, particularly on which plants that indigenous Amazonians use for medicinal purposes, as well as on the nomenclature and classification of those plants in Amazonian languages (e.g., Alexiades 2009; Balée 1994; Bennett and Prance 2000; Boom 1987; Duke and Vasquez 1994; Prance et al. 1987; Vickers and Plowman 1984; Zent and Zent 2004; Zent 2009; Anderson and Posey 1989; Balée 1994; Coimbra-Junior 2004; Elisabetsky and Posey 1989; Erickson 2006; Erickson and Balée 2006; Macia 2008; Posey and Plenderleith 2002; Stearman 1989).

The western areas of Amazonia, particularly the pre-Andean region, have been identified as centres of diversity for a number of significant crop species (Clement 1999). Globally important staples including manioc (*Manihot esculenta*), peanuts, pineapple, cacao, runner beans, peach palm ('pupunha' in Brazil 'pijuayo' in Peru), tree cotton, coca, tomatoes, and *Capsicum frutense* chiles (which include some remarkably hot cultivars such as the habanero), are believed to be among the most important plants to have been domesticated in the region.

While there are several valuable studies of forest management practiced by indigenous peoples, the plant knowledge of caboclo, ribereño, and other 'non-indigenous' Amazonian peoples has received considerably less attention (although see Amorozo and Gély 1988; Shanley and Rosa 2004; Silva et al. 2007). Many researchers have remarked on the richness of Amazonian plant knowledge, especially of local pharmacopeias. For instance, the Joti of Venezuela are said to use at least 180 species of plants as medicines (Zent 2009), while the Esse-Eja of Peru and Bolivia use some 143 species, caboclos along Brazil's Rio Negro know and use 193 plant species (Silva et al. 2007), and the Tacana of Bolivia use 150 species as medicines. Each of these groups, of course, also uses plants and plant products for a large number of other purposes including food, fibre, ritual, construction, hunting, fishing, and other utilitarian and technological needs, as well as for firewood. A number of ethnobotanists have also researched what Amazonians know about the plants that animals prefer as foods; the Joti, for instance know more than 500 species that are sought out by animals. This plant knowledge presumably helps the Joti locate places where animals can be found at particular times of the year or the day, and thus are particularly good places to hunt (Zent 2009; McKey et al. 2001; Rival 2002).

Quantifying and comparing ethnobotanical knowledge of Amazonians in a different fashion, 'quantitative ethnobotanists' have determined the percentage of plants found in a typical hectare of forest that are known to particular Amazonian groups as useful for meeting particular needs. Using this approach, Prance et al. (1987) found that the Chacobo of Bolivia had uses for 78.7% of the tree species inventoried in 1 ha, while the Ka'apor of Brazil used 76.8%, the Tembe also of Brazil used 61.3%, and the Panare of Venezuela used 48.6% (Prance et al. 1987). Of all the plant families that these groups used, palms were found to be most frequently used by all four of the groups consulted (Prance et al. 1987). Palms more generally are salient features in the landscapes, and the cultural ecology of both indigenous and peasant production systems (Henderson 1995; Kahn and Granville 1992; Porro 2005). Medical ethnobotany has been another especially important area of recent research, although controversial because of the history and potential for biopiracy (Bertani et al. 2005; Jullian et al. 2006).

Apart from documenting the surprising diversity of plants used by and the high percentage of forest plants that are given a particular use by Amazonian peoples, some recent ethnobotanical work in the Amazon basin has focused on how that plant knowledge may be changing. Although traditionally indigenous knowledge has been presented as centuries-old, accumulative, and fast-disappearing, more recent work has focused on how learning, adaptation, innovation, borrowing, and hybridization continues to occur within the various Amazonian plant knowledges, including pharmacopeias and adoption of useful plants from European sources as well as the African diaspora, such as watermelons, calabash trees, pigeon pea, and rice (e.g., Alexiades 1999, 2009; Zent 2009; Carney and Rosomoff 2009; Carney 2001; Crosby 1986; Ganson 2003; Price 1983; Schiebinger and Swan 2005; Voeks 1997). These works and others suggest that adaptation and resilience are processes that characterize local and indigenous forest knowledge and that the frequent focus on the erosion of that knowledge and its imminent demise does not tell the whole story.

Fig. 4.4 Small portable sawmills process cheap materials of fast-growing timbers produced by smallholders (Photo: Christine Padoch)

Some articles, for example have also focused on the 'hybridization' of traditional or local Amazonian knowledge with science-based or other 'modern' knowledges, including the information that indigenous and other farmers may receive from the many development, extension, and conservation projects that have been directed at Amazonian communities—from farmers' attendance at schools and training courses, from the broadcast and other media, and frequently from involvement in wage labour or participation in other market-based activities (Ayres 1996; Ellen et al. 2000; Laurie et al. 2005). For instance, working as loggers and in sawmills, many Amazonians, especially young men, acquired specialized knowledge about the particular industrial uses of forest species and their desirability in local, regional, and even international timber markets (Sears et al. 2007). In the estuarine floodplain forests of Amapá, caboclo smallholders and family-run sawmills are enjoying economic success today because they formed a new 'hybrid' forest product industry and fashioned a niche in a market left open after the regional timber boom ended in the 1970s. The phenomenon started with a diversified resource management system based on traditional knowledge of forest species, combined with the knowledge and technology gained during employment with boom-time large-scale timber operations and markets (Fig. 4.4). With this new, combined knowledge, caboclo families transformed the local and regional timber markets to include many more locally grown and locally milled timber species (Sears et al. 2007).

4.4   Modern Settlement

Politically, present-day Amazonia reaches into eight countries and one political dependency; Brazil, with almost 70% of the basin dominates; Peru has 13%; and smaller territories are found in Bolivia, Ecuador, Colombia, Venezuela, Guyana, Suriname, and French Guiana.[1] The residents and communities of this complex region are also extremely diverse. Many of them are urbanites, but regionally the level of urbanization varies, with Amazonian Venezuela, Brazil, and Peru each having about 70% or more of their populations living in urban spaces, while the Amazonian areas of Ecuador and Guyana are still predominantly rural (GEOAmazonia 2009). Several recent researchers examining urbanization trends have argued that the conventional division of Amazonian individuals or households into 'urban' and 'rural' has become increasingly difficult as a pattern of dispersed or 'multi-sited' households has emerged throughout the region (Padoch et al. 2008; Brondizio 2006; WinklerPrins and Barrios 2007). In many areas, most families and individuals are highly mobile and maintain household economies that depend simultaneously on both rural and urban incomes and resources. Such emerging patterns have been shown to affect both the extent and the composition of forests (Padoch et al. 2008). Such multi-sitedness may have deeper roots in indigenous cultural practices, which featured seasonal trekking and periodic dispersion and convergence of groups (Redmond 1998; Rival 2002; Whitehead 2003).

Ethnically Amazonia is also extremely complex. People and communities who are strictly considered indigenous by their governments are now a small minority in most of the region. Estimates by the indigenous rights organization Cultural Survival puts the number of indigenous people in the Amazon Basin today at about 2-½ million, and the number of identifiable indigenous groups at 400, or a mere 20% of what they estimate were once 2,000 cultural groups in Amazonia (Castro et al. 1993; da Cunha 1984, 1992). The Brazilian anthropologist Darcy Ribeiro indicated that 82% of the native groups recorded at the turn of the twentieth century were extinct by the middle of the 20th. This was a conflictive and desperate process, as John Hemming, the pre-eminent historian of Amazonian indigenous peoples, has shown in relentless detail (Hemming 1978, 1987, 2003).

While native populations have justifiably received a great deal of attention, many other populations continue to maintain complex relations with forests. These other populations—which include Brazil's caboclos and quilombolas, Peru's ribereños, and others—were actually outcomes of various diasporas that affected the region as a result of detribalization processes, black runaway slave communities, the rubber tappers, and colonization programmes that unfolded at various times throughout out the region, and whose descendents continue to live in Amazonia (Auricchio et al. 2007; de Mello e Souza 1996; Dos Santos Gomes 2002; Gomes 1999b; Hecht 2010;

---

[1] Again, as mentioned above, the precise numbers and percentages depend on the particular delimitation of Amazonia that is used. For some recent alternative numbers from the Amazon Cooperation Treaty Organization, see GEOAmazonia 2009.

Hecht and Cockburn 1989; Karasch 1996; Nugent and Harris 2004). These communities also have multiple and complex relationships with Amazonian forests and continue to maintain, evolve, and create knowledge of how to manage, create, conserve, and use the forests and other resources of the Basin.

Among these communities are African-Amazonians, whose ancestors first came as slaves from large and important rural and urban communities especially in the eastern Amazon of Brazil, Suriname, and Guyana. In Brazil many of these are designated quilombos or settlements descended from independent black communities (often of 'maroons' or slaves who escaped into the forests); these communities number in the hundreds in Brazil's eastern Amazon. In colonial Guyana, Suriname, and French Guiana, where Africans made up the main work force during the colonial period, many of them escaped slavery and also settled in the maroon communities in the interior. In Suriname these descendants of slaves who found freedom in Amazonian forests formed a variety of separate ethnic groups including the present-day Saramaca, Djuka, Paramaka, Matawai, Aluku, and the Kwinti (Escalante 1981; Funes 1995, 2004; Gomes 1999a; 2005; Price 1975; Stedman et al. 1992). Known as 'palenques' in Venezuela, Ecuador, Bolivia, and Colombia, these maroon societies can be found throughout Amazonia. Black slaves were used in agriculture as well as gold mining, and thus gold-mining areas, such as Mato Grosso (Brazil), French Guiana, Ecuador, and Colombia, had extensive fugitive settlements.

People from many other regions and nations also came to Amazonia, and their contributions figure in the management practices and the state of its forests. Guyana saw massive immigration from India, where workers were recruited to replace African labour after slavery was abolished. In many other countries, internal migration flows brought workers from impoverished areas outside of the lowland humid tropics. The rubber boom and many subsequent economic opportunities brought large populations of Nordestinos, or peasants from Brazil's arid Northeast into the Amazon to tap rubber, gather a variety of forest and agricultural products, and eventually to settle (Barham and Coomes 1994; Dean 1987; Domínguez and Gómez 1990; Hecht 2010; Reis 1953; Santos-Granero and Barclay 2000; Santos 1980; Taussig 1986; Wolff 1999). In Peru, residents of Andean and sub-Andean communities have been descending the mountains and becoming Amazonians for millennia. Other important additions to the populations of Amazonia include the Chinese who came to work on Peruvian railroads before settling in Amazonian towns and as 'coolie' labour in gold mines and railroad development. Japanese farmers were also settled in areas of the Brazilian and Bolivian Amazon. Middle Easterners, largely from Syria and Lebanon, became an important part of the commercial class with the disruption of the Ottoman Empire and the various upheavals in the Mediterranean.

The various economic booms and busts that periodically lifted and sank Amazonian economies and recurrent political changes and crises, further enriched the ethnic diversity of Amazonia throughout the eighteenth, nineteenth, and twentieth centuries. For example, the task of building the never-completed Madeira-Mamore railway in western Amazonia reportedly involved people of about 50 nationalities (Craig 1907; Ferreira 2005). Regionally, migrants came from Bolivia, Brazil, Colombia, Ecuador, Peru, and Venezuela; outside the region, they arrived from Cuba,

Granada, Ireland, Sweden, Belgium, China, Japan, India, Turkey, Russia, and many others (Hardman-Foot 1988, quoted in GEOAmazonia 2009). The earlier attempts at building the rails involved chartering a company, PC Phillips, a construction company from Philadelphia, so the region was also filled with American engineers and adventurers; many of these Americans provided important information on labour practices (Craig 1907), including denunciations of the horrors of the Casa Arana treatment of native populations, which resulted in the denunciations of terror slavery (Hardenburg and Enock 1913; Taussig 1984, 1986).

With these recurrent immigration flows, population growth rates have continually swelled and ebbed. The population of all Amazonia is estimated to have grown at an average annual rate of 2.3% from 1990 to 2007 (GEOAmazonia 2009), although there is considerable variation among regions in population dynamics. The recent growth of the Amazonian population is associated both with natural demographic increase as well as with continuing immigration that reflects many factors, including a variety of national policies that have deliberately stimulated and supported Amazon colonization and fluctuating expansion and contraction of productive activities—including agriculture, ranching, mining, timber extraction, and in some areas, coca production (Almeida 2004; Barbieri et al. 2009; Etter et al. 2008; Hardman-Foot 1988; Frost 2005; Hecht 2005; Hecht and Cockburn 1989; Hecht and Mann 2008; Hecht et al. 1988; Jepson 2006; Lavalle 1999; Pacheco 2006; Schmink and Wood 1984; Smith 1982). Population growth also reflects immigration from outside the region to the Amazon spurred by the poverty, political instability, and violence prevalent at various times in areas such as the Andean highlands.

4.5 Knowledge and Management of Amazonian Environments: Dynamism and Evolution

There continues to be considerable controversy about the exact extent and continuing importance of the management or disturbance of Amazonia's forests. However, there is little doubt that many of the forests that are now found throughout the Amazon Basin do bear the mark of management by pre-historic, historic, and contemporary peoples, probably often superimposed in a complex mixture that cannot be disentangled. These patterns of forest disturbance, management, or manipulation take many forms. Amazonians through the ages and belonging to various communities have developed highly diversified, complex, and dynamic ways of enhancing the economic value of their environments, including the production of timber and non-timber forest products (Guariguata et al. 2010). In many ways perhaps the most useful way to think about them is as 'silvo-agrarian populations' since simple categories of gathering, or settled agriculturalists, simply don't fit the realities of Amazonian livelihood strategies, which involve a complex mixture of forest-agrarian shaping and interactions, including urban-based strategies.

Although many of these management practices are subtle and 'invisible or illegible' to conventional foresters, they have been used to transform forests, to create new types of forests, and to maintain them for generations. Fire is probably the most frequently cited and most effective management tool that past generations as well as today's small farmers appear to wield for changing forests and other components of their environments. Fire is still widely used in swidden (or shifting cultivation) systems and in the management of grasslands for enhancing game production, and more recently the management of pastures for livestock production. Fire not only affects standing vegetation but also, as a plethora of recent research efforts have uncovered, the soils upon which those forests stand (Blate 2005; Hammond et al. 2007; Hecht 2009; McDaniel et al. 2005; Nepstad et al. 2001; Ramos-Neto and Pivello 2000; Rodriguez 2007; Vieira 1999).

Knowledge of the practices and methods by which Amazonian landscapes, including their topography and soils, have been and continue to be altered, has grown rapidly in the past few decades. There also recently has been considerable progress in understanding that both contemporary and past Amazonians engaged in the active manipulation of watercourses as well. In several locations of Amazonia, most notably in the estuarine floodplain, the 'course, dynamics, volume, and significance of local rivers and streams are being changed' (Raffles and WinkerPrins 2003). Raffles, for instance has described how a community of caboclos in the Brazilian state of Amapá have enlarged the main channel of a river in the estuarine floodplain and have dug a series of canals leading from the river into the neighbouring forest (Raffles 2002). This manipulation was done largely to ease access to valuable forest products, including timber, as well as to reach hunting grounds and agricultural lands. The work was done by groups of villagers working over several seasons with the help of water buffalos (Raffles 2002).

Farther upstream on the Amazon near the city of Santarém, farmers have reportedly been manipulating sedimentation processes for many years in order to fill in swampy areas, making them usable for agriculture (Raffles and Winkler-Prins 2003). Using a 'slackwater' process, farmers rely on the fact that the heavily sediment-laden Amazon waters release much of their load when their flow is slowed, thus filling in low-lying areas behind the natural levee. Raffles and Winkler-Prins (2003), cite research that shows that sediment deposition could reach rates of about 3 cm/day, 'producing a potentially significant sedimentation of up to 2 m in any flood season.' This method eventually results in substantially changing the vulnerability to floods of certain floodplain sites, raising them above the zone that is apt to be inundated most years and thus allowing for the production of flood-intolerant crops, as well as changing the forests that can grow on the sites. It is worth noting that Padoch and Pinedo-Vasquez observed other methods of using sediments to 'make land' used for farming and agroforestry that did not involve the digging of channels, but rather the spreading of palm fronds on the ground to slow the velocity of floodwaters and ensure the deposition of flood-borne sediments (Padoch and Pinedo-Vasquez 2000).

Raffles and Winkler-Prins cite many other instances of anthropogenic changes to Amazonian watercourses, including speculation that part of the 'lower reaches of the famous Casiquiare Canal linking the Amazon (via the Rio Negro) and Orinoco river systems may have been opened by Arawak labor' (Raffles and Winkler-Prins 2003, page 177). A recent particularly dramatic example of the effects that deliberate modifications of river channels can have, is provided by Coomes et al. (2009), who recount a human-initiated process that resulted in a substantial shift in the course of the Ucayali River upstream of Peruvian city of Pucallpa. The change began in the flood season of 1997 when a 71 km meander loop of the Ucayali, the major stem of the Amazon, was cut through at a site called Masisea. By 1998, the cutoff gave rise to the most dramatic change in course of the Ucayali in at least 200 years (Coomes et al. 2009). It reduced the length of the Ucayali River by about 64 km, creating a large lake and altering local flooding regimes and sediment deposition patterns all along the floodplain; it even produced a significant shift in the river that is now threatening the port of the city of Pucallpa. Coomes et al. (2009, page130) also report that due to this change, 'long-term shifts are expected in local edaphic and ecological conditions, particularly with respect to vegetation, fish, and other aquatic life in the area'. Their research also indicated that agricultural conditions and general livelihoods up and downstream from the site of the cutoff have been profoundly affected. The massive Masisea cutoff, Coomes et al. (2009, page 132) report, 'resulted from the work of a small number of people who, using simple tools in an incremental but steady manner, took advantage of natural large river dynamics;' their interviews reveal that local residents dug ditches, cleared brush, and engaged in other seemingly small earthmoving activities, thereby changing the river to facilitate their own river transport. The result was far larger in scale and effect than they had anticipated, reflecting the power of the river currents that were unleashed. In a broader historical perspective, this recent example is particularly important as it suggests that

> 'the course and floodplains of major meandering rivers can be strongly influenced by small groups of people working with simple tools, with important consequences—both upstream as well as downstream—for people and the environment; such is the case today and perhaps also in prehistoric times, when rural riverine Amazonia was much more densely populated' (Coomes et al. 2009, page 133).

One can only speculate on how many watercourses and their surrounding forests that are now accepted as completely natural may have resulted from pre-Columbian fluvial manipulation. Those processes are now lost to history and hidden from view.

4.6   Managing Amazon Forests

Present-day methods of directly managing forest vegetation may also give clues to the past. The methods and tools that are being used by today's small farmers are only now being discovered. Apart from the use of fire, the Amazon farmer's and forester's toolbox includes various forms of planting and transplanting (often within forest stands), as well as selective weeding, and a plethora of silvicultural treatments including maintenance of seed trees and other forms of managing natural regeneration, protection of seedlings and saplings, and pruning of trees. Amazonian forests abound in patches and swaths that are particularly rich in economically valuable species, including timbers, fruits, and others. Many of these forest patches are almost

Table 4.1 List of some important types of manchales

Remnant managed forests	Characteristics
1. Caobales (Spanish) or Mogonais (Portuguese)	Forests that include clusters of mahogany trees and were found all over Amazonia (Balée 2003)
2. Cacaois (Portuguese) or Cacauales (Spanish)	Forests where the sub-canopy is dominated by dense stands of cacao (Peters et al. 1989)
3. Zapotales (Spanish) or Zapotais (Portuguese)	Forest where the dominant canopy is the zapote (*Quararibea cordata*) fruit species called (Pinedo-Vasquez 1995)
4. Babaçuais	Forest where the most dominant species is the palm babaçu. These forests are the product of fire
5. Castanhal (Portuguese) or Castañales (Spanish)	Dominated by Brazil-nut trees forming dense stands (Almeida 2004)
6. Siringal (Portuguese) or Shiringal (Spanish)	Dominated by rubber trees and very low-density sub-canopy vegetation (Almeida 2004)
7. Açaizais	Flooded forests dominated by dense stands of açaí palms (Brondizio 2008)
8. Sacha manguales	Forest where the sub-canopy is dominated by the fruit species known as sacha mangua (Peters et al. 1989)

certainly remnants of intensively managed gardens that have now been largely abandoned, or of planted or protected wayside or campsite vegetation that has been periodically manipulated by seasonal trekkers, travellers, or migrants (Alexiades 2009; Anderson and Posey 1989; Kerr and Posey 1984; Rival 2002).

There is a great diversity of such remnants or patches in forests throughout Amazonia; their ages and those of the vegetation that surrounds them vary from the very old (and often believed to have been created by long-dead ancestors), to recently abandoned swiddens. They therefore are essentially 'cultural forests.' In Brazil and Peru most of these forests are named after their most abundant and valuable tree species (e.g., 'siringal' in Brazil to indicate a forest where rubber trees are abundant; 'sacha mangual' in Peru to indicate a forest rich in sacha mangua or *Grias peruviana*). In many parts of Amazonia these patches or remnants fall into the general category of 'manchales,' which locally differentiates them from unmanaged forests. Table 4.1 lists several kinds of manchales that are commonly found in the forests of lowland Peruvian Amazon and neighbouring areas. A similar naming system is found in Brazil: 'bacabal' *(Oenocarpus bacaba)* 'buritizal' *(Mauritia flexuosa),* 'acaizal' *(Euterpe oleracea),* and 'castanhal' *(Bertholetta excelsa)*, just to mention a few. Toponyms are also useful for indicating other aspects of the landscape that suggest historical occupation as well: 'Piquia dos negros,' 'Kalunda,' and 'macumbinha' all suggest historical Afro-Brazilian settlements, even if such sites today might not be inhabited.

Ribereños of the Ucayali River often consider zapotales (forests of zapote trees) to be especially ancient. The zapote (*Quararibea cordata*) is a tall emergent forest tree that reportedly can survive for hundreds of years. It yields a large, sweet fruit with bright orange pulp. Zapotales are most frequently found along paths that have been used for centuries by indigenous and non-indigenous people to collect forest resources or along varaderos, the important forest pathways that link one river system or watershed to another. These zapotales are believed to have originated centuries ago, but Amazonians today continue to manage them by either intentionally or accidentally dispersing the seeds (while eating or processing food), protecting the seedlings and juveniles in the forests through selective weeding, and occasionally by transplanting seedlings from forests to the edges of pathways, agricultural fields, or fallows or into house gardens. Amazonians tend to clear vines away from any zapote trees they encounter in forests. People not only value zapotes as fruit, but they also know that the fruits attract a variety of game animals, ranging from monkeys to tapirs. Since zapotales are known to be rich in wildlife, Amazonians will use them as preferred hunting sites. Zapote trees are believed to protect people during high winds and storms, because they do not easily break. Travellers therefore have yet another reason to choose zapotales as good places to spend the night when they are journeying through the forest and to contribute to their continual, informal management. Zapotales are particularly abundant in the forests of western Amazonia.

Numerous other fruit and multi-purpose tree species that are found in manchales can be considered indicators of past human residence or presence. The forest disturbances that shifting cultivators typically cause (such as creation of gaps, burning, selective weeding, and others) may also stimulate the growth and therefore creation of manchales of several valuable timber species. For instance, mahogany (*Swietenia macrophylla*), which is one of the most valuable Amazon timbers, is known to successfully reproduce only in clearings (Shono and Snook 2006; Snook and Negreros-Castillo 2004). Areas with dense stands of mahogany, such as some parts of the upper Purus River basin in Peru, may well have been cleared for agriculture in the past by indigenous farmers or during the active phase of the rubber period, creating ideal conditions for regeneration and managed recruitment of mahogany and other valuable species. Virola (*Virola surinamensis*), another highly valued timber species, also appears to thrive with human disturbance of forests. Scholars have suggested that abundant stands in areas such as the Rio Preto in Brazil's estuarine zone may show virola's positive response to fires. More specifically, the now well-known practice of managing swidden fallows or areas that were intensively farmed by slash-and-burn methods for a year or two and then managed as tree-dominated gardens of combined planted and spontaneous vegetation may well be the origin of many fruit as well as timber-rich forest plots (Coomes and Burt 1997; DeJong 1996; Denevan and Padoch 1987; Denevan et al. 1984; Freire 2007; Posey and Balick 2006; Posey and Plenderleith 2002). Amazonian swiddens are frequently interplanted with a great variety of crops that may include annuals such as maize and manioc, semi-perennials such as bananas and plantains, and trees including a great variety of fruits, palms, and other utilitarian species (Padoch et al. 1988; Coomes 2010) that develop over time into dense and diverse forest-gardens. As annual swidden cropping fades with time; mature swidden fallows typically feature an abundance of planted tree crops and managed natural regeneration. Some of the larger trees—such as fruits including Brazil nuts and zapote, or timbers such as tropical cedar (*Cedrela odorata*)—typically persist for many decades within the regrowing forest. The management of swidden fallows both by indigenous groups and by ribereños or caboclos have been described in a very large literature on New World tropical agricultures of which we cite only a few examples (Adams 2009; Balée 1994; Erickson and Balée 2006; Cormier 2006; Denevan 2001; Denevan and Padoch 1987; Denevan and Turner 1974; Domínguez and Gómez 1990; Erickson 2006; La Rotta Cuellar 1989; Pardo et al. 2004; Posey and Balée 1989).

Effects of human activity may be somewhat difficult to ascertain in some other unusual, low-diversity or 'oligarchic' (Peters et al. 1989) forests of the Amazon. For instance, the extensive forests dominated by various native bamboo species (*Guadua* spp.) that cover large areas especially in the state of Acre in the western Brazilian Amazon are believed by some experts to have originated from the clearing of the area by indigenous groups (Balée 2003), or to have specifically resulted from repeated burning. In cattle ranches in southern Para and Northern Mato Grosso, *Guadua* forests appeared in pastures that were repeatedly burned. These, however, could have reflected a soil seed stock derived from indigenous uses (for arrow and artisanal products). Under normal native uses, successional processes would suppress the rapid extension of *Guadua*, but the repeated burning and the cutting away of other successional species favoured *Guadua*, causing it to become a severe pasture pest that required abandonment of some pastures. A similar dynamic seems to be in play for babassu (*Orbignea phalerata*) invasions of pastures as well. However, some recent research has questioned those assumptions suggesting other mechanisms for the bamboos' dominance. It is, of course, difficult to determine with any certainty what initial disturbances might have put these forests first on the road to bamboo dominance.

On the other hand, the importance of human management is clearly evident in the estuarine forests that are dominated by the valuable açai (*Euterpe oleracea*) palm. The fruit of the açai has long been a staple of the caboclo diet in Amazonian Brazil. It has recently also become an important source of cash as the market, because açai has expanded to urban areas and into cities beyond Amazonia and now even overseas. Açaí palms are native to and abundant in the floodplain forests of the Amazon estuary. Both the density and the distribution of the palm in estuarine forests depend on environmental and anthropogenic factors (Brondizio et al. 2002). The application of a variety of management and planting strategies is increasingly transforming the varzea forests of the estuary and beyond into açaí agroforests or 'açaizais' in a process that has been called the açaization of the Amazon estuary (Hiraoka 1994; Brondizio 2008). Açai agroforests include stands under different types and intensities of management, with varying population densities, structures, species diversity and composition. Brondizio (2008) suggests that 'while at the plot level one may observe a decline in tree species diversity in the açaizais when compared to unmanaged floodplain forest, a broader landscape view including plots at different levels of management may show an increase of tree species diversity.'

From his research in the great estuarine island of Marajó, Brondizio (2008), describes three main methods of managing and/or creating açai agroforests: (1) management of native stands; (2) planting of açai stands following an annual cropping phase; and (3) combined management and planting in native stands (Table 4.2).

Table 4.2 Three management systems to produce açaí fruits in house gardens, fallows, and forests

Management system	Techniques	Comments
1. Selective forest enrichment system	1. Seed gathering of best phenotypes 2. Seedlings production in jirau (raised beds) 3. Cleaning under trees and planting seedlings	This system is mainly used to manage açaí in house gardens. Fruits are produced in 3 years.
2. Seed broadcasting forest management system	1. Broadcasting of seeds in newly open agricultural fields, jogo de semente 2. Select weeding, samear 3. Thinning, afastamento	This system is used to manage açaí in fallows. Fruit are produced in 5 years.
3. Forest stand management system	1. Thinning, afastamento 2. Cleaning, limpieça das toras 3. Removal of vines and shrubs, roça	This system is used to manage açaí in mature forests. Fruit production increases after 3 years of management.

Management of natural stands is done at both the stand and individual levels. As is typical in the management of other fruit-rich forest stands described above, açaí agroforesters often make openings in the canopy and thin the understory to promote the growth of açaí palms. They continue to thin and selectively weed the stands and may enrich them further by planting or transplanting additional seedlings. Managing at the individual level, açaí producers prune their palms to achieve an optimal number of stems per multi-stemmed palm clump.

When an açaizal is created from swidden or agricultural field, many of the same techniques are used. Fields are first cleared from forest, usually using fire, and in the ashes some combination of annual and perennial crops are planted. As annuals are harvested and gradually eliminated from the farmed plot, açaí seedlings (and other economic interesting plants like cacau and cupuassu) are planted in. Weeding and pruning are used to encourage the development of açaí seedlings and to control the density of stems per clump. As the plot matures these techniques continue to be used. Brondizio et al. (2002, pages 74–75) report that 'in unmanaged floodplain forests, açaí stems contribute less than 15% to total stand basal area, and represent less than 20% of total individuals. As management proceeds, this contribution tends to increase to a level of 50% of total biomass, and up to 90% of total number of individual in 5 years. It is interesting to note how the maintenance of basal area is at similar levels in both floodplain forests and açaí agroforestry. Management does not radically change stand biomass but instead influences which species contribute to it. Forest inventory data shows that basal area ranges from 29 to 31 m^2/ha in different types of floodplain forests (Pinedo-Vasquez and Zarin 1995; Peters et al. 1989). In açaí agroforestry, it ranges widely from 22 to 41 m^2/ha (Brondizio et al. 2002). It is important to emphasize that açaí stands vary very broadly in the type and intensity of their management, so that no 'standard' açaizal can be described. In the estuarine

Fig. 4.5 A highly productive açai stand in a house garden in Foz de Mazagão, Amapá, Brazil (Photo: Christine Padoch)

floodplain of the state of Amapá, caboclo households are increasingly producing large amounts of açai fruits in the extensive gardens that surround their houses (Fig. 4.5).

With the recent extraordinary expansion of the international açai market, various, more conventional plantation approaches to the planting and management of the palm are being developed and tried in the estuary. These 'scientific' or 'modern' plantings, which typically feature far less diversity of plant species than do traditionally managed stands, still constitute only a very small portion of all açai production in the Amazon estuary. Brondizio has noted that with the increasing value of the açai market the adequacy of local caboclo methods of managing as well as of marketing their traditional resource have been questioned, and caboclo açai management techniques either ignored or disparaged (Brondizio 2008). The substitution of unproven but purportedly scientific approaches to açai planting and production undoubtedly reflect this trend.

The booming urban and export market appears also to have stimulated considerable innovation in the system by promoting the selection and promotion of particular varieties of açai, notably a 'white' (*Açai branco*) variety, perhaps signalling an incipient domestication of the palm and more intensive manipulation of açai on a population or genetic level.

4.7   Conventional Silviculture in Amazonia

Silviculture for timber in the way people think about it in the temperate zone hardly exists in Amazonia. However, there have been some large pulp plantations on savanna lands such as those in Amapá. In this Amazonian region, the Jari Project embarked upon by Daniel Ludwig was based largely on the Asian pulp species, *Gmelina arborea,* and was a spectacular failure at the end of the day. In the 1970s and 1980s, there were some demonstration or experimental forest management and timber processing projects in Peru, Bolivia, Colombia, and Ecuador. For instance, BOLFor a sustainable forestry project in Bolivia, was carried out with funding by USAID. Similarly, the Pichis-Palcazu forest project, also financed by USAID, has experimented with low-damage timber extraction systems in the Upper Peruvian Amazon. There have also been many small community forest projects that have had diverse and overall dubious social, silvicultural, and ecological outcomes (De Pourco et al. 2009; Hoch et al. 2009; Larson et al. 2007; McCarthy 2005; Medina et al. 2009; Menzies 2007; Pacheco et al. 2010; Stearman 2006). On the whole the sector remains compromised by extremely high levels of corruption everywhere in the basin, with the World Wildlife Fund (WWF) estimating that more than 80% of timber in commerce is illegally harvested (Tacconi 2009).

There is a vastly growing literature on management of Amazonian forests for fruit (Miller and Nair 2006) and other non-timber products, but very little attention has been paid to the methods and techniques of management of forests for timber or construction materials by local residents (Guariguata et. al. 2010). Furthermore, little has been recorded of the knowledge that they may have of the phenology, regeneration, seed dispersal, and growth characteristics of tree species used for construction. In the next Sect. 4.8 we outline the informal forestry practices that perhaps are responsible for much of the wood production in current commerce.

4.8   Informal Forestry

Both indigenous Amazonians and their caboclo or ribereño counterparts are known to use a great variety of technologies for producing timber, fruits, and other forest products in their fields, fallows, forests, and house gardens (e.g., Padoch and Pinedo-Vasquez 2006; Sears et al. 2007). Although the specifics of each system may well be different and each deserves detailed study, a number of generalizations could be made that describe most management systems that have been observed. One characteristic of Amazonian silvicultural systems is that they tend not to segregate agriculture, agroforestry, and forestry products or practices, and they do not separate planted from spontaneously occurring useful plants. Timber species are often managed in agricultural fields; many stands rich in timber originate as agricultural plots (Fig. 4.6).

Fig. 4.6 Amazonian indigenous and non-indigenous farmers use selected weeding techniques to protect natural regeneration of valuable timber species in their agriculture fields (Photo: Miguel Pinedo-Vasquez)

Timber is also managed in diverse, mature forests, and still other timber trees are planted in house gardens. A diversity of species, purpose, origin, and age characterizes most systems. Secondly and related to the above, all useful species— crops, 'wild' trees, planted trees, etc.—in a particular plot tend to be managed concurrently through a series of complex and complementary processes (Padoch and Pinedo-Vasquez 2006; Pinedo-Vasquez and Zarin 1995; Sears and Pinedo-Vasquez 2004). The management techniques used by Amazonian foresters are typically part of elaborate systems that use and build upon the natural dynamics of vegetation and stand development and use the favourable aspects of each stage of land use. When a field is first cleared for agriculture, typically trees of useful species are spared and even protected from any burning that is done. These often include palm species used for thatch (like Inaja) and other fire-resistant cultivars of crops such as manioc. Subsequently, when crops are planted and weeded, seedlings of economic species encountered in the field are again spared and protected. Crops that are due to be harvested within 3 months or so are managed together with spontaneously occurring trees that may not be harvested as timbers for another 30 years or more. In indigenous production systems, researchers have noted circular plantings with different cultivation zones within which different planting (and burning) strategies are carried out.

This permits more targeted approaches while still maintaining polycropping patterns and diversity in the types of cultivars (Hecht and Posey 1989).

After 2 years or so, intensive weeding tends to be reduced as annual crops are eliminated and semi-perennial and perennial crops come to predominate. At this point the farmers typically leave more natural vegetation to regenerate, which, among its other functions, helps recuperate fertility of the soil. The growth of timber and other useful woody species is encouraged during this transition from field to fallow, particularly during the final season of intensive cropping when farmers encourage or assist in the development of valuable tree seedlings that establish through natural regeneration or are planted, by weeding around the seedlings or freeing them from vines. Fallow management for timber production focuses on juveniles and on the removal of selected vines, shrubs, and trees to create the gradients of light and humidity necessary for the natural regeneration of a diversity of timber species.

Timber management in fields and house gardens also focuses on the protection of seed producer trees and seedlings. Selected seeds and seedlings from these trees are planted or transplanted into more hospitable environments. Seedlings are managed by weeding and are protected from insects, rodents, and floods. The house garden is also an important component of the system, serving as nursery and experimental plot where farmers test selected varieties and species of plants, including timber species. This function of dooryard gardens has been widely documented but very much underappreciated.

This process of fallow enrichment allows farmers to increase the utility and option value of their landholdings while allowing for the necessary recuperation of soil fertility. When one annual crop field is allowed to fallow, Amazonian farmers typically open another field, usually from an old fallow. A trend noted of late in some regions, including the açai-growing estuarine floodplain, however, is a decrease in the area dedicated to annual crops and an increase in area of production forest (Brondizio 1999).

The various smallholder traditional silvicultural systems for managing the seedlings, juveniles, and adult trees of valuable species vary significantly from species to species. Valuable species that are fast-growing are managed differently from the one ones that grow more slowly. Similarly, valuable species that are light-tolerant are managed differently from species that prefer shade. When managing fast-growing timber species, smallholders encourage competition among seedlings because a small number of seedlings (approximately 1 for every 100 seedlings) actually reach the juvenile stage. Amazonians actively thin stands of juveniles of fast-growing species by removing the bark around the stem. This technique is known in the Peruvian Amazon as anillado. The anillado technique helps to control gap formation and sunlight to avoid the establishment of vines and other weedy species. Smallholders manage adult individuals of valuable fast-growing species by repeatedly removing vines and termite nests; local farmer-foresters report that vines and termite nests are the main causes of damage to both wood and fruit production in fast-growing valuable species.

In managing slow-growing species, smallholders selectively weed to protect seedlings that have naturally regenerated or were transplanted in fields, fallows, or forests.

In most cases, the seedlings of slow-growing species cannot compete successfully with seedlings of fast-growing species; therefore, smallholders tend to remove all competing vegetation (including seedlings of fast-growing species) from around the seedlings of the slow growers. Since the natural regeneration of slow-growing species is far lower than that of fast-growing trees, Amazonian farmer-foresters tend to transplant seedlings of the slow growers to enrich their fields, fallows, and forests. Smallholders tend to weed around the juveniles of fast-growing species, especially to control the growth of vines and other fast-growing weedy species. On the whole, Amazonians tend to put much more effort into managing the juveniles of slow-growing timber species than those of fast-growing species. Smallholders annually selectively clear around adult trees of the slow-growing timbers, largely to reduce competition for the natural regeneration of the seedlings of these valuable species.

Amazonians encourage the regeneration of light-tolerant species in their forests by removing emergent trees with particular types of canopies, especially those that are umbrella-shaped; regeneration of those timber species that are not light-tolerant is promoted by leaving two or three seedlings of *Cecropia* or other pioneer species nearby to provide the requisite amount of shade. In some cases small-holders protect the seedlings of light-intolerant species by keeping banana or other perennial or semi-perennial crops in the field specifically for the shade that they provide.

We have observed that Amazonian smallholders encourage stand development by removing vines, opening gaps, and removing any individual trees that are mis-shapen or have large branches. In forest stands where species of fast-growing timber predominate, Amazonian farmer-foresters do the thinning by harvesting some individuals that can be used for any number of purposes, and/or by eliminating individuals that show any deformations or are not producing fruits. In forest stands where slow-growing species are dominant, smallholders tend to thin by eliminating any individual trees that are deformed or, in the case of fruit trees, those that have been less productive. In most cases stand management includes the simultaneous management of seedlings, juveniles, and adult trees of many species.

In the Peruvian Amazon, farmers manage seedlings in forest stands using two principal forest management techniques: huactapeo and jaloneo.

- Huactapeo is a technique that consists of three operations: selective weeding, cleaning (including the removal of the roots and stems of species that regenerate by sprouting), and controlled burning (including the elimination of regeneration materials such as the roots and stems of sprouting species by burning).
- Jaloneo is a thinning technique that includes the selection of healthy, well-formed seedlings, and uprooting of seedlings with any defects. By uprooting seedlings, managers reduce the probability of their resprouting.

Amazonians also manage the juveniles of valuable timber species along with any planted and protected fruit and other valuable species. While in Brazil smallholders use the afastamento system (enrichment), in Peru farmers use the manchal system (enrichment) to manage juveniles of timber and non-timber species.

The manchal system includes three specific techniques, termed huactapeo, raleo, and mocheado.

- Huactapeo is essentially selective weeding with a focus on removing vines, particularly woody vine species that are believed by farmer-foresters to be the main killers of juveniles.
- Raleo is a thinning technique that consists of removing deformed and diseased juveniles. In communities near urban centres farmers also use raleo to harvest some juveniles to sell as poles in urban markets.
- Mocheado is a pollarding technique that consists of cutting back the top branches of individuals of species that have umbrella-type branching, such as many *Inga* species. Mocheado is done to open the canopy of the stand and let more light into the stand and speed the development in height of juveniles of valuable species.

Overall the three techniques involve selection, marking, pollarding, pruning, liberation, and thinning activities; and they serve to remove undesirable individuals. By removing less desirable individuals from the stands, space and light are made available for growth in height and diameter of stems of the juveniles of valuable tree species. Farmers also use a technique called huahuancheo. This technique consists in cleaning and pruning of the best juveniles or adult trees that are then harvested to sell in the market.

In many cases, farmers begin harvesting fast-growing trees in 4 year fallows and again harvest in the 6th and 8th years. We have observed that in some cases, for instance in stands rich in capirona (*Calycophyllum spruceanum*) or bolaina (*Guazuma crinita*), two marketable fast-growing species, immediately after harvest farmers make new fields to plant agricultural crops, but also to encourage natural regeneration of the tree species (Fig. 4.7). This case is more common in regions where cheap poles and small boards are in high demand (Sears et al. 2007; Padoch et al. 2008).

Two management techniques commonly used to manage adult trees are anillado and desangrado. These techniques are applied on an average of once every 6–8 years.

- Anillado involves the killing of selected stems of competitor species using girdling and fire. Its application causes the tree to die rapidly and avoids resprouting from the roots or stem. Amazonians use it to kill stems of large species that are difficult to control.
- Desangrado is a commonly used girdling technique. Using desangrado, Amazonian farmers remove small stems of competitor species and individuals of stranglers and woody climbing vines. Desangrado involves two operations: selection, in which all individuals of vine and other species that are climbing or strangling the trunk and/or covering the canopy of valuable species are selected for removal; and girdling, which involves the removal of bark, cambium, and sapwood in a ring extending around the selected individual at the bottom of its trunk. From the ring fissure, sap, resins and water are lost. The abundance of resin or sap attracts ants, termites, and other insects that not only extract the sap or resin but also damage any new sprouts. The infestation of insects thus controls the sprouting of vines and helps kill them.

Fig. 4.7 Managed swidden fallows rich in fast-growing timbers (e.g. *Calycophyllum spruceanum* and *Guazuma crinita*) exist throughout Amazonia (Photo: Miguel Pinedo-Vasquez)

The suite of traditional management techniques outlined above is used to increase the value of forests. The application of these techniques results in a significant increase in the commercial volume per hectare of timber. For instance, the mean commercial volume of managed stands rich in capinuri (*Maquira coreacea*), known as capinurales, was 81 m³/ha for areas that had been managed as mature capinural forests for 8 years, 89 m³/ha for those managed for 16 years, and 85 m³/ha for those managed for 24 years. All the above values were significantly higher than the estimated mean of 54 m³/ha for unmanaged forests in the Peruvian Amazon (Pinedo-Vasquez 1995).

The management of forests for the production of both timber and non-timber products is cyclically linked to agricultural management. When the rate of growth of the timber trees begins to decline, the stand is slashed and a new agricultural phase begins. This cyclical characteristic of forestry practiced by indigenous and non-indigenous Amazonians, makes it particularly different from conventional forestry.

The combination or 'hybridization' of conventional forestry with Amazonian practices is little discussed and has been little explored in the literature. The example we described above (Sect. 4.4) of local farmers who successfully integrated exogenous technology and knowledge gained while working for the large-scale timber industry, with local knowledge of the environment and their silvicultural practices to develop a lucrative local industry in Amapá (Sears et al. 2007), is certainly not unique. The study of such examples, perhaps like the practices themselves, has been neglected because it falls between forestry research categories.

4.9   Forests, Forestry, and the Future of Amazonia

Forest types in Amazonia are a good deal more complex that the standard literature would suggest, and local knowledge systems show a much more elaborate set of classification systems that take in more environmental and ecological complexity than non-native classification systems. This complexity of forest types reflects the degree of understanding of forest histories. Recent studies in archaeology, pollen analysis, ethnography, environmental history, and soil science is leading to an understanding of Amazonian environmental history that views the region as having been substantively modified by human agency in Pre Colombian times and dramatically affected by different booms. In addition, a number of diasporas as well as economic pressures have shaped patterns of exploitation, production techniques, and resource strategies. What is emerging from the ethnographic approaches to resource uses is a sense of an array of techniques ranging from individual species to landscapes that has framed Amazonia's biota in ways that make it very difficult to untangle the 'wild' from the tame, and in many cases, these are not even useful categories. What is clear is that there are suites of management techniques that provide for income and resilience, and protect and enhance diversities, while maintaining biomass through successional processes at the landscape level. These knowledge systems certainly merit greater attention for the longer term, especially given the uncertainties about the impacts of climate change in tropical regions and their pivotal role in global climate mediation.

Outside observers often assume that local tropical people do not know how to manage forests and that the logs and veneers were harvested from 'wild trees' that presumably were in place from time immemorial. We have shown the diversity of knowledge about the management of trees and landscapes in Amazonia, and this huge library of vernacular knowledge could go far in improving forestry in Amazonia. The history of modern forestry with its emphasis on monocultures and rapid yields may yet be successful in some areas of Amazonia, but overall the successes of the formal forestry sector have been open to question. Local knowledge systems that are capable of producing timber and many other commodities, as well as environmental services, merit much more attention. The forms of knowledge and the capacity for innovation in vernacular science deserve to be viewed with a great deal of thoughtfulness rather than dismissing the complexity of such knowledge because it doesn't match the frameworks of conventional forestry, which for Amazonia has largely a history of failure.

Timber is without question the most valuable biotic commodity coming out of Amazonia at this time, and given the prevalence of local extinctions and expanding demand, rethinking forestry in Amazonia is overdue. The education of new foresters in the value of the current practices, along with analyses of the dynamics that produce economically diverse and ecologically complex forest systems, suggests a recasting of pedagogy as well as practice.

The process of integrating the principles, methods, and practices of traditional forest management into training a new generation of foresters is still fragmented despite the existing volume of technical information. Some Amazonian universities have integrated agroforestry, ecology, and other related training courses in their syllabi, but most of these courses are based on conventional principles, methods, and practices rather than on the plethora of knowledge and experiences of smallholder forestry practices. Lack of integration of traditional forest management practices is expressed by the use of conventional techniques in reforestation programmes, rather than the rich diversity of techniques that are equally practiced by indigenous and non-indigenous Amazonians. Similarly, forest management plans designed by smallholders are still not recognized as technical documents by the public forest sector in most Amazonian countries. The above examples are only two of the reasons why a new generation of foresters needs to be trained to use and apply traditional forest management principles, methods, and practices. An accumulating body of knowledge and emerging insights into Amazonian patterns of forest management should help foresters, in both Amazonia and elsewhere, to liberate their thinking and their training from long and complete domination by foreign, colonial ideas of what forestry means and how it needs to be done.

References

Adams C (2009) Amazon peasant societies in a changing environment political ecology, invisibility and modernity in the rainforest. Springer, London

Alexiades MN (1999) Ethnobotany of the Ese Eja: plants, health, and change in an Amazonian society. Dissertation, City University of New York, New York

Alexiades MN (2009) Mobility and migration in indigenous Amazonia: contemporary ethnoecological perspectives. Berghahn, Oxford

Almeida MWdB (2004) Direitos à Floresta e Ambientalismo: os seringueiros e suas lutas. Rev Bras Ci Soc 19(55):35–52

Amorozo MCVM, Gély A (1988) Uso de plantas medicinais por caboclos de Baixo Amazonas, Barcarena, PA, Brasil. Boletim do Museu Paraense Emilio Goeldi. Ser Bot 4(1):47–131

Anderson AB, Posey DA (1989) Management of a tropical scrub savanna by the Gorotire Kayapo. Adv Econ Bot 7:159–173

Auricchio M, Vicente JP, Meyer D, Mingroni-Netto RC (2007) Frequency and origins of hemoglobin S mutation in African-derived Brazilian populations. Hum Biol 79:667–677

Ayres MJ (1996) Mamirauá management plan. SCM/CNPq/MCT, Brasília

Balée WL (1994) Footprints of the forest: Ka'apor ethnobotany—the historical ecology of plant utilization by an Amazonian people. Columbia University Press, New York

Balée WL (2003) Native views of the environment in Amazonia. In: Selin H (ed) Nature across cultures: views of nature and the environment in non-western cultures. Kluwer, Dordrecht, pp 277–279

Balée WL, Erickson CL (2006) Time and complexity in historical ecology: studies in the neotropical lowlands. Columbia University Press, New York

Barbieri AF, Carr DL, Bilsborrow RE (2009) Migration within the frontier: the second generation colonization in the Ecuadorian Amazon. Popul Res Policy Rev 28:291–320

Barbosa RI, Fearnside PM (2005) Fire frequency and area burned in the Roraima savannas of Brazilian Amazonia. For Ecol Manag 204:371–384

Barham BL, Coomes OT (1994) Reinterpreting the Amazon rubber boom—investment, the state, and Dutch disease. Lat Am Res Rev 29:73–109

Bennett BC (1992) Plants and people of the Amazonian rainforests: the role of ethnobotany in sustainable development. Bioscience 42(8):599–607

Bennett BC, Prance GT (2000) Introduced plants in the indigenous pharmacopoeia of northern South America. Econ Bot 54:90–102

Bertani S, Bourdy G, Landau I, Robinson JC, Esterre P, Deharo E (2005) Evaluation of French Guiana traditional antimalarial remedies. J Ethnopharmacol 98:45–54

Blate GM (2005) Modest trade-offs between timber management and fire susceptibility of a Bolivian semi-deciduous forest. Ecol Appl 15:1649–1663

Boom BM (1987) Ethnobotany of the Chacobo Indians, Beni, Bolivia, vol 4, Advances in economic botany. New York Botanical Garden Press, New York

Brondizio ES (1999) Agroforestry intensification in the Amazon estuary. In: Granfelt T (ed) Managing the globalized environment: local strategies to secure livelihoods. IT Publications, London, pp 88–113

Brondizio ES (2006) Landscapes of the past, footprints of the future: historical ecology and the analysis of land use change in the Amazon. In: Balée W, Erickson C (eds) Time and complexity in historical ecology: studies in the neotropical lowlands. Columbia University Press, New York, pp 365–405

Brondizio ES (2008) The Amazonian caboclo and the açaí palm: forest farmers in the global market. New York Botanical Garden Press, New York

Brondizio ES, Safar CAM, Siqueira AD (2002) The urban market of açaí fruit (*Euterpe oleracea* Mart.) and rural land use change: ethnographic insights into the role of price and land tenure constraining agricultural choices in the Amazon estuary. Urban Ecosyst 6(1–2):67–97

Carneiro RL (1978) The knowledge and use of rain forest trees by the Kuikuru Indians of Central Brazil. In: Hemming J (ed) Change in the Amazon Basin, vol 1. Manchester University Press, Manchester, pp 70–89

Carneiro RL (1983) The cultivation of manioc among the Kuikuru of the upper Xingu. In: Vickers W, Harnes R (eds) Adaptive responses of Native Amazonians. Academic, New York, pp 65–111

Carney JA (2001) Black rice: the African origins of rice cultivation in the Americas. Harvard University Press, Cambridge

Carney J, Rosomoff R (2009) Seeds of slavery. University of California Press, Berkeley

Castro EBVD, da Cunha MC, Dreyfus S (1993) Núcleo de história indígena e do indigenismo. Fundação de Amparo à Pesquisa do Estado de São Paulo, São Paulo

Church GE (1912) Aborigines of South America. Chapman & Hall, London

Clement CR (1998) A center of crop genetic diversity in western Amazonia: a new hypothesis of indigenous fruit-crop distribution. Bioscience 39(9):624–631

Clement CR (1999) 1492 and the loss of Amazonian crop genetic resource: the relation between domestication and human population decline. Econ Bot 53(2):188–202

Coimbra-Junior CEA (2004) The Xavâante in transition: health, ecology, and bioanthropology in central Brazil. University of Michigan Press, Ann Arbor

Coomes OT (2010) Of stakes, stems and cuttings: the importance of local seed systems in traditional Amazonian societies. Prof Geogr 62(3):323–334

Coomes OT, Burt GJ (1997) Indigenous market-oriented agroforestry: dissecting local diversity in western Amazonia. Agrofor Syst 37:27–44

Coomes OT, Abizaid C, Lapointe M (2009) Human modification of a large meandering Amazonian river: genesis, ecological and economic consequences of the Masisea cutoff on the central Ucayali, Peru. Ambio 38(3):130–134

Cormier L (2006) Between the ship and the bulldozer: Guaja subsistence, sociality and symbolism after 1500. In: Balée WL, Erickson C (eds) Time and complexity in historical ecology: studies in the neotropical lowlands. Columbia University Press, New York, pp 341–364

Craig NB (1907) Recollections of an ill-fated expedition to the head-waters of the Madeira River in Brazil. Lippincott, Philadelphia

Crosby AW (1986) Ecological imperialism: the biological expansion of Europe, 900–1900. Cambridge University Press, Cambridge

Da Cunha MC (1984) Sobre os silêncios da lei: lei costumeira e positiva nas alforrias de escravos no Brasil do século XIX. Unicamp, Campinas

Da Cunha MC (1992) História dos índios no Brasil. Fundação de Amparo à Pesquisa do Estado de São Paulo, Companhia das Letras, Secretaria Municipal de Cultura, São Paulo

De Mello e Souza L (1996) Violencia e practicas culturais no cotidiano de uma expedicao contra quilambolas. In: Reis JJ, Gomes FdS (eds) Liberdad por um fio. Companhia das Lettras, São Paulo, pp 193–212

De Pourco K, Thomas E, Van Damme R (2009) Indigenous community-based forestry in the Bolivian lowlands: some basic challenges for certification. Int For Rev 11:12–26

De Toledo MB, Bush MB (2007) A mid-Holocene environmental change in Amazonian savannas. J Biogeogr 34:1313–1326

De Toledo MB, Bush MB (2008a) A Holocene pollen record of savanna establishment in coastal Amapa. An Acad Bras Cienc 80:341–351

De Toledo MB, Bush MB (2008b) Vegetation and hydrology changes in eastern Amazonia inferred from a pollen record. An Acad Bras Cienc 80:191–203

Dean W (1987) Brazil and the struggle for rubber: a study in environmental history. Cambridge University Press, New York

DeJong W (1996) Swidden-fallow agroforestry in Amazonia: diversity at close distance. Agrofor Syst 34:277–290

Denevan WM (1964) The native population of the Americas in 1492. University of Wisconsin Press, Madison

Denevan WM (1970) Aboriginal drained-field cultivation in the Americas. Science 169:647–654. doi:10.1126/science.169.3946.647

Denevan WM (2001) Cultivated landscapes of Native Amazonia and the Andes. Oxford University Press, Oxford

Denevan WM, Padoch C (1987) Swidden-fallow agroforestry in the Peruvian Amazon. New York Botanical Garden, New York

Denevan WM, Turner BL (1974) Forms, functions and associations of raised fields in old world tropics. J Trop Geogr 39:24–33

Denevan WM, Treacy JM, Alcorn JB, Padoch C, Denslow J, Paitan SF (1984) Indigenous agroforestry in the Peruvian Amazon—Bora Indian management of swidden fallows. Interciencia 9:346–357

Domínguez C, Gómez A (1990) La economía extractiva en la Amazonia colombiana, 1850–1930. TROPENBOS, Bogota

Dos Santos GF (2002) A 'safe haven': runaway slaves, Mocambos, and borders in colonial Amazonia, Brazil. Hisp Am Hist Rev 82:469–498

Duke JA, Vasquez R (1994) Amazonian ethnobotanical dictionary. CRC Press, Boca Raton

Eden MJ, Bray W, Herrera L, McEwan C (1984) Terra-preta soils and their archaeological context in the Caqueta basin of southeast Colombia. Am Antiq 49:125–140

Elisabetsky E, Posey DA (1989) Use of contraceptive and related plants by the Kayapo Indians (Brazil). J Ethnopharmacol 26:299–316

Ellen RF, Parkes P, Bicker A (2000) Indigenous environmental knowledge and its transformations: critical anthropological perspectives. Harwood Academic, Amsterdam

Erickson CL (2006) Domesticated landscapes of the Bolivian Amazon. In: Balée WL, Erickson CL (eds) Time and complexity in historical ecology: studies in the neotropical lowlands. Columbia University Press, New York, pp 235–278

Erickson CL, Balée WL (2006) Historical ecology of complex landscapes of the Bolivian Amazon. In: Balée WL, Erickson CL (eds) Time and complexity in historical ecology: studies in the neotropical lowlands. Columbia University Press, New York, pp 187–233

Escalante A (1981) Palenques in Colombia. In: Price R (ed) Maroon societies. Johns Hopkins, Baltimore

Etter A, McAlpine C, Possingham H (2008) Historical patterns and drivers of landscape change in Colombia since 1500: a regionalized spatial approach. Ann Assoc Am Geogr 98:2–23

Ferreira MR (2005) A ferrovia do diabo. Melhoramentos, São Paulo

Fleck DW, Harder JD (2000) Matses Indian rainforest habitat classification and mammalian diversity in Amazonian Peru. J Ethnobiol 20(1):1–36

Food and Agriculture Organisation [FAO] (2001) Global forest resource assessment, 2000, vol 140, FAO Forestry Paper. FAO, Rome

Frechione J, Posey DA, da Silva LF (1989) The perception of ecological zones and natural resources in the Brazilian Amazon: an ethnoecology of lake Coari. Adv Econ Bot 7:260–282

Freire GN (2007) Indigenous shifting cultivation and the new Amazonia: a Piaroa example of economic articulation. Hum Ecol 35:681–696

Frost H (2005) De-coca-colonization: making the globe from the inside out. Antipode 37:384–387

Funes E (1995) Nasci na matas nuca tive senhor: historia e memoria de mocambos do Baixo Amazonas. Universidade de São Paulo, São Paulo

Funes E (2004) Mocambos do Trombetas: historia memoria e identidade. EAVirtual (Barcelona) 1(1):5–25

Furley PA, Ratter JA, Procter J (1992) Nature and dynamics of forest-savanna boundaries. Chapman a& Hall, New York

Ganson BA (2003) The Guaraní under Spanish rule in the Río de la Plata. Stanford University Press, Stanford

GeoAmazonia (2009). Environment outlook in Amazonia. United Nations Development Programme [UNDP], Nairobi

Gilmore MP (2005) An ethnoecological and ethnobotanical study of the Maijuna Indians of the Peruvian Amazon. Dissertation, Miami University, Miami

Gomes FdS (1999a) Fronteiras e Mocambos. In: FdS G (ed) Nas terras do Cabo Norte: fronteiras, colonização, escravidão na Guyana Brasiliera XVIII-XIX. UFPA, Belém, pp 225–318

Gomes FdS (1999b) Outras paisagens coloniais: notas sobre desertores militares na Amazonia Setecentista. In: FdS G (ed) Nas terras do Cabo Norte: fronteiras, colonização, escravidão na Guyana Brasiliera XVIII-XIX. UFPA, Belem, pp 196–223

Gomes FdS (2005) A hidra e os pântanos: mocambos, quilombos e comunidades de fugitivos no Brasil (séculos XVII-XIX). UNESP, São Paulo

Guariguata MR, Garcia-Fernandez C, Sheil D, Nasi R, Herrero-Jauregui C, Cronkleton P, Ingram V (2010) Compatibility of timber and non-timber forest product management in natural tropical forests: perspectives, challenges, and opportunities. For Ecol Manag 259:237–245

Halme KJ, Bodmer RE (2007) Correspondence between scientific and traditional ecological knowledge: rain forest classification by the non-indigenous ribereños in Peruvian Amazonia. Biodivers Conserv 16:1785–1801

Hammond DS, ter Steege H, van der Borg K (2007) Upland soil charcoal in the wet tropical forests of central Guyana. Biotropica 39:153–160

Hardenburg WE, Enock CR (1913) The Putumayo, the devil's paradise; travels in the Peruvian Amazon region and an account of the atrocities committed upon the Indians therein. Unwin, London

Hardman-Foot F (1988) Trem fantasma: a modernidade na selva. Cia das Letras, São Paulo

Harris DR (1980) Human ecology in savanna environments. Academic, London

Hecht SB (2003) Indigenous management and the creation of Amazonian dark earths: implications of Kayapo practices. In: Lehman J, Kern D, Glaser B, Woods W (eds) Amazonian dark earths: origins, properties, management. Kluwer, Dordrecht, pp 355–373

Hecht SB (2005) Soybeans, development and conservation on the Amazon frontier. Dev Chang 36:375–404

Hecht SB (2009) Kayapó savanna management: fire, soils, and forest islands in a threatened biome. In: Woods WI, Teixeira WG, Lehmann J, Steiner C, WinklerPrins AMGA, Rebellato L (eds) Amazonian dark earths: wim sombroek's vision. Springer, Heidelberg, pp 143–162

Hecht SB (2010) Amazon oddysey: Euclides da Cunha and the scramble for the Amazon. University of Chicago Press, Chicago

Hecht SB, Cockburn A (1989) The fate of the forest: developers, destroyers, and defenders of the Amazon. Verso, London

Hecht SB, Mann CC (2008) How Brazil outfarmed the American farmer. Fortune 157:92–105

Hecht SB, Posey D (1989) Preliminary results on Kayapo soil management techniques. Adv Econ Bot 7:174–188

Hecht SB, Norgaard RB, Possio G (1988) The economics of cattle ranching in eastern Amazonia. Interciencia 13:233–240

Heckenberger M (2005) The ecology of power: culture, place, and personhood in the southern Amazon, A.D. 1000–2000. Routledge, New York

Heckenberger MJ, Peterson JB, Neves EG (1999) Village size and permanence in Amazonia: two archaeological examples from Brazil. Lat Am Antiq 10:353–376

Heckenberger MJ, Russell JC, Fausto C, Toney JR, Schmidt MJ, Pereira E, Franchetto B, Kuikuro A (2008) Pre-columbian urbanism, anthropogenic landscapes, and the future of the Amazon. Science 321:1214–1217

Hemming J (1978) Red gold: the conquest of the Brazilian Indians. Harvard University Press, Cambridge

Hemming J (1987) Amazon frontier: the defeat of the Brazilian Indians. Macmillan, London

Hemming J (2003) Die if you must: Brazilian Indians in the twentieth century. Macmillan, London

Henderson A (1995) The palms of the Amazon. Oxford University Press and World Wildlife Fund, New York

Hill JD (1988) Rethinking history and myth: indigenous South American perspectives on the past. University of Illinois Press, Urbana

Hill JD, Santos-Granero F (2002) Comparative Arawakan histories: rethinking language family and culture area in Amazonia. University of Illinois Press, Urbana

Hiraoka M (1994) Mudanças nos padrões econômicos de uma população ribeirinha do estuário do Amazonas. In: Furtado L, Mello AF, Leitão W (eds) Povos das Águas: realidade e perspectivas na Amazônia. MPEG/Universidade Federal do Para, Belém, pp 133–157

Hoch L, Pokorny B, de Jong W (2009) How successful is tree growing for smallholders in the Amazon? Int For Rev 11:299–310

Jackson J (1999) The politics of ethnographic practice in the Colombian Vaupes. Identities Global Stud Cult Power 6:281–317

Jepson W (2006) Private agricultural colonization on a Brazilian frontier, 1970–1980. J Hist Geogr 32:839–863

Jullian V, Bourdy G, Georges S, Maurel S, Sauvain M (2006) Validation of use of a traditional antimalarial remedy from French Guiana, Zanthoxylum rhoifolium Lam. J Ethnopharmacol 106:348–352

Kahn F, Granville J-Jd (1992) Palms in forest ecosystems of Amazonia. Springer, Berlin

Karasch M (1996) Os quilombos do ouro na Capitania de Goias. In: Gomes FdS, Reis JJ (eds) Liberdad por um fio. Companhia das Lettras, São Paulo, pp 240–262

Kerr WE, Posey DA (1984) New information on Kayapo agriculture. Interciencia 9:392–400

La Rotta CC (1989) Estudio ethnobotánico sobre las especies utilizadas por la comunidad indigena Miraña, Amazonas-Colombia. World Wildlife Fund, FEN-Colombia, Bogotá

La Torre-Cuadros MD (2008) One hundred twelve years of scientific research on ethnic groups in the Peruvian Amazon. Boletin Latinoamericano y Del Caribe De Plantas Medicinales y Aromaticas 7:171–179

Larson AM, Pacheco P, Toni F, Vallejo M (2007) Trends in Latin American forestry decentralisations: legal frameworks, municipal governments and forest dependent groups. Int For Rev 9:734–747

Laurie N, Andolina R, Radcliffe S (2005) Ethnodevelopment: social movements, creating experts and professionalising indigenous knowledge in Ecuador. Antipode 37:470–496

Lavalle B (1999) Frontiers, colonization and Indian manpower in Andean Amazonia (16th-20th centuries): the construction of Amazon socioeconomic space in Ecuador, Peru and Bolivia (1792–1948). Caravelle-Cahiers du Monde Hispanique et Luso-Bresilien 2:315–316

Lehmann J, Kern DC, Glaser B, Woods WI (eds) (2006) Amazonian dark earths: origin, properties, management. Springer, New York

Liang B, Lehmann J, Solomon D, Sohi S, Thies JE, Skjemstad JO, Luizao FJ, Engelhard MH, Neves EG, Wirick S (2008) Stability of biomass-derived black carbon in soils. Geochimica Et Cosmochimica Acta 72:6069–6078

Lima HN, Schaefer CER, Mello JWV, Gilkes RJ, Ker JC (2002) Pedogenesis and pre-Colombian land use of "Terra Preta Anthrosols" ("Indian black earth") of western Amazonia. Geoderma 110:1–17

Macia MJ (2008) Woody plants diversity, floristic composition and land use history in the Amazonian rain forests of Madidi National Park, Bolivia. Biodivers Conserv 17:2671–2690

Malleux JO (1983) Inventario forestales en bosques tropicales. Universidad Nacional Agraria La Molina, Lima

Mann CC (2005) New revelations of the Americas before Columbus. Knopf, New York

Martínez LJ (1998) Suelos de la Amazonia. Ministerio de Educación Nacional, Santa Fe de Bogotá

McCarthy J (2005) Devolution in the woods: community forestry as hybrid neoliberalism. Environ Plann A 37:995–1014

McDaniel J, Kennard D, Fuentes A (2005) Smokey the tapir: traditional fire knowledge and fire prevention campaigns in lowland Bolivia. Soc Nat Resour 18:921–931

McEwan C (2001) Unknown Amazon: culture in nature in ancient Brazil. British Museum Press, London

McKey D, Emperaire L, Elias M, Pinton F, Robert T, Desmouiliere S, Rival L (2001) Local management and regional dynamics of varietal diversity of cassava in Amazonia. Genet Sel Evol 33:S465–S490

Medina G, Pokorny B, Campbell B (2009) Loggers, development agents and the exercise of power in Amazonia. Dev Chang 40:745–767

Meggers BJ (1954) Environmental limitation on the development of culture. Am Anthropol 56(5):801–824

Meggers BJ (1971) Amazonia: man in a counterfeit paradise. Aldine-Atherton Publisher, Chicago

Menzies NK (2007) Our forest, your ecosystem, their timber: communities, conservation, and the state in community-based forest management. Columbia University Press, New York

Miller RP, Nair PKR (2006) Indigenous agroforestry systems in Amazonia: from prehistory to today. Agrofor Syst 66:151–164

Mistry J (2001) Savannas. Prog Phys Geogr 25:552–559

Mistry J, Berardi A (2006) Savannas and dry forests: linking people with nature. Ashgate, Aldershot

Moran E (1993) The human ecology of Amazonian populations. University of Iowa Press, Iowa City

Nepstad D, Carvalho G, Barros AC, Alencar A, Capobianco JP, Bishop J, Moutinho P, Lefebvre P, Silva UL, Prins E (2001) Road paving, fire regime feedbacks, and the future of Amazon forests. For Ecol Manag 154:395–407

Nugent S, Harris M (2004) Some other Amazonians: perspectives on modern Amazonia. Institute for the Study of the Americas, London

Pacheco P (2006) Agricultural expansion and deforestation in lowland Bolivia: the import substitution versus the structural adjustment model. Land Use Policy 23:205–225

Pacheco P, de Jong W, Johnson J (2010) The evolution of the timber sector in lowland Bolivia: examining the influence of three disparate policy approaches. Forest Policy Econ 12:271–276

Padoch C, Pinedo-Vasquez M (2000) Farming above the flood in the várzea of Amapa: some preliminary results of the Projeto Várzea. In: Padoch C, Ayres JM, Pinedo-Vasquez M, Henderson A (eds) Varzea: diversity, development, and conservation of Amazonia's Whitewater Floodplain. New York Botanical Garden Press, New York, pp 345–354

Padoch C, Pinedo-Vasquez M (2006) Concurrent activities and invisible technologies: an example of timber management in Amazonia. In: Posey DA (ed) Human impacts on the Amazon: the role of traditional ecological knowledge in conservation and development. Columbia University Press, New York, pp 172–180

Padoch C, Chota Inuma J, de Jong W, Unruh J (1988) Marketed oriented agroforestry at Tamshiyacu. In: Denevan WM, Padoch C (eds) Swidden-fallow agroforestry in the Peruvian Amazon, vol 5, Advances in Economic Botany. New York Botanical Garden, New York, pp 90–96

Padoch C, Brondizio E, Costa S, Pinedo-Vasquez M, Sears RR, Siqueira A (2008) Urban forest and rural cities: multi-sited households, consumption patterns, and forest resources in Amazonia. Ecol Soc 13(2):2, Available via http://www.ecologyandsociety.org/vol13/iss2/art2/

Pardo M, Mosquera C, Ramírez MC (2004) Panorámica afrocolombiana: estudios sociales en el Pacífico. Instituto Colombiano de Antropología e Historia, Universidad Nacional de Colombia, Bogotá

Peters CM, Balick MJ, Kahn F, Anderson AB (1989) Oligarchic forests of economic plants in Amazonia: utilization and conservation of an important tropical resource. Conserv Biol 3(4):341–349

Peterson J, Neves E, Heckenberger M (2001) Gift from the past: terra preta and the prehistoric occupation of the Amazon. In: McEwan C, Barretos C, Neves E (eds) Unknown Amazon. British Museum, London, pp 86–105

Pinedo-Vasquez M (1995) Human impact on varzea ecosystems in the Napo-Amazon, Peru. Dissertation, Yale University School of Forestry and Environmental Science, New Haven

Pinedo-Vasquez M, Zarin D (1995) Local management of forest resources in a rural community of Northeast Peru. In: Nishizawa T, Uitto J (eds) Fragile tropics of Latin America: changing environments and their sustainable management. United Nations University Press, Tokyo

Pinedo-Vasquez M, Zarin D, Jipp P (1993) Economic returns from forest conversion in the Peruvian Amazon. Ecol Econ 6:163–173

Pinedo-Vasquez M, Padoch C, McGrath D, Ximenes-Ponte T (2002) Biodiversity as a product of smallholder response to change in Amazonia. In: Brookfield H, Padoch C, Parson H, Stocking M (eds) Cultivating biodiversity: understanding, analysing and using agricultural diversity. ITDG Publishing, London, pp 167–178

Porro R (2005) Palms, pastures, and swidden fields: the grounded political ecology of "agro-extractive/shifting-cultivator peasants" in Maranhao, Brazil. Hum Ecol 33:17–56

Posey DA (1996) Traditional resource rights: international instruments for protection and compensation for indigenous peoples and local communities. International Union for the Conservation of Nature [IUCN], Gland

Posey DA, Balée WL (1989) Resource management in Amazonia: indigenous and folk strategies. New York Botanical Garden, New York

Posey DA, Balick MJ (2006) Human impacts on Amazonia: the role of traditional ecological knowledge in conservation and development. Columbia University Press, New York

Posey DA, Plenderleith K (2002) Kayapó ethnoecology and culture. Routledge, London

Prance GT, Balée WL, Boom BM, Carneiro RL (1987) Quantitative ethnobotany and the case for conservation in Amazonia. Conserv Biol 1:296–310

Price R (1975) Saramaka social structure: analysis of a maroon society in Surinam. University of Puerto Rico, Institute of Caribbean Studies, Río Piedras

Price R (1983) First-time: the historical vision of an Afro-American people. Johns Hopkins University Press, Baltimore

Raffles H (2002) In Amazonia: a natural history. Princeton University Press, Princeton

Raffles H, WinklerPrins AMGA (2003) Further reflections on Amazonian environmental history: transformations of rivers and streams. Lat Am Res Rev 38(3):165–187

Ramos-Neto MB, Pivello VR (2000) Lightning fires in a Brazilian savanna National Park: rethinking management strategies. Environ Manage 26:675–684

Redmond EM (1998) Chiefdoms and chieftaincy in the Americas. University Press of Florida, Gainesville

Reis ACF (1953) O seringal e o seringueiro. Ministério da Agricultura Servico e InformaçãoAgrícola, Rio de Janeiro

Rival LM (2002) Trekking through history: the Huaorani of Amazonian Ecuador. Columbia University Press, New York

Rodriguez I (2007) Pemon perspectives of fire management in Canaima National Park, southeastern Venezuela. Hum Ecol 35:331–343

Roosevelt AC (1991) Moundbuilders of the Amazon: geophysical archaeology on Marajo Island, Brazil. Academic, San Diego

Saatchi SS, Houghton RA, Dos Santos Alavala RC, Soares JV, Yu Y (2007) Global change. Biology 13(4):816–837

Santos R (1980) História econômica da Amazônia (1800–1920). T.A. Queiroz, São Paulo

Santos RV, Flowers NM, Coimbra CEA, Gugelmin SA (1997) Tapirs, tractors, and tapes: the changing economy and ecology of the Xavante Indians of central Brazil. Hum Ecol 25:545–566

Santos-Granero F, Barclay F (2000) Tamed frontiers: economy, society, and civil rights in upper Amazonia. Westview Press, Boulder

Schiebinger LL, Swan C (2005) Colonial botany: science, commerce, and politics in the early modern world. University of Pennsylvania Press, Philadelphia

Schmink M, Wood CH (1984) Frontier expansion in Amazonia. University of Florida Press, Gainesville

Sears R, Pinedo-Vasquez M (2004) Axing the trees, growing the forest: smallholder timber production on the Amazon Várzea. In: Zarin D (ed) Working forests in the neotropics: conservation through sustainable management? Columbia University Press, New York, pp 258–275

Sears R, Padoch C, Pinedo-Vasquez M (2007) Amazon forestry transformed: integrating knowledge for smallholder timber management in eastern Brazil. Hum Ecol 35(6):697–707

Shanley P, Rosa NA (2004) Eroding knowledge: an ethnobotanical inventory in Eastern Amazonia's logging frontier. Econ Bot 58(2):135–160

Shepard G Jr, Yu DW, Lizarralde M, Italiano M (2001) Rain forest habitat classification among the Matsigenka of the Peruvian Amazon. J Ethnobiol 21(1):1–38

Shono K, Snook LK (2006) Growth of big-leaf mahogany (Swietenia macrophylla) in natural forests in Belize. J Trop For Sci 18:66–73

Silva AL, Tamashiro J, Begossi A (2007) Ethnobotany of riverine populations from the Rio Negro, Amazonia (Brazil). J Ethnobiol 27(1):46–72

Smith NJH (1982) Rainforest corridors: the Transamazon colonization scheme. University of California Press, Berkeley

Snook LK, Negreros-Castillo P (2004) Regenerating mahogany (Swietenia macrophylla King) on clearings in Mexico's Maya forest: the effects of clearing method and cleaning on seedling survival and growth. For Ecol Manag 189:143–160

Stearman AM (1989) Yuqui foragers in the Bolivian Amazon—subsistence strategies, prestige, and leadership in an acculturating society. J Anthropol Res 45:219–244

Stearman AM (2006) One step forward, two steps back: the Siriono and Yuqui community forestry projects in the Bolivian Amazon. Hum Organ 65:156–166

Stedman JG, Price R, Price S (1992) Stedman's Surinam: life in eighteenth-century slave society. Johns Hopkins University Press, Baltimore

Tacconi L (2009) Illegal logging. Earthscan, London

Taussig M (1984) Culture of terror—space of death: Roger Casement's Putumayo report and the explanation of torture. Comp Stud Soc Hist 26:467–497

Taussig MT (1986) Shamanism, colonialism, and the wild man: a study in terror and healing. University of Chicago Press, Chicago

Tuomisto H, Ruokolainen K, Kalliola R, Linna A, Danjoy W, Rodriguez Z (1995) Dissecting Amazonian biodiversity. Science 269(5220):63–66

Vickers WT, Plowman T (1984) Use plants of the Siona and Secoya Indians of eastern Ecuador. Fieldiana Bot 15:1–63

Vidal SM (2000) Kuwe Duwakalumi: the Arawak sacred routes of migration, trade, and resistance. Ethnohistory 47:635–667

Vieira EM (1999) Small mammal communities and fire in the Brazilian Cerrado. J Zool 249:75–81

Voeks RA (1997) Sacred leaves of Candomblé: African magic, medicine, and religion in Brazil. University of Texas Press, Austin

Whitehead NL (ed) (2003) Histories and historicities in Amazonia. University of Nebraska Press, Lincoln

Whitmore T, Thomas M, Turner BL II (2001) Cultivated landscapes of middle America on the eve of conquest. Oxford University Press, Oxford

WinklerPrins A, Barrera-Bassols N (2004) Latin American ethnopedology: a vision of its past, present, and future. Agricult HumValues 21:139–156

WinklerPrins AMGA, Barrios E, WinklerPrins AMGA, Barrios E (2007) Ethnopedology along the Amazon and Orinoco Rivers: a convergence of knowledge and practice. Rev Geogr 142(Julio-Diciembre):111–129

Wolff CS (1999) Mulheres da floresta: uma história—Alto Juruá, Acre, 1890–1945. Editora Hucitec, São Paulo

Woods WI, Teixeira WG, Lehmann J, Steiner C, WinklerPrins AMGA, Rebellato LE (2009) Amazonian dark earths: wim sombroek's vision. Springer, New York

Zent S (2009) Traditional ecological knowledge (TEK) and biocultural diversity: a close-up look at linkages. In: Bates P, Chiba M, Kube S, Nakashima D (eds) Learning and knowing in indigenous societies today. UNESCO, Paris

Zent S, Zent E (2004) Ethnobotanical convergence, divergence, and change among the Hotï of the Venezuelan Guayana. In: Carlson T, Maffi L (eds) Ethnobotany and conservation of biocultural diversity, vol 15, Advances in Economic Botany. New York Botanical Garden Press, Bronx, pp 37–78

Chapter 5
North America

Ronald L. Trosper, Fred Clark, Patrica Gerez-Fernandez,
Frank Lake, Deborah McGregor, Charles M. Peters, Silvia Purata,
Teresa Ryan, Alan Thomson, Alan E. Watson, and Stephen Wyatt

Abstract The colonial history of North America presents a contrast between Mexico and the two predominantly English-speaking countries, the United States and Canada. In Mexico, indigenous and other local communities own considerable forested lands, a consequence of the Mexican Revolution of the early twentieth century. In the United States, forest land is now primarily in private or federal hands, while in Canada forest land is primarily managed by the provinces. In all three countries, traditional knowledge had little effect upon forestry until the end of the twentieth century. In Mexico and the United States, the central government retained control over forested lands ostensibly held by communities. Policy changes in those two countries have

R.L. Trosper (✉) • T. Ryan
University of British Columbia, Vancouver, BC, Canada
e-mail: rltrosper@email.arizona.edu; ryantl@shaw.ca

F. Clark
U.S. Forest Service, Washington, DC, USA
e-mail: fclark@fs.fed.us

P. Gerez-Fernandez
Instituto de Biotechnologia y Ecología Aplicada, Universidad Veracruzana,
Xalapa, Veracruz, Mexico

F. Lake
U.S. Forest Service, Arcata, CA, USA
e-mail: flake@fs.fed.us

D. McGregor
University of Toronto, Toronto, ON, Canada
e-mail: d.mcgregor@utoronto.ca

C.M. Peters
The New York Botanical Garden, Bronx, NY, USA
e-mail: cpeters@nybg.org

S. Purata
People and Plants International, Xalapa, Veracruz, Mexico

J.A. Parrotta and R.L. Trosper (eds.), *Traditional Forest-Related Knowledge:*
Sustaining Communities, Ecosystems and Biocultural Diversity,
World Forests 12, DOI 10.1007/978-94-007-2144-9_5,
© Springer Science+Business Media B.V. (outside the USA) 2012

decentralized control to indigenous peoples, and their ideas have started to affect forestry. In Canada, although traditional management of lands in remote regions persisted until the middle of the twentieth century, provincial policies have generally been displacing indigenous control; First Nations knowledge, which has survived well in some areas, is only recently being applied to forest management, and in only a few examples.

Keywords Canada • Cultural diversity • Forest history • Ejidos • Forest management • Forestry education • Indigenous peoples • Mexico • Traditional knowledge • United States

5.1 Introduction

The chapter begins with a brief survey of ecological and cultural diversity in North America and Mexico before proceeding to a summary of the history of aboriginal people since the arrival of European settlers. The rest of the chapter addresses the contribution of traditional forest-related knowledge to modern forest management, and to good practices in the utilization of traditional knowledge for management and research. A number of boxes provide specific examples from North America.

5.2 Context and History

5.2.1 Regional Overview of Cultural and Ecological Diversity

This section provides a general overview of the great cultural diversity of three countries of North America: the United States, Canada, and Mexico (Fig. 5.1). The view that the North American continent was pristine and untouched prior to settlement by Europeans has been discarded. Forest researchers and managers now recognize that widespread epidemics of communicable diseases greatly reduced the population of indigenous peoples, creating a 'widowed land.' European settlement was aided in many regions as a result of fields that had already been cleared and forests that had been managed to produce products useful to humans.

A. Thomson
Canadian Forest Service, Victoria, BC, Canada
e-mail: ajthomson@shaw.ca

A.E. Watson
Aldo Leopold Wilderness Research Institute, Missoula, MT, USA
e-mail: awatson@fs.fed.us

S. Wyatt
Université de Moncton, Campus d'Edmundston, Edmundston, NB, Canada
e-mail: swyatt@umce.ca

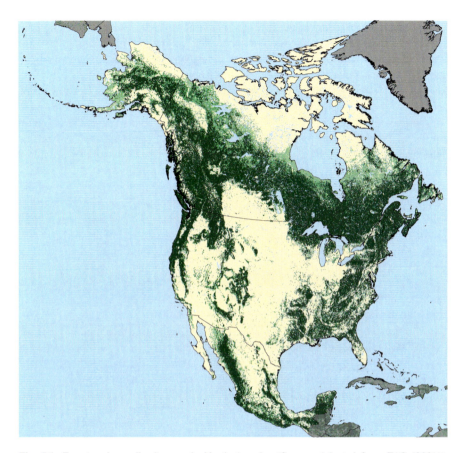

Fig. 5.1 Forest and woodland cover in North America (Source: Adapted from FAO (2001)). Key: *Dark green* closed forest, *light green* open or fragmented forest, *pale green* other wooded land, *yellow* other land

In the east of what became the United States, the indigenous peoples of the northern hardwood forest had a territorial system that included agriculture. The Haudonosaunee, Huron, and neighbouring tribes, for instance, were organized in villages and towns that moved from place to place within the forest, based on a system of shifting agriculture having long fallow periods for the forests to grow back (Fenton 1998; Trigger 1969). A town would clear the forest and establish agriculture based on the corn-bean-squash complex and managed by women, who held title to the farmed land. Fire was used to clear the forest for agriculture, and also used within the forest to favour particular plants and wildlife. Confinement of these peoples to reservations in the United States and Canada has removed that system of forest management.

In the Southeast, the Cherokee, prior to relocation by the U.S. government, also had agriculture and used fire to enhance the productivity of forested lands (Chapman et al. 1989; Delcourt and Delcourt 1997, 2004). In New England, the landscape changed after contact and the displacement of indigenous peoples as European settlers established farms using plants and animals new to America (Cronon 1983).

In Canada, the Algonquin peoples of the boreal forest followed a semi-nomadic lifestyle with distinctive family or clan territories. In these regions, land ownership was by territory, with overall authority for access and management of the territory held by an elder, steward, titleholder, or chief (Feit 1992). For purposes of this discussion, we will refer to them as stewards. The steward was expected to know his or her area well, and to pass along the knowledge to successors. The territories were not the personal property of the steward; the use of the land belonged to a group, always some kind of kinship-defined association. The stewards also had systems of relationships among each other, which served as higher level governing structures (Feit and Beaulieu 2001; Feit 2010; Scott 2001; Tanner 1988).

Similar but more elaborate territorial systems occurred among the peoples of the west coast from Alaska to northern California (Kroeber 1939). For instance, titleholders in and near the west coast held periodic feasts to recognize each other's authority and profiled efficient stewardship through sharing goods (Mills 1994, 2005). Knowledge of how to use plants was widespread (Deur and Turner 2005).

Complex resource management systems were reinforced by highly developed socio-political and religious organizational structures. Tribal clans or village families with chiefs or headmen often regulated access and use of terrestrial and aquatic flora, fauna, fungi, and geologic properties (see references in Suttles (1990b)). Tribal groups of different linguistic and ethnic origins intermarried and maintained socioeconomic trade relations. Along the Northwest coast, from Southeast Alaska to northern Oregon, many tribal village systems existed. In the southern coastal range from central Oregon to northwestern California, a more decentralized, village-family headman structure regulated commerce and management of natural resources (Trosper 2009).

Common among all nations, tribes, and bands were individuals' inherited rights and responsibilities to own, access, manage, and use resources or perform ceremonial practices that were reinforced by transmission of intellectual and spiritual properties. Diverse and productive coastal marine and riverine resources of mammals, birds, fish, and shellfish enabled larger and more stable tribal populations. Coastal to interior forests and grasslands were influenced by cultural burning practices (Boyd 1999). Vegetable, berry, nuts/seeds, and plants used for basketry were enhanced and benefited from periodic burns (Anderson 2005; Boyd 1999; Lewis 1993). Indigenous terrestrial, aquatic, and marine management systems accentuated existing geologic and ecological diversity and adapted to influences of ecological processes such as volcanism, earthquakes, floods, fires, and landslides (Suttles 1990a). The temperate rainforests of the coastal regions contain significantly high biodiversity with climax forests of western hemlock (*Tsuga hetreophylla*) and western redcedar (*Thuja plicata*) from the sea water edges to the steep mountain subalpine mix of spruces and firs. Crossing the mountainous terrain into the interior basin plateaus, the forest transitions from Douglas-fir (*Pseudotsuga menziesii*) into ponderosa pine slopes edged with oak (*Quercus garyyana*) at lower latitudes.

Interior forests and grasslands—primarily composed of Douglas-fir, pines, and oaks—were systematically burned by tribal groups to improve access to and quality of food plants; to enhance range forage quality; to aid hunting; to promote growth structure of plants used for basketry, implements, or cordage; to reduce pests; to lessen fuel loads and threats of catastrophic fires; to increase water yield at springs; and to facilitate travel (Anderson 2005; Boyd 1999). Along the Cascades and Rocky Mountain regions, fire was used primarily to maintain camas meadows, to open understory pine/fir forests for hunting, to maintain huckleberry and other berry fields, and to improve quality of forage or medicinal plants (Boyd 1999; Turner and Peacock 2005). Tribal groups in California's interior valleys, foothills, and mountains used fire primarily to enhance acorn productions, root-bulb crops, basketry plants; to reduce seed and parasite pests; and to maintain ecological diversity of various habitats for access and use of plants, animals, and minerals (Anderson 2005; Lewis 1993).

In the southwestern region of the United States and northern Mexico, mountains and mesas converge with alluvial desert flood plains where cycles of drought, fire, and water inundation transform the landscape and plant communities. Southwest forests contain the largest contiguous ponderosa pine forest in the world. The intense elevation gradients from desert to alpine tundra enable exceptional diversity and support the greatest numbers of endemic species in North America. For thousands of years, indigenous cultural groups tended careful use of forest resources and practised agriculture in seasonal rounds between valley deserts and mountain forests with reciprocal exchange of disturbance and restoration at sites based on fire ecology and water characteristics. The White Mountain Apache tribal stories describe stewardship practices and often do not differentiate between crops and cultivated wild plants (Long et al. 2003; Wilkinson 2005).

In Mexico, indigenous communities have been using and managing forest resources since pre-Columbian times (Barrera et al. 1977; Gómez-Pompa 1987; Peters 2000). For several thousand years, both temperate and tropical forests throughout the country have been periodically cleared and burned as part of the 'milpa' cycle, a traditional system of growing corn, beans, and squash that is a fundamental part of Mexican culture (Coe 1984; Harrison and Turner 1978). A milpa produces food, but perhaps more importantly, it also represents an essential link between the community, the land, the plants, and the universe (Hernandez Xolocotizi 1985). Cutting and burning the forest to plant corn has, as a result, traditionally been somewhat of a divine obligation for Mexican farmers. In most regions, the milpa is not abandoned after 2 or 3 years, but is enriched with an assortment of fruit trees, construction materials, and economic plants in sophisticated managed-fallow, agroforestry systems (Alcorn 1984, 1990). This practice is especially well-developed among Mayan communities who plant, spare, graft, or coppice dozens of useful tree species in their managed fallows (Gómez-Pompa and Kaus 1987; Harrison and Turner 1978). Totonac communities in northern Veracruz add allspice trees and vanilla after the last maize crop (Medellín Morales 1986); farmers in more arid regions frequently plant agave for making mescal, *Opuntia* cactus for their fruits, and assorted leguminous trees for improving soil quality (Messer 1975).

Managed fallows are either cycled back into milpas after a decade or two, or maintained and gradually converted into managed forests. The latter form of landuse is one of the most invisible, poorly understood, and potentially valuable forms of resource management in Mexico. Managed forests, which look identical to "undisturbed" forests, contain several hundred species of useful trees, shrubs, and herbs, many of which are shade tolerant, native plants adapted for growth and regeneration under a closed canopy (Alcorn 1983; Gómez-Pompa et al. 1987). When abandoned, these forests maintain themselves, with much of the original species composition introduced by the traditional farmers who created them. Nowhere is the imprint of indigenous forest management more visible than in the structure and composition of the forests of Mexico (Lundell 1937).

Indigenous communities in Mexico also manage their forest commercially for the production of timber (Bray et al. 2003). Most of the communities managing tropical forests (selva mediana subperrenifolia, or medium sub-deciduous forest) for wood products are located in the states of Campeche and Quintana Roo (Bray et al. 2005); an even greater number of community forestry operations are located in the temperate pine and oak forests of the states of Chihuahua, Durango, Michoacán, Guerrero, Puebla, and Oaxaca. A few of the indigenous management programmes in Quintana Roo have been certified by the Forest Stewardship Council (FSC) for more than a decade and are some of the oldest certified tropical forests in the world. Taking temperate and tropical forests together, Mexico has more certified community forest operations than any other country in the world (Gerez-Fernández and Alatorre-Guzmán 2005).

5.2.2   History of the Interaction of Traditional Societies and Modern Forest Management

Although both Canada and the United States primarily use English legal systems, each country has a different history of the relationship between colonizers and indigenous peoples. Also, the timing of colonization and the development of modern forestry are also different. When settlers harvested trees on the eastern side of the continent, modern forestry had not yet been created. By the late nineteenth century, as modern forestry was established, indigenous people were in a period of political weakness and dispossession. Subsequently, as indigenous political strength has improved, so has their influence on contemporary forest management decisions. But even when indigenous power was low, the peoples had some effect on forest management (by encouraging uneven-aged management) or participated in forestry operations as workers.

In Mexico, colonization occurred much earlier. By the time industrial forestry began to be practised, in the aftermath of the Mexican Revolution, indigenous peoples had acquired title to much forested land. In spite of this, the federal government maintained the right to grant forest concessions on both ejido and indigenous community lands until the 1980s, when forestry laws were changed.

A common view, shared by many indigenous peoples, was the existence of reciprocal relationships among humans, plants, and animals. The land required respect, and humans had a responsibility to take care of the land so it could continue to provide resources for their use. Other species were seen as having powers of their own that humans needed to recognize and consider in managing the relationships among different entities occupying the land. These peoples illustrated Bruno Latour's observation that indigenous people did not live 'in harmony with nature' because the separate idea of 'nature' did not exist (Latour 2004). The world was composed of humans and others interacting in ways that required humans to behave properly (McGregor 2004). Enrique Salmon describes the view as kincentric ecology, a viewpoint consistent with Gregory Cajete's description of native science (Cajete 1999; Salmón 2000).

European colonists introduced and later imposed a fundamentally different view of the forest. Since it was thought that forests were frightening places, 'much of the forest was an enemy to be eradicated as quickly as possible' (Lambert and Pross 1967). Later, forests began to take on a different, more utilitarian value and were exploited on an increasing scale. In the early years of colonization, this exploitation involved little in the way of management. Acquisition of land was an important policy objective of the colonial governments. Aboriginal people, like forests, were regarded as impediments to the path of colonial progress and subsequently were systematically removed through treaties, legislation, and policies (e.g., British North America Act, Indian Act) of the Dominion and later Canadian governments. (Alfred 1999; McGregor 2011). In the early colonial period, indigenous peoples had sufficient military and political power to insist on treaties that recognized their status as self-governing entities. These peace treaties were later replaced by treaties of cession. In this period, treaties seemed necessary as a way to remove people from the path of 'progress,' including settlements and other developments (Lambert and Pross 1967; RCAP 1996b).

In time, forests became valued for their timber, and management of the forests began to be seen as important. The dominant form of human interaction with the forest thus rapidly shifted from systems of aboriginal stewardship to 'management' as practised by Europeans and their entrepreneurial descendants. The territories upon which aboriginal people depended for their survival were wrested from their control without their consent, and in many cases, without their knowledge until facts on the ground revealed what had occurred.

5.2.2.1 United States

Control over indigenous forested lands by indigenous people occurred rarely in the United States in the early years of modern forestry. After the period of treaty-making ended in 1871, the United States moved to establish control on reservations. With pressure from states and economic interests, the federal government passed the General Allotment Act in 1887 (Dawes Act). The Act gave the Indian Office the authority to parcel out tribal land to individual Indians (usually 160 acres to the head of the family, with less to wives and children). Because the allotment policy emphasized agricultural lands, forested lands were not allotted to a great degree; the Quinault Reservation in western Washington state is a famous exception because its extensive rain forest was divided into individual parcels, which made forest management difficult. Allotment similarly affected the Hoopa-Yurok Reservation in California.

The Allotment Act is one of the most devastating policies in the history of American Indians. The act resulted in the loss of millions of acres because of fraudulent or coercive land transactions or the failure to pay taxes after the land passed out of trust. On many reservations, unallotted lands were open to homesteading by non-Indians. Between 1887 and 1934, Indian land holdings fell from 138 million acres to 48 million acres (Deloria and Lytle 1983). This was a loss of land whose magnitude was the size of the state of Montana, one of the largest states in the United States. The allotment and homesteading also created the checkerboard pattern of land ownership on reservations outside of Arizona and New Mexico; this checkerboard of land holdings and the existence of multiple heirs to allotments complicate federal and tribal jurisdiction, and create obstacles to land management (McDonnell 1991).

The Allotment Act was one of several federal policies that were intended to break up tribes and assimilate tribal members into non-Indian society. Boarding schools were established, traditional culture was suppressed, and local Indian agents essentially governed tribes on reservations. The original 'trust responsibility' of the federal government was implemented by the Allotment Act, which determined that individual allotments would be held in trust by the federal government until an individual was declared competent. Similarly, the federal government held tribal land in trust and asserted its management authority over that land.

Early in the twentieth century, the Forestry Branch of the Office of Indian Affairs asserted control over forest policy on reservations, assisted for a brief period by the U.S. Forest Service, which was established during the period of the allotment policy. On most reservations, loss of self-government and subsequent control by the BIA prevented indigenous people from having much influence on forestry management (McQuillan 2001; Sassaman and Miller 1986). In spite of their low power, however, the practices of American Indians may have influenced the BIA, particularly regarding the use of fire and uneven-aged management.

The story of the Yurok in California presents a well-documented example of the removal of indigenous people from their lands (Huntsinger and McCaffrey 1995), with consequences both for the land and for the people. The land changed because the shift in power was from people who valued oak savannas for food for themselves and browse for animals they hunted, to people who valued Douglas-fir for making lumber. The indigenous people burned to exclude Douglas-fir; the colonists excluded fire to encourage Douglas-fir and enclose the openings with timber. Tribal people became impoverished as their ability to obtain livelihood from the land was removed, at the same time as they were also increasingly excluded from their fisheries. Eventually the Yurok obtained rights to land adjoining the Hoopa Valley Indian Reservation. That reservation represents a different type of history, in which some Indians were confined to a reservation, which assured them some land but remained under the control of the BIA (Harris et al. 1995).

In contrast to the Yurok and most other tribes under BIA control, the Menominee of Wisconsin were able to avoid allotment and exert some influence on forest management. But they had to confront both the U.S. Forest Service and the Forestry Branch of the BIA to change forestry practices and policy. The Menominee had pressed for a law in 1908 to govern the management of their forests based on their principles of sustained yield, and had to pursue legal court action to enforce the 1908 law. Their opponents in court were lawyers defending the decisions of foresters trained in contemporaneous forest management (Davis 2000; Trosper 2007). The Menominee were awarded judgment with compensation because of federal government harvest rates that had exceeded the 1908 authorization.

The allotment policy ended with the passage of the Indian Reorganization Act (IRA) in 1934. Although the IRA allowed tribes to establish their own governments, forests remained under the control of the Bureau of Indian Affairs, which succeeded the older Office of Indian Affairs. A change in power in the U.S. Congress in 1952 led to the policy of termination, which was a policy to remove the trust relationship between the federal government and Indian tribes, and place the tribes under the control of states. Forty California Tribes and about 15 other tribes were singled out for termination. The Menominee were on the list; their victory in court over the 1908 Act created a vulnerability under termination, because they were told that in order to obtain the funds they had to agree to termination.

Many tribes fought the termination policy, which came to an end when self-determination became federal policy under the Nixon Administration in 1975. The Menominee Reservation that had been terminated was returned to federal status (Davis 2000; Huff and Pecore 1995; Pecore 1992), as were many other tribes that had been terminated. One exception was the Klamath of Oregon, whose land remained in the Winema-Fremont National Forest; the Klamath did retain some aboriginal rights, such as hunting rights in that forest. The Taos Pueblo in New Mexico, which had been fighting to protect the land and water of the watershed above their village, successfully had those lands returned to their control (Gordon-McCutchan 1991).

After the start of the self-determination movement, a group of Northwest Indian tribes organized the Intertribal Timber Council (ITC) in 1976. In an effort to change the thinking of the BIA, as well as to share ideas among the tribes, the ITC began a series of annual symposia, and published the proceedings of each of them. These proceedings are excellent source material for tracing the gradual change in BIA policy as well as gradual change of the direction of the management of forests on Indian reservations (http://www.itcnet.org/).

Several acts at the end of the twentieth century changed the relationship between the federal government and tribes regarding forest management. Most significant of these were the Indian Self-Determination Act of 1975 and the National Indian Forest Resources Management Act, which restructured the relationship with the BIA; and the Tribal Forest Protection Act of 2004. Following that act, the U.S. Forest Service and the Bureau of Land Management established rules that allow tribes to create cooperative relationships with the Forest Service on lands of significance to the tribes. Federally recognized Indian tribes in the United States now manage more than 18 million acres of forested landscape, with the Bureau of Indian Affairs involved to varying degrees (the Second Indian Forest Management Assessment

Team [Second IFMAT] 2003). Some tribes contract all management functions under the authority of the Self-Determination Act, reducing the BIA's role to oversight. For tribes that aren't contracting or compacting, the BIA plays a greater role in forest management.

5.2.2.2 Canada

Canada had three different general patterns of dispossession, and the consequences for forest management also varied. The timing of colonial control also had a different relationship to the timing of modern forest management and settlement. For instance, as the forests of the Great Lakes states were being cleared in the nineteenth century to support the growth of cities in the United States, the forests of British Columbia remained untouched. Harvest of those forests occurred a century later. Parts of the boreal forest in Canada remain unharvested even at the start of the twenty-first century (Hayter 2003; Korber 1997; Lendsay and Wuttunee 1999; McGregor 2000, 2002; Scott 2001; Smith 1998, 2001; Stevenson and Webb 2003).

The three general patterns of dispossession can be classified by type of treaty (Mann 2003; Richardson 1993; RCAP 1996c). Early treaties in the east were treaties of peace, not dispossession of land. The treaties in the middle of the country (which were numbered from 1 to 11) were viewed as treaties of cession by the Canadian state, leaving indigenous people with small reserves; however, indigenous people thought the treaties were arrangements to share the land. The extent of control over ceded lands became subject to treaty interpretation (Dickason 1997). The interpretation of treaty rights between the contradictory ideas of sharing compared to cession presents a major ongoing concern in Canada, particularly in relation to issues around access to resource rights (Macklem 1997; Smith 1998; Venne 1997). As a government commission found, 'the representatives of the Crown had come to see treaties merely as a tool for clearing Aboriginal people off desirable land' (RCAP 1996d).

Despite legally binding treaties, aboriginal people have for centuries been relegated to the fringes of Canadian society. More often than not, they have been seen as 'irrelevant to present-day concerns' (Berger 1991). There is a long history in Canada of oppression and colonization directed specifically at aboriginal people (Berger 1991; Boldt 1993; Borrows 2010; Little Bear et al. 1984; Miller 1989). Colonization was institutionalized and legislated in the Indian Act of 1876, which continues in many ways to regulate the lives of registered 'Indians' in Canada. Ongoing colonial policies and legislation have undermined aboriginal peoples access to their territories (McDonald 2003; Notzke 1994; RCAP 1996d).

In Quebec, British Columbia, Newfoundland and Labrador, Yukon, and the Northern Territories, no treaties were signed in the early years of settlement. Disputes over land ownership in those regions became very heated in the late twentieth century, and 'modern treaties' or rather comprehensive agreements are currently being contemplated through negotiation processes. Rather than one or two pages, such agreements are long and detailed, as illustrated by the James Bay and Northern

Quebec Convention signed in 1975 and the Nisga'a Final Agreement, ratified in 2001 (Hayward 2001). See http://www.gov.bc.ca/arr/firstnation/nisgaa/default.html.

In the early years of colonization, there was minimal regulation of forest use, mostly because resources were assumed to be bountiful and conservation was not an issue for some time (Lambert and Pross 1967). Later, the need for conservation was recognized and forest policy and management frameworks were developed. Beyers and Sandberg (1998) and Levy (1994) provide a more thorough historical account of forest policy in Canada and Ontario. On-reserve forests are an important source of livelihood for First Nations. Unfortunately, the forests tend to be in poor condition because of a long history of Crown mismanagement. The 1867 British North America Act (BNA Act) divided powers among the federal and most provincial governments. In relation to aboriginal people and their territories, the BNA Act gave jurisdiction of Indians and lands reserved for Indians to the federal parliament (Erasmus 1989). The responsibility for the management of natural resources fell to the provinces; thus the traditional territory that aboriginal groups enjoyed since time immemorial came under provincial jurisdiction (Bombay et al. 1996). This arrangement of confederation without the consent of aboriginal people has been a source of problems ever since. Aboriginal groups have effectively been stripped of their authority and jurisdiction over the land upon which they relied (National Aboriginal Forestry Association 1993).

The Indian Act was enacted after the BNA Act to provide for federal control of aboriginal groups. An important consideration in the Indian Act is that the Crown retains authority and vested interest in assets, which are held and managed in trust for the Indians, thus complicating the use of land for collateral in business enterprises. Although the act is inadequate in scope in terms of forest management, in remains in force (Auditor General of Canada 1994).

Recognition and protection of aboriginal rights in the 1982 Constitution Act has provided some leverage for indigenous peoples, now recognized to include Indians (or First Nations), Inuit, and Métis peoples. Major court decisions have clarified aboriginal rights and government obligations, leading to some changes in resource management regimes. Colonial legislation and policies, however, continue to permeate conflicts and resource management with changes occurring in isolation as provincial governments attempt to narrowly interpret court decisions. In some localized areas of Canada, resource management is coordinated through negotiation, including forest management (McGregor 2000, 2002; Ross and Smith 2002; Scott 2001; Smith 2001, 2007). The localized nature of these co-management arrangements have not precipitated broad national or provincial legislation or policy changes to date.

Although there are exceptions, the legislative and policy frameworks that govern Canada's forest industry continue to alienate and exclude aboriginal people from forest management. This involves restricting access to forest resources (e.g., harvesting timber and non-timber resources) and denying access to decision-making such that aboriginal cultural and traditional uses and values continue to be unaccounted for (McGregor 2011; National Aboriginal Forestry Association 1993). There has been in the past considerable conflict over forest resources between aboriginal and non-aboriginal society (Notzke 1994). Aboriginal assertions of rights and court decisions in their favour have recently led to a somewhat more favourable climate for aboriginal involvement in decisions affecting their lands. Despite these small inroads into the current system, the state of aboriginal forestry in Canada is unfortunately still characterized by exclusion.

Perhaps Canada will move into a period that is similar to the shift in control that has occurred as a result of self-determination in the United States. While the recognition of aboriginal rights in the 1982 Constitution Act is providing some impetus in that direction, reluctance by provinces to comply with federal court rulings makes the outcome uncertain.

5.2.2.3   Mexico

The lands occupied by the indigenous communities in Mexico were abruptly transformed during the colonial period by the introduction of livestock and new cultivars such as wheat, barley, and sugar cane. The coastal zones and plains were the first areas to be opened for intensive agriculture, and, in response, local indigenous communities took refuge in the mountains.

During the second half of the nineteenth century, the Leyes de Reforma de la Constitución Federal profoundly changed access to the land and natural resources of Mexico by promoting the privatization of national territories to increase local and foreign investment, by granting concessions for the commercial exploitation of forests, mines, and petroleum to American and British companies, and by funding the construction of railroads and an extensive network of roads. The net effect of these policies on indigenous communities in forested regions was the loss of their rights to harvest and sell forest resources and the geographic isolation caused by the placement of roads. Numerous indigenous communities saw the land titles granted by the vice-royalty during the Spanish Colonial period become invalid. During the dictatorship of Porfirio Diaz (1880–1909), it is estimated that the indigenous groups in Mexico lost control over 90% of their lands (Klooster 1996).

The Constitution of 1847 recognized the absolute right of private property over land and natural resources. No resource-use regulations of any kind were established, such that the extraction of forest products was done in the same way as mining—i.e., to obtain the largest benefit possible in the shortest amount of time, with no attention paid to regeneration or forest recuperation. Timber extraction during this period could best be described as 'forest mining.' New regulations that recognized the sovereignty of the Mexican Nation over all the land and water within its territory were not put into place until the Constitution of 1917 at the end of the Mexican Revolution.

The re-allocation of territory following the Mexican Revolution provided a more equitable distribution of agricultural lands and promoted productive farming and livestock management. However, this re-allocation effort did not grant absolute property rights to the ejidos (farming cooperatives), as the Mexican government maintained the right to grant concessions on communal lands. As a result, the ejidos were essentially tenants on the land that could make use of local resources; i.e., they were given usufruct rights but could not sell the land.

These regulations permitted the state to grant concessions on ejido and community lands for the use and exploitation of certain resources, such as minerals, oil, and wood products. In the case of indigenous communities that had been given deeds by the vice-royalty and could prove it, the re-allocation of lands restored their ownership to these territories.

The Forestry Law of 1926 established the inalienable character of community forests and stipulated that harvesting in communal forests could be conducted only by ejido cooperatives, and, as a result, numerous cooperatives were formed in forested areas. Unfortunately, this law was not accompanied with a policy to train and support the communities for managing their forests. The new owners of these forests were left without funds, markets, or the technical expertise for managing forest resources (Merino-Perez and Segura-Warnholtz 2005).

During the 1930s, the government of Lázaro Cárdenas realized an even greater re-distribution of lands, and at the conclusion of his presidency over 6.8 million ha of forest had been handed over to ejidos. This area represents about 18% of all the forests in Mexico (Merino-Pérez 2004).

Between 1934 and 1940, Lázaro Cárdenas promoted the organization of forestry cooperatives to facilitate the exploitation of these new community forests. The lack of investment policies, adequate training, and oversight, however, resulted in a pattern of uncontrolled, indiscriminate logging with little concern about long-term sustainability. This situation, together with the existing policies to promote the spread of agriculture and cattle, had a devastating effect on the most valuable forest in the country. The Mexican government responded by establishing national parks in forested areas above 3,000 m above sea level to protect the forests growing on the slopes of major mountains and to prohibit the extraction of wood products in these regions; people continued to live in the designated areas. The majority of the national parks in Mexico were established during this period.

From 1940 to 1970, the extraction of forest products was completely banned in selected areas of different states to try and stop the illegal logging and deforestation that was rampant. Logging, for example, was banned in parts of Veracruz, Chiapas, Puebla, Hidalgo, Distrito Federal, Morelaos, Durango, and Jalisco. At the same time, government policies regarding community and ejido forests were concentrated on establishing concessions with private companies, initially for exploiting timber and later with government-run companies to strengthen the link with the overall industrial development of the country and the export market for forest products. These types of concessions were operated in the forests of Quintana Roo, Oaxaca, Michoacán, Guerrero, Jalisco, Durango, and Chihuahua, the principal producers of forest products in the country.

Although the re-allocation of agrarian lands had recognized and re-instated the rights of indigenous communities over their territories with respect to forests, the federal government continued to grant logging concessions to private companies without consulting local communities or involving them in harvest activities. These concessions covered large extensions of forest, frequently including the territory of several indigenous communities, and were operable for 25 years. It is important to note that the concession holders invested large amount of money in these areas, building an extensive network of roads to the most productive parts of the forest. It was frequently the case that these roads did not service the local villages, which continued to be isolated from existing markets.

With the enactment of the Forestry Law in 1943, selected forest areas were designated as national supply zones in an attempt to integrate certain forested regions with local private industries known as Forest Exploitation Industries (Unidades Industriales de Explotación Forestal). The ejido and indigenous communities included in these areas were allowed to harvest timber, but they could only sell it to the local industries. At the same time, the technicians required for forest operations were usually supplied by the state or the local industry. Although the initial idea was that these industries would serve as a source of employment for indigenous groups, in practice each industry usually had its own group of trained technicians (Bray and Merino-Perez 2004).

Overall, the economic benefit of these activities to indigenous communities was nominal. The company paid a fixed stumpage price for the wood that was established by the government, and these fees were received directly by the Secretary of Agriculture who used part of these funds to cover the costs of public services for the inhabitants of the region.

In the 1950s, the Instituto Nacional Indigenista established a centre in the Tarahumara zone of Chihuahua to promote the training and education of the local forest stewards and to act as an intermediary among the indigenous communities, the ejidos, and timber buyers. At the end of this decade, the Fondo Nacional para la Promoción Ejidal (FONAFE) was created to use a portion of the forest taxes revenue to develop the productive capacity of local communities.

During this period there were attempts by several indigenous groups to regain control over the use of their forests, but none were successful. Several communities managed to have roads built to their communities and to negotiate permission for community members to work in timber harvesting; the latter allowed them to get trained in various timber-related activities. These advances were not trivial, and implied long struggles, imprisonment, and the occasional assassination of local leaders. After continued and increasingly vociferous pressure from indigenous groups, the first community forestry enterprise was established in Durango in 1965.

The period from 1971 to 1986 was characterized by the spread of forestry concessions to logging companies and state forestry enterprises. There was also, however, much interest during these years on the part of professional forestry groups to promote the development of indigenous communities and forestry ejidos by enhancing their technical and productive capabilities. It was hoped that this would help increase the overall productivity of the forestry sector in Mexico. With this in mind, formal partnerships were brokered between the forest industry and several indigenous communities, but the majority of these associations were inequitable and inefficient and produced local conflicts. In spite of this, a number of community forest enterprises were created during this period, initially with financial support from FONAFE. By 1975, 21% of the total volume of wood produced in the country came from these community-based enterprises. Community forestry enterprises were formed in 15 states, including those with the greatest amount of forest cover.

Although the first community forestry enterprises were dedicated exclusively to harvesting and extraction, as revenues increased and the level of technical expertise improved, several communities started sawmills to capture a higher percentage of the actual value of the resource. These changes were reflected in the Forestry Law of 1986, which stopped granting concessions to private enterprises and initiated a process to dismantle government-run forest industries. The law also established the right of ejidos and forestry communities to contract technical forestry services to assist them in the development of management plans. This opened a whole new phase in the management of forests in Mexico. The Mexican government relinquished control over the technical aspects of forestry, and permitted the owners of the forests, in partnerships or by themselves, to actually manage forest resources and take an active role in the production and commercialization of these products. Numerous indigenous communities in the mountains, however, had to go on strike, block roads, and disrupt work to exercise this right because most of the local forest industries were in collusion with government officials and forest technician groups. In contrast, the Forestry Law of 1986 was rigid with respect to the type of forest management that had to be implemented, and stipulated the exact silvicultural system that had to be used.

Community forestry enterprises that were able to gain access to markets and form local groups of technicians started to consolidate during the end of the 1980s. Other groups were unable to achieve this level of success and remained suppliers of raw material; i.e. they sold stumpage or sawlogs to various buyers. The administration of President Carlos Salinas de Gortari (1988–1994) was noteworthy for introducing programmes of decentralization and deregulation in the Mexican countryside. A new Forestry Law (1992) was drafted that greatly simplified all aspects of forest exploitation from extraction to milling. The law also eliminated technical forestry services from the government, opening the door for the creation of new private companies of professional foresters to provide these services.

In the background to all of these changes was the total economic transformation of the Mexican forestry sector caused by the General Agreement on Tariffs and Trade in 1986. Once the commercial borders of Mexico were opened to forest products from the United States, Canada, and Chile, the prices for locally produced timber plummeted. Yet, in spite of this, none of the community forestry enterprises in Mexico went bankrupt or stopped operating (Bray and Merino-Perez 2004).

Since the creation of the Comision Nacional Forestal (CONAFOR) in 2003, a significant percentage of the total investment in the forestry sector in Mexico has been designated for plantation establishment. The economic impact of these investments has yet to be realized as none of these plantations are of harvestable size.

5.3   Contribution of Traditional Knowledge to Modern Forest Management

Given the great power inequality during colonial dispossession of indigenous peoples, those people were not able to contribute their ideas to the development of forestry management. The idea that fire could be used to manage forests was derided as 'Piute [Paiute] Forestry' by the Assistant Chief of the U.S. Forest Service as his agency sought to exclude all fire from forests in the United States (Greeley 1920). Both fire and indigenous ideas were suppressed and excluded. The anthropologist Omar Stewart was unable to publish his manuscript documenting the use of fire by native peoples, for obscure reasons possibly related to the fact that fire suppression was at its height in the 1950s when he tried to publish the book (Lewis 2002; Stewart 2002).

Serious consideration of indigenous knowledge in forest management did not occur until indigenous peoples were able to assert their authority over forest lands, and this situation is a pattern of the late twentieth century, as described in the previous section that summarized colonization. Inequity still exists regarding Indian sacred sites. The indigenous view that sacred sites must be protected in order to protect the integrity of the natural and spiritual world is disregarded by the land management agencies. Western managers also still do not make the connection between indigenous control and uses of traditional resources on their traditional landscapes, and the health and well-being of those communities. Some see a direct connection between alienation from spiritual landscapes and the high levels of violence, intoxicants, and suicide that so mar Native communities in North America (Parkins et al. 2006).

Once indigenous people are able to manage their own lands, as on reservations in the United States, they still may not be contributing ideas to modern forest management in general. A measure of such contribution could be the citation of indigenous ideas in current textbooks; such citations are hard to find. For instance, the third edition of Kimmins' *Forest Ecology: A Foundation for Sustainable Forest Management and Environmental Ethics in Forestry* (Kimmins 2004) confines indigenous contributions to slash-and-burn agriculture, a reference to Omar Stewart, and recognition of 'local people with experienced-based wisdom' (p. 610). The author regards 'experience-based' knowledge as less valuable than scientific knowledge, in spite of referring to it as 'wisdom.' The fourth edition of *Forest Management* (Davis et al. 2000) also does not include references to indigenous contributions, according to its index. In Quebec, the 2008 edition of the legally mandated forest management manual *Manual d'aménagement forestier du Québec* now includes a short mention of the necessity to address indigenous values and knowledge in forest management, in regard to FSC certification requirements. In Smith (1997), *The Practice of Silviculture: Applied Forest Ecology*, 9th edition, there is no mention of indigenous forest management practices or of fire-use by tribal groups. 'In most parts of the world the most common natural disturbance is fire…which was kindled by lightning and volcanic eruptions long before people put it to use and abuse' (p. 162, Kinds of Natural Regenerative Disturbance). The text describes reforestation following agricultural land use, yet does not recognize those centuries-old abandoned fields were the result of native-white conflicts in eastern North America. In Kohm and Franklin (1997), *Creating a Forestry for the 21st Century: The Science of Ecology Management,* very few contributors reference Native American management or utilization of forest resources other than use of special forest products (Molina et al. 1997). In *Forest Ecosystems* (Perry et al. 2008), the historical and modern role of indigenous knowledge and forest stewardship is recognized and may be an improvement in text content.

Recognition in textbooks follows recognition in the professional literature, and in the professional literature, the contributions are expanding. The Second Indian Forest Management Assessment Team (Second IFMAT), for instance, recognized that '. . .the condition of Indian forests can also yield valuable lessons for society in general; indeed, Indian forests have the potential to be models of integrated resource management and forest sustainability from which we can all learn.' There is also growing recognition that Indians used fire for beneficial purposes, and that the resulting knowledge might be helpful. Even with improved self-determination of tribes in the United States, the record is mixed on reservations according to the Second Indian Forest Management Assessment Team (Second IFMAT 2003). The winter 2005–2006 edition of the journal Evergreen provides summaries of efforts on some reservations.

In Canada, widespread recognition of the potential contribution of indigenous peoples and their knowledge to sustainability first began to manifest itself in various international commissions, conferences, protocols, and conventions. For example, the 1992 Convention on Biological Diversity makes explicit recognition of indigenous peoples and their knowledge (Higgins 1998; NAHO 2007a; Scientific Panel for Sustainable Forest Practices in Clayoquot Sound 1995a). As signatory to the Convention on Biological Diversity (CBD), Canada has shown substantially increased interest in traditional forest-related knowledge in recent years (Battiste and Henderson 2000; Ellis 2005; MacPherson 2009; Manseau et al. 2005; McGregor 2004). Traditional forest-related knowledge in environmental and resource management is thus now emerging as a field of study, complete with theory, research approaches, models, and applications (Berkes 1999, 2008; Grenier 1998; Houde 2007; Inglis 1993; Johnson 1992; McGregor 2002).

In Canada, the use of traditional knowledge in sustainable forest management has been influenced by the Convention on Biological Diversity (CBD), a legally binding international agreement, and has found expression in national forest policy in Canada since 1992. The Canadian Council of Forest Ministers' National Forest Strategy in 1992 makes clear the importance of incorporating native values in sustainable forest management and planning in Canada. Aboriginal contributors to the 1992 Strategy stressed the importance of the forest to aboriginal peoples and the need to have native and non-native parties work together to protect cultural and spiritual forest values (CCFM 1992, p. iii). In 1998, the National Forest Strategy was renewed. The updated version more specifically identified the goal of increasing engagement between aboriginal peoples and the rest of the forest community in the area of 'traditional forest values and modern Aboriginal aspirations and needs' (CCFM 1998, p. 35).

In 1998, the National Forest Strategy was renewed for another 5 years. The National Forest Strategy Coalition specifically mentions the role of the CBD. A key action item is to 'incorporate traditional knowledge in managing forest lands and resources in accordance with the Convention on Biological Diversity' (NFSC 2003, p. 15). Traditional forest-related knowledge gained further recognition in the following Strategy in 2003, before being ignored in the most recent version, released in 2008. Canada has reported on progress of traditional forest-related knowledge in forest management in international forums (Bombay 1996; Brubaker 1998). In 1999, the Canadian Forest Service conducted a review of the case studies funded by the First Nation forestry programme for case studies on the implementation of Article 8(j) and related provisions under the Convention on Biological Diversity (Canadian Forest Service 1999).

In spite of this limited recognition, traditional knowledge and the views of Indian people are contributing in four areas in particular: the use of fire as a management tool, uneven-aged silviculture, long-term monitoring, and developing integrated resource management.

5.3.1 Fire As a Management Tool

The idea that indigenous people in North America used fire as a management tool has gradually become more and more accepted. Recognition of the importance of human manipulation of the pre-Columbian landscape has allowed forest historians to reinterpret data, and scientists to improve their understanding of forest ecosystems. Those interested in ecosystem management and restoration have particularly used the growing literature on the indigenous use of fire. Prominent goals of indigenous management, respecting the diversity and productivity of the landscape and maintaining balance, is similar to the goals of non-indigenous management to restore ecosystem functioning.

Kimmerer and Lake (2001) provide a useful summary of the contributions that indigenous fire management can make in changing the management of forests. They focus on five aspects of fire management: seasonality, frequency, extent, site, and outcome. Indigenous burning tended to occur in cool seasons rather than the heat of summer, allowing the beneficial effects of fire to outweigh the catastrophic consequences. Fire frequency also could be shifted in comparison to a regime based on ignition by lightning; generally, and increase in frequency accompanied by a decrease in fire extent and severity. Sites that would normally not burn, such as riparian areas with plants important for basket making, would be burned to improve outcomes of interest to humans (Kimmerer and Lake 2001).

Harold Weaver recognized the importance of fire for ponderosa pine when working on reservations, and he established control plots on the Colville Indian Reservation which have been maintained since their establishment. Under the direction of Wallace Covington, Victor Morfin revisited the plots and provided analysis of the effects of different fire regimes. This study provides the quantitative support that might not be available from other data sources regarding the impact of fire as a management tool (Morfin 1997).

Recently, contributions have expanded rapidly. After Stephen Pyne summarized the evidence of the use of fire by indigenous peoples (Pyne 1982), other researchers have documented contributions (Lake 2007; Miller et al. 2010), and the U.S. Forest Service published an extensive bibliography(Williams 2003). Archaeologists

(Delcourt and Delcourt 1997, 2004; Dods 2002) and ecologists (Nowacki and Abrams 2008) are also contributing to this literature. The Confederated Salish and Kootenai Tribes reclassified their forest based on four fire regimes (Confederated Salish and Kootenai Tribes 2000). The Ontario Department of Natural Resources has started to work with the Pikangikum First Nation on fire in the boreal forest (Miller et al. 2010).

5.3.2 Uneven-Aged Silviculture

Because their clients had some influence over forest management, the BIA has used uneven-aged management systems for forest harvest to a greater extent than used by other forest land owners. As a consequence, forests on Indian reservations in the United States are a fertile site for learning about the consequences of uneven-aged silviculture on forest structure (Becker and Corse 1997; McTague and Stansfield 1994, 1995).

5.3.3 Long Term Monitoring

The Continuous Forest Inventory (CFI) programme of the Bureau of Indian Affairs provides measurement of forest growth on a 10-year interval for all forested lands on reservations in the United States, using a system of fixed plots. This system, which originated on the Menominee Indian Reservation and spread to the rest of the Bureau, provides an excellent basis for long-term monitoring of forest conditions. McTague and Stansfield (1994) used the continuous forest inventory data for the Fort Apache Indian Reservation in Arizona to estimate growth-and-yield equations for ponderosa pine managed with an uneven-aged system. This study would not have been possible without both uneven-aged management and the CFI data (McTague and Stansfield 1994, 1995).

5.3.4 Integrated Resource Management

The management of forest lands for multiple purposes has become a reality on Indian reservations as tribes have implemented self-determination policy. Terminology has varied, with 'total resource management' being popular with the Intertribal Timber Council, which used that term as a theme for one of its annual timber symposia. The BIA forestry division has a small programme that supports tribes if they wish to engage in integrated resource management planning.

The Flathead Indian Reservation in Montana provides an example of planning for integrated resource management. In 1985, when the Salish and Kootenai Tribes

Box 5.1 Combining Traditional Forest-Related Knowledge and Ecological Science on the Flathead Indian Reservation

As tribes in the United States worked to implement the federal policy of self-determination, the Confederated Salish and Kootenai Tribes of the Flathead Indian Reservation were able to assert de facto decision-making authority when their dependence on forest revenue was removed by the signing of a hydroelectric power licence that gave an annual rental twice the average value of timber harvest. As a result, the Tribal Council was able to endure delays that the BIA required for complying with the wishes of the Tribes' Tribal Council. The Council had refused to approve a forest management plan written by the BIA, with an approved annual cut of 54 million board feet; it later modified the annual cut to 38 million board feet. In the 1990s, the tribes set out to write their own plan. They changed the classification system of the forest from cover type to type of fire regime, using four different fire regimes defined by return frequency ('nonlethal,' fires every 5–30 years; 'mixed,' fires 30–100 years; 'lethal,' 70–500 years; 'timberline,' 30–500 years). Silvicultural prescriptions were rewritten to accommodate the characteristics of fire regimes; principles of ecosystem management were used throughout the plan, with the pre-contact fire regimes as the first step in defining desired conditions. A substantial public participation process examined five different management strategies; the tribes selected the 'Modified Restoration' option, which had an annual cut of 19 million board feet (Confederated Salish and Kootenai Tribes 2000; Tecumseh Professional Associates 1999). Restoration refers to creation of conditions under the pre-contact fire regimes, modified to take account of current desired outcomes. The reduction in timber harvest from previous BIA recommendations is due to consideration of 10 other factors, such as wildlife, forest health, fisheries, and culture.

   Another innovation by the Salish and Kootenai Tribes was the establishment of a tribal wilderness area in the Mission Mountains, which form the eastern boundary of the Flathead Indian Reservation. Subsequently, a substantial public involvement process has been undertaken to plan for fire and fuels management in the buffer zone at the foot of the mountains in order to allow restoration of fire in the wilderness and to address healthy ecosystem concerns for adjacent lands (Carver et al. 2009; Krahe 2001; Watson et al. 2008).

became joint licensees for Kerr Dam, located on the Flathead Indian Reservation, their dependence on forest income fell and they were able to insist that the BIA establish interdisciplinary teams for the evaluation of all timber sale proposals. They were able to do this because the BIA's threat to delay income if timber sales had to be re-planned no longer carried force. Subsequently, they wrote their own management plan, taking over that function from the BIA (Box 5.1) (Confederated Salish and Kootenai Tribes 2000).

5.4 Development of Good Practices

5.4.1 Good Practices in Education

Some universities and tribal colleges have established forestry programmes to support education of indigenous people in contemporary forestry management; most of these programmes acknowledge traditional knowledge. Programmes exist at the University of British Columbia in Vancouver, British Columbia; Lakehead University in Thunder Bay, Ontario; Salish Kootenai College in Pablo, Montana; College of the Menominee Nation in Keshena, Wisconsin; Northern Arizona University in Flagstaff, Arizona; the State University of New York; and Humboldt State University in Arcata, California. Tribal colleges, being closer to their communities, can develop stronger links. Universities, especially those sponsored by states or provinces, have to overcome considerable suspicion resulting from years of difficulties resulting from research practices and public policy. These barriers are not easy to overcome, because of their strong connection to the colonial practices of their government sponsors, which reflect differences in power. Profound differences between traditional knowledge holders and universities regarding epistemology also creates barriers, as does the attempt by many scientists to draw a strong line separating knowledge, ethics, and religion (Drew and Henne 2006; James 2001).

Differences in approaches to epistemology—the study of the origins of knowledge—affect relationships between educational organizations and indigenous knowledge holders. Several authors have explored the differences in assumptions about the source of knowledge and the best methods for passing knowledge across generations (Bala and Joseph 2007; Davidson-Hunt and O'Flaherty 2007; Houde 2007; Moller et al. 2004; Shackeroff and Campbell 2007). Because these profound differences are not easy to accommodate without substantial changes to educational institutions, building links in education between Western science and traditional knowledge in education remains challenging (Kimmerer 2002).

For instance, many traditional knowledge holders regard a long personal experience with particular areas as essential to knowing about that land; such people have little tolerance for theories based on knowledge from elsewhere. Many scientists, however, seek generalizations and place high regard for knowledge of written materials that describe a variety of places (McQuat 1998). While recognizing that universal theories remain undeveloped in forest ecology, scientists nonetheless regard knowledge of particular places as incomplete (Davidson-Hunt and O'Flaherty 2007; Nadasdy 2003b). This affects education as well as other aspects of good practices.

In universities, young people are able to obtain their PhDs and become licensed as experts. In traditional communities now disconnected from their traditional territories, few young people can accumulate the experience needed to become recognized as experts. Just the difference in the ages of professionals in the two cultures can create problems. Without the certification of a professional degree, traditional elders cannot become professors at universities, although special arrangements can recognize their knowledge. With a PhD but without experience, young scientists working with elders risk being unable to attain the respect needed to truly share knowledge (Snively 2006).

In spite of these problems, colleges and schools of forestry have been successful in educating indigenous foresters. In the United States, the School of Forestry at Northern Arizona University, the University of Montana, Oregon State University, and the University of Idaho have all had American Indians complete degrees from bachelor to doctorate. The University of British Columbia, University of Alberta, Université Laval, University of New Brunswick, Université de Moncton, and University of Toronto have all graduated First Nations students in forestry. Northern Arizona University's multi-resource approach to forestry has proved to be especially appealing to indigenous students.

5.4.2 Good Practices in Research

While changes to educational practices may overcome the barriers of the structure of educational institutions, changes to practices in research also require adjustments and can be carried out in good measure by individuals engaged in research (Battiste and Henderson 2000; McGregor 2010; Roots 1998; Smith 1999). Several guides are available (Battiste and Henderson 2000; Crowshoe 2005; Grenier 1998; Roots 1998; Smith 1999) to include the consideration of traditional forest-related knowledge in environmental and resource management. Many academic and federal agency researchers must comply with Institutional Review Board standards and follow guidelines established for ethical conduct of human subjects (Amdur and Bankert 2002). Some First Nations and American Indian tribes have established and others are beginning to require similar reviews of research methods and agreements with outside researchers. Approvals of research are generally granted by tribal councils, cultural committees, or tribal department leaders after discussion. The John Prince Research Forest provides an example of successful co-management of research (Box 5.2).

Good research practices involve development of community-based research in which those holding traditional knowledge are able to formulate research questions and control research methods in conjunction with scientists (Brant Castellano 2004; Davidson-Hunt and O'Flaherty 2007; Huntington 2000; Lewis and Sheppard 2005; Long et al. 2003; Macaulay et al. 1998; MacPherson 2009; Menzies 2001, 2004; Piquemal 2000; Sheppard et al. 2004). Participatory action research (PAR) methods can foster closer relationships between indigenous groups and scientists, although parties need to negotiate terms and conditions of research practices and who is responsible for data collection, storage, and reporting of results (Caldwell et al. 2002). Challenges can arise between Western scientific approaches to research and what is acceptable to indigenous groups (Davis and

Box 5.2 Co-Management of Research on the John Prince Research Forest

Founded in 1999, the John Prince Research Forest is a joint venture between the Tl'azt'en Nation and the University of Northern British Columbia. It is located on the traditional territory of the Tl'azt'en Nation, north of the city of Fort St. James, British Columbia. A board of directors with equal numbers of representatives from both parties governs the research forest, which operates on Crown land with a special use permit from the BC government. The vision statement for the research partnership states:

'Internationally recognized, the John Prince Research Forest is well known for both its ecological approach to forest stewardship and its leadership in building success-ful partnerships between Aboriginals and non-Aboriginals. Integrating traditional and current scientific approaches into resource management and research has achieved long term sustainable and sound management.
 The co-management approach between the University of Northern BC and the Tl'azt'en Nation serves as a model for effective partnerships. Professional capacity and high respect for both partners has been built through innovative educational approaches. The John Prince Research Forest, together with its founding partners, is recognized in Canada and beyond, for its vision, leadership and research on the cooperative management of natural resources.'

The vision has been implemented with a number of research projects with funding from research sources in the governments of British Columbia and Canada. The board of directors also has a policy of sharing benefits with the families who hold keyohs, the traditional form of land tenure, within the research forest. The holder of the provincially recognized traplines in the for-est is also a participant. Research projects have involved the people of the Tl'azt'en Nation, with research outputs such as a set of aboriginal criteria and indicators of sustainable forest management (Grainger et al. 2006; Karjala 2001; Karjala and Dewhurst 2003).

Reid 1999). The Whitefeather Forest provides an example of successful integra-tion of different knowledge systems (Box 5.3).

With support from the Ford Foundation, the Community Forestry Research Fellowship programme at the University of California, Berkeley, supported many PhD students in conducting research with communities, including aboriginal com-munities, in the United States. The book edited by Wilmsen et al. (2008) provides many insightful case studies of participatory research. Wulfhorst and others provide a set of criteria for the evaluation of participatory research. They discuss the impor-tance of three criteria: community-centred control, reciprocal production of knowl-edge, and attention to the distribution of benefits (Wulfhorst et al. 2008).

Community-centred control is their first criterion. Because community members participate as researchers, they become actual owners of the research, likely to use

Box 5.3 Joint Learning on the Pikangikum Land Base: The Whitefeather
Forest Initiative

In 1996, leaders of the Pikangikum First Nation of northern Ontario, Canada,
were concerned that forest harvesting south of their homeland was going to
extend into their lands. They engaged the Ontario Ministry of Natural
Resources of the Province of Ontario (OMNR) in a land use planning process.
They established the Whitefeather Forest Management Corporation (WFMC)
to put them in the driver's seat as they built relationships with university
researchers interested in traditional knowledge, environmental nongovernmental organizations, other First Nations, and other non-Pikangikum organizations (Smith 2007).

As a consequence of their efforts, they created a land use plan,
*Cheekahnahwaydaymungk Keetahkeemeenann—Keeping the Land: A Land
Use Strategy for the Whitefeather Forest and Adjacent Areas* (PFN and OMNR
2006). The strategy combined Anishinaabe knowledge with Western science,
and was a joint effort with the Ontario Ministry of Natural Resources. In
2009, the WFMC and the OMNR were working together to comply with environmental assessment requirements.

Working with university researchers, the Pikangikum elders have produced
a number of publications explaining their worldview and desired approach to
caring for the land in their traditional territory. Utilizing both their approach
and that of Western science has proved to be a challenge (O'Flaherty et al.
2008, 2009). Cooperative research has produced unique research results, such
as their 'cultural landscape framework,' which explains the Pikangikum
approach to interpreting change in their territory (Shearer et al. 2009).
University researchers have articulated the joint learning process (Davidson-
Hunt 2006; Davidson-Hunt and O'Flaherty 2007). They have provided guidance for the relationship between humans and woodland caribou and they
have also articulated the current Pikangikum view of the role of fire in the
boreal forest (Miller et al. 2010).

the results. Community control also increases credibility, particularly when all
groups in a community participate. Good community-centred control also builds
trust and shared development of locally applicable best available science.

Reciprocal production of knowledge begins by recognizing the different goals of
communities and most researchers. When a research project meet the goals of both,
then research can proceed successfully. Recognizing the goals of communities
involve recognition as well of all groups in the community, including those usually
marginalized. Knowing that political agendas matter in different ways to communities and researchers aids in improving joint knowledge production.

A third strength of participatory research is that action benefiting a community is
more likely, because the community has participated in formulating and prioritizing
research questions and in generating answers. A result is that 'extension,' getting

results to users, is much less important; the research process is a type of extension. The same results may appear differently to the community, which desires action, and to the researcher, who wishes to contribute to academic knowledge.

Wilmsen and Krishnaswamy (2008) consider the many challenges to participatory research, stemming primarily from funding agencies and universities. Funding agencies tend to separate community development from research; therefore, participatory research projects that do both do not fit well in either category. Universities have a rhythm of study for graduate students that goes rather quickly from the viewpoint of communities who need to work on learning the importance of research for development and other desires (Wilmsen and Krishnaswamy 2008).

Two of the chapters in the book edited by Wilmsen et al. (2008) deal specifically with Native American issues. Long and others describe the consequences when a stream restoration project was threatened by water runoff following a catastrophic fire on the Fort Apache Reservation in Arizona (Long et al. 2008). One consequence of the research project was that the community had obtained the ability to force compromise from officers of the BIA in efforts to protect a bridge. Hankins and Ross consider how the general issues of participatory research work for Native American scholars, who are more inclined to accept community knowledge as valid. They find that participatory research methods address most of the key issues Native American communities raise when dealing with researchers (Hankins and Ross 2008).

Problems in research can occur because of different ethical systems. Destructive sampling, for instance, shows disrespect for the tree if it is not subsequently utilized. Yet transporting such trees to a mill may not appear to be cost-effective to a growth-and-yield researcher (McTague and Stansfield 1994, 1995). While radio-collaring is not an acceptable technique for studying human migration, it is acceptable for scientific study of moose or caribou; traditional elders may see such collars as disrespectful of animal autonomy. Kluane elders objected to the disruption caused by helicopters in counting Dall sheep, in addition to complaining that such counts were not accurate, a conflict over both ethics and methodology (Nadasdy 2003b). Additional conflicts about sampling methods and adequate statistical rigour to confidently draw conclusions can arise between indigenous groups and scientists who have fundamentally different approaches for research methods, data analysis, and determination of causal factors (Ford and Martinez 2000).

Another example of good practices in tribal forest research that serves both parties' management objectives and interest was a joint research project between the Confederated Tribes of Warm Springs and Oregon State University in northern Oregon, which examined the various silvicultural treatments by seasonality (winter compared to summer) of ground-based harvesting systems on huckleberry rejuvenation and productions (Anzinger 2002). Another example is the effects of landscape-level wildlife damage, e.g., black bear cambium consumption, on tree regeneration linked to forest composition and structure on the Hoopa Indian Reservation in northwestern California (Matthews et al. 2008).

Aboriginal organizations are developing their own research protocols to protect traditional knowledge (Assembly of First Nations of Quebec and Labrador 2005; CIHR 2007) Important principles include OCAP—ownership, control, access, and possession—by aboriginal peoples of the research process (NAHO 2007b).

The overarching paradigm of research with respect to aboriginal people includes a shift to a paradigm of research *with* aboriginal peoples rather than *on* them (McNaughton and Rock 2003).

5.4.3 Good Practices in Forest Management

In Canada, the National Aboriginal Forestry Association has recommended good practices (Bombay 1992; Bombay et al. 1996; Rekmans 2002). Aboriginal forestry is being developed as a concept for education (Parsons and Prest 2003) and management (Wyatt 2008). Traditional forest-related knowledge can contribute to improving forest management practices in several distinct ways.

First, traditional knowledge can be used as an additional source of information in forest management planning, the essential first step of controlling practices. Contemporary forest planners typically use computer-based geographic information systems (GIS) to integrate a wide variety of data sources, including forest types and inventories, soil and geology maps, fauna habitats, visual landscapes, and social values. It is now common across Canada for indigenous communities to be requested to provide traditional forest-related knowledge for inclusion in plans, and a number of books are available describing data collection and mapping methods. Information provided frequently includes camp sites, travel routes, areas of spiritual or cultural significance, and fauna habitats. This information is then used in forest plans to identify areas that should be protected from logging or other management activities, or areas where logging may be permitted with modified practices to respect particular values. However, this approach expects that indigenous communities will turn over their traditional information to forestry companies or government agencies, something that many groups are reluctant to do. Similarly, traditional knowledge exists within a cultural context. Collecting information for inclusion in a GIS strips traditional forest-related knowledge of associated knowledge that cannot be recorded on a map, such as cultural norms concerning how a resource may be used or the ecological or spiritual importance of a particular site. Traditional forest-related knowledge can certainly be a source of useful information for good practices, but this approach limits the extent of the potential contribution (Wyatt et al. 2010).

Involving indigenous people, as holders of traditional forest-related knowledge, in determining forest management practices can be seen as a logical response to the limitations of treating such knowledge simply as another source of information. Public consultations have become widespread in forest management across Canada, and distinct processes are now obligatory under provincial law in Québec, British Columbia, and Ontario. Consultation can provide a forum where indigenous elders, or other holders of traditional knowledge can speak directly with forest managers and planners, providing information as they judge necessary and appropriate. Consultation processes may encourage participants not only to provide information, but also to contribute to modifying forest plans and practices on the basis of this information, as occurred in Clayoquot Sound (Box 5.4).

Box 5.4 Scientific Panel for Sustainable Forest Practice in Clayoquot Sound

During the period prior to 1993, a large environmental conflict developed over the logging of timber in Clayoquot Sound, on the west coast of Vancouver Island in British Columbia, Canada. Much of the island had been logged; not only was Clayoquot Sound one of the last unlogged areas, it was also important for recreation and fishing. International environmental organizations became involved along with local organizations, leading to protests and a logging bridge blockade resulting in arrests and negative worldwide publicity. As a means to resolve the issue, the provincial government appointed a panel of scientists to evaluate the situation.

The indigenous peoples of the Sound, five Nuu-Chah-Nulth First Nations, had been pursuing land claims and an interim measures agreement prior to treaty settlement. The government appointed one of the traditional leaders of the Nuu-Chah-Nulth as co-chair of the scientific panel, along with three other elders as members. The elders urged the panel to adopt Nuu-Chah-Nulth inclusive process for negotiations, and the panel agreed. This meant that everyone would be listened to with respect, and all decisions made by consensus. It also meant that ethical concerns would receive a fair hearing (Lertzman and Vredenburg 2005).

The panel recognized and incorporated two key Nuu-Chah-Nulth concepts: *hishuk ish ts'awalk* ('everything is one') and *hahuulhi* (the system of control of traditional territories by recognized hereditary leaders). The first of these recognizes the sacredness of all life forms; the second provides a basis for meaningful co-management of the resources of Clayoquot Sound. Because of this decision and the inclusive process, key elements of Nuu-Chah-Nulth traditional knowledge informed a great many of the overall recommendations of the panel. A special report on First Nations' perspectives provided 27 specific action recommendations to support the panel's general position that the Nuu-Chah-Nulth be significantly involved in all decisions regarding resources in Clayoquot Sound . Among the recommendations was explicit recognition both of traditional ecological knowledge and of the traditional system of leadership and decision-making in traditional territories. (Scientific Panel for Sustainable Forest Practices in Clayoquot Sound 1995a).

Informed by both science and traditional knowledge, the recommendations of the Scientific Panel in 1995 provided innovative ideas for ecosystem management. Among them was the silvicultural concept of variable retention, meaning that the previous pattern of clear-cutting was modified to leave significant areas of uncut forest. This recommendation was part of the general recommendation that planning focus on the condition of the ecosystem rather than the products removed from the ecosystem. This focus on the condition of the ecosystem implements the Nuu-Chah-Nulth concept of hishuk ish *ts'awalk*. (Scientific Panel for Sustainable Forest Practices in Clayoquot Sound 1995b).

(continued)

Box 5.4 (continued)

Concurrent with the work of the panel, an Interim Measures Agreement created a decision-making structure for ecosystem management in Clayoquot Sound; central to the structure was a Clayoquot Sound Central Region Board with significant First Nations participation; no final decisions could be undertaken without agreement from the Nuu-Chah-Nulth. The decisions of the Board, however, were advisory to the relevant provincial ministers; the province retained final formal control, but usually, the decisions of the Central Region Board were confirmed by the province. Subsequent studies of the co-management arrangements revealed that the statutory situation (with the province having final say) and the utilization of scientific ecosystem terminology limited the full implementation of equal power in management decisions. Significant changes, however, had been made in comparison to previous levels of integration of the ideas of traditional knowledge and Western science (Goetze 2005; Lertzman 2006; Mabee and Hoberg 2006). The Interim Measures Agreement, however, had expired as of 2010.

There are, however, many difficulties associated with consultation, for both the general public and indigenous peoples, and consultation techniques need to be chosen carefully. Differences in knowledge and world-view may block effective exchanges; specific mandates may limit the relevance of the process; inappropriate techniques may result in poor communication or mistrust; and hierarchies of power and decision-making responsibility may limit the value of participation for an indigenous community. Feit and Beaulieu noted that Cree participation in consultation processes in northern Quebec seemed to be aimed at legitimizing existing practices (Feit and Beaulieu 2001). Perhaps in response to this situation, the Cree of northern Québec have developed their own process for appropriate and effective consultations between community members and forestry companies and for the use of traditional forest-related knowledge (Waswanipi Cree Model Forest 2007). Marsden found that the Province of British Columbia oriented its consultation policies toward justifying infringement rather than seriously taking aboriginal concerns into account (Marsden 2005).

Partnerships between indigenous communities and the forest industry have also developed in Canada, with considerable attention to evaluation of the successes and failures (Anderson 1997, 1999; Anderson and Bone 1999; Beckley and Korber 1996; Brubacher et al. 2002; La Rusic 1995; Larsen 2003; McKay 2004; Merkel et al. 1994; Trosper et al. 2008). Such partnerships are usually aimed at obtaining a share of economic benefits of forestry development or at obtaining a measure of control or influence on forestry practices. As a means of applying traditional knowledge to forest management, a partnership can enable indigenous people (either as a community or individually) to conduct forestry operations in a way that is based on their own knowledge and in respect of their own values and culture. Meadow Lake

Tribal Council in Saskatchewan and the Atikamekw of Opiticiwan in Québec are among the communities who have established timber processing plants in partnership, while numerous nations across the country have various forms of agreements or contractual relationships with industry.

In the United States, the recommendations of the Indian Forest Management Assessment Team (IFMAT)(IFMAT 1993) and the Second IFMAT (Second IFMAT 2003) provide one set of good practices. The first report documented that Indian people and the foresters in the BIA had different priorities for the management of forests. Indian people tended to prefer 'protection' and the use of forests for 'subsistence.' The non-Indians in the BIA, and to an extent the Indian employees, believed their clients preferred the use of the forests to provide income (IFMAT 1993, pp. III-3 and III-4). The groups converged by the time of the Second IFMAT, with the BIA moving toward Indian values. In their summary of the Second IFMAT, the team stated:

> 'The timber-production focus of the past has begun to give way to integrated resource management to better fit the visions of tribal communities' (Second IFMAT 2003, p. 102).

They also say the following:

> ... the conditions of Indian forests can also yield valuable lessons for society in general; indeed, Indian forests have the potential to be models of integrated resource management and forest sustainability from which we can all learn (Second IFMAT 2003, p. 102).

The IFMAT recommendations are tailored for the situation on reservations in the United States, where tribes should clearly be the managers of their forests and the beneficiaries of the 'trust relationship,' based on the federal government's holding title to reservation lands. The two studies occurred because tribes and the Intertribal Timber Council advocated the passage of the National Indian Forest Resources Management Act (Title II of Public Law 101–630, November 28, 1990).

With tribes compacting and contracting management of forestry departments because of the Indian Self Determination Act, many reservation forests are now managed substantially in the way that their tribes want them managed. The second IFMAT report reveals a wide range of ways in which management is shared between the BIA and tribes.

The relationship between tribes and the U.S. Forest Service has also been changing. In the United States, many federal agencies, including the Forest Service, do not legally recognize 'co-management' with American Indian tribes, but they implement similar outcomes with contract, grants and agreements, and compliance with policies and legal mandates. The Tribal Forest Protection Act (TFPA) provides a mechanism for tribes to work with federal agencies, especially the Forest Service and the Bureau of Land Management, to protect their lands from threats arising on federal lands. The threats of wildfire, insect and disease infestations, etc. stem largely from the federal lands not being managed to as high a standard as the tribal lands, due for the most part to combined effects of litigation and reduced funding levels for fuels reduction. The TFPA is one of the few laws that recognizes the value of traditional knowledge and supports its application in TFPA project proposals. Memoranda of Understanding and Memoranda of Agreement are also important tools, as are contracts by which agencies hire tribes to do work in accord with traditional values. Legal tools include treaties, legislation such as TFPA, executive orders, and regulations that require consultation between tribes and federal agencies in the United States (Clinton 2000).

In Canada, some attempts have been made in the area of co-management to include aboriginal perspectives; sometimes this is not successful (Gardner 2001; Grainger et al. 2006; Greskiw 2006; Mabee and Hoberg 2004, 2006; Nadasdy 2003a, b, 2005; Natcher 1999, 2000; Natcher et al. 2005; Peters 2003; RCAP 1996a; Sneed 1997; Stevenson 1998; Wanlin 1999; Witty 1994).

Canadian authors have developed a framework for consideration of different types of co-management; Plummer and FitzGibbon (2004) have summarized this literature. They propose that three dimensions need to be examined: the form of power distribution between communities and government (on a scale from mere informing of the community of decisions made to full community control of decisions); the nature of representation that defines the community in question (commercial, national, local, indigenous groups); and the process for negotiations (whether formal or informal in nature). While their focus is on environmental management generally, they include forestry examples from Canada (Plummer and FitzGibbon 2004). After reviewing a variety of experiences in Quebec, Rodon (2003) concluded that it was more useful to think of co-management as a process and a structure by which aboriginal peoples negotiate power with governments, than simply as an organizational model. Jason Forsyth has applied a power scale to different levels of forest management—operational, tactical, and strategic—in arrangements between First Nations and the Ministry of Forests and Range in British Columbia (Forsyth 2006).

The role of traditional land use studies has been assessed (Horvath et al. 2002; Markey 2001; Natcher 2001; Wyatt et al. 2010). Understanding how and why indigenous practitioners used forest resources is important to implementing effective management practices. Providing access to and fostering an adequate supply of forest resources is critical to survival of indigenous groups and the perpetuation of traditional knowledge and management practices.

Different systems of criteria and indicators attempt to include indigenous priorities, with several levels of comparison being possible. First, national-scale criteria-and-indicator systems may not easily translate to local-scale criteria and indicators (Beckley et al. 2002; Duinker et al. 1994; MacKendrick et al. 2004; Sherry et al. 2005). Second, local criteria and indicators may not include indigenous priorities (Karjala and Dewhurst 2003; Marlor et al. 1999; Nilsson and Gluck 2001; Parkins et al. 2001; Sherry et al. 2005). Several studies have showed ways to include indigenous priorities in criteria and indicators (Bombay et al. 1995; Collier et al. 2002; Karjala and Dewhurst 2003; Karjala et al. 2003; Lewis and Sheppard 2005; Natcher and Hickey 2002). Smith and others have recently summarized the use of criteria and indicators in Canada (Smith et al. 2010).

Non-timber forest products are often very important for indigenous priorities regarding types of forest management (Berkes et al. 1995; George et al. 1996; Korber 1997; Kuhnlein 1991; Natcher 2004; Natcher et al. 2004; Parlee et al. 2005; Turner 2001). Many Western-scientific forestry studies can have results that are of importance to or have implications for indigenous groups who use non-timber forest products (for huckleberries and forest stand condition, see Jones and Lynch 2007; Kerns et al. 2004). Although it is important that agency regulations and policies recognize and acknowledge indigenous rights and interests in non-timber forest products (cf the Forest Service Handbook (FSH) 2409.18—Timber Sale Preparations Handbook, Chapter 80—Uses of Timber other than Commercial Timber Sales, Special Forest Products-Botanical forest products), in the United States giving access to aboriginal people for botanical products on terms different from other forest users is controversial.

The goals of community forestry overlap in many ways with ideas of co-management. In the United States, those advocating community forestry recognize the importance of traditional and local knowledge. Based on extensive interviews and workshops, Mark Baker and Jonathan Kusel have summarized the community forestry efforts throughout the country. Their introductory chapter recognizes the importance of aboriginal practices in determining the state of forests prior to European settlement. They also recognize the Hispanic communities in New Mexico as important sources of traditional knowledge and management practices (Baker and Kusel 2003). The Seventh American Forestry Congress in February 1996 was important in demonstrating broad interest in reforming forestry in the United States; American Indians participated in that event, which was organized to give voice to all participants.

In Canada as well, discussion of community forestry often includes indigenous perspectives (Bagby and Kusel 2003; Booth and Skelton 2008; Curran and M'Gonigle 1999; Duinker et al. 1994; Greskiw 2006; McGregor 2002; Smith 2006; Stevenson and Natcher 2010). Cree people have been involved with the Waswanipi Cree Model Forest (Box 5.5 and Fig. 5.2).

5.5   The Present Role of Traditional Forest-Related Knowledge

Given the many developments described in the previous section, the role of traditional forest-related knowledge in forestry in Mexico, the United States, and Canada seems to be increasing. In Mexico, the 1986 change in the national forestry law gave communities more control over their forests, thus allowing them to use their own knowledge to a greater degree. Forests managed using traditional systems to produce timber, fruits, latex, and medicines extend throughout the tropical, sub-tropical, and temperate biomes of Mexico. The milpa system is still a ubiquitous feature of the Mexican landscape, and species-rich home gardens are found outside the back door of many contemporary Mayan households in Quintana Roo. Mexico has more certified community forestry enterprises than any other country in the world, each of these involving to varying degrees the application of traditional forest-related knowledge.

In the United States, the increasing implementation of self-determination on reservations has also provided tribes opportunities to increase their control of forestry operations (Miller et al. 2010, Second IFMAT 2003). This increase has led to more efforts in the area of integrated forest management. It has also allowed more attention to non-timber forest products. Community-based research, such as by the White

Box 5.5 The Ndhoho Itschee Process—Sharing Knowledge and
Development Management Strategies

Faced by expanding forestry activities on their traditional lands, the Cree of
Waswanipi in northern Quebec developed their own process for bringing their
traditional knowledge to forest management consultations.

> Cree land users felt that they cannot be expected to detail such a complex system of
> knowledge to a foreigner who they see, at most, once or twice a year. They would
> rather enter into a system that recognizes their stewardship role about the land and
> values their body of knowledge. They expect that their voice will be heard because
> they have this knowledge. (Waswanipi Cree Model Forest 2007)

There are two principal parts of the process. First, members of the community document their past, present, and projected future land use. This 'family
map' is considered confidential; the goal is not just to document the information, but also to involve land users in reviewing their use and in deciding what
should happen in the future. The family map is used to prepare a separate map
of conservation values (Fig. 5.2) that can subsequently be shared with industry
and governments. At the same time, an assessment is made of the state of each
trapline (traditional area for Cree resource management), enabling trappers,
elders, and other users to focus on critical issues for each part of the land.

In the second part of the process, Cree representatives and planners from the
forest industry and governments meet to review the values and issues identified
by the Cree and to explore management strategies that can provide solutions
to these. 'This is where imagination, innovation and experience come into play.

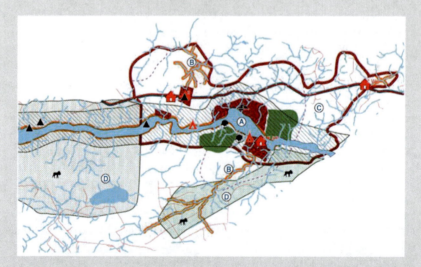

Fig. 5.2 A conservation value map. A planning support tool accompanied by Cree objectives and desired forest conditions (Source: Waswanipi Cree Model Forest 2007)

(continued)

Box 5.5 (continued)

The goal is to find possible solutions to the issue' (Waswanipi Cree Model Forest. 2007). Cree elders and land users, as well as outside experts, may be asked to provide insights and ideas at this point. Once solutions and strategies have been determined, then they need to be accepted by the community (as well as by government and industry). Finally, the community also joins the monitoring process to ensure that goals are attained and to bring information back to users of the land for future planning cycles.

Mountain Apache regarding riparian restoration, has involved elders in directing the questions asked and has involved tribal youth in carrying out the projects (Long et al. 2008).

Another development in the United States has been the ability of Indians to make agreements with the U.S. Forest Service through stewardship projects. The Maidu Stewardship Project was an example of a local aboriginal community, in this case not a federally recognized group, being able to implement some of their ideas in a small area (Cunningham and Bagby 2004).

Because of the strong role Canadian provinces have in land management, and the resistance from those governments to significant aboriginal participation, use of traditional forest-related knowledge in Canadian forestry has been more problematic. In spite of the problems, however, some significant examples exist. In 1995, the Scientific Panel for Sustainable Forest Practices in Clayoquot Sound completed a large study that gave traditional forest-related knowledge a key role (Scientific Panel for Sustainable Forest Practices in Clayoquot Sound 1995b). The Waswanipi Model Forest has used traditional knowledge (Waswanipi Cree Model Forest 2007). The planning document *Keeping the Land* resulted from a collaboration of the Pikangikum First Nation and the Ontario Department of Natural Resources, with significant use of traditional knowledge (PFN and OMNR 2006).

Many Canadian aboriginal communities have been active in addressing systems of criteria and indicators, in an effort to have their concerns included in the lists of criteria and monitored using indicators (Smith et al. 2010). The Sustainable Forest Management Network during its existence funded a number of research projects with significant participation by communities. The work with the Little Red River Cree Nation led to the publication of a book that has been useful as a textbook for at least one course in aboriginal forestry (Natcher 2008). Some of the research carried out by the Whitefeather Forest Management Corporation was funded by the Sustainable Forest Management Network. Two edited volumes report the results of these research projects, many of which considered issues of the relevance of traditional forest-related knowledge to forest management (Stevenson and Natcher 2009, 2010). While these examples show recognition of traditional forest-related knowledge in Canada, they are exceptions to the general rule that official forestry departments hardly use such knowledge.

References

Alcorn JB (1983) El Te'lom huasteco: presente, pasado y futuro de un sistema de silvicultura indigena. Biotica 8:315–331

Alcorn JB (1984) Huastec Mayan ethnobotany. University of Texas Press, Austin

Alcorn JB (1990) Indigenous agroforestry systems in the Latin American tropics. In: Altieri M, Hecht S (eds) Agroecology and small farm development. CRC Press, Boca Raton, pp 203–218

Alfred T (1999) Peace, power, righteousness: an indigenous manifesto. Oxford University Press, Don Mills

Amdur RJ, Bankert EA (2002) Institutional review board: management and function. Jones and Bartlett, Boston

Anderson RB (1997) Corporate/indigenous partnerships in economic development: the First Nations in Canada. World Dev 25:1483–1503

Anderson RB (1999) Economic development among the aboriginal peoples in Canada. Captus Press, Concord

Anderson MK (2005) Tending the wild: Native American knowledge and the management of California's natural resources. University of California Press, Berkeley

Anderson RB, Bone RM (1999) First Nations economic development: the Meadow Lake Tribal Council. J Aboriginal Econ Dev 1:13–34

Anzinger DL (2002) Big huckleberry (*Vaccinium membranaceum*) ecology and forest succession, Mt. Hood National Forest and Warm Springs Indian Reservation, Oregon. Thesis, Oregon State University, Corvallis

Assembly of First Nations of Quebec and Labrador (2005) First Nations of Quebec and Labrador research protocol. First Nations of Quebec and Labrador, Health and Social Services Commission, Ottawa

Auditor General of Canada (1994) Report of the auditor general of Canada to the House of Commons. Government of Canada, Ottawa

Bagby KA, Kusel J (2003) Partnering with underserved communities in community forestry: some lessons from North American experience. In: XII World forestry congress, Quebec City, Canada, United Nations. Food and Agriculture Organization [FAO], Rome

Baker M, Kusel J (2003) Community forestry in the United States: learning from the past, crafting the future. Island Press, Washington, DC

Bala J, Joseph G (2007) Indigenous knowledge and western science: the possibility of dialogue—race and class. Inst Race Relations 49(1):39–61

Barrera A, Gómez-Pompa A, Vázquez-Yanes C (1977) El manejo de las selvas por las Mayas: sus implicaciones silvícolas y agrícolas. Biotica 2:47–61

Battiste M, Henderson JSY (2000) Protecting indigenous knowledge and heritage: a global challenge. Purich Publishing, Saskatoon

Becker RR, Corse TS (1997) Resetting the clock with uneven-aged management. J For 95(11):29–32

Beckley TM, Korber D (1996) Clear cuts, conflict, and co-management: experiments in consensus forest management in northwest Saskatchewan. Northern Forestry Centre, Canadian Forest Service, Natural Resources Canada, Edmonton

Beckley T, Parkins J, Stedman R (2002) Indicators of forest-dependent community sustainability: the evolution of research. For Chron 78:626–636

Berger TR (1991) A long and terrible shadow: white values, native rights in the Americas, 1492–1992. Douglas and McIntyre, Vancouver

Berkes F (1999) Sacred ecology: traditional knowledge and resource management. Taylor & Francis, Philadelphia

Berkes F (2008) Sacred ecology, 2nd edn. Routledge, New York

Berkes F, Hughes A, George PJ, Preston RJ, Cummins BD, Turner J (1995) The persistence of aboriginal land use: fish and wildlife harvest areas in the Hudson and James Bay lowland, Ontario. Arctic 48:81–93

Beyers J, Sandberg A (1998) Canadian federal policy: present initiatives and historical constraints. In: Sandberg A, Sorlin S (eds) Sustainability the challenge: people, power and the environment. Black Rose Books, Montreal, pp 99–107

Boldt M (1993) Surviving as Indians: the challenge of self-government. University of Toronto Press, Toronto

Bombay H (1992) An aboriginal forest strategy. National Aboriginal Forestry Association [NAFA], Ottawa

Bombay H (1996) Aboriginal forest-based ecological knowledge in Canada: discussion paper. National Aboriginal Forestry Association [NAFA], Ottawa

Bombay H, Smith P, Wright D (1995) An aboriginal criterion for sustainable forest management. National Aboriginal Forestry Association [NAFA], Ottawa

Bombay H, Smith P, Murray A (1996) Aboriginal forest-based ecological knowledge in Canada. National Aboriginal Forestry Association [NAFA], Ottawa

Booth A, Skelton N (2008) Indigenous community values and commercial forestry: a case study of community forestry in the Tl'azt'en Nation. Natural resources and environmental studies institute occasional paper no. 3, University of Northern British Columbia, Prince George. Available via http://www.unbc.ca/assets/nres/nres_op_03_booth_2008.pdf. Cited 4 Feb 2011

Borrows J (2010) Canada's indigenous constitution. University of Toronto Press, Toronto

Boyd R (ed) (1999) Indians, fire and the land in the Pacific Northwest. Oregon State University Press, Corvallis

Brant Castellano M (2004) Ethics of aboriginal research. J Aboriginal Health 1(1):98–114

Bray DB, Merino-Perez L (2004) La experiencia de las comunidades forestales en México. SENARNAT, INE, CCMSS, Mexico City

Bray DB, Merino-Perez L, Negreros-Castillo P, Segura-Warnholtz G, Torres-Rojo JM, Vester H (2003) Mexico's community-managed forests as a global model for sustainable landscapes. Conserv Biol 17(3):672–677

Bray DB, Merino-Perez L, Barry D (cds) (2005) The community forests of Mexico. University of Texas Press, Austin

Brubacher D, Gladu JP, Bombay H (2002) First Nations governance and forest management: a discussion paper. National Aboriginal Forestry Association [NAFA], Ottawa

Brubaker E (1998) The common law and the environment: the Canadian experience. In: Hill PJ, Meiners RE (eds) Who owns the environment? Rowman and Littlefield, Lanham, pp 97–118

Cajete G (1999) Native science: natural laws of interdependence. Clear Light Publishers, Santa Fe

Caldwell JY, Jamie JD, Du Bois B, Echo-Hawk H, Erickson JS, Goins RT, Hill C, Hillabrant W, Johnson SR, Kendall E et al (2002) Culturally competent research with American Indians and Alaska Natives: findings and recommendations of the first symposium of the work group on American Indian research and program evaluation methodology American Indian and Alaska Native Mental Health Research. J Natl Center Univ Colo Health Sci Center USA 12(1):1–21

Canadian Council of Forest Ministers [CCFM] (1992) Sustainable forests: a Canadian commitment. National forest strategy. Canadian Council of Forest Ministers, Hull

Canadian Council of Forest Ministers [CCFM] (1998) National forest strategy 1998–2003: Sustainable forests, a Canadian commitment. Canadian Council of Forest Ministers, Ottawa

Canadian Forest Service (1999) Traditional ecological knowledge within the government of Canada's First Nation forestry program. Canadian Forest Service, Natural Resources Canada, and Indian and Northern Affairs Canada, Ottawa

Canadian Institutes of Health Research CIHR (2007) CIHR guidelines for health research involving aboriginal people. CIHR, Ottawa

Carver S, Watson A, Waters T, Matt R, Gunderson K, Davis B (2009) Developing computer-based participatory approaches to mapping landscape values for landscape and resource management. In: Geertman S, Stillwell J (eds) Planning support systems: best practices and new methods. Springer, New York

CCFM (1992) and CCFM (1998): See: Canadian Council of Forest Ministers

Chapman J, Delcourt HR, Delcourt PA (1989) Strawberry fields, almost forever. Nat Hist 9:50–59

Clinton WJ (2000) Order 13175, consultation and coordination with Indian tribal governments. Fed Regist 65(218):67249–67250

Coe MD (1984) Mexico. Thames and Hudson, New York

Collier R, Parfitt B, Woollard D (2002) A voice on the land: an indigenous peoples' guide to forest certification in Canada. National Aboriginal Forestry Association and Ecotrust Canada, Ottawa

Confederated Salish and Kootenai Tribes (2000) Flathead Indian Reservation forest management plan: an ecosystem approach to tribal forest management. Confederated Salish and Kootenai Tribes, Pablo, Montana

Cronon W (1983) Changes in the land: Indians, colonists, and the ecology of New England, 1st edn. Hill and Wang, New York

Crowshoe C (2005) Sacred ways of life: traditional knowledge. National Aboriginal Health Organization, Ottawa

Cunningham F, Bagby K (2004) The Maidu stewardship project: blending of two knowledge systems in forest management. Pacific West Community Forestry Center, Taylorsville. Available via http://www.sierrainstitute.us/PWCFC/publications/2004/MCDG_Final_Report_5_04.doc.pdf. Cited 20 Dec 2010

Curran D, M'Gonigle M (1999) Aboriginal forestry: community management as opportunity and imperative. Osgoode Hall Law J 37:711–774

Davidson-Hunt IJ (2006) Adaptive learning networks: developing resource management knowledge through social learning forums. Hum Ecol Interdiscip J 34:593–614

Davidson-Hunt IJ, O'Flaherty RM (2007) Researchers, indigenous peoples, and place-based learning communities. Soc Nat Resour 20:291–305

Davis T (2000) Sustaining the forest, the people, and the spirit. State University of New York Press, Albany

Davis SM, Reid R (1999) Practicing participatory research in American Indian communities. Am J Clin Nutr 69:755S–759S

Davis LS, Johnson KN, Bettinger PS, Howard TE (2000) Forest management: to sustain ecological, economic, and social values., 4th edn, McGraw-Hill series in forest resources. McGraw Hill, Dubuque

Delcourt HR, Delcourt PA (1997) Pre-Columbian native American use of fire on Southern Appalachian landscapes. Conserv Biol 11(4):1010–1014

Delcourt PA, Delcourt HR (2004) Prehistoric Native Americans and ecological change: human ecosystems in eastern North America since the Pleistocene. Cambridge University Press, Cambridge

Deloria V Jr, Lytle CM (1983) American Indians, American justice. University of Texas Press, Austin

Deur D, Turner NJ (2005) Keeping it living: traditions of plant use and cultivation on the northwest coast of North America. University of Washington Press, Seattle

Dickason OP (1997) Canada's First Nations: a history of founding peoples from earliest times, 2nd edn. Oxford University Press, Toronto

Dods RR (2002) The death of Smokey Bear: the ecodisaster myth and forest management practices in prehistoric North America. World Archaeol 33(3):475–847

Drew JA, Henne AP (2006) Conservation biology and traditional ecological knowledge: integrating academic disciplines for better conservation practice. Ecol Soc 11(2):34

Duinker PN, Matakala PW, Chege F, Bouthillier L (1994) Community forests in Canada: an overview. For Chron 70:711–720

Ellis S (2005) Meaningful consideration? A review of traditional knowledge in environmental decision making. Arctic 58(1):66–77

Erasmus G (1989) A native viewpoint. In: Hummel M (ed) Endangered spaces: the future for Canadian wilderness. Key Porter Books, Toronto

Feit HA (1992) Waswanipi Cree management of land and wildlife: Cree ethno-ecology revisited. In: Cox BA (ed) Native people, native lands: Canadian Indians, Inuit and Metis, vol 142, Carleton Library Series. Carleton University Press, Ottawa, p. 75

Feit HA (2010) Neoliberal governance and James Bay Cree governance: negotiated agreements, oppositional struggles, and co-governance. In: Blaser M (ed) Indigenous peoples and autonomy: insights for a global age. University of British Columbia Press, Vancouver, pp 49–79

Feit HA, Beaulieu R (2001) Voices from a disappearing forest: government, corporate, and Cree participatory forestry management practices. In: Scott CH (ed) Aboriginal autonomy and development in northern Quebec and Labrador. University of British Columbia Press, Vancouver, p 119

Fenton WN (1998) The great law and the longhouse: a political history of the Iroquois Confederacy. University of Oklahoma Press, Norman

Food and Agriculture Organisation [FAO] (2001) Global forest resource assessment, 2000. FAO forestry paper 140, FAO, Rome

Ford J, Martinez D (2000) Traditional ecological knowledge, ecosystem science and environmental management. Ecol Appl 10(5):1249–1250

Forsyth JP (2006) The balance of power: assessing conflict and collaboration in aboriginal forest management. Thesis, University of British Columbia, Vancouver

Gardner J (2001) First Nations cooperative management of protected areas in British Columbia: tools and foundations. Canadian Parks and Wilderness Society-BC Chapter and Ecotrust Canada, Vancouver

George P, Berkes F, Preston RJ (1996) Envisioning cultural, ecological and economic sustainability: the Cree communities of the Hudson and James Bay lowland, Ontario. Can J Econ 29:356–360

Gerez-Fernández P, Alatorre-Guzmán E (2005) Challenges for forest certification and community forestry in Mexico. In: Bray DB, Merino-Perez L, Barry D (eds) The community forests of Mexico. University of Texas Press, Austin, pp 71–87

Goetze TC (2005) Empowered co-management: towards power-sharing and indigenous rights in Clayoquot Sound, BC. Anthropologica 47(2):247–265

Gómez-Pompa A (1987) On Maya silviculture. Mexican Studies 3:1–33

Gómez-Pompa A, Kaus A (1987) Traditional management of tropical forests in Mexico. In: Anderson AB (ed) Alternatives to deforestation: steps toward sustainable use of the Amazon rain forest. Columbia University Press, New York, pp 45–63

Gómez-Pompa A, Flores JS, Sosa V (1987) The 'pet kot': a man-made tropical forest of the Maya. Interciencia 12:10–15

Gordon-McCutchan RC (1991) The Taos Indians and the battle for Blue Lake. Red Crane Books, Santa Fe

Grainger S, Sherry E, Fondahl G (2006) The John Prince Research Forest: evolution of a co-management partnership in northern British Columbia. For Chron 82(4):484–495

Greeley WB (1920) 'Piute forestry' or the fallacy of light burning. The Timberman. Reprinted in forest history today, Spring 1999. pp 33–37

Grenier L (1998) Working with indigenous knowledge: a guide for researchers. International Development Research Centre, Ottawa

Greskiw GE (2006) Communicating 'forest': co-managing crises and opportunities with Northern Secwepemc First Nations and the province of British Columbia. Dissertation, University of British Columbia, Vancouver

Hankins DL, Ross J (2008) Research on native terms: navigation and participation issues for native scholars in community research. In: Wilmsen C, Elmendorf W, Fisher L, Ross J, Sarathy B, Wells G (eds) Partnerships for empowerment: participatory research for community-based natural resource management. Earthscan, London, pp 239–257

Harris RR, Blomstrom G, Nakamura G (1995) Tribal self-governance and forest management at the Hoopa Valley Indian Reservation, Humboldt County, California. Am Indian Cult Res J 19(1):1–38

Harrison PD, Turner BL II (eds) (1978) Pre-Hispanic Maya agriculture. University of New Mexico Press, Albuquerque

Hayter R (2003) 'The war in the woods': post-Fordist restructuring, globalization, and the contested remapping of British Columbia's forest economy. Ann Assoc Am Geogr 93:706–729

Hayward DC (2001) The Nisga'a final agreement and its impact on forest management. Lakehead University, Faculty of Forestry and the Forest Environment, Thunder Bay

Hernandez Xolocotizi E (1985) La agricultura en la Peninsula de Yucatán. In: tomo I (ed) Xolocotzia: obras de E. Hernandez Xolocotzi, Revista de Geografia Agricola. Universidad Autonoma, Chapingo, pp 371–410

Higgins C (1998) The role of traditional ecological knowledge in managing for biodiversity. For Chron 74:323

Horvath S, Dickerson MO, MacKinnon L, Ross MM (2002) The impact of the traditional land use and occupancy study on the DeneTha' First Nation. Can J Nativ Stud 22:361–398

Houde N (2007) The six faces of traditional ecological knowledge: challenges and opportunities for Canadian co-management arrangements. Ecol Soc 12(2):34

Huff PR, Pecore M (1995) Case study: menominee tribal enterprises. The Institute for Environmental Studies and the Land Tenure Centre, Menominee

Huntington HP (2000) Using traditional ecological knowledge in science: methods and applications. Ecol Appl 10(5):1270–1274

Huntsinger L, McCaffrey S (1995) A forest for the trees: forest management and the Yurok environment, 1850 to 1994. Am Indian Cult Res J 19(4):155–192

Indian Forest Management Assessment Team [IFMAT] (1993) An assessment of Indian forests and forest management in the United States. Intertribal Timber Council, Portland

Inglis J (ed) (1993) Traditional ecological knowledge: concepts and cases. In: Common property conference, international workshop on indigenous knowledge and community based resource management. International Program on Traditional Ecological Knowledge, International Development Research Centre (Canada), Ottawa

James K (ed) (2001) Science and Native American communities: legacies of pain, visions of promise. University of Nebraska Press, Lincoln

Johnson M (1992) Lore: capturing traditional environmental knowledge. Dene Cultural Institute; International Development Research Centre, Yellowknife

Jones ET, Lynch KA (2007) Nontimber forest products and biodiversity management in the Pacific Northwest. For Ecol Manag 246:29–37

Karjala MK (2001) Integrating aboriginal values into strategic level forest planning on the John Prince Research Forest, central interior, British Columbia. Thesis University of Northern British Columbia, Prince George

Karjala MK, Dewhurst SM (2003) Including aboriginal issues in forest planning: a case study in central interior British Columbia, Canada. Landsc Urban Plan 64:1–17

Karjala MK, Sherry EE, Dewhurst SM (2003) The aboriginal forest planning process: a guidebook for identifying community-level criteria and indicators. University of Northern British Columbia, Prince George

Kerns BK, Alexander SJ, Bailey JD (2004) Huckleberry abundance, stand condition, and use in western Oregon: evaluating the role of forest management. Econ Bot 58(4):668–678

Kimmerer RW (2002) Weaving traditional ecological knowledge into biological education: a call to action. BioScience 52(5):432–438

Kimmerer R, Lake FK (2001) Maintaining the mosaic: the role of indigenous burning in land management. J For 99:36–41

Kimmins JP (2004) Forest ecology: a foundation for sustainable forest management and environmental ethics in forestry, 3rd edn. Pearson Prentice Hall, Upper Saddle River

Klooster D (1996) Como no conservar el Bosque: la marginalización del campesino eh la historia forestall Mexicana. Cuadernas Agrarias 14(6):144–156

Kohm KA, Franklin JF (eds) (1997) Creating a forestry for the 21st century: the science of ecology management. Island Press, Washington, DC

Korber D (1997) Measuring forest dependence: implications for aboriginal communities. Thesis, University of Alberta, Edmonton

Krahe DL (2001) A sovereign prescription for preservation: the Mission Mountains tribal wilderness. In: Clow RL, Sutton I (eds) Trusteeship in change: toward tribal autonomy in resource management. University Press of Colorado, Boulder, pp 195–221

Kroeber AL (1939) Cultural and natural areas of Native North America. University of California Press, Berkeley

Kuhnlein HV (1991) Traditional plant foods of Canadian indigenous peoples: nutrition, botany, and use—food and nutrition in history and anthropology. Gordon and Breach, New York

La Rusic IE (1995) Managing mishtuk: the experience of Waswanipi band in developing and managing a forestry company. In: Elias PD (ed) Northern aboriginal communities: economies and development. Captus Press, North York, pp 53–87

Lake FK (2007) Traditional ecological knowledge to develop and maintain fire regimes in the northwestern California, Klamath-Siskiyou bioregion: management and restoration of culturally significant habitats. Dissertation Oregon State University, Corvallis

Lambert RS, Pross AP (1967) Renewing nature's wealth; a centennial history of the public management of lands, forests and wildlife in Ontario, 1763–1967. Ontario Department of Lands and Forests, Toronto

Larsen SC (2003) Promoting aboriginal territoriality through interethnic alliances: the case of the Cheslatta T'en in northern British Columbia. Hum Organ 62:74–84

Latour B (2004) Politics of nature: how to bring the sciences into democracy. Harvard University Press, Cambridge

Lendsay KJ, Wuttunee W (1999) Historical economic perspectives of aboriginal peoples: cycles of balance and partnership. J Aboriginal Econ Dev 1:87–101

Lertzman DA (2006) Bridging traditional ecological knowledge and western science in sustainable forest management: the case of the Clayoquot scientific panel. University of Calgary, Haskayne School of Business, Calgary

Lertzman DA, Vredenburg H (2005) Indigenous peoples, resource extraction and sustainable development: an ethical approach. J Bus Ethics 56:239–254

Levy M (1994) The policies and politics of forestry in Ontario. Faculty of Environmental Studies, York University, Ontario

Lewis HT (1993) Patterns of Indian burning in California: ecology and ethnohistory. In: Blackburn TC, Anderson K (eds) Before the wilderness: environmental management by native Californians. Ballena Press, Menlo Park, pp 55–116

Lewis HT (2002) An anthropological critique. In: Lewis HT, Anderson MK (eds) Forgotten fires: Native Americans and the transient wilderness. University of Oklahoma Press, Norman, pp 17–36

Lewis JL, Sheppard SRJ (2005) Ancient values, new challenges: indigenous spiritual perceptions of landscapes and forest management. Soc Nat Resour 18(10):907–920

Little Bear L, Boldt M, Long JA (1984) Pathways to self-determination: Canadian Indians and the Canadian state. University of Toronto Press, Toronto

Long J, Tecle A, Burnette B (2003) Cultural foundations for ecological restoration on the White Mountain Apache Reservation. Ecology Society 8(1): Article 4. Available via www.ecologyandsociety.org/vol8/iss1/art4/inline.html. Cited 15 Jan 2011

Long J, Endfield MBD, Lupe C (2008) Battle at the bridge: using participatory approaches to develop community researchers in ecological management. In: Wilmsen C, Elmendorf W, Fisher L, Ross J, Sarathy B, Wells G (eds) Partnerships for empowerment: participatory research for community-based natural resource management. Earthscan, London, pp 217–237

Lundell CL (1937) The vegetation of Peten. Publication 478. Carnegie Institution of Washington, Washington, DC

Mabee HS, Hoberg G (2004) Protecting culturally significant areas through watershed planning in Clayoquot Sound. For Chron 80(2):229–240

Mabee HS, Hoberg G (2006) Equal partners? Assessing comanagement of forest resources in Clayoquot Sound. Soc Nat Resour 19(10):875–888

Macaulay A, Delormier T, McComber A, Cross E, Paradis G (1998) Participatory research with Native community of Kahnawake creates innovative code of research ethics. Can J Public Health 89(105):108

MacKendrick N, Parkins J, Northern Forestry Centre (2004) Frameworks for assessing community sustainability: a synthesis of current research in British Columbia. Information report. Northern Forestry Centre, Edmonton

Macklem P (1997) The impact of treaty 9 on natural resource development in northern Ontario. In: Asch M (ed) Aboriginal and treaty rights in Canada: essays on law, equality, and respect for difference. University of British Columbia Press, Vancouver, pp 97–134

MacPherson NE (2009) Traditional knowledge for health. Thesis, University of British Columbia and Siska Traditions Society, Vancouver

Mann M (2003) Capitalism and the dis-empowerment of Canadian aboriginal people. In: Anderson R, Bone R (eds) Natural resources and aboriginal people in Canada: readings, cases and commentary. Captus Press, Concord, pp 18–29

Manseau M, Parlee B, Ayles G (2005) A place for traditional ecological knowledge in resource management. In: Berkes F, Huebert R, Fast H, Manseau M, Diduck A (eds) Breaking ice: renewable resource and ocean management in the Canadian north. University of Calgary Press, Calgary

Markey NM (2001) Data 'gathering dust': an analysis of traditional use studies conducted within aboriginal communities in British Columbia. Thesis, Simon Fraser University, Vancouver

Marlor C, Barsh R, Duhaylungsod L (1999) Comment on 'defining indicators which make sense to local people: intra-cultural variation in perceptions of natural resources. Hum Organ 58:216–219

Marsden T (2005) From the land to the Supreme Court, and back again: defining meaningful consultation with First Nations in northern British Columbia. Thesis, University of Northern British Columbia, Prince George

Matthews SM, Golightly RT, Higley JM (2008) Mark-resight density estimation for American black bears in Hoopa, California. Ursus 19(1):13–21

McDonald M (2003) Aboriginal forestry in Canada. In: Anderson R, Bone R (eds) Natural resources and aboriginal people in Canada: readings, cases and commentary. Captus Press, Concord, pp 230–256

McDonnell JA (1991) The dispossession of the American Indians, 1887–1934. Indiana University Press, Bloomington

McGregor D (2000) From exclusion to co-existence: aboriginal participation in Ontario forest management planning. Dissertation, University of Toronto, Faculty of Forestry, Toronto

McGregor D (2002) Indigenous knowledge in sustainable forest management: community-based approaches achieve greater success. For Chron 78:833–836

McGregor D (2004) Coming full circle: indigenous knowledge, environment, and our future. Am Indian Q 28(3/4):385–410

McGregor D (2010) Traditional knowledge, sustainable forest management and ethical research involving aboriginal peoples. In: White JP (ed) Exploring voting, governance and research methodology, vol 10, Aboriginal Policy Research. Thompson Educational Publishers, Toronto, pp 229–246

McGregor D (2011) Aboriginal/non-aboriginal relations and sustainable forest management in Canada: the influence of the Royal Commission on Aboriginal Peoples. J Environ Manag 92:300–310

McKay RA (2004) Kitsaki management limited partnership: an aboriginal economic development model. J Aboriginal Econ Dev 4:3–5

McNaughton C, Rock D (2003) Opportunities in aboriginal research: results of SSHRC's dialogue on research and aboriginal peoples. Social Sciences and Humanities Research Council (SSHRC), Ottawa

McQuat G (1998) What is western science? In: Manseau M (ed) Traditional and western scientific environmental knowledge. Institute for Environmental Monitoring and Research, Goose Bay-Labrador, pp 7–10

McQuillan AG (2001) American Indian timber management policy: its evolution in the context of U.S. forest history. In: Clow RL, Sutton I (eds) Trusteeship in change: toward tribal autonomy in resource management. University Press of Colorado, Boulder, pp 73–104

McTague JP, Stansfield WF (1994) Stand and tree dynamics of uneven-aged ponderosa pine. For Sci 40(2):289–302

McTague JP, Stansfield WF (1995) Stand, species, and tree dynamics of an uneven-aged, mixed conifer forest type. Can J For Res 25(5):803–812

Medellín Morales SG (1986) Uso y manejo de las especies vegetales comestibles, medicinales, para construcción y combustibles en una comunidad Totonaca de la costa (plan de Hidalgo, Papantla). Instituto Nacional de Investigaciones sobre Recursos Bioticos, Xalapa

Menzies C (2001) Reflections on research with, for, and among aboriginal people. Can J Nativ Educ 25(1):19–36

Menzies CR (2004) Putting words into action: negotiating collaborative research in Gitxaala. Can J Nativ Educ 28(1, 2):15–32

Merino-Pérez L (2004) Conservación o deterioro: el impact de las políticas públicas en las instituciones comunitarias y en los usos de los bosques en México. SEMARNAT, INE, CCMSS, Mexico City

Merino-Perez L, Segura-Warnholtz G (2005) Forests and conservation policies and their impact on forest communities in Mexico. In: Bray DB, Merino-Perez L, Barry D (eds) The community forests of Mexico. University of Texas Press, Austin, pp 49–69

Merkel G, Osendarp F, Smith P (1994) For seven generations: an information legacy of the royal commission on aboriginal peoples. In: Sectoral study: forestry—an analysis of the forest industry's views of aboriginal participation. Royal Commission on Aboriginal Peoples, Ottawa

Messer E (1975) Zapotec plant knowledge: classification, uses, and communication about plants in Mitla. Dissertation, University of Michigan, Ann Arbor

Miller JR (1989) Skyscrapers hide the heavens: a history of Indian-white relations in Canada. University of Toronto Press, Toronto

Miller AM, Davidson-Hunt IJ, Peters P (2010) Talking about fire: Pikangikum First Nation elders guiding fire management. Can J For Res 40:2290–2301

Mills A (1994) Eagle down is our law: Witsuwit'en law, feasts, and land claims. University of British Columbia Press, Vancouver

Mills A (ed) (2005) 'Hang onto these words': Johnny David's Delgamuukw evidence. University of Toronto Press, Toronto

Molina R, Vance N, Weigand JF, Pilz D, Amaranthus MP (1997) Special forest products: integrating social, economic, and biological considerations into ecosystem management. In: Kohm KA, Franklin JF (eds) Creating a forestry for the 21st century: the science of ecosystem management. Island Press, Washington, DC, pp 315–336

Moller H, Berkes F, Lyver PO, Kislalioglu M (2004) Combining science and traditional ecological knowledge: monitoring populations and co-management. Ecology and Society 9(3): Article 2. Available via http://www.ecologyandsociety.org/vol9/iss3/art2/. Cited 15 Jan 2011

Morfin VL (1997) Changes in composition and structure in ponderosa pine/Douglas-fir stands on the Colville Indian Reservation. Thesis, Northern Arizona University, Flagstaff

Nadasdy P (2003a) Reevaluating the co-management success story. Arctic 56:367–380

Nadasdy P (2003b) Hunters and bureaucrats: power, knowledge, and aboriginal-state relations in the southwest Yukon. University of British Columbia Press, Vancouver

Nadasdy P (2005) The anti-politics of TEK: the institutionalization of co-management discourse and practice. Anthropologica 47:215–232

NAHO (2007a) and NAHO (2007b): See National Aboriginal Health Organization

Natcher DC (1999) Co-operative resource management as an adaptive strategy for aboriginal communities. Sustainable Forest Management Network, Edmonton

Natcher DC (2000) Constructing change: the evolution of land and resource management in Alberta, Canada. Int J Sustain Dev World Ecol 7(4):363

Natcher DC (2001) Land use research and the duty to consult: a misrepresentation of the aboriginal landscape. Land use Policy 18:113–122

Natcher DC (2004) Implications of fire policy on native land use in the Yukon flats, Alaska. Hum Ecol Interdiscip J 32(4):421–442

Natcher DC (2008) Seeing beyond the trees: the social dimensions of aboriginal forest management. Captus Press, Concord

Natcher DC, Hickey CG (2002) Putting the community back into community-based resource management: a criteria and indicators approach to sustainability. Hum Organ 61:350–363

Natcher DC, Hickey CG, Davis S (2004) The political ecology of Yukon forestry: managing the forest as if people mattered. Int J Sustain Dev World Ecol 11(4):343–355

Natcher DC, Davis S, Hickey CG (2005) Co-management: managing relationships, not resources. Hum Organ 64(3):240

National Aboriginal Forestry Association (1993) Forest lands and resources for aboriginal people: an intervention submitted to the Royal Commission on Aboriginal Peoples. National Aboriginal Forestry Association, Ottawa

National Aboriginal Health Organization [NAHO] (2007a) Handbook and resource guide to the convention on biodiversity. NAHO, Ottawa

National Aboriginal Health Organization [NAHO] (2007b) OCAP: ownership, control, access and possession. First Nations Centre, Ottawa

National Forest Strategy Coalition [NFSC] (2003) National forest strategy, 2003–2008: a sustainable forest, the Canadian commitment. Canadian Council of Forest Ministers, Ottawa

Nilsson S, Gluck M (2001) Sustainability and the Canadian forest sector. For Chron 77:39–47

Notzke C (1994) Forestry. In: Notzke C (ed) Aboriginal peoples and natural resources in Canada. Captus Press, North York, pp 81–108

Nowacki GJ, Abrams MD (2008) The demise of fire and 'mesophication' of forests in the eastern United States. BioScience 58(2):123–138

O'Flaherty R, Davidson-Hunt I, Manseau M (2008) Indigenous knowledge and values in planning for sustainable forestry: Pikangikum First Nation and the Whitefeather Forest Initiative. Ecol Soc 13(1):6–16

O'Flaherty RM, Davidson-Hunt IJ, Miller A (2009) Anishinaabe stewardship values for sustainable forest management of the Whitefeather Forest, Pikangikum First Nation, Ontario. In: Stevenson MG, Natcher DC (eds) Changing the culture of forestry in Canada: building effective institutions for aboriginal engagement in sustainable forest management, vol 60, Occasional Publications Series. CCI Press, Edmonton, pp 19–34

Parkins JR, Stedman RC, Varghese J (2001) Moving towards local-level indicators of sustainability in forest-based communities: a mixed method approach. Soc Indic Res 56:43–52

Parkins JR, Stedman RC, Patriquin MN, Burns M (2006) Strong policies, poor outcomes: longitudinal analysis of forest sector contributions to aboriginal communities in Canada. J Aboriginal Econ Dev 5(1):61–73

Parlee B, Berkes F, Council Teel'it Gwich'in Renewable Resources (2005) Health of the land, health of the people: a case study on Gwich'in berry harvesting in northern Canada. EcoHealth 2:127–137

Parsons R, Prest G (2003) Aboriginal forestry in Canada. For Chron 79:779–784

Pecore M (1992) Menominee sustained yield management: a successful land ethic in practice. J For 90:12–16

Perry DA, Oren R, Hart SC (2008) Forest ecosystems, 2nd edn. The John Hopkins University Press, Baltimore

Peters CM (2000) Precolumbian silviculture and indigenous management of neo-tropical forests. In: Lentz DL (ed) Imperfect balance: landscape transformation in the Pre-Columbian Americas. Columbia University Press, New York, pp 203–224

Peters EJ (2003) Views on traditional ecological knowledge in co-management bodies in Nunavik, Quebec. Polar Record 39(208):49–60

Pikangikum First Nation [PFN] and Ontario Ministry of Natural Resources [OMNR] (2006) Keeping the land: a land use strategy for the Whitefeather Forest and adjacent areas. PFN and OMNR, Pikangikum and Red Lake. Available via http://www.whitefeatherforest.com/pdfs/land-use-strategy.pdf. Cited 4 Feb 2011

Piquemal N ((Dec 2000) Four principles to guide research with aboriginals. Policy Options, pp 49–50

Plummer R, FitzGibbon J (2004) Co-management of natural resources: a proposed framework. Environ Manag 33(6):876–885

Pyne SJ (1982) Fire in America: a cultural history of wildland and rural fire. Princeton University Press, Princeton

Rekmans L (2002) Aboriginal people, science and innovation. For Chron 78(1):101–102

Richardson B (1993) People of the terra nullius: betrayal and rebirth in aboriginal Canada. Douglas and McIntyre, Vancouver

Rodon T (2003) En partenariat avec l'état; les expériences de cogestion des autochtones du Canada. Les Presses de l'Université Laval, Quebec

Roots F (1998) Inclusion of different knowledge systems in research. In: Manseau M (ed) Traditional and western scientific environmental knowledge. Institute for Environmental Monitoring and Research, Goose Bay-Labrador, pp 42–49

Ross M, Smith P (2002) Accommodation of aboriginal rights: the need for an aboriginal forest tenure. In: Synthesis report prepared for the Sustainable Forest Management Network. University of Alberta, Edmonton

Royal Commission on Aboriginal Peoples [RCAP] (1996a) Appendix 4B. Co-management agreements, report of the Royal Commission on Aboriginal peoples, Ottawa

Royal Commission on Aboriginal Peoples [RCAP] (1996b) Lands and resources: report of the Royal Commission on Aboriginal Peoples, vol 2, Restructuring the relationship. Canada Communications Group Publishing, Ottawa, pp 421–685

Royal Commission on Aboriginal Peoples [RCAP] (1996c) People to people, nation to nation: highlights from the report of the Royal Commission on Aboriginal Peoples. Minister of Supply and Services, Ottawa. Available via http://www.ainc-inac.gc.ca/ap/pubs/rpt/rpt-eng.asp. Cited 25 Feb 2011

Royal Commission on Aboriginal Peoples [RCAP] (1996d) Report of the Royal Commission on Aboriginal Peoples. Vol 2, part 2. Ottawa

Salmón E (2000) Kincentric ecology: indigenous perceptions of the human nature relationship. Ecol Appl 10(5):1327–1332

Sassaman RW, Miller RW (1986) Native American forestry. J For 84(10):26–31

Scientific Panel for Sustainable Forest Practices in Clayoquot Sound (1995a) First Nations' perspectives relating to forest practices standards in Clayoquot Sound. Report 3, Clayoquot Scientific Panel, Victoria. Available via http://www.cortex.org/dow-cla.html. Cited 4 Feb 2011

Scientific Panel for Sustainable Forest Practices in Clayoquot Sound (1995b) Sustainable ecosystem management in Clayoquot Sound: planning and practices. Report 5, Clayoquot Scientific Panel, Victoria

Scott CH (ed) (2001) Aboriginal autonomy and development in northern Quebec and Labrador. University of British Columbia Press, Vancouver

Second Indian Forest Management Assessment Team [Second IFMAT] (2003) An assessment of Indian forests and forest management in the United States. Intertribal Timber Council, Portland

Shackeroff JM, Campbell LM (2007) Traditional ecological knowledge in conservation research: problems and prospects for their constructive engagement. Conserv Soc 5(3):343–360

Shearer J, Peters P, Davidson-Hunt IJ (2009) Co-producing a Whitefeather Forest cultural landscape monitoring framework. In: Stevenson MG, Natcher DC (eds) Changing the culture of forestry in Canada: building effective institutions for aboriginal engagement in sustainable forest management, vol 60, Occasional Publications Series. CCI Press, Edmonton, pp 63–84

Sheppard S, Lewis JL, Akai C (2004) Landscape visualization: an extension guide for First Nations and rural communities. Sustainable Forest Management Network, Edmonton

Sherry E, Halseth R, Fondahl G, Karjala M, Leon B (2005) Local-level criteria and indicators: an aboriginal perspective on sustainable forest management. Forestry 78(5):513–539

Smith DM (1997) The practice of silviculture: applied forest ecology, 9th edn. Wiley, New York

Smith P (1998) Aboriginal and treaty rights and aboriginal participation: essential elements of sustainable forest management. For Chron 74:327–333

Smith LT (1999) Decolonizing methodologies: research and indigenous peoples. Zed Books, University of Otago Press, London

Smith P (2001) Indigenous peoples and forest management in Canada. In: Rolfe T (ed) The nature and culture of forests: implications of diversity for sustainability, trade and certification. University of British Columbia, Institute for European Studies, Vancouver

Smith P (2006) Community-based framework for measuring the success of indigenous people's forest-based economic development in Canada. In: Merino L, Robson J (eds) Managing the commons: indigenous rights, economic development and identity. Instituto de Ecologia [NE], Mexico City

Smith P (2007) Creating a new stage for sustainable forest management through co-management with aboriginal peoples in Ontario: the need for constitutional-level enabling. Dissertation, University of Toronto, Toronto

Smith P, Symington E, Allen S (2010) First Nations' criteria and indicators of sustainable forest management: a review. In: Stevenson MG, Natcher DC (eds) Planning co-existence: aboriginal issues in forest and land use planning, research and insights from the aboriginal program of the sustainable forest management network, vol 64, Occasional Publications Series. CCI Press, Edmonton, pp 225–264

Sneed PG (1997) National parklands and northern homelands: toward co-management of national parks in Alaska and the Yukon. In: Stevens S (ed) Conservation through cultural survival: indigenous peoples and protected areas. Island Press, Washington, DC, p 135

Snively G (2006) Honoring aboriginal science knowledge and wisdom in an environmental education graduate program. In: Menzies CR (ed) Traditional ecological knowledge and natural resource management. University of Nebraska Press, Lincoln, pp 195–220

Stevenson MG (1998) Traditional knowledge in environmental management: from commodity to process. Sustainable Forest Management Network, Edmonton

Stevenson MG, Natcher DC (eds) (2009) Changing the culture of forestry in Canada: building effective institutions for aboriginal engagement in sustainable forest management, vol 60, Occasional Publications Series. CCI Press, Edmonton

Stevenson MG, Natcher DC (eds) (2010) Planning co-existence: aboriginal issues in forest and land use planning, research and insights from the aboriginal program of the sustainable forest management network, Occasional Publications Series. CCI Press, Edmonton

Stevenson M, Webb J (2003) Chapter 3: Just another stakeholder? First Nations and sustainable forest management in Canada's boreal forest. In: Burton P, Messier C, Smith DW, Adamowicz WL (eds) Towards sustainable management of the boreal forest. National Research Council of Canada Press, Ottawa, pp 65–112

Stewart OC, Lewis HT, Anderson MK (2002) Forgotten fires: Native Americans and the transient wilderness. University of Oklahoma Press, Norman

Suttles W (1990a) Introduction. In: Suttles W (ed) Handbook of North American Indians, volume 7, northwest coast. Smithsonian Institution, Washington, DC, pp 1–15

Suttles W (ed) (1990b) Handbook of North American Indians, vol 7. Northwest coast. Smithsonian Institution, Washington, DC

Tanner A (1988) The significance of hunting territories today. In: Cox BA (ed) Native people, native lands: Canadian Indian, Inuit and Métis. Carleton University Press, Ottawa, pp 60–74

Tecumseh Professional Associates I (1999) Flathead Indian Reservation forest management plan: final environmental impact statement. Bureau of Indian Affairs and Confederated Salish and Kootenai Tribes, Pablo

Trigger BG (1969) The Huron: farmers of the north. Holt, Rinehart and Winston, New York

Trosper RL (2007) Indigenous influence on forest management on the Menominee Reservation. For Ecol Manag 249:134–139

Trosper RL (2009) Resilience, reciprocity and ecological economics: sustainability on the Northwest Coast. Routledge, New York

Trosper R, Nelson H, Hoberg G, Smith P, Nikolakis W (2008) Institutional determinants of profitable commercial forestry enterprises among First Nations in Canada. Can J For Res 38(2):226–238

Turner NJ (2001) 'Doing it right': Issues and practices of sustainable harvesting of non-timber forest products relating to First Peoples in British Columbia. BC J Ecosyst Manage 1:1–11

Turner NJ, Peacock S (2005) Solving the perennial paradox: ethnobotanical evidence for plant resource management on the northwest coast. In: Deur D, Turner N (eds) Keeping it living: traditions of plant use and cultivation on the Northwest Coast of North America. University of Washington Press/University of British Columbia Press, Seattle/Vancouver, pp 101–150

Venne S (1997) Understanding treaty 6: an indigenous perspective. In: Asch M (ed) Aboriginal and treaty rights in Canada: essays on law, equality, and respect for difference. University of British Columbia Press, Vancouver, pp 173–207

Wanlin M (1999) A voice and a plan: key steps toward economic development. J Aboriginal Econ Dev 1:44–48

Waswanipi Cree Model Forest (2007) Ndoho istchee: an innovative approach to aboriginal participation in forest management planning. Waswanipi, Quebec

Watson A, Matt R, Waters T, Gunderson K, Carver S, Davis B (2008) Mapping tradeoffs in values at risk at the interface between wilderness and non-wilderness lands. In: González-Cabán A (ed) Proceedings of the international symposium on fire economics, planning, and policy: common problems and approaches, 29 Apr–2 May 2008, Carolina, Puerto Rico. General Technical Report PSW-GTR-227. U.S. Department of Agriculture [USDA], Forest Service, Pacific Southwest Station, Albany, pp 375–388

Wilkinson CF (2005) Blood struggle: the rise of modern Indian nations, 1st edn. Norton, New York

Williams GW (2003) References on the American Indian use of fire in ecosystems. U.S. Department of Agriculture [USDA], Forest Service, Washington, DC. Available via http://www.blm.gov/heritage/docum/Fire/Bibliography%20-%20Indian%20Use%20of%20Fire.pdf. Cited 21 Dec 2010

Wilmsen C, Krishnaswamy A (2008) Challenges to institutionalizing participatory research in community forestry in the US. In: Wilmsen C, Elmendorf W, Fisher L, Ross J, Sarathy B, Wells G (eds) Partnerships for empowerment: participatory research for community-based natural resource management. Earthscan, London, pp 47–67

Wilmsen C, Elmendorf W, Fisher L, Ross J, Sarathy B, Wells G (2008) Partnerships for empowerment: participatory research for community-based natural resource management. Earthscan, London

Witty D (1994) The practice behind the theory: co-management as a community development tool. Plan Canada 44:22–27

Wulfhorst JD, Eisenhauer BW, Gripne SL, Ward JM (2008) Core criteria and assessment of participatory research. In: Wilmsen C, Elmendorf W, Fisher L, Ross J, Sarathy B, Wells G (eds) Partnerships for empowerment: participatory research for community-based natural resource management. Earthscan, London, pp 23–46

Wyatt S (2008) First Nations, forest lands, and 'aboriginal forestry' in Canada: from exclusion to comanagement and beyond. Can J For Res 38(2):171–180

Wyatt S, Natcher DC, Smith P, Fortier J (2010) Aboriginal land use mapping: what have we learned from 30 years of experience? In: Stevenson MG, Natcher DC (eds) Planning co-existence: aboriginal issues in forest and land use planning, research and insights from the aboriginal program of the sustainable forest management network, Occasional Publications Series. CCI Press, Edmonton, pp 185–198

Chapter 6
Europe

Elisabeth Johann, Mauro Agnoletti, János Bölöni, Seçil Yurdakul Erol, Kate Holl, Jürgen Kusmin, Jesús García Latorre, Juan García Latorre, Zsolt Molnár, Xavier Rochel, Ian D. Rotherham, Eirini Saratsi, Mike Smith, Lembitu Tarang, Mark van Benthem, and Jim van Laar

Abstract Forests and other wooded lands cover about a third of the European land area and are therefore a characteristic element of the continent's natural and cultural landscape. Woodland has always provided people with economic, social and environmental products and services. Indeed the history of Western civilisation would

E. Johann (✉)
Austrian Forest Association, Vienna, Austria
e-mail: elisabet.johann@aon.at

M. Agnoletti
Dipartimento di Scienze e Teconolgie Ambientali Forestali,
Università di Firenze, Facoltà di Agraria, Florence, Italy
e-mail: mauro.agnoletti@unifi.it

J. Bölöni • Z. Molnár
Institute of Ecology and Botany, Hungarian Academy of Sciences, Vacratot, Hungary
e-mail: jboloni@botanika.hu; molnar@botanika.hu

S.Y. Erol
Faculty of Forestry, Istanbul University, Istanbul, Turkey
e-mail: sechily@hotmail.com

K. Holl
Forest Research, Northern Research Station, Bush, Midlothian, UK
e-mail: Kate.Holl@snh.gov.uk

J. Kusmin
State Forest Management Centre, Tallinn, Estonia
e-mail: jurgen@velise.ee

J. García Latorre
Federal Ministry of Agriculture, Forestry, Environment and Water Management, Vienna, Austria
e-mail: Jesus.Garcia-Latorre@lebensministerium.at

J. García Latorre
Association for Landscape Research in Arid Zones, Almería, Spain
e-mail: jglatorre2@gmail.com

J.A. Parrotta and R.L. Trosper (eds.), *Traditional Forest-Related Knowledge:*   203
Sustaining Communities, Ecosystems and Biocultural Diversity,
World Forests 12, DOI 10.1007/978-94-007-2144-9_6,
© Springer Science+Business Media B.V. (outside the USA) 2012

be dramatically different without the multiple benefits forests provided to society. However, the current distribution and composition of forests in most parts of Europe reflect the profound cumulative impacts of many centuries of land use change and forest management. While in many cases the loss of biodiversity of cultural landscapes we observe today is closely related to modern exploitation strategies, very often this situation is connected with changes in traditional agricultural systems and the abandonment of traditional land management practices.

In this chapter, we examine the principal factors responsible for the development of locally adapted technologies and traditional forest management practices used historically to sustain the long-term availability of forest resources through generations. The chapter also considers the influences of science and modern forestry on the development of cultural landscapes. A central part of the chapter considers the relevance of traditional forest-related knowledge to current debates about sustainable forest management. Inclusion of traditional forest-related knowledge within formal scientific forestry is considered a necessary step to maintain an important part of the European cultural heritage. Such knowledge is also regarded vital for the development of an effective approach to maintaining the ecological balance of European forests and for securing the sustainable development of rural areas in Europe.

Keywords Cultural heritage • Europe • Forest history • Forest management • Forest policy • Multiple-use forestry • Science-based forestry • Sustainable forest management • Traditional knowledge • Woodland history

X. Rochel
Département de Géographie, Université Nancy 2, Nancy, France
e-mail: Xavier.Rochel@univ-nancy2.fr

I.D. Rotherham
Sheffield Hallam University, Sheffield, UK
e-mail: I.D.Rotherham@shu.ac.uk

E. Saratsi
Department of Politics, Centre for Rural Policy Research, University of Exeter, Exeter, UK
e-mail: E.Saratsi@exeter.ac.uk

M. Smith
Scottish Natural Heritage, Edinburgh, UK
e-mail: Mike.Smith@forestry.gsi.gov.uk

L. Tarang
Estonian Society of Foresters, Haapsalu, Estonia
e-mail: letarang@hot.ee

M. van Benthem
Probos Foundation, Wageningen, The Netherlands
e-mail: mark.vanbenthem@probos.net

J. van Laar
Forest and Nature Conservation Policy Group, Wageningen University,
Wageningen, The Netherlands
e-mail: Jim.vanlaar@wur.nl

6.1 Introduction

The European landscape is characterized by richly varied climate, topography, population density, and political and social conditions. The long relationship between nature and human activity has produced distinct landscapes shaped by cultural differences among societies and reflecting how humans have responded to their natural environments (Kirby and Watkins 1998). In the same way, European wooded and forested landscapes bear witness to cultural processes and developments, showing evidence of impacts through numerous constantly changing human needs. The shaping and re-shaping of these landscapes over the centuries is the result of both private and public land management decisions, in national and local levels; decisions often influenced by predominant cultural, ethical, and societal trends of the time. Understanding this legacy, acquiring more information about it, and developing awareness of it through interdisciplinary research, will lead to new and important insights. Information generated by this approach will potentially foster the essential awareness required for more appropriate assessments and evaluations to be made in order to influence present-day cultural accountability and environmental decision-making.

Social and environmental factors have come together to create the forests that today constitute part of Europe's identity and its cultural heritage. The sustainable management and conservation of this unique cultural heritage requires consideration of social and cultural values related to forests and forestry. Historically, traditional knowledge of woodland management to provide multiple goods and services, food, raw material, and energy sources, secured the livelihoods of local populations. However, social and cultural values change over time as societies develop. From the early nineteenth century onwards, rural depopulation and changes in agricultural practices (retreat or intensification) proceeded alongside socioeconomic development and science-based forestry. These processes have deeply altered the relationship between society and forest resources, interrupting the transmission of traditional forest-related knowledge between generations and reducing its contribution to the cultural identity of European regions. With the introduction of the science of forestry, cultural wooded landscapes were managed differently based on expert knowledge and professional skills. In recent decades discussions about sustainability have placed great emphasis on the integration of broader criteria, which incorporate a historical understanding about the development of such areas. The breakdown in transmission of traditional knowledge has made sustaining the diversity of cultural wooded landscapes and their local communities more difficult because valuable information related to landscape management has been lost. Yet international agreements, processes, and programmes increasingly recognize wooded landscapes as cultural heritage. Examples include the Convention on Biological Diversity (CBD), the United Nations Convention of Compact Desertification (UNCCD), UNESCO's World Heritage Convention, the United Nation Forum on Forests (UNFF), and the fourth Ministerial Conference on the Protection of Forest in Europe (Saratsi et al. 2007; Agnoletti et al. 2007).

Some decades ago also foresters rarely considered the many traditional forest functions as relevant to economically driven extraction. They did not see them as a credible option for broadening the circle of beneficiaries or stakeholders, substituting declining income from timber, or gaining new social engagement with forests and forestry in post-industrial societies. Qualities such as recreation, culture, beauty, and water protection were seen as by-products of a system for timber production. They were assumed to occur almost spontaneously, and costs (if any at all) were included in timber production (Anko 2005). However, there is now a growing awareness of the forest's cultural and environmental functions and their importance for environmental and ecological quality. The forest in its entirety is increasingly regarded as an essential component of the landscape. It is not only integrated, as in the past, in a grid of productive structures, but is now also defined in a network of linguistic and cultural relations (Zanzi Sulli 1997).

Based on these grounds a series of questions arise. What kind of social and economic structures contributed to the evolution of diversity of the cultural landscapes, and what factors were mainly responsible for the development of locally adapted technologies? What kind of forest management was practised in previous times and what measures were taken by local populations in order to secure the available natural resources over the long-term and maintain them for future generations? In what way and dimension did increasing industrial demands influence traditional forest management and the practice of locally adapted skills and techniques, and what was the contribution of modern forestry and forest science to this development? To deepen our understanding and to be able to approach the questions presented above, we conducted a historical review of traditional forest- related knowledge.

6.2   Europe and Its Forests

The region covered in this chapter includes the 45 European countries participating in the Ministerial Conference on the Protection of Forests in Europe (MCPFE) (Fig. 6.1). Because of the variety and heterogeneity of European cultural landscapes, it is not possible to provide detailed information on the history and preservation of traditional forest-related knowledge in all countries. Therefore case studies from the different eco-regions have been selected to illustrate the general trends in history and contemporary importance of traditional forest-related knowledge in detail. In selecting the countries to be included, attention was paid to the different climatic conditions, forest composition and distribution of tree species (conifers or broad-leaved forests), and the different socioeconomic and cultural importance of forests. As the Russian Federation is a separate chapter in this book, its data, apart from some general information mentioned below, are not included in this chapter. The intention is to present a broad overview for the European region but illustrated by pertinent specific examples from individual areas and countries.

Within the territory of the MCPFE countries, forests cover more than one billion ha, approximately 25% of the world's total forest area. This is equivalent to 44.3%

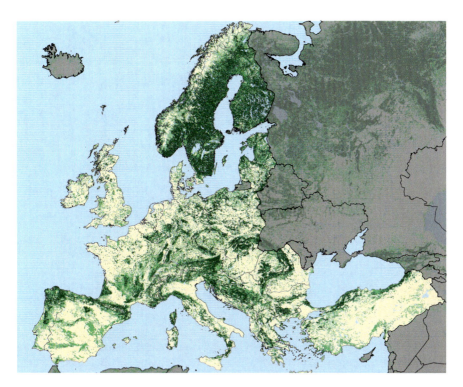

Fig. 6.1 Forest and woodland cover in Europe (Source: Adapted from FAO (2001)). *Key*: *Dark green* closed forest, *light green* open or fragmented forest, *pale green* other wooded land, *yellow* other land

of the region's total land area. In addition, other wooded lands constitute 5.0% of the total land area. The distribution of forests varies substantially among countries, with the percentage of forest cover being twice as high in the Russian Federation and in the Nordic countries than in the northwestern and southeastern European countries. Per capita forest area is highest in the Russian Federation (5.7 ha/capita), followed by Finland and Sweden (4.2 and 3.1 ha/capita, respectively). However, there are only 0.5 ha or less of forest per capita in most other countries. The area of other wooded land per capita is less than 0.6 ha/capita in all countries. In the past 15 years European forests have increased by almost 13 million ha due mainly to planting of new forests and natural expansion of forests onto abandoned former agricultural land (MCPFE et al. 2007).

Europe's temperate forests occur over a broad bioclimatological spectrum, ranging from oceanic to continental forests, and from floodplain to mountain forests up to the alpine timberline. The Atlantic climate in the west turns into a continental climate with decreasing humidity and wider temperature variation in the eastern part of Europe. At present in some countries of the temperate zone—such as in Denmark, the United Kingdom, Ireland, and the Netherlands—forest cover is only about 10% of the total area, while in Austria, Estonia, Latvia, Liechtenstein, Slovakia, and Slovenia,

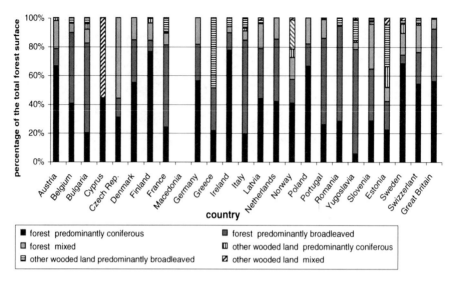

Fig. 6.2 Forest and other wooded land by forest types (species groups) in several European countries (Source: MCPFE 2003d)

forest cover exceeds 40%. In Austria, Ireland, Poland, and the United Kingdom, more than 60% of the forest area is dominated by conifers, while in Bulgaria, Croatia, France, Hungary, Luxembourg, Romania, and the Republic of Moldova, more than 60% of the forest area is dominated by broadleaved trees (Fig. 6.2).

High forests—generally of seed or seedling origin, mainly even-aged, and normally characterized by a high closed canopy—are the most common silvicultural structure category. In Bulgaria, France, Liechtenstein, Italy, the Republic of Moldova, and the Ukraine, coppice forests and coppice-with-standards constitute more than 40% of the forest area. In a few countries—namely Croatia, Germany, Liechtenstein, Slovakia, Slovenia, and Switzerland—the area of uneven-aged forest amounts to 15% or more of the total forest area (Parviainen 2007). About 70% of the European forests are classified as semi-natural, 4% as plantations, and the remaining 26% (located mainly in eastern and northern European countries) are considered undisturbed (Fig. 6.3).

Woodland cultures survive wherever knowledge and skills accumulate through the continuous application of practice and theory. Cultural heritage and traditional forest-related knowledge can either be reflected in human activity related to the forest or be features or activity situated within the forest. For example, extraction of various forestry related resources creates a physical heritage in the forest, such as sawpits and slag heaps; in other cases, different land uses in forested areas have led to more of a cultural heritage in the forest (Svensson 2006). Knowledge of the history of forests comes from diverse places: the forest itself (archaeological sites and the forest structure), field names indicating past uses, species composition of forests, archives, libraries, and memories. Ancient monuments are often considered to be the only evidence about past societies and can be an important part of a national

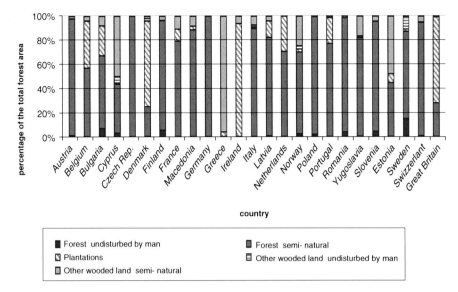

Fig. 6.3 Forest and other wooded land by categories of naturalness in several European countries (Source: MCPFE 2003d)

identity; however, the woodlands were places where former people lived and worked, and forests also contain historical records and other evidence. Therefore as far as sustainable forest management is concerned, one has to acknowledge history and adopt a regional approach.

This study draws on research presented at several international conferences such as Vienna, Austria, 2004 (Johann 2005); Sunne, Sweden, 2005 (MCPFE 2006); Florence, Italy, 2006 (Parrotta et al. 2006); and Thessaloniki, Greece, 2007 (Saratsi et al. 2009). We organized information derived from these sources to present and reveal the current status, presence, or absence of traditional forest-related knowledge in different parts of Europe, and to make general comment and comparison possible. In addition, this essay is based on data from a questionnaire developed to obtain a regional approach to information gathering. The information was derived from reviews of historical documents and other literature related to forest and forest management history and presented in 11 reports carried out by the co-authors of this study. These encompass detailed information gathered along a gradient across Europe from northwest to southeast and from northeast to southwest, and encompassing the following countries: Great Britain (including Scotland), the Netherlands, France, Germany, Austria, Greece, Turkey, Estonia, Hungary, and Spain. Together these accounts illustrate the very heterogenic picture of European forestry with regard to the safeguarding and implementing cultural heritage in scientific-based sustainable forest management. The study also takes into account international agreements that have been signed by most of the European Union (EU) member countries and the EU such as the *Fourth Ministerial Conference on the Protection of Forest in Europe* (MCPFE) 2003, Vienna, Austria (MCPFE 2003c); the *Convention for the Safeguarding of the Intangible Cultural Heritage* (UNESCO 2003); and the *Declaration Population and Culture of the Alpine Convention.*[1] Additional information was gathered from two *COST* Actions (Glück and Weber 1998; Frank et al. 2007).

6.3   History of Traditional Forest-Related Knowledge with Respect to Forest Management

Forests have a long memory, telling stories of past mismanagement but also of conscious human efforts to preserve their natural structure. They also reveal traces of long-ago abandoned forest uses such as charcoal burning, pollarding, and coppicing. In a very special way, the allocation and spatial distribution patterns of forest matrices, patches, and corridors in a forest also represent a valuable historic record (Forman 1996). Forests, indeed, are the shadow of civilization. They reflect wars and economic crises as well as the rise and fall of industries and markets. Cultural landscapes, as a product of human intervention in natural processes, have always depended on socioeconomic evolution and have adapted to changing needs and evolving technologies (Anko 2005). Forests are entities and agencies that invoke a wealth of knowledge, which is historically accumulated and embedded in their features and management history. Socioeconomic processes that have taken place mainly during the past two centuries have deeply affected the flow of this lay knowledge from generation to generation.

6.3.1   Forest Management from Prehistoric to Medieval Times

6.3.1.1   Early Records of Forest Use

The time when the forest surface first became altered by human activities has many uncertainties. However, Bohn and Neuhäusl (2000/2003) have proved that the distribution of tree species of European forests of the present day deviates remarkably from the potential natural vegetation of Europe, considering that the climate has remained fairly steady over the past 4,500 years (except for relatively wet and cold

---

[1] The Convention is a framework that sets out the basic principles of all the activities of the Alpine Convention and contains general measures for the sustainable development in the Alpine region. It was signed by the Federal Republic of Germany, the French Republic, the Italian Republic, the Republic of Slovenia, the Principality of Liechtenstein, the Republic of Austria, the Swiss Confederation, and the European Economic Community, and entered into force on March 1995. http://www.alpconv.org/NR/rdonlyres/E0A45E2E-5540-4C6B-8708-C7E173DC983D/0/PopCult_en.pdf

events such as experienced about 2,600, 1,400, and 700–200 years ago). Agriculture seems to have been present in Turkey and eastern Greece as early as 8,000–9,000 years ago (Whittle 1985; Barker 1985), but its impact on vegetation is believed to have been quite localized. However, by 5,000 years ago agriculture had been spread to most parts of Europe. This human influence on prehistoric vegetation is evident from pollen analysis and dendrochronological studies, such as in the Netherlands where recent remote sensing techniques enabled the discovery of Celtic fields under ancient woodlands— sites previously thought to have been forest since prehistoric times (Kooistra and Maas 2008).

Forests generally were part of agro-sylvo-pastoral systems, and they formed vital economic resources for local populations. Ground flora was grazed, wood and non-wood products were collected, and fields were opened up and cultivated wherever the topography allowed it (Saratsi 2003, 2005). Pollen diagrams give evidence of such silvo-pastoral systems, such as in the evergreen oak forests (*Quercus rotundifolia*) of southwestern Spain, in Spanish known as dehesas, dating back to the Bronze Age (4,500 BP) (Stevenson and Harrison 1992). Studies in Pindos Mountain, Greece, suggest that leaf fodder collection has continued from prehistoric times until present days (Halstead 1990a, b, 1998). Charcoal analysis gives evidence of the use of oak and ash wood in settlement construction in Greece since Neolithic times (Dinou and Badal 2001). Prehistoric carpentry tools in the early Bronze Age indicate the existence of a highly skilled and specialized woodworking industry during this period in the Aegean (Downey 2001). In Great Britain, artefacts from lowland fens such as prehistoric causeways and some "bog oaks" provide evidence of early tree management and coppice. In the eastern part of the Netherlands, roads made from thousands of poles from oak (*Quercus*), alder (*Alnus*), birch (*Betula*), and lime tree (*Tilia*) have been excavated in former peat land areas in the province of Drenthe. These trees had been cut by axes around 2,430 BC. Again it has been suggested that forest management took place in the Neolithic period because the trees had about the same age and even a cutting regime might be identified from the annual ring patterns of the wood sections (Casparie 1995). Track-ways in raised bogs made from timber and coppice wood were also constructed during the Bronze Age and Iron Age but without evidence of silviculture. Further, willow branches were used for prehistoric road construction as found near Emmen. This particular road dates back to 200 BC. The branches seem to have been harvested from willow stools, a prehistoric coppice system with a 4 year rotation; and the excavated road constructions are comparable to present-day constructions of willow branches used in waterworks. Excavations along the Via Claudia in Tyrol, Austria and Italy (Nikolussi 1998), give evidence of the use of the local available species, diameters, and lengths of wood for the construction of Roman roads. Excavations in Lermoos, Austria, demonstrate that the species used—fir (*Abies*), beech (*Fagus*), and spruce (*Picea*)—had to be replaced after 30 years because of deterioration. Even today, a life-span of 30–50 years is estimated for wood-built constructions located in open situations.

6.3.1.2 Forest Management as Recorded by Roman Authors

During the Roman Empire the number and size of permanent settlements increased and stable economic structures were established. The Roman Period is of particular interest because it is characterized not only by the expansion of colonization but also by multiple branches of economy with a high demand for wood, thus for the first time influencing forest management activities on a European-wide scale. The high demand for wood could only be satisfied by planned management of the available resources. This holds true for areas surrounding big settlements as well as for mining districts.

The long-term maintenance of timber supplies was already an aspiration of Roman legislation, which particularly emphasized management regulations for coppice forests and the regeneration of high forests (Johann 2008). The historical importance of coppice forests derives from the fact that for the first time the idea of a sustainable management was put into practice. Thus a given forest area was divided into certain sections that could be harvested annually according to the planned rotation period (4–30 years). Archive material gives evidence of the continuation of this management system beyond the Roman era. It was practised during the Middle Ages and has been locally maintained until the present day. This is well-documented for specific regions of the study area (for example, see Rackham (1976, 1980, 2000); Smout et al. (2005), and Rotherham and Jones (2000) for Western Europe; Szabó (2005) for Hungary; or Johann (1996, 2008) for the Austrian Alpine Region).

From Greek (since 400 BC) and Roman (since 200 BC) writings we can derive useful information about tools and techniques used in forest utilization and management. This epoch of early history is marked by the transition from orally transferred traditional knowledge to written sources. Thus, for the first time, Roman authors such as Plinius Caius Secundus[2] or Columella[3] provide information about contemporary forest utilization practices, the domestication of trees, and the cultural and socioeconomic relation of local people and authorities. Plinius, for instance, described use and properties of and ways to grow tree species such as beech, oak, pine, and chestnut (*Castanea*). This source provided also information for woodlands outside the Mediterranean region. Triangulations of written sources from the Roman era, pollen analysis, archaeological excavations, and relicts from charcoal confirm forest utilization practices such as clearing for farming, woodland pasture, pollarding, barking, cutting of timber and firewood, wood for tools, and harvest of hazelnuts and acorns (Seidensticker 1886, p. 303). Also, Palladius[4] who lived about 300 years after Plinius tackled the sowing of pine including the appropriate time for harvesting (Herr 1538).

---

[2] Plinius Caius Secundus, 23–79 AD. *Naturalis Historia.* 37 volumes.

[3] Lucil Columella, 50–60 AD. *Liber de Arboribus,* translated into German by Herr (1538).

[4] Rutilius Taurus Aemilianus Palladius, 330 AD. *Opus Agriculturae De Re Rustica.* 13 volumes. Palladius' work became well-known in German-speaking countries when Michael Herr, Doctor of Medicine in Basel, published it in 1538.

Plinius stated in his *Naturalis Historia,* 'Acorns form the wealth of many peoples'; in times when cereals are lacking people live on these fruits'.[5] Virgil describes the happiness of pigs in a year with good yield of acorns, which also influenced the mood of the people (Radkau 2000, p. 77). When flour from grains was not available, acorns were dried, milled, and baked as bread, sometimes after roasting in hot ash. They were also used to feed goats and oxen and in particular pigs from time to time. Beech nuts also belonged to the fruits that were highly appreciated as fodder for pigs because they produced soft meat. Further edible forest products were honey, fruits, mushrooms, roots, berries, and leaves. Some products had medicinal effects, such as the needles from fir and larch or the ash of beech nuts. When grass or herbs were not available in a sufficient amount, green leaves were cut from trees and shrubs and used as leaf fodder. Plinius also reported the drying and storing of leaves as stock for wintertime. Leaves from mountain ash were particularly suited for the feeding of sheep and goats (Seidensticker 1886, II, pp. 335–347). The value of leaf litter as fertilizer was already known in ancient times. Cato mentioned the usefulness of leaf litter for the stable and field and that this was the reason why it was collected from places where it accumulated. Columella recommended to cut fern and mix it with the manure of henhouses or to strew it in sheep stables (Herr 1538).

Roman writers differentiated between softwood and broadleaved timber. They knew about the importance of the weight and density, tenacity, elasticity, moisture content, and burning qualities for the potential application of wood. They were also aware that these qualities were closely related to tree species, the stand, stand density, age and tree health, the logging period, and drought (Seidensticker 1886, I, pp. 256–257; 1886, II, p. 66). Oak, ash (*Fraxinus*), elm (*Ulmus*), poplar (*Populus*), spruce, pine (*Pinus*), fir, and larch (*Larix*) trees were in popular use for the construction of houses. Walnut (*Juglans*) and maple (*Acer*) trees were used for beams, larch for the construction of bridges and water pipes, and beech (*Fagus*) and oak trees for manufacturing pots and bowls.

There was a high demand for firewood for cooking and heating, and wood for craftwork as well as for the production of charcoal, which was necessary for the melting of iron or copper. Large quantities of wood were also demanded for the construction of barns and stables for breeding cattle and poultry to nourish the troops. Roman farmsteads and estates consumed a remarkable amount of various kinds of wood for different uses, which they called *lignum utile*. Examples are wheels, furniture, pulley blocks, and poles. Additionally, there were many agricultural tools that were completely or partly made out of wood, such as barrels, dippers, baskets, ladles, or bottles (Johann 2004a). The bast from the lime tree was used for the manufacture of ropes. While bark from oak was needed for the tanning of leather, the bark of beech, lime tree, spruce, and fir were used for manufacturing different kinds of containers and pots. Apples, grapes and other fruits were stored in sawdust from fir and poplar. Wood was also an important part of bows, fishing rods, and shafts for tools, lances, and spears. Coppice wood was also used for the

---

[5]Plinius Caius Secundus, 23–79 AD. *Naturalis Historia*, volume 16 (botany, forest trees); German/ Latin edition 1994 published by Artemis & Winkler Verlag Zürich.

production of besoms, fences, and fascine revetments (Hughes and Thirgood 1982; Johann 1996). There were many everyday objects such as walking sticks made out of beech, elm or oak; toothpicks, prams, combs, and flutes were made out of box-tree. Roman literature also indicates the collection of sap and tar.

6.3.1.3 Safeguarding of Trees and Forests

The particular ambiences of woods and trees provoke people's imagination; therefore, very often people give forests meanings and create myths. According to Celtic and ancient-Roman views, each natural grown up tree was inhabited by a tutelary god (*numen*)[6] as long as it was green (Strack 1968). In a broader sense each natural tree, respective of each natural wood, was a residence of gods and goddesses and therefore prohibited from general use (Seidensticker 1886, II, p. 144). These kinds of trees became early temples. According to Columella, holy woodlands occurred in two different types, such as *sacra nemora* and *luci* (Herr 1538). The latter experienced a higher level of protection. In general these woodlands were mixed forests of variable size. Only very limited grazing of sacrificial animals took place in them. In holy woodlands private ownership was not allowed (Mantel 1990, pp. 113–117). Trees could grow until they died naturally, and only dead wood or stems and branches were sanctioned to be harvested. In Central Europe, particularly from the eighth century onwards, Christian missionaries made great efforts to exterminate these holy trees and groves by cutting them because they considered them to be symbols for pagan rites.

6.3.2 Forest Management of Medieval and Early Modern Times

The medieval and early modern forest management system was fairly varied and flexible. Most farms were autonomous economic units and tied to the market to only a limited extent. The farmer acquired the bare necessaries of life by diligence, skill and traditional knowledge related to fields, woodland, cattle, and Alpine pasture. Cattle breeding was the centre of a mixed economy including livestock, cultivation of crops, collecting activities, forest use, and the production of several wooden tools for daily use. Evidence shows that this pastoral lifestyle, which involved seasonal transhumance, existed in the Mediterranean countries (Saratsi 2003; Di Martino et al. 2006) as well as in the Alpine regions (Johann 2004b) or right across the Scottish uplands (Bil 1990). The most essential wooden products were firewood, charcoal, timber, and structural wood for various farm items (such as fences, barrels, wheels). Hence historical records refer between 50 and 90 different kinds of wood products (Johann 2004c). The coexistence of different types of forest use and different kinds of forest features was typical and formed a mosaic landscape of various

---

[6]Plinius X: 64.

types of woods and other habitats like meadows, grasslands, and arable lands across Europe. This utilization resulted in a high diversity handicraft and culture with balances of different uses.

Already the Bavarian common laws (*Lex bajuvariorum*), written down under Roman and West-gothic influences, tackled fundamental regulations concerning the utilization of woodland and pastures. From the sixth century onwards, the amount of livestock farmers could drive into the forest and Alpine pastures depended on the available resources of the farm to feed it during winter time (Mantel 1990, p. 61). This regulation aimed at avoiding overuse of the local woodland and proved to be very effective and successful. In remote areas of several Central European countries such as Slovenia, Slovakia, Czech Republic, and Austria, this approach was maintained and still in use until the nineteenth century. In some of the earliest written documents such as the Freisinger documents from 793 AD, 937 AD, and 957 AD (Bavaria, Germany), the right of forest use is mentioned as part of the farmstead. Thereby they differentiated between woodland that was suited for conversion into fields, and woodland that was not yet considered for clearance. These documents already differentiated between woodland in personal ownership and woodland for common use (Sturm 1937).

In several European countries commons were particular important resources for the poorest of the poor. Predominantly in mountainous areas, villagers depended basically on the use of forest resources, with agriculture a mere complement to their self-supporting livelihood. Among the applied management techniques were pollarding, pruning, shredding, and coppicing (Valdés 1996; García Latorre and García Latorre 2007). This was for instance the case in Spain during medieval times, where the communal use strongly enforced the relationship between peasant communities and forests (De la Cruz Aguilar 1994). During the 'Ancient régime' a complex ownership system existed including village commons, commons belonging to small groups, landlord grounds, royal grounds, etc. There was indeed a sort of municipal forester called 'caballeros de sierra' who was commissioned to observe and monitor the use of communal forests. Another example of a commonly developed system of governance and land use was the corporations of users within local communities, so called marken or maalschappen in the Netherlands (Sloet 1911), which continued to the nineteenth century. They were established after the overexploitation of woodlands and grazing lands, when drifting sands sometimes threatened even villages and agricultural fields on a wide scale. In medieval Europe the way these corporations were organized and the power relationships surrounding the use of natural resources within the corporations differed from region to region. During meetings, which regularly took place at historic sites in the forest, the number of domestic animals that were allowed to graze in the forest and on common pastures, the quantities of timber that could be harvested and by whom, as well as other uses of the natural resources were determined either by the commoners or by administrative structures dominated by nobility.

It can be assumed that many of the old management systems survived the course of time. Traditional forest-related knowledge was passed orally from generation to generation to become part of the customary law of the villages. In several countries

Fig. 6.4 Charcoal making in Pindos Mountains, northwest Greece. From the *right*: kilns in the second, sixth and tenth day of burning. June 2001 (Photo: Eirini Saratsi)

of Central Europe these laws were written down from the thirteenth century onwards. Attempts were made to regulate the natural resources of the villages (water, woodland) in a sustainable way. In Austria, for instance, these laws were collected and published by the Austrian Academy of Science in the second half of the nineteenth century (Siegel and Tomaschek 1870; Bischoff and Schönbach 1881). Accounts from other parts of Europe such as Greece or Transylvania document that traditional forest management regulations were written down from the seventeenth and eighteenth centuries onwards (see Saratsi 2003 for Greece; Imreh 1983 for Transylvania).

From the beginning of early modern times onwards, mining and industrial growth, as well as population, increased in many parts of Europe, requiring a high amount of charcoal, fuel wood, pit props, and timber (Figs. 6.4–6.6). In Western Europe the high demand resulted in significant pressures on the remaining domestic woodlands such as in the Netherlands, where a high demand gave rise to woodland reduction on a large scale. In the industrial areas or England, Wales, and Scotland the management particularly of lowland woodland was for intensive industrial coppice during the period 1500 AD to 1900 AD (Rotherham and Jones 2000; Rotherham 2007). In Central Europe, regulations and forest laws were implemented from the sixteenth century to ensure the sustainable supply of the mining industry with the required energy (Mantel 1980). In this way the amount of wood that was allowed to be annually cut depended on the result of a pre-determined inventory. The legally prescribed harvesting methods were diverse clear-cut systems; reforestation was to come about by natural regeneration.

Köhlerei in der Hölle.

Fig. 6.5 Charcoal burning for iron production in Austria (nineteenth century). Richard Püttner (Source: Rosegger et al. (s.d.) (s.d. means sine datum = without datum) published about 1880)

Fig. 6.6 Charcoal burner's hut, Sheffield Old Park Wood, Sheffield, Great Britain (Source: Addy 1898)

However, already in the sixteenth century, methods of sowing seeds from coniferous trees, which had been practised 200 years earlier in Nuremberg and other parts of Germany, increased in several European countries such as the Netherlands, Austria, and Hungary. They probably were encouraged by Roman authors, whose translated books became available in Europe in the Middle Ages. Authors such as

Petrus de Crescentiis (1538) began to write of forests and forestry; his treatises about silviculture, forest management, and trees and woodland were based on information from the time of antiquity. The most important books are volume II about nature, V about trees, VII about forests, and VII about gardens (Mantel 1980, pp. 569–570). De Crescentiis's handwritten manuscript turned up first in Latin language around 1305 and was translated into Italian, German, and French already in the fourteenth century. It was the first printed book in German Language, published just after the Bible; 12 editions followed. These and other publications, such as from the French naturalist Pierre Belon (1517–1564), formed the basis of European literature on forestry practices until the eighteenth century. However, not before the second half of the eighteenth century was artificial regeneration from seeds and seedlings generally practised (Mantel 1980; Johann et al. 2004; Bendix 2008).

6.4 The Development of Science-Based Forestry and Its Influence on Traditional Forest-Related Knowledge and Management Practices (Nineteenth–Twentieth Centuries)

There is no doubt that the forest area of Europe was at a minimum around 1800. In addition, many former agricultural lands had been severely degraded, transformed to heathland, karst, eroded mountain slopes, and drifting sand dunes Thus the land remaining in forest did not exceed 2–3% in Denmark and 3–4% in the Netherlands around 1800. A similar development took place in almost all Western European countries such as in Ireland (land remaining in forest 1%) and Great Britain (5%) in 1900, Belgium (14%), and Portugal (7%) in the middle of the nineteenth century (Johann et al. 2004). The lack of wood-based energy was in contrast with the high demand of industry and population. In most areas no suitable resource other than charcoal was available for many industrial processes. Experiments with fossil fuels in Great Britain and peat in the Netherlands, Great Britain, and Germany, began to address these issues. But at the beginning of the nineteenth century it seemed that there was only one solution to get out of this dilemma: the afforestation of the wasteland and the implementation of new techniques with regard to forest management. Because forest science contributed remarkably to reach this ambitious goal, it was highly appreciated by policy.

6.4.1 Development of Science-Based Forestry

Under the influence of the German scientist Justus von Liebig (1803–1873), father of agricultural chemistry, state-aided agricultural research and extension took off on quite a large scale, with the aim to increase crop production. There was a high demand for scientific-based research at the time when forest science became a subject of teaching at the newly established forest academies in Banská Štiavnica, Slovakia (1808), Tharandt, Germany (1816), Nancy, France (1824), Skoplje, FYR of Macedonia (1847), Zurich, Switzerland (1855), Istanbul, Turkey (1857), Evo, Finland (1860s), Vallombrosa, Italy (1869), Vienna, Austria (1872), Wageningen, Netherlands (1883), Cham Koria, Bulgaria (1896), Helsinki/Finland (1907), and Athens, Greece (1917) (Welzholz and Johann 2007; Saratsi 2009).

The German classical school of forestry developed by Cotta (1763–1844) and Hartig (1764–1837), among others, influenced forestry practice in Germany and, later on, European forestry on a wider scale. Thereby the German forest became an archetype for imposing on disorderly nature the neatly arranged constructs of science. Practical goals had encouraged mathematical utilitarianism, which seemed, in turn, to promote geometric perfection as the outward sign of well-managed forests. In turn, the rationally ordered arrangement of trees offered new possibilities for controlling nature. Human intervention should be visible. It followed the ideal of designed nature. Thus clear-cutting with subsequent planting became common practice from 1820 onwards.[7] Conifer species were highly favoured because the demand for softwood was high, the rotation period was short, the intermediate cutting of stems with a rather small diameter also brought some money, and the growing stock was high when final cutting took place. The successful regeneration with spruce or pine converted the former mixed forest into large areas of monocultures of conifers and even-age forests (Johann et al. 2004).

The developments in German forestry provided the scientific and intellectual basis for advancements in many other European countries. Moreover, with regard to afforestation and the increasing of growth and yield, this doctrine was so successful that it became a model for forestry that was transferred to many parts of the world such as for colonial forestry in Burma (Myanmar) and India and in the Americas from the second half of the nineteenth century onwards. For more than 200 years Central European forests were managed according to the principles mentioned above.

The particular way of Central European forestry emphasizing timber primacy, sustainable yields, and optimization of forest functions (Glück 1987) might to a certain extent be derived from the fact that countries such as Germany, Switzerland, or the Austrian Empire were forced to rely on their local timber sources, whereas European countries situated along the Atlantic coast—such as Great Britain, France, and the Netherlands—relied on their colonial territories in the Americas, Africa, and Asia to supply timber. This was the reason why Central Europe became the classical region of afforestation of high forest stands from the beginning of the nineteenth century onwards (Bendix 2008). In contrast to German forestry, the characteristic French forestry system—which was also designed during the first half of the nineteenth century by the School of Forestry of Nancy, France (especially by Bernard Lorentz and Adolphe Parade)—relied on a shelterwood system, based on natural regeneration.

---

[7]Cotta, H. 1817 Anweisung zum Waldbau. Dresden. Hartig, G.L. 1808. Lehrbuch für Förster und die es werden wollen. Cotta'sche Buchhandlung Tübingen.

Regions with a high demand for charcoal and timber were the first to shift to clear-cutting systems with artificial regeneration. European afforestation activities were inspired by either the French or German management systems, although the German classical school of forestry seems to have dominated. Economic considerations, the question of the highest financial yield, played an important role. Afforestation activities were driven by political and strategic considerations, the aim being to improve environmental and living conditions by the re-cultivation of 'wasteland.' This included areas located in the plains of Germany, the Netherlands, and Hungary; the mountain region of Switzerland, France, Italy, Austria, and Slovenia; and the coastal dunes in Denmark and the Netherlands (Johann et al. 2004; Kossarz 1984).

In European countries with a low percentage of forest cover—such as in Spain, Portugal, or Great Britain—afforestation represented the most important forest activity even in the twentieth century. Here clear-cut and re-planting became norm, and many areas were planted with coniferous, often exotic species (see Fig. 6.2) (Johann et al. 2004). Often the expansion of forest area was driven by official reforestation policies, such as the British government's establishment of the Forestry Commission in 1919. In France the transformation occurred with the reforestation of the Landes area of Gascony, the development of protection forests under the Mountain Land Restoration Programme, and the establishment of the National Forest Fund in 1946 (Morin 1996).

However, already from the second half of the nineteenth century onwards an increasing urban population required various other goods and services apart from wood and timber, such as recreation and nature protection. Changes in scientific forestry or in the view of governments towards forests occurred because of the recognition of ecosystem function and because democratic processes were giving local people more say. Thus in some regions experiments were carried out concerning 'silviculture close to nature,' initiating various scientific disputes about the advantages of different silvicultural systems (Johann 2006).

Since the beginning of the twentieth century, silvicultural systems have aimed for continuous forest cover and the abandonment of clear-cuttings and monocultures, particularly in Central Europe. These methods have been practised, for instance, in Switzerland, Slovenia, and parts of France, Germany, and Austria. They became known as 'jardinage' in the French and 'Plenterwald' in the German literature. The development differed with regard to space and time, often depending on varying framework conditions of policy, economy, and society. The main influential factors were World Wars I and II, inflation and depression in the 1920s, the obligation of forestry to contribute to national budgets, tax policies, the stock market crash of 1929, legislation, political developments, and changing forest ownership structures (Johann 2006). In Sweden, for example, particularly in the 1920s and 1930s, an increasing number of foresters endorsed the principle of selective felling as the supreme way of regeneration forests. 'Shame on the forester who cannot regenerate with the axe' was one telling and commonly quoted slogan (Hofsten 1996).

For many centuries forest management focussed on the traditional multiple demands of residents had been considered as archaic and useless; science and progress had to triumph over that outdated folklore. But the rebirth of continuous cover forestry shows us that there is an unmistakable continuum between the traditional management systems and the most up-to-date-techniques (Rochel 2006). There are various 'close to nature' forest management methods, such as 'Pro Silva.' The Pro-Silva movement originated from southeastern and Central European deciduous forests since 1989 (in Slovenia), but it has also attracted supporters from the boreal forest zone and Western European countries such as the Netherlands. As a result of the discussions, the current concept of silviculture includes, apart from wood production, an emphasis on maintaining forest biodiversity, recreational, landscape, protective, socio-economic, and cultural factors. The 'natural forest living community' is accepted generally as a model for the realization of nature oriented silviculture (Mlinsek 1996).

6.4.2   The Influence of Forest Science-Based Forestry: Evaluation of Conflicts, Detriments and Benefits

Present-day cultural developments are not just adding contemporary layers of decisions affecting forests and landscapes on top of the previous ones. They also recall earlier meanings and thus evoke the hidden memory of the past and make it relevant to decision-making. Decision-making can be based on new ideas and intellectual concepts, but the natural foundations of forests and landscapes involve pre-existing historic layers. Post-industrial decision-making processes—being generally participatory, democratic, and decentralized—are currently shaping local life-worlds on the basis of global values (Schmithuesen and Seeland 2006). To understand landscapes requires interaction among different actors in society. Does the paradigm 'industrial forestry' compete with the paradigm 'multiple use'? This raises the issue of scale in an economic sense. At what spatial and temporal scale is traditional forestry for wood production beneficial? At the national level, policy instruments include legislation, information, subsidies, monitoring, education, and other instruments. However, safeguarding natural and cultural heritage cannot be maintained by institutions alone, but rather by local people acting in different governance systems (Angelstam and Elbakidze 2006). Whilst many conflicts occurred in regions where traditional agriculture and pastoralism were still important economic assets, in other regions, where industry was already well-developed and farmers' subsistence did not rely on forest resources any more, there were very few forest-related conflicts at the same time.

6.4.2.1   Common Rights

Particularly in the nineteenth century, conflicts with rural populations were provoked by forest administrations of several European countries when striving for a monopoly in the regulation of villages' commons and/or when traditional forest uses were restricted or abolished by law due to political or economic interest (De la Cruz Aguilar 1994; Sala 2000). Some examples help illustrate similar developments under the influence of the German classical school of forestry, whereby modern forestry favoured the utilization of forests for primarily timber production.

Spanish authors demanded a rational and scientific approach to natural resource management; disliked traditional practices applied by peasants; and perceived the cultural landscape as degraded. Agustín Pascual and Esteban Boutelou—who were trained as foresters in Heinrich Cotta's (1763–1844) institution in Tharandt, Germany—brought to Spain not only the techniques of forestry but German foresters' cultural values, too: the necessity of state intervention in forest management, the professional group of foresters as the only accredited people for forest management, and a positive exaltation of forest stands (Casals Costa 1996). In Spain the new professional group exaggerated the degraded situation of traditional forests as well as the ineptitude of villages in order to carry out a rational management. According to Sala (2000) there was an assault of new forestry directives upon villages, usurping their regulatory capacity. Furthermore, requests to link customary forestry practices with scientific modern forestry management at local and county levels were disregarded. Consequently, conflicts between the public forest administration and rural communities have been very frequent during the nineteenth and twentieth centuries (De la Cruz Aguilar 1994). An example is Extremadura, Spain, a region that was characterized by the predominance of the open oak parkland (dehesa), which offered simultaneously livestock, forestry, and agricultural production by transhumance. When the forest service attempted to apply the German principles, this intent had to be continuously reviewed to update it to local economic and ecological settings. However, planning at least gave scientific support not to the forest intervention itself, but rather to the traditional potentiality of the dehesa system (Linares 2006).

In countries that were part of the Napoleonic Empire, such as in Veneto (northeastern Italy), from the nineteenth century conflicts arose with traditional forest management as forest science's dictates clashed with the customs and popular beliefs that had evolved in previous centuries within local communities (Lazzarini 2006). In Turkey the historical right to unlimited use of forests free of charge (apart from obeying some rules) was curtailed for the first time in 1870. This was by legal arrangements and later on by law during the National Independence War in 1920 (Yurdakul and Ekizoğlu 2005) (Fig. 6.7). Aiming at the protection of the state forest, it revoked the former utilization rights in these forests and obliged the villagers to supply their demands from the coppice given to them. However, this 'Coppice Law,' as well as the 'Law on Villagers' Right of Benefiting from State Forests' from 1924, stipulating that the villagers must pay for their benefits deriving from the State Forest, did not improve relationships between forest and villagers or provide effective protection of the forests. In 1937, the Forest Law revoked the right of free-of-charge utilization from state forests by forest villagers. However they were granted the right of picking up forest products under a free-of-charge permission (Özdönmez 1973). Today, there are still 20,974 villages and 7.8 million forest villagers who live in or around forests, and their livelihoods depend on utilization of available forest resources nearby (DPT 2007).

Fig. 6.7 Carrying of firewood by forest villager women in the initial years of Turkish Republic (Source: Forest History Archive of Istanbul University, Faculty of Forestry Department of Forest Policy and Administration)

Ownership rights on private, communal, or state forests were responsible for long periods of hostilities between the state and its representatives and communities in many parts of Europe, particularly in mountainous regions (Fig. 6.8). In Greece such hostilities were also maintained because of the nature of the supervision the Greek State used and the way this was applied in practice. As a matter of emergency the Greek State after the 1820s emphasized the need for all forested and wooded areas to be owned and managed by the state. The influence of forestry ideas originating in Central Europe encouraged the perception of the existing cultural landscapes as being degraded ecosystems that urgently needed rational management. The ultimate target was forested areas to be used as a national resource. In addition, in the eyes of educated intellectuals, the dryness of the Greek landscape was a great disagreement compared with the idyllic landscapes of Antiquity often portrayed as evergreen, paradise-like gardens (Saratsi 2009). The Greek Law never gave local populations the right to utilize the forests under their own management, not even for their private needs. Despite the eventual recognition of a significant part of the Greek forests (mainly those belonging to the village communities) as communal, in many cases the debates were sustained until the first years after the end of the dictatorship in 1974 (Damianakos et al. 1997). This approach to the management of Greek forests had negative effects because it instigated illegal uses of the resource and in many cases led to the destruction rather than the protection of forested areas.

Fig. 6.8 Collecting fuelwood in a coppice forest in Great Britain. The Burnham Beeches, by Birket Foster (1800) (Source: I. Rotherham)

6.4.2.2 Traditional Skills and Tools

Since the nineteenth century, always-increasing industrial mass production has resulted in gradual abandonment of handicrafts and a subsequent decline in the use of traditional materials. With regard to woodland utilization, because many substitutes for tree and forest products became available, there was a decreasing demand for traditional woodland products and a loss of traditional knowledge and skills. This was caused by market changes and intensified by migration of local population to the industrial centres. The influence of science in general was direct where forestry promoted the introduction of monocultures and indirect where wood substitutes and other forest products such as iron and plastics, artificial turpentine, and chemical dyes entered the markets (Pantera et al. 2009). Almost unnoticed by society, the declining use of these mainly renewable materials signified the loss of valuable knowledge associated with the required handicraft work for their production.

However, a body of traditional knowledge persisted until the early decades of the twentieth century in some regions linked to the relatively slow development of technology in agriculture and forestry (Figs. 6.9 and 6.10). In these regions traditional techniques were still used in the 1960s (Agnoletti 2006). Cultural heritage was still evident at the beginning of the twentieth century in these landscapes, which were not yet strongly affected by timber production but were sustained by agro-silvo-pastoral societies, for whom forest resource served multiple functions. In these regions the disappearance of traditional forest uses is a relatively recent process, triggered by the

Fig. 6.9 Harvesting of resin
in the Alpine region of
Austria late nineteenth
century (Source: Johann
2002)

Fig. 6.10 Non-wood forest
products: storax-production
by forest villagers in Turkey,
1930s (Source: Forest History
Archive of Istanbul
University, Faculty of
Forestry Department of
Forest Policy and
Administration)

Fig. 6.11 Initial experiments concerning the re-forestation of cleared woodland (Source: Agricolae 1772)

development of industrial forestry. This is, for instance, the case of transhumance system in the middle south of Apennines. It has shaped over the centuries one of the most important historical and cultural landscapes in Italy; in particular the web of roads of transhumance representing in the past a wide connection between Abbruzzian pastures, wood pastures, and Apulia lowland pastures is now disappearing (Di Martino et al. 2006).

6.4.2.3 Ecology and Diversity of Forests

Scientific-based forest management applied extensively throughout Europe during the nineteenth century had major impacts on ecology, hydrology, and soils. Particularly the afforestation of former wasteland or unproductive pastures and the transformation of coppice forests into high forest stands of monocultures caused an increase in spruce and pine plantations, thus contributing to either an intended or an uncontrolled decrease of species diversity (Fig. 6.11) (Johann et al. 2004).

Although former coppices had a relatively low wood production, their structure and composition were diverse. This diversity changed radically by the shift in forest management techniques. When traditionally managed coppice woods were converted to conifer plantations especially in the lowland zones, replacement of native vegetation and tree species with exotic ones as well as cessation of traditional management took often place (Johann 2007). In Hungary as elsewhere in Europe the promotion of even-aged high forests caused the disappearance of giant, veteran trees, which became extremely rare. Not many coppiced areas of woodland remain

Fig. 6.12 Remnants of
former coppice forests in the
Carpatian Basin nearby
Negreni/Romania, 2001
(Photo: János Bölöni)

today (Fig. 6.12). However, coppice management has been reintroduced locally, for
example in Great Britain and in the Netherlands, mainly to favour biodiversity.

The main effects in the landscape caused by science-based forestry of the nineteenth century are what Professor Jean-Jacques Dubois called géométrisation (the
introduction of geometric shapes in forest landscapes) and the appearance of homogenous compartments, wide-range extension of conifers (e.g., 31% of French forests), and deep changes in forest soils and flora. In general terms, from the middle
of the twentieth century even remote European regions became influenced by
encompassing processes of mechanization, intensive farming, specialization, and
rationalization. The removal of trees and shrubs, the introduction of larger machin-
ery, the application of fertilizers inducing widespread of nutrient enrichment, and
the use of pesticides, led to an escalating loss of landscape biodiversity from the
1950s to the 1990s (Welzholz and Johann 2007; Galán et al. 2003).

6.4.2.4 Forest Area

From the second part of the nineteenth century onwards, afforestation constituted
one of the principal tasks carried out by European forest administrations. In general,
forest area increased remarkably (Johann et al. 2004). This development can be

observed in various European countries, particularly those having been without significant forest cover. Afforestation changed cultural landscapes by introducing non-native plantation forestry, particularly when heathland was re-cultivated (e.g., in the United Kingdom, Belgium, Denmark, Germany, and the Netherlands) (Fig. 6.2). In Great Britain during the twentieth century the area of productive coppice dropped from 230,000 to 40,000 ha, much of which was ancient semi-natural woodland. But within this period the area of productive conifer high forests increased threefold, with particularly extensive planting programmes in the uplands. Since 1919 the forest area of Great Britain has more than doubled from 5% to 12% (Henderson-Howat 1996). Few changes in Scottish land-use have been so marked in physical terms, or so compressed in time. The increase in area under wood has been from about 5% of the land surface in 1900, to about 17% around 2000. Three-quarters of this change were compressed into about 40 years of the second half of the century.

In Mediterranean countries large reforestation programmes also took place. For example, in Greece in 1950s and 1960s this occurred across the country but particularly in mountainous areas (Saratsi 2003). In Turkey afforestation activities have been carried out since 1937. By now approximately two million have been afforested (DPT 2007). Also in Spain, an area of 3.5 million ha was afforested from 1940 to the middle of the 1980s, mostly with pines (Groome 1988; Gómez de Mendoza and Mata Olmo 1992). In Eastern Europe, such as in Hungary, from 1960 onwards the reforestation programme launched by the government aimed at a minimum yearly increase of forest land of 25,000 ha (with 50% as poplar plantations). Thus the forest land share increased from 12% (1920) to about 20% today, but with 12–13% being plantations of both native and alien tree species. In Poland from 1950 to 1980, the forest area increased around 10,000 ha per year due to government-induced reforestation policy. Finally, in Bulgaria 1.5 million ha of eroded and devastated land were re-cultivated as forest between 1945 and 1984 (Kossarz 1984).

6.4.2.5 Socioeconomic Effects

Scientific-based forest management led to increases in the forest area all over Europe (Bendix 2008). This generated an increasing availability of valuable timber as a renewable resource, and a recovery in the supply of timber to the markets. This was an important development and evolution at a time when charcoal and firewood had been replaced by fossil fuels as energy supplies. Thus the coppice forest had lost its importance for firewood and charcoal production (Johann et al. 2004), and serious declines had followed. Disadvantages of the increased forest area were the change from formerly broadleaved to coniferous forests, (often monocultures), the loss of biodiversity, and the simplification of the forest structure (Johann 2003). However, afforestation activities improved some environmental conditions (e.g. soil, climate, and livelihood) in several parts of Europe. They also provided working places in rural areas, such as the afforestation programmes in the karst region in Slovenia since the 1880s (Johann 2001), in Scotland and Wales in the 1960s, and in Spain particularly during the hard decades following the Spanish Civil War (Sánchez

Martínez and Araque Jiménez 1993). Twentieth century forestry displaced traditional work in the woods and on estates, but replaced this by limited provision of local jobs in remote areas. For many decades, this was an economic and social justification for government-funded forestry policies in Great Britain. In recent years, this approach has been reviewed and greater emphasis is now placed on conservation, on associated tourism and recreational benefits, and on wider ecosystem services (carbon sequestration and water/soil management).

6.5 The Role of Traditional Forest-Related Knowledge in European Policy and Its Influence on Forest Management at Present

In Europe the present forest management concept is based on a multifunctional approach, with forests being used simultaneously for economic, recreational, and other purposes, integrated with the maintenance of biodiversity. Sustainable forest management (SFM) is an inspiration in continuous development, the interpretation of which in Europe varies greatly over time and among countries, regions, and even localities. As a consequence, the necessary knowledge to comprehend the requirements of SFM is heterogeneous and dependent on sets of values with different spatial and temporal dimensions. The development of the Pan-European forest policy process reflects this. Moving into the post-industrial society, ecological principles started to be included in the definition of SFM in the 1990s (Angelstam and Elbakidze 2006). Implementing SFM policies requires practical experience, engineering, and science. The relative roles of these dimensions vary with the type of forest and woodland goods and services, and among regions. Preserving and using traditional knowledge today does not mean only reapplying the techniques of the past but rather understanding the underlying bases of these models of knowledge (Agnoletti 2006). Angelstam and Elbakidze (2006) argue that Europe's diverse forest and woodland landscapes can be used to better understand the perceptions of SFM among actors at different levels of governance and the role of traditional knowledge for sustainable landscapes. The historical development of forest use within a region usually goes through more or less distinct phases: (1) local use and exploitation of natural dynamic forest ecosystems, (2) development of sustained yield forestry, and (3) efforts to satisfy ecological and socio-cultural dimensions. Thus because different regions are located in different phases in this development, one can 'travel in time' and learn from both past and future phases of development. As tradition is rapidly disappearing along with the cultures and landscapes where this knowledge lives, the task to document the role of traditional knowledge for sustainable development becomes an urgent matter.

Because of their fundamental and multifunctional characters, cultural and spiritual aspects create an important bridge for humans between forestry and the society's other functions and needs. In raising awareness of forestry related cultural and spiritual issues and their interrelationships in society, dissemination of information, communication, and education are extremely important. Through interactive discussion on cultural and spiritual aspects between various interest groups, new intellectual, cultural, and socio-cultural innovations based on multifunctional values of forests can be promoted (Parviainen 2006).

In recent years there has been much debate about the functions of forests. The old certainties have been replaced by a state of fluidity and uncertainty about the role of forests and, more importantly, the weighting of the different roles. According to Slee (1998) this is perceived differently by diverse scientific communities. To the social scientist, rooted in the reflexive tradition, this is a fertile research field, made more interesting as governments, agencies, and policy makers design new instruments and create new institutions to address these new challenges. To the positivistic social scientist, new research agendas may arise as, for example, new questions are asked about non-market values of forests. To the biological scientist the challenge is huge. The traditional sets of knowledge, often accumulated over a long period of time, are no longer adequate to address the multiplicity of demands placed on the biologist, in this cases the silviculturalist. Furthermore, the analytical closure, which is both necessary and desirable in the pursuit of improvement of practices with respect to a single function—be it avalanche protection, cultural heritage, or maximizing the yield of useable timber—is rendered massively more complex by the inclusion of different functions. This is especially so when these must be negotiated rather than taken as given. Social assessment is a subject often mentioned in recent forest resource management. It is used to determine variables in the planning environment and to evaluate alternate management regimes concerning social conditions. Although modern planning increasingly takes cultural values into consideration, they often only refer to tangible components of culture. However, intangible elements are often hidden (Okan and Ok 2006). These are often connected with knowledge related to work and wood processing or to the non-wood products having been passed down from generation to generation. The knowledge of historical development and about the relationships between people and forests is necessary to understand the present situation and to adapt and modify forest management practices to reflect society's needs.

6.5.1   The Present Role of Traditional Forest-Related Knowledge in European Policy

There is a growing awareness of the significance and relevance of local, indigenous knowledge about forests and traditional ways of forest utilization amongst the international community of forest science and forest-related policy. There is also a need to incorporate such knowledge in the development of political strategies that aim for sustainable forest management. The protection, documentation, and utilization of forest-related, tradition-based knowledge are the focus of numerous political discussions held within regional, national, and international organizations and forums. Because of this development, the demands for changes in silvicultural practices have been particularly strong. The content of traditional silviculture has changed and the terminology has had to be reassessed. Parallel to the term silviculture, taking care of forests and wooded areas can mean ecosystem management, biodiversity-oriented silviculture, close-to-nature silviculture, continuous cover silviculture, landscape management, or landscape biological planning.

The material and spiritual development of societies is tightly connected with forests. In many countries the cultural legacy of forests means tradition and the way of life, and these are connected with values, ethics, moral rights, aesthetics, legitimacy responsibility, and religion. The fundamental role of cultural heritage provided by forests was acknowledged beforehand at the first MCPFE, where the signatory states affirmed that 'forests in Europe make up an ecological, cultural and economic heritage that is essential to our civilization' (Strasbourg 1990, Resolution S1) (MCPFE 2000). Since 1990 the Ministerial Conferences on the Protection of Forests in Europe (MCPFE), now called Forest Europe,[8] have provided a forum for cooperation towards common principles of sustainable forest management throughout the continent, developing common strategies for its 46 member countries and the European Union. The MCPFE commitments (Strasbourg 1990, Helsinki 1993, Lisbon 1998, Vienna 2003, Warsaw 2007, Bergen 2009) promote a balance between the economic, ecological, social, and cultural dimension of sustainable forest management, clearly defined at the Helsinki Conference 1993 (Helsinki 1993, Resolution H1) (MCPFE 2000). On the whole the Helsinki Resolutions and the General Declaration constitute a joint response of the European countries to several forest decisions taken at global level through the ministerial commitments to stimulate and to promote the implementation of the Rio Declaration, Agenda 21, the Statement of Forest Principles, and the Convention on Biological Diversity. These cover a whole range of issues on cultural and spiritual facts in international conventions, agreements, and processes. Encompassing entire Europe, it can be regarded as an example of cross-border cooperation stimulating mutual awareness and understanding of ecological, economic, and socio-cultural dimension of forests. Thus the specific relationships between society and forest as well as a variety of market and non-market forest socio-cultural services—for example, the acknowledgment of forest recreational and aesthetic values, and the protection of areas with spiritual and cultural heritage, including the role of traditional forest-related knowledge—have all been taken into account (Gaworska and Kornatowska 2006).

Moreover, *The EU-Forest Action Plan 2007-2011*[9] acknowledged that cultural landscapes, traditional practices, and other cultural values of forests were some of the ways to achieve local and regional sustainable development. Landscape values are also included in the new Common Agricultural Policy (CAP) and the European Landscape Convention, presently ratified by 29 and signed by 7 countries.[10] The

---

[8] http://www.foresteurope.org/eng/, website visited 2010-07-12.

[9] European Commission Agriculture and Rural Development. *The EU-Forest Action Plan 2007–2011.* http://ec.europa.eu/agriculture/fore/publi/2007_2011/brochure_en.pdf, website visited 2010-07-12.

[10] Status 2009-01-02.

Alpine Convention and its protocols[11] put attention on the fact that there is a demand to constitute a people-centred, sustainable development policy that focuses on the needs, wishes, and opinions of the people who live in the Alpine area. Research, maintenance, and development of the existing physical and non-physical cultural heritage and traditional knowledge are addressed with regard to safeguarding the tangible and intangible cultural heritage. This applies in particular to traditional methods of landscape management, architectural and artistic heritage, forestry, handicrafts, and industrial production. The maintenance and safeguarding of knowledge and practices concerning nature and the universe developed and perpetuated by communities in interaction with their natural environment is also targeted by the Convention for the Safeguarding of the Intangible Cultural Heritage passed 2003 by the 37th assembly of UNESCO and presently signed by 114 states (UNESCO 2003).

In spite of these initiatives, traditional cultural landscapes are still threatened by socioeconomic changes in agriculture and forestry, and by some nature protection strategies. As noted by MCPFE, applied research on the social and cultural aspects of SFM to support the development of solutions to these challenges requires a multidisciplinary approach where human and natural science approaches are used (Angelstam and Elbakidze 2006, p. 350). As noted by Rotherham and Ardron (2006) and Rotherham (2007), such multi- and cross-disciplinary approaches are often sadly lacking both in recognizing issues and especially in coordinating practical actions.

6.5.2 Examples of Landscape Assessments

Forest management planning is an essential tool in applying cultural and spiritual aspects in practical forestry operations. Recognizing the cultural and spiritual values shows that forest management is not only production or protection, but also a mean of maintaining the relationship between people and forests (Parviainen 2006).

Although forest planning and sustainable forest management have to take into account the history of the types of former utilization and the intensity, and techniques applied, adequate studies are often lacking. This holds true for the historic elements present as a physical part of cultural heritage, but also for immaterial cultural heritage such as knowledge also being addressed by the Convention for the Safeguarding of the Intangible Cultural Heritage[12] or the Alpine Convention.[13] Until today, only forest managers with a personal interest in the history of their forest and the historic forest elements of their area have paid attention to the preservation of these elements. However there are attempts all over Europe to improve knowledge about cultural heritage and to see cultural heritage grow as an integral part of forest management plans. As a first step, investigations are being carried out in several

---

[11] http://www.convenzionedellealpi.org/page5_de.htm AC_IX_11_1_e.n.

[12] http://unesdoc.unesco.org/images/0013/001325/132540e.pdf

[13] http://www.alpconv.org/NR/rdonlyres/E0A45E2E-5540-4C6B-8708-C7E173DC983D/0/PopCult_en.pdf

European countries such as in England, Scotland, or the Netherlands, where forest policy is at the moment in favour of protecting this cultural heritage. In Scotland the RCAHMS (Royal Commission of the Ancient and Historical Monuments of Scotland) is actively working with the Forestry Commission Scotland and SNH (Scottish National Heritage) to improve the recording of this evidence more widely, while the Historic Land-use Assessment (HLA)[14] is identifying areas of woodland that have been managed in the past. This is an important aspect of the cultural landscape, and it is clearly an area where cultural and natural heritage and woodland management come together. In the Netherlands the results of a 3 year research project carried out by Jansen and van Benthem (2005) concerning historic elements in the forests was published and communicated to forest managers. Also in several federal provinces of Germany such as Bavaria or Hessen, there are increasing activities carried out by forest historians and public forest administrations to document historical sites and artefacts in the forest (Hamberger et al. 2008). Concerning cultural heritage related to traditional land use and management, including traditional skills and tools as well as folklore, a project called Agrarkulturerbe has been carried out in Germany since the 2000s.[15]

In Sweden a documentation project called Forest and History is carried out in cooperation among organizations of cultural environments and forestry. Local people educated and aided by trained archaeologists conduct the surveys. The main objectives are: (1) to increase knowledge about ancient monuments and other cultural remains in the forest; (2) to demonstrate measures to protect and maintain cultural environments in the forest; (3) to increase knowledge and consideration of cultural remains in forestry and other land use sectors; and (4) to focus on cultural remains as a resource in rural development. The latter, for instance, includes the tourist sector and strengthened feeling for the 'home area' and 'one's roots' (Marntel and Wagberg 2006).

The Swedish case proves that an important step in creating more initiatives on the subject of preservation is in the elaboration of appropriate methodologies for the analysis and management of cultural landscapes. The evaluation of cultural landscapes has to consider their dynamics and the role of time in creating or erasing values. The roles of forests cannot be assessed by the evaluation of forest units alone, but need the consideration of their wider landscape context. For Mediterranean countries this has been confirmed by the outcome of a project including 13 studies in the Tuscan territory, each representing different landscape features (Agnoletti 2006). Moreover, at present a Forestry History documentation and archive project is in its starting phase planned by the Department of Forest Policy and Administration of the Faculty of Forestry, Istanbul University.

New trends in Estonia during the past decade have highlighted a problem that has not been obvious before: the problem of continuous safeguarding of land culture. Estonia only regained its independence in 1991, and over recent centuries Estonian

---

[14] HLA: a joint initiative between Historic Scotland and RCAHMS to describe the historic origin of current land-uses.

[15] http://www.agrarkulturerbe.de/

forest cover ranged from 30% to 60%. Consequently, the signs of human activity or cultural heritage have been variably distributed over the forest areas, depending on the time of their origin. Due to the initiatives of Tarang and others, the identification, registration, and exposition of these objects in the natural landscapes (mainly in the forest) have started to receive more attention (Tarang et al. 2006). The presently ongoing inventory will be finalized in Estonia by 2011, and is also influencing similar activities in Finland and Latvia.

6.5.3 Best Practices for Including Traditional Knowledge into Forest Management

In Europe's intensively managed forests, where wood production is still a dominating goal of management, nature conservation values and the safeguarding of the material and immaterial cultural heritage are becoming more important. It is the response to a society strongly expressing a need for 'nature', which has shifted its interest from the tree as a source of wood, to the forest as an ecological system. This change is debated vigorously. In some cases, particularly in countries with a low forest cover but a high population density, the environmental and cultural functions of the forest have grown to the point where its productive function is declining. Also in the 'new' member-countries of the EU, where the importance of forests apart from wood production has generally already been acknowledged, there are nonetheless new plantations being established, supported by EU and national policy, to get poor agricultural land out of production (England, Scotland, France, Portugal, and Spain). This tendency is not a European peculiarity, but can be observed on a global scale, particularly in the countries of the South (e.g., Australia, New Zealand, South Africa, and South America).

As already noted, two different attitudes of relevance for heritage management are noticeable. One approach is to use the forest as a resource in a wide sense (cultural heritage of the forest). This celebrates activities such as hunting, iron production, tar production, forest grazing, and logging, as well as protecting remains such as iron production sites, charcoal pits, shielings, and logging huts. The other approach is to evaluate the replacement of forests by something else. In this case the forest has been cleared to give way to what people once wanted to create. Examples of such activities are expanding and establishing new settlement, cereal cultivation, and leaving remains of houses and fields. When located in today's forest, these remains just happened to be overgrown with trees (cultural heritage in the forest). In a long-term heritage management perspective, and above all in a sustainable development perspective, the distinction is vital because the different human relationships with the forest in times past, call for different and reflected strategies of how we manage the cultural heritage of the forest for the future (Svensson 2006).

Recent studies indicate that human impact on the forest landscape has been more intense than it has been realized before. They also state the important fact that knowledge of forest history is a necessary tool to make nature preservation more efficient. The step of taking the cultural values and traditional ecological knowledge from ancestors and integrating them into management plans differs among European countries and ecoregions. Some examples, which are given bellow, may illustrate this.

6.5.3.1 Western Europe

Probably because of the low proportion of forest or wooded land, combined with high population density, interest in cultural heritage started first in Western European countries such as the Netherlands. In the case of the Netherlands, after 1800, when the land was heavily deforested, only 100,000 ha were left, mainly coppiced woodlands and shrubs. Since then the forest area has increased more than threefold. As a consequence the majority of historic elements now in the forest came into being outside the forest and are mostly not related to the forest. The following, quite often recent, evidences of traditional forest-related knowledge can be found: (former) coppiced woodlands and relicts of their exploitation, such as forges and charcoal burning sites (e.g., for iron works); and pits dug for several purposes such as sawpits, man-made wells, and springs and waterways (Jansen and van Benthem 2005).

Over a number of years archaeological elements were identified, preserved, and in some cases restored. Yet only a few woodlands in the Netherlands have been thoroughly surveyed for historical objects. The few that have been examined have shown a surprising number of elements, many of which were not known before the survey and inventory. Once finished, the results of the inventories are published for a wider audience. In some cases routes are signposted along the different objects, explaining the objects in the field. These routes are attractive destinations for both tourists and locals. However, even in the Netherlands, traditional forest-related knowledge is only taken into account in current forest management to a very limited extent. Coppiced woodlands are perhaps the best example of such knowledge incorporated in current forest management, although a more detailed knowledge of the specific management techniques is lacking and the majority of the former or outgrown coppiced woodlands are not now managed as such (van Benthem and Jansen 2006).

In many cases, actions concerning forest cultural heritage make an important contribution to sustainable rural development, such as in Great Britain, where woods and forests remain perhaps the most evocative and iconic features of the landscape. Many skills and knowledge of traditional forest uses and management were almost entirely lost when from the 1960s onwards there were increasing concerns about lost skills, and then sporadic and often localized attempts to document and safeguard cultural knowledge started. Current evidence of traditional forest-related knowledge has obviously been maintained in areas with traditional subsistence communal uses such as coppicing, pollarding, and wood pastures; and the remains of built structures used by the wood workers, sawpits, tracks for transport, ponds and dams, charcoal hearths, white-coal and potash pits, kilns, forges, metal working (smelting) sites, wood banks, and park pales (Rotherham and Ardron 2006; Rotherham 2007). Archaeological sites also include Iron Age enclosures, major regional boundaries, cultivation terraces (from pre-Roman times), deserted villages, former early industrial sites and workings, stone-getting and quarrying sites, medieval field systems, packhorse routes, Bronze Age features such as cup and ring marked stones, and Stone Age evidence such as flints tools. Major initiatives such as the New National Forest in the English Midlands, and numerous Community Forest Projects give evidence of the importance placed on this resource by politicians, planners, and decision makers.

In Scotland interest in the veteran trees and how they have been used—looking at evidence for coppicing, for example—is fairly well-established in the context of designed landscapes. However, it is an expanding area of interest in a wider landscape context, and there are efforts to improve the recording of this evidence more widely, while the Historic Land-use Assessment is identifying areas of woodland that have been managed in the past. The conservation of archaeological and historical features within woodlands to protect these are in place within the Scottish Forestry Strategy and the Forestry Commission's Archaeological Guidelines.[16] These features often have no direct connection to the woodland itself, such as prehistoric burial mounds, although some are associated with past management of the woodland or the use and exploitation of timber, such as plantation boundary banks and charcoal clamps. However, activities are necessary to ensure that traditional building materials and associated craft skills are not lost, particularly because there is a continuing need for these in the repair of ancient monuments and historic buildings

6.5.3.2 Northern Europe

Wood production was the main objective in Finland and the other Nordic countries until the late 1980s. Since then emphasis is also placed on biodiversity aspects (Parviainen 2006). As long as work in the forest was mainly done by manpower and horse, it had been construed as meaning little harm would be done to cultural heritage. However, from 1970 onwards, modern forestry has mainly been based on the use of big forestry machines, and cultural heritage might be endangered. This was the reason why in 1999 the Swedish Parliament adopted 15 national environmental quality objectives. In Sweden the government has stated strongly that 'nature' and 'culture' are two sides of the same coin. One of the targets points for sustainable forests is that 'by 2010 forest land will be managed in such a way as to avoid damage to ancient monuments and to ensure that damage to other known valuable cultural remnants is negligible.' To this end a list of valuable cultural environments (e.g., abandoned crofts, overgrown pastures and meadows with trees, clearing cairns, and remains of sawmills) is included in the Swedish Forestry Act. The physical cultural heritage described above is very obvious to most people. But to make dead stones speak, one has to explain them and tell stories about them. The immaterial cultural heritage includes many different phenomena, names of places, knowledge of handicrafts, songs, poems, tales, and proverbs (Svensson 2006).

---

[16] Forestry Commission Anon. 1995. Forests and archaeology guidelines. http://www.forestry.gov. uk/fr/INFD-5W2FZT, website visited 2010-07-12.

Experiences from Sweden and other Nordic countries in Europe indicate that safeguarding cultural heritage can be improved through protection by the owner. It gives additional value to his property, distinguishes the property from the neighbouring land, and improves the self-image of the owner. In a wider sense, the exhibited objects of cultural heritage add value to the whole region and improve the image of the community. Therefore, in Estonia where the concept of cultural heritage is one of the factors shaping the general forest policy, the inventory of cultural heritage carried out in 2003 aimed not only to highlight cultural heritage but also to motivate landowners to protect the objects. Consequently Tarang et al. (2006) recommended that registering and protecting these objects should be driven primarily by forestry officials themselves with the aim to include them in forest management plans. However, even today there are still examples of accidentally destroyed funeral sites and demolished ruins of historic limestone kilns. At the same time several unique stone fences have been used as reinforcement for the roads used for timber transportation. This has happened even after Estonia regained its independence and began the process of heritage conservation.

6.5.3.3 Central Europe

Even though academic interest has been oriented towards valuing cultural landscapes and traditional management practices, apart a few exceptions there is generally no integration of traditional forest-related knowledge into contemporary forest management. Currently, only a few countries have started to investigate the complex processes underlying forest bio-cultural heritage and to propose conservation methods and principles. For forest authorities and other forest stakeholders, preserving and enhancing social and cultural dimensions of sustainable forest management is often still a new task.

Mountain forests of the Alpine Arc such as in Bavaria, Northern Italy, France, Switzerland, or Austria show evidence of former efforts towards sustainable forest management. Many traces of traditional forest utilization practices are still visible in the present-day landscape. Site-specific strategies, based on multi-disciplinary consultation, and mobilizing resources and political support from a wide range of stakeholders, are promoting the sustainable development of rural landscapes. Accordingly the strengthened cooperation of agriculture and forestry is of essential importance. Thereby, local people are considered to be the main pillar of a self-determining economic development of the Alpine region. The association with tourism is obvious. Within the past few years several initiatives have emerged to promote the production of alternative products by launching forestry and cultural initiatives (traditional use in connection with tourism). The long-term goal of the activities is to generate initiatives to raise additional sources of income for rural population apart from wood and timber. This can be done by creating markets for the multiple services the forest can offer in harmony with regional concepts of development. There are different initiatives promoting this goal such as 'Alliance in the Alps', which was established in 1997 and joins communities and regions of seven states of the Alpine Arc. Its goal is to develop the Alpine area for the future in a sustainable and participatory way.

6.5.3.4 Mediterranean Countries

Several corporation woodlands—a cultural and natural heritage from traditional rural collectivism dating back to the Middle Ages—have survived in remote areas of, for example, Greece, Spain, or Italy, thanks to the permanence of traditional cultures in undeveloped rural societies where social relationships remain at local level. They are an interesting asset for rural development through sustainable forest management (Montiel Molina 2006). However, land abandonment is evident in Italy, Spain, Portugal, and Turkey and across all parts of Greece. Natural regeneration has long been taking over and many former agricultural areas have been re-forested. However agricultural terraces, threshing floors, and stone huts and shredded trees are to some extent still evident to witness a previous agro-silvo-pastoral system of utilization of recourses. In remote areas Greek people are still practising these traditional techniques but only on a small scale (Saratsi 2003, Halstead 1990ab, Forbes 2000). Recent changes in Greek environmental law emphasized the importance of cultural landscapes and heritage, including the preservation of traditional management practices in protected areas along with the protection of natural heritage and nature conservation.[17] Although this law will affect directly only part of the country, it can provide motivation for further initiatives like that. As large protected areas are in forested or wooded areas, this development offers for the first time the opportunity for forest-related cultural values and traditional knowledge in Greece to be protected.

Also, in Turkey one of the main purposes of the 'nature protection system' is to protect not only natural beauty but also any kind of historical, cultural, and natural richness within the natural structure (Akesen 1983). Today villagers who live in and around state forests have some rights concerning forest use such as obtaining timber and firewood, picking up seeds and fruits, and pasturing. Because these rights were granted under the Forest Law of 1937, coppicing has continued up to the present. With the attempt of the government to dedicate several forests to protection, the traditional knowledge of the villagers will be retained. Thereby the central policy of multifunctional forest management includes the planning of forest areas on an integrated district basis and the participation and cooperation of the forestry organization and other interest groups (Turkish Ministry of Environment and Forestry 2004). Accordingly, several historical and archaeological objects in the forests as well as abandoned old villages in forests, if they are of archaeological and historical value, are included in the protected areas.

6.5.3.5 Eastern Europe

In a range of eastern European countries, the location, type, and number of cultural heritage sites are yet not well known. So a necessary precondition to action is a good knowledge of the size of the issue. Concepts and definitions are also not clear. This

---

[17]Government gazette, 14th June 2005.

is the case for instance in Hungary, where, also because of the influence of policy after World War II until the beginning of the 1990s, the traces of culture of the life of many generations lie hidden in the forest. Because most national cultural heritage in forest areas is still unidentified, its importance has not been recognized by the people and it is unprotected against conscious or unconscious destruction. In Hungary after World War II the traditional forest management system was stopped and the new forest management approach transformed the former woodland into high-yielding homogenous forests. This is the reason why almost nothing was left from former traditional forest management. Important elements of former forest management such as coppicing, pollarding, and grazing disappeared in the twentieth century. However, some remnants of former traditional forest use can be found first of all in protected areas and in the core of large woodlands. In some places remnants of formerly grazed open woodland can be found, characterized by trees with huge, wide crowns that are well over 100 years old. There is evidence of former pollarding of beech, lime, or hornbeam trees *(Carpinus betulus)* in wooded pastures. Former mounds can frequently be recognized, and charcoal places and lime kilns are common in the large woods. Tree species composition often suggests the former influence of charcoal and lime burning, and sometimes the tree species composition reflects the former human impact (Molnár et al. 2008).

In Europe's post-socialist former Eastern Bloc, the relationship between traditional knowledge and ecosystems has become increasingly vital during the transition period from central-planned to market economies with limited societal support. In the Ukrainian Carpathian Mountains, local people are still using their traditional knowledge as a base for maintaining a distinctive local culture. Livelihoods of local communities depend directly on traditional knowledge regarding whole landscapes including in traditional village systems once common throughout Europe (Elbakidze and Angelstam 2006). The collapse of the Soviet Union has revived traditional knowledge as an important tool for rural development and the maintenance of biodiversity and cultural landscape values (Angelstam and Elbakidze 2006). In countries such as Romania or Bulgaria, traditional forest farming still plays an ecological, economic, and social role for regional sustainable development. At the same time, however, there are conflicts between the needs of the local people and limitations on the use of forest resources that belong to the state.

6.6   Priorities and Demands for Scientific Studies About Traditional Forest-Related Knowledge and Forest Management

It is obvious that traditional forest-related knowledge has been developed within specific cultural groups over a specific period of time and within specific environmental and social settings. At the same time, history has demonstrated how knowledge has been actively shared and exchanged among societies, and in this matter, holders of traditional knowledge do not differ. The experience of the past half century reveals a variety of relationships between science and traditional forest-related knowledge, in which the general trend has moved from the initial disapproval towards appreciation of traditional knowledge. However, a number of constraints must also be recognized, particularly regarding the inability to access traditional knowledge efficiently and poor communication between holders and potential users.

Since the 1980s, researchers started to recognize the significance of indigenous knowledge for sustainable development (both for environmental conservation and for agricultural technological improvements). Policies, projects, and programmes have also been developed to create synergies between traditional and modern scientific knowledge. This may be for exploring solutions to common problems such as the Declaration Population and Culture of the Alpine Convention (2006), or the European Thematic Network on Cultural Landscapes and their Ecosystems (PAN), which gathers active research groups to develop an overall understanding of man as a landscaping factor.

In recent decades the ways we measure and evaluate the relationship between people and their rural environments have undergone significant advances. Changes of human impacts on forests can be measured and evaluated with indicators such as: naturalness; fragmentation; areas of forest for landscape management and protection; lists of potential sites of cultural and spiritual values; urban forest areas for recreational purposes; or the use of non-wood forest products. However, there are a few existing sources that provide specialized information on cultural and spiritual factors in forestry. In most cases those indicators are not compiled from various statistical sources, such as lists of archaeological sites and sites of ceremonies or customs or through special research or surveys. Some separate indicators for measuring cultural and spiritual factors include both descriptive and quantitative aspects. Parviainen (2006) on the occasion of the Seminar of Forestry and Our Cultural Heritage in Sunne, Sweden, 2005, pointed out the need for further research, means, and efforts in order to develop approaches for including, when relevant, cultural and spiritual aspects in national forest policies, in legislation, and in national forest programmes; and to develop measurable indicators for cultural and spiritual aspects.

On the occasion of the IUFRO European Congress 2007, Forests and Forestry in the Context of Rural Development, organized in conjunction with the EFI Annual Conference 2007 and the Forestry Faculty of the Warsaw Agricultural University, panellists noted from the presentations on the future research agenda that non-wood products, environmental services, and social values are presently not adequately addressed by forest science. Thus the interaction between forest management and rural development should be seen from a social perspective. It was noted that issues such as biodiversity could not be solved without understanding the social and economic factors that influence human decisions. However, while two-thirds of issues identified in terms of European forests need socioeconomic research, only 10% of the existing research is presently dedicated to it. Rural development needs to be engaged in the community of place, and to look for new ways to engage these communities as stakeholders in research.

Concerning the demand for future research it is important to note that the signatory states and the European Community committed themselves in the Vienna

Resolution 3 from 2003 to furthering this work MCPFE (2003b). They agreed to identify, assess, and encourage the conservation and management of significant historical and cultural objects and sites in forests and related to forests in collaboration with relevant institutions. To translate this resolution into more specific and action oriented guidelines, a document has been prepared by Agnoletti et al. (2007) aiming to facilitate its implementation. Although several countries such as the Netherlands, Scotland, Sweden, Estonia, Finland, and Latvia have already carried out or are presently carrying out assessments with regard to archaeological and cultural sites, there is a demand for adequate inventories in other parts of Europe.

The existing diversity of the safeguarding of cultural heritage has geographical, historical, and socioeconomic roots also influenced by ownership structures, which have to be understood and respected. In order to get a reliable and comparable picture of the status of the maintenance of forest-related cultural values, further research is required. According to the suggestions of the MCPFE-countries, amongst others the following objectives should be tackled: (1) recognition of cultural values as essential elements of the diversity and richness of national cultural heritage; (2) identification of the requirements and provision of legislation to incorporate cultural values into local plans for forest and woodland management; (3) promotion of cultural values by including them in educational programmes and development of training courses on conservation and management of cultural values in forestry; (4) identification of cultural values in the territory defining their significance, integrity, and vulnerability; (5) monitoring of the processes of transformation; (6) definition of criteria and indicators for their management; (7) definition of planning tools and management techniques; (8) ensuring research development in order to increase knowledge and gather evidence to limit actual and potential negative impacts on cultural heritage; (9) management of the processes of data collection and collation; and (10) promotion of interdisciplinary studies for the identification, inventory, and documentation of local cultural heritage related to forestry and woodland landscapes (MCPFE 2003a).

6.7   Conclusion

Traditional forest management focused on the local availability of materials. Production was adapted to the natural resources and to the demands of the local market. Aiming at various non-wood products in addition to timber could be carried out as long as the industrial demand for energy, wages, and population density were low. The increasing demand for wood for an expanding industry and growing population, as well as liberalism, caused a shortage of the renewable resource of wood and resulted in conflicts with regard to participation in the use of woodland. Wood thereby became the most demanded product, and other uses were of political minor interest. As growth and yield of forests could not be increased by traditional forest management because it focused also on other products that were important for the subsistence of farms (e.g., grazing and litter harvesting), a general over-utilization came into being. This development was forced by conflicts with regard to ownership and utilization rights, and by both globalization of the market and science.

At present the old skills are disappearing very rapidly. Knowledge that was passed down to descendents for centuries cannot easily be rekindled at the present time. From the viewpoint of the 'modern' market, neither traditional handicraft-products nor traditional forest management have a chance to compete successfully with cheap industrial mass production and plantation forestry. Forest science has contributed significantly to this development by focusing on the production of a maximum of valuable timber (mostly softwood) only and by a one-sided interpretation of the term sustainability. The historical tie between the sustainable use of woodland and the maintenance of its social and cultural values does not exist any more. However, present-day 'close-to-nature forest management' and several policies also include social and cultural traditional values (apart from economic and ecological targets).

There is no doubt that sustainable development must include people and the importance of both their pasts and their futures. What is the value of a 'restored nature in balance' with neither history nor future? It will be just another kind of museum or reserve. The European forests and wooded landscapes are, and will always be, products of both natural and cultural processes. Human impacts on forests and woods have been there since the dawn of humankind, resulting in both monuments and new or changed environments. If the forest is comprehended as 'wilderness' and not as a cultural landscape, it places the people living there as of a low status as cultural beings, and even a problem in attainment of sustainability. They become people living in an eternal periphery, being completely alien to our urban cultural conception. However, the cultural landscapes of the European forests tell different stories. In some parts of Europe these are stories of the use of forest resources for trade, and the achievement of wealth in the forested local societies. In older times these local societies were far from being the marginalized entities we consider them today. Svensson (2006) points out that an acknowledged history and an upgrading of the importance of the cultural heritage of the forest would be an act of contributing to a regional sustainable development.

In this context there is an urgent need to exchange knowledge on a European level. However, it is also most important to place the site in its landscape and cultural contexts. Europe is at the beginning of a working process where cultural and social dimensions of sustainable forest management are to be integrated in relevant policies such as MCPFE and the Alpine Convention. In order to more fully and effectively achieve the necessary inputs of enthusiasm and data, extensive and dedicated networking is essential. In this work it will be important to encompass the widest possible spectrum of stakeholders. There is much to be done and the challenges are great. However, for the wider European community, the cultural and heritage costs of failure are beyond calculation, and the benefits of success—to people, the environment and its biodiversity, the economy, and to sustainability—are enormous.

References

Addy SO (1898) The evolution of the English house. Swan Sonnenschein, London

Agnoletti M (2006) Traditional knowledge and the European Common Agricultural Policy (PAC): the case of the Italian national Rural Development Plan 2007–2013. Parrotta J, Agnoletti M, Johann E (eds) Cultural heritage and sustainable forest management: the role of traditional knowledge. Proceedings of the conference 8–11 June 2006, Florence, Italy. MCPFE Liaison Unit, Warsaw, pp 11–14, 17–25

Agnoletti M, Anderson S, Johann E, Klein R, Kulvik M, Kushlin AV, Mayer P, Montiel Molina C, Parrotta J, Patosaari P, Rotherham ID, Saratsi E (2007) Guidelines for the implementation of social and cultural values in sustainable forest management: a scientific contribution to the implementation of MCPFE–Vienna Resolution 3. IUFRO occasional paper no. 19. International Union of Forest Research Organizations (IUFRO), Vienna

Agricolae GA (1772-1784) Versuch einer allgemeinen Vermehrung aller Bäume, Stauden und Blumengewächse anjetzo auf ein neues Übersehen, mit Anmerkungen und einer Vorrede. 2 vols. Verlag Montag und Gruner, Regensburg

Akesen A (1983) Ulusal doğayı ve doğal kaynakları koruma sistemimizin temel öğeleri (The main elements of our national nature and natural resources conservation system). Tabiat ve İnsan (Ankara) 7(1):16–18

Angelstam P, Elbakidze M (2006) Sustainable forest management in Europe's East and West: trajectories of development and the role of traditional forest knowledge. In: Parrotta J, Agnoletti M, Johann E (eds) Cultural heritage and sustainable forest management: the role of traditional knowledge. Proceedings of the conference 8–11 June 2006, Florence, Italy. MCPFE Liaison Unit, Warsaw, pp 349–357

Anko B (2005) Woodlands as cultural heritage—yet another challenge for contemporary and future forestry. News of forest history 36/37 (III), 2 vol. vol 1/2005. Bundesministerium für Land- und Forstwirtschaft, Umwelt und Wasserwirtschaft, Vienna, pp 57–65

Barker G (1985) Prehistoric farming in Europe. Cambridge University Press, Cambridge

Bendix B (2008) Geschichte der Forstpflanzenanzucht in Deutschland von ihren Anfängen bis zum Ausgang des 19. Jahrhunderts. Verlag Kessel, Remagen

Bil A (1990) The shieling 1600–1840: the case of the central Scottish highlands. John Donald Publishers Ltd., Edinburgh

Bischoff F, Schönbach A (eds) (1881) Steirische und Kärnthnerische Taidinge. Bd. 6. Kaiserliche Akademie der Wissenschaften. Verlag Braumüller, Vienna

Bohn U, Neuhäusl R, Gollub G, Hettwer C, Neuhäuslová Z, Schlüter H, Weber H (2000/2003) Karte der natürlichen Vegetaton Europas. Maßstab 1:2500000. Teil 1 Erläuterungstext mit CD-Rom, Teil 2 Legende, Teil 3 Karten. Landwirtschaftsverlag, Münster

Casals CV (1996) Los ingenieros de montes en la España contemporánea, 1848–1936. Ediciones del Serbal, Barcelona

Casparie WA (1995) Houtgebruik van Paleolithicum to Middeleeuwen: voorbeelden uit de archeologie. In: Helfrich K, Benders JF, Casparie WA (eds) Handzaam hout uit Groninger grond: houtgebruik in de historische stad. Stichting Monument and Materiaal, Groningen, pp 212–219

Damianakos S, Zakopoulou E, Casimis C, Nitsiakos B (1997) Eksousia, ergasia kai mnimi se tria xoria tis Epirou: i topiki dunamiki tis epiviosis. (Power, work and memory in three villages of Epiros: the local dynamic of survival). Plethron/EKKE, Athens

De Crescentiis P (1538) De agricultura. omnibusque plantarum et animalium generibus libri XII. Westhemer, Basel

De la Cruz AE (1994) La destrucción de los montes (claves histórico-jurídicas). Universidad Complutense de Madrid, Madrid

Di Martino P, Di Marzio P, Giancola C, Ottaviano M, Di Martino P, Di Marzio P, Giancola C, Ottaviano M (2006) The forest landscape of transhumance in Molise, Italy. In: Parrotta J, Agnoletti M, Johann E (eds) Cultural heritage and sustainable forest management: the role of traditional knowledge. Proceedings of the conference 8–11 June 2006, Florence, Italy. MCPFE Liaison Unit, Warsaw, pp 196–208

Dinou M, Badal E (2001) Charcoal analysis: application of the method and preliminary results from the Neolithic settlement of Makri, Thrace. In: Bassiakos Y, Aloupi E, Facorellis Y (eds) Archaeometry issues in Greek prehistory and antiquity. Hellenic Society for Archaeometry and Society of Messenean Archaeological Studies, Athens, pp 125–138

Downey CJ (2001) Prehistoric tools and the Bronze Age woodworking industry. In: Bassiakos Y, Aloupi E, Facorellis Y (eds) Archaeometry issues in Greek prehistory and antiquity. Hellenic Society for Archaeometry and Society of Messenean Archaeological Studies, Athens, pp 791–800

DPT (2007) Dokuzuncu Kalkınma Planı. 2007–2013. Ormancılık özel ihtisas komisyonu raporu (Ninth five-year development plan: forestry expert commission report). Devlet Planlama Teşkilatı (TR. Prime Ministry State Planning Organization) Publications, Ankara, ISBN 978-975-19-4031-5 (in Turkish)

Elbakidze M, Angelstam P (2006) Role of traditional villages for sustainable forest landscapes: a case study in the Ukrainian Carpathian Mountains. In: Parrotta J, Agnoletti M, Johann E (eds) Cultural heritage and sustainable forest management: the role of traditional knowledge. Proceedings of the conference 8–11 June 2006, Florence, Italy. MCPFE Liaison Unit, Warsaw, pp 371–380

Food and Agriculture Organisation (FAO) (2001) Global forest resource assessment 2000: Main report, Forestry Paper 140. Food and Agriculture Organization of the United Nations, Rome

Forbes H (2000) Security and settlement in the medieval and post-medieval Peloponnese, Greece: 'hard' history versus oral history. J Mediterr Archaeol 13(2):204–224

Forman RTT (1996) Land mosaics. Cambridge University Press, Cambridge

Frank G, Parviainen J, Vandekerhove K, Latham J, Schluck A, Little D (2007) COST Action E27. Protected forest areas in Europe—analysis and harmonisation: results, conclusions and recommendations. BFW, Vienna

Galán R, Segovia C, Martínez MA, Alés E, Coronilla R, Barrera M (2003) La colonia de buitre negro de Sierra Pelada. Quercus 211:26–33

García Latorre J, García Latorre J (2007) Almería: hecha a mano: una historia ecológica. Fundación Cajamar, Almería

Gaworska M, Kornatowska B (2006) MCPFE commitments—political framework for social and cultural dimension of SFM. In: Parrotta J, Agnoletti M, Johann E (eds) Cultural heritage and sustainable forest management: the role of traditional knowledge. Proceedings of the conference 8–11 June 2006, Florence, Italy. MCPFE Liaison Unit, Warsaw, pp 243–249

Glück P (1987) Social values in forestry. Ambio 16(2/3):158–160

Glück P, Weber M (1998) Mountain forestry in Europe—evaluation of silvicultural and political means, vol 35. Publication Series of the Institute for Forest Sector Policy and Economics, Vienna

Gomez de Mendoza J, Mata Olmo R (1992) Actuaciones forestales públicas desde 1940. Objetivos, criterios y resultados. Agric Soc 65:15–64

Groome H (1988) El desarrollo de la política forestal en el Estado español: desde la guerra civil hasta la actualidad. Arbor XXIX:65–110

Halstead P (1990a) Present to past in the Pindos: specialisation and diversification in mountain economies. Rivista di Liguri 56:61–80

Halstead P (1990b) Waste not, want not: traditional responses to crop failure in Greece. Rural Hist Econ Soc Cult 1:147–164

Halstead P (1998) Ask the fellows who lop the hay: leaf-fodder in the mountains of northwest Greece. Rural Hist 9(2):211–234

Hamberger J, Irlinger W, Suhr G (2008) In Boden und Stein. Denkmäler im Wald Bayerische Landesanstalt für Wald und Forstwirtschaft und Zentrum Wald Forst Holz Weihenstephan (eds). Lerchl Druck, Freising

Henderson-Howat DB (1996) Great Britain. In: Morin G-A, Kuusela K, Henderson-Howat DB, Efstathiadis NS, Orozi S, Sipkens H, Hofsten E, MacCleery DW (eds) Long-term historical changes in the forest resource: case studies of Finland, France, Great Britain, Greece, Hungary, The Netherlands, Sweden and the United States of America, vol 10, Geneva Timber and Forest Study Papers. United Nations, New York, pp 23–26

Herr M (1538) Das Ackerwerk (De re rustica) Lucil Columella und Palladius (Rutilius Taurus Aemilianus Palladius) zweyer hocherfarner Römer. Verteutscht durch Michael Herren. Melchior Rabus, Strassburg

Hofsten Ev (1996) Sweden. In: Morin G-A, Kuusela K, Henderson-Howat DB, Efstathiadis NS, Orozi S, Sipkens H, Hofsten Ev, MacCleery DW (eds) Long-term historical changes in the forest resource: case studies of Finland, France, Great Britain, Greece, Hungary, The Netherlands, Sweden and the United States of America. United Nations, New York, pp 43–51

Hughes JD and Thirgood JV (1982) Deforestation, erosion, and forest management in ancient Greece and Rome. J For Hist and American Society for Environmental History 26(2):60–75

Imreh I (1983) A rendtartó székely falu. Faluközösségi határozatok a feudalizmus utolsó évszázadából (Szekler Dorfordnung. Dorfgemeinschaften aus den letzten Jahrhunderten des Feudalismus). Kriterion, Bukarest

Jansen P, van Benthem M (2005) Historische boselementen: geschiedenis, herkenning en beheer. Waanders, Zwolle

Johann E (1996) Zur Geschichte der Waldbewirtschaftung im Alpenvorland. Österreichische Forstzeitung 12:8–11

Johann E (2001) Zur Geschichte des Natur- und Landschaftsschutzes in Österreich: Historische, Ödflächen“ und ihre Wiederbewaldung”. In: Weigl N (ed) Faszination Forstgeschichte, vol 42. Schriftenreihe des Instituts für Sozioökonomik der Forst- und Holzwirtschaft, Vienna, pp 41–60

Johann E (2002) Zukunft hat Vergangenheit. 150 Jahre Österreichischer Forstverein. Österreichischer Forstverein, Wien

Johann E (2003) More about diversity in European forests: the interrelation between human behaviour, forestry, and nature conservation at the turn of the 19th century. In: Jelecek L et al (eds) Dealing with diversity. 2nd International conference of the European society for environmental history, Prague. Charles University in Prague, Faculty of Science, Department of Social Geography and Regional Development, Prague, pp 202–205

Johann E (2004a) Forest history in Europe. In: Werner D (ed) Biological resources and migration. Springer, Berlin/Heidelberg, pp 73–82

Johann E (2004b) Wald und Mensch: Die Nationalparkregion Hohe Tauern (Kärnten). Verlag des Kärntner Landesarchivs, Klagenfurt

Johann E (2004c) Landscape changes in the history of the Austrian alpine regions: ecological development and the perception of human responsibility. In: Honnay O, Verheyen K, Bossuyt B, Hermy M (eds) Forest biodiversity: lessons from history for conservation. CABI Publishing, Wallingford, pp 27–40

Johann E (ed) (2005) Woodlands—cultural heritage. Proceedings of the international IUFRO–conference, Vienna. News of forest history Nr. III (36/37) 1/2005, 2/2005. Bundesministerium für Land- und Forstwirtschaft, Umwelt und Wasserwirtschaft, Wien

Johann E (2006) Historical development of nature-based forestry in Central Europe. In: Diaci J (ed) Nature-based forestry in Central Europe: alternatives to industrial forestry and strict preservation. CIP Katalozni zapis o publikacij, Narodna in Universitetna Knjiznica, Ljubljana

Johann E (2007) Traditional forest management under the influence of science and industry: the story of the alpine cultural landscapes. For Ecol Manag 249(2007):54–62

Johann E (2008) Wirtschaftsfaktor Wald. Am Beispiel des österreichischen Alpenraumes. In: Vavra E (ed) Der Wald im Mittelalter: Funktion–Nutzung–Deutung: das Mittelalter. Zeitschrift des Mediävistenverbandes 13/2:12–27

Johann E, Agnoletti M, Axelsson AL, Bürgi M, Östlund L, Rochel X, Schmidt EU, Schuler A, Skovsgaard JP, Winiwarter V (2004) History of secondary Norway spruce in Europe. In: Spiecker H, Hansen J, Klimo E, Skovsgaard JP, Sterba H, von Teuffel K (eds) Norway spruce conversion—options and consequences, vol 18, European Forest Institute Research Report. Brill, Leiden, pp 25–62

Kirby K, Watkins C (eds) (1998) The ecological history of European forests. CAB International, Wallingford

Kooistra MJ, Maas GJ (2008) The widespread occurrence of Celtic field systems in the central part of the Netherlands. J Archaeol Sci 35:2318–2328

Kossarz W (1984) Der Wald in den Volksrepubliken des Donauraumes. Verlagsunion Agrar, Österreichischer Agrarverlag, Wien

Lazzarini A (2006) Relationships between local communities and forest administration: the Cadore region after the Napoleonic laws. In: Parrotta J, Agnoletti M, Johann E (eds) Cultural heritage and sustainable forest management: the role of traditional knowledge. Proceedings of the conference 8–11 June 2006, Florence, Italy. MCPFE Liaison Unit, Warsaw, pp 65–68

Linares AM (2006) Forest science and local experience in the management of the woodland: the case of Extremadura's dehesa. In: Parrotta J, Agnoletti M, Johann E (eds) Cultural heritage and sustainable forest management: the role of traditional knowledge. Proceedings of the conference 8–11 June 2006, Florence, Italy. MCPFE Liaison Unit, Warsaw, pp 69–77

Mantel K (1980) Forstgeschichte des 16. Jahrhunderts unter dem Einfluss der Forstordnungen und Noe Meurers. Paul Parey, Hamburg

Mantel K (1990) Wald und Forst in der Geschichte. Verlag M&Schaper, Alfeld

Marntel A, Wagberg C (2006) Cultural remains of the forests—a resource in rural development. In: MCPFE (ed) Forestry and our cultural heritage. MCPFE Liaison Unit, Warsaw, pp 82–86

Ministerial Conference on the Protection of Forests in Europe (MCPFE) (2000) General declaration and resolutions adopted at the ministerial conferences on the protection of forest in Europe, Strasbourg 1990–Helsinki 1993–Lisbon 1998. MCPFE Liaison Unit, Vienna

Ministerial Conference on the Protection of Forests in Europe (MCPFE) (2003a) Improved Pan-European indicators for sustainable forest management as adopted by MCPFE expert level meeting 7–8 Oct 2002, Vienna, Austria. MCPFE Liaison Unit, Vienna

Ministerial Conference on the Protection of Forests in Europe (MCPFE) (2003b) Vienna declaration and Vienna resolutions. Adopted at the fourth ministerial conference on the protection of forests in Europe, 28–30 Apr 2003, Vienna, Austria. MCPFE Liaison Unit, Vienna

Ministerial Conference on the Protection of Forests in Europe (MCPFE) (2003c) Fourth ministerial conference on the protection of forests in Europe conference proceedings, 28–30 Apr 2003 Vienna, Austria. MCPFE Liaison Unit, Vienna

Ministerial Conference on the Protection of Forests in Europe (MCPFE) (2003d) State of Europe's forests 2003. The MCPFE report on sustainable forest management in Europe. MCPFE Liaison Unit and UNECE/FAO, Vienna

Ministerial Conference on the Protection of Forests in Europe (MCPFE) (ed) (2006) Forestry and our cultural heritage. Proceedings of the seminar 13–15 June 2005, Sunne, Sweden. MCPFE Liaison Unit, Warsaw

Ministerial Conference on the Protection of Forests in Europe (MCPFE), Liaison Unit Warsaw, UNECE, FAO, (eds) (2007) State of Europe's forests. The MCPFE report on sustainable forest management in Europe. MCPFE Liaison Unit, UNECE, FAO, Warsaw

Mlinsek D (1996) From clear-cutting to a close-to-nature silvicultural system. IUFRO News 25:6

Molnár Z, Bartha S, Babai D (2008) Traditional ecological knowledge as a concept and data source for historical ecology, vegetation science and conservation biology: a Hungarian perspective. In: Szabó P, Hédl R (eds) Human nature: studies in historical ecology and environmental history. Institute of Botany of the ASCR, Brno, pp 14–27

Montiel Molina C (2006) Cultural heritage, sustainable forest management and property in inland Spain. In: Parrotta J, Agnoletti M, Johann E (eds) Cultural heritage and sustainable forest management: the role of traditional knowledge. Proceedings of the conference 8–11 June 2006, Florence, Italy. MCPFE Liaison Unit, Warsaw, pp 417–423

Morin GA (1996) Common points. In: Morin GA, Kuusela K, Henderson-Howat DB, Efstathiadis NS, Orozi S, Sipkens H, Hofsten E, MacCleery DW (eds) Long-term historical changes in the forest resource: case studies of Finland, France, Great Britain, Greece, Hungary, The Netherlands, Sweden and The United States of America, vol 10, Geneva Timber and Forest Study Papers. United Nations, New York, pp 1–2

Nikolussi K (1998) Die Bauhölzer der Via Claudia. In: Walde E (ed) Via Claudia—Neue Forschungen. Hörtenbergdruck, Telfs

Okan T, Ok K (2006) Opportunities in Turkish cultural heritage for conservation of natural values. In: Parrotta J, Agnoletti M, Johann E (eds) Cultural heritage and sustainable forest management: the role of traditional knowledge. Proceedings of the conference 8–11 June 2006, Florence, Italy. MCPFE Liaison Unit, Warsaw, pp 424–431

Özdönmez M (1973) Devlet ormanlaından köylülerin faydalanma hakları (Utilization rights of viligers from public forests). İstanbul Üniversitesi Orman Fakültesi Dergisi (Review of the Faculty of Forestry University of İstanbul), B Series, İstanbul, Sermet Matbaası, 23(2):62–77

Pantera A, Fotiadis G, Aidinidis E (2009) History and current distribution of valonia oak in Greece. In: Saratsi E, Buergi M, Johann E, Kirby K, Moreno D, Watkins C (eds) Woodland cultures in time and space: tales from the past, messages for the future. Embryo, Athens, pp 228–235

Parrotta J, Agnoletti M, Johann E (eds) (2006) Cultural heritage and sustainable forest management: the role of traditional knowledge. Proceedings of the conference 8–11 June 2006, Florence, Italy. 2 vols. MCPFE Liaison Unit, Warsaw

Parviainen J (2006) Forest management and cultural heritage. In: MCPFE (ed) Forestry and our cultural heritage. MCPFE Liaison Unit, Warsaw, pp 67–75

Parviainen J (2007) Virgin and natural forests in the temperate zone of Europe. In: Commarmot B, Hamor F (eds) Natural forests in the temperate zone of Europe—values and utilisation. Proceedings of the conference 13–17 Oct 2003 Mukachevo, Ukraine. Swiss Federal Research Institute WSL, Birmensdorf, pp 9–18

Rackham O (1976) Trees and woodland in the British landscape: a complete history of Britain's trees, woods and hedgerows. Phoenix Press, New York

Rackham O (1980) Ancient woodlands, its history, vegetation and uses in England. Edward Arnold, London

Rackham O (2000) The history of the countryside: the classic history of Britain's landscape, flora and fauna. Phoenix, New York

Radkau J (2000) Natur und Macht: eine Weltgeschichte der Umwelt. Beck, München

Rochel X (2006) Selection forestry between tradition and innovation: five centuries of practice in France. In: Parrotta J, Agnoletti M, Johann E (eds) Cultural heritage and sustainable forest management: the role of traditional knowledge. Proceedings of the conference 8–11 June 2006, Florence, Italy. MCPFE Liaison Unit, Warsaw, pp 85–90

Rosegger PK, Pichler F, Rauschenfels AV (about 1880) Wanderungen durch Steiermark und Kärnten. Druck und Verlag Gebrüder Kröner, Stuttgart

Rotherham ID (2007) The implications of perceptions and cultural knowledge loss for the management of wooded landscapes: a UK case-study. For Ecol Manag 249:100–115

Rotherham ID, Ardron PA (2006) The archaeology of woodland landscapes: issues for managers based on the case-study of Sheffield, England and four thousand years of human impact. Arboricult J 29(4):229–243

Rotherham ID, Jones M (2000) The impact of economic, social and political factors on the ecology of small English woodlands: a case study of the ancient woods in South Yorkshire, England. In: Agnoletti M, Anderson S (eds) Forest history: international studies in socio-economic and forest ecosystem change. CAB International Wallingford, Oxford, pp 397–410

Sala P (2000) Modern forestry and enclosure: elitist state science against communal management and unrestricted privatisation in Spain, 1855–1900. Environ Hist 6(2):151–168

Sánchez Martínez JD, Araque Jiménez E (1993) El impacto de la política de repoblación forestal de postguerra. Dos ejemplos municipales en la Sierra de Segura (Jaen). In: Congreso Forestal Español, Tomo IV, Lourizán, pp 471–476

Saratsi E (2003) Landscape history and traditional management practices in the Pindos Mountains of northwest Greece c 1880–2000. University of Nottingham, School of Geography, Nottingham

Saratsi E (2005) The cultural history of 'Kladera' in the Zagori area of Pindos Mountain, NW Greece. News of Forest History III (36/37). Bundesministerium für Land und Forstwirtschaft Umwelt und Wasserwirtschaft, Wien, pp 107–117

Saratsi E (2009) Conflicting management cultures: an historical account of woodland management in Greece during the 19th and 20th centuries. In: Saratsi E, Buergi M, Johann E, Kirby K, Moreno D, Watkins C (eds) Woodland cultures in time and space: tales from the past, messages for the future. Embryo, Athens, pp 57–64

Saratsi E, Agnoletti M, Watkins C (2007) What the past can do for the future: towards the inclusion of social and cultural values in sustainable forest management. IUFRO WP 6.07.00 Scientific Summary related to IUFRO News 10, 2007. International Union of Forest Research Organizations (IUFRO), Vienna

Saratsi E, Buergi M, Johann E, Kirby K, Moreno D, Watkins C (eds) (2009) Woodland cultures in time and space: tales from the past, messages for the future. Embryo, Athens

Schmithuesen H, Seeland K (2006) European landscapes and forests as representations of culture. In: Parrotta J, Agnoletti M, Johann E (eds) Cultural heritage and sustainable forest management: the role of traditional knowledge. Proceedings of the conference 8–11 June 2006, Florence, Italy. MCPFE Liaison Unit, Warsaw, pp 217–224

Seidensticker A (1886) Waldgeschichte des Alterthums. Ein Handbuch für akademische Vorlesungen. 2 Bde. Trowitzsch and Sohn, Frankfurt a. d. Oder

Siegel H, Tomaschek H (eds) (1870) Salzburgische Taidinge. Bd. 1. Kaiserliche Akademie der Wissenschaften, Wien

Slee B (1998) Epilogue. In: Glück P, Weber M (eds) Mountain forestry in Europe—evaluation of silvicultural and political means, vol 35. Publication Series of the Institute for Forest Sector Policy and Economics, Vienna, pp 295–297

Sloet JJS (1911) Gelderse markerechten. Oud-vaderlandse rechtsbronnen. Werken der Vereeniging tot Uitgaaf der Bronnen van het Oud-Vaderlands Recht, Utrecht.Tweede reeks no. 12. 's-. Martinus Nijhoff, Gravenhage

Smout TC, MacDonald AR, Watson F (2005) A history of the native woodlands of Scotland 1500–1920. Edinburgh University Press, Edinburgh

Stevenson AC, Harrison RJ (1992) Ancient forest in Spain: a model for land-use and dry forest management in south-west Spain from 4000 BC to 1900 AD. Proc Prehist Soc 58:227–247

Strack CFL (1968) Plinius Caius Secundus. Naturalis Historiae. Unchanged reprint of the first edition Bremen 1853–1855. Translated by Strack M.E.D.L. Darmstadt

Sturm J (1937) Der Wald in den Freisinger Traditionen. Kommission für bayerische Landesgeschichte bei der Bayerischen Akademie der Wissenschaften in Verbindung mit der Gesellschaft für fränkische Geschichte München, editor. Zeitschrift für bayerische Landesgeschichte 10:311–373

Svensson E (2006) Cultural heritage of the forest and cultural heritage in the forest. In: MCPFE (ed) Forestry and our cultural heritage. MCPFE Liaison Unit, Warsaw, pp 106–110

Szabó P (2005) Woodland and forests in medieval Hungary. Archaeopress, Oxford

Tarang L, Kusmin J, Pommer V, Matila A, Külvik M (2006) Forest landscape cultural heritage inventory: an Estonian model. In: Parrotta J, Agnoletti M, Johann E (eds) Cultural heritage and sustainable forest management: the role of traditional knowledge. Proceedings of the conference 8–11 June 2006, Florence, Italy. MCPFE Liaison Unit, Warsaw, pp 268–271

Turkish Ministry of Environment and Forestry (2004) Türkiye ulusal ormancılık programı. (National forestry programme of Turkey), Ankara

United Nations Educational Scientific and Cultural Organization (UNESCO) (2003) Convention for the safeguarding of the intangible cultural heritage. General conference Paris, 29 Sept to 17 Oct 2003, 32 session. Available via http://unesdoc.unesco.org/images/0013/001325/132540e.pdf. Cited 24 Mar 2009

Valdés CM (1996) Tierras y montes públicos en la sierra de Madrid (sectores central y meridional). Ministerio de Agricultura, Pesca y Alimentación, Madrid

Van Benthem M, Jansen P (2006) Management of historic elements in Dutch forests. In: MCPFE (ed) Forestry and our cultural heritage. MCPFE Liaison Unit, Warsaw, pp 77–81

Welzholz JC, Johann E (2007) History of protected forest areas in Europe. In: Frank G, Parviainen J, Vandekerhove K, Latha J, Schluck A, Little D (eds) COST Action E27. Protected forest areas in Europe—analysis and harmonisation: results, conclusions and recommendations. BFW, Vienna, pp 17–49

Whittle A (1985) Neolithic Europe: a survey. Cambridge University Press, Cambridge

Yurdakul S, Ekizoğlu A (2005) The historical development of one of the forest villagers' utilization rights from the public forests: sale rights of wood material in the market. News of forest history 36/37 (III), 2 vol. vol 1/2005. Bundesministerium für Land- und Forstwirtschaft, Umwelt und Wasserwirtschaft, Vienna, pp 142–152

Zanzi Sulli A (1997) Notes on the concept of 'multiple use forestry'. In: Johann E (ed) Multiple use forestry from the past to present times. Proceedings of the symposium organized by IUFRO research group 6.07. Forest history 2–4 May 1996 Gmunden. News of forest history 25/26. Österreichischer Forstverein, Wien, pp 11–17

Chapter 7
Russia, Ukraine, the Caucasus, and Central Asia

Vladimir Bocharnikov, Andrey Laletin, Per Angelstam, Ilya Domashov, Marine Elbakidze, Olesya Kaspruk, Hovik Sayadyan, Igor Solovyi, Emil Shukurov, and Tengiz Urushadze

Abstract Interconnection and interaction of human beings and forests have shaped traditional forest-related knowledge (TFRK) over centuries in the vast region of Eurasia, where forestry practices of Russian and other former Soviet origin prevailed during the past century. There are significant differences in forestry practices across this region, a result of geographic and cultural diversity as well as historical differences in social, economic, and political conditions. In this chapter, the diversity of traditional forest-related knowledge and associated practices—as well as the problems and the prospects for their preservation—are introduced by focussing on

V. Bocharnikov (✉)
Pacific Institute of Geography, Russian Academy of Science,
Vladivostok, Russia
e-mail: vbocharnikov@mail.ru

A. Laletin
Friends of the Siberian Forests, Krasnoyarsk, Russia
e-mail: laletin3@gmail.com

P. Angelstam
Faculty of Forest Sciences, School for Forest Engineers, Swedish University of
Agricultural Sciences, Skinnskatteberg, Sweden
e-mail: per.angelstam@smsk.slu.se

I. Domashov
"BIOM" Ecological Movement, Bishkek, Kyrgyzstan
e-mail: idomashov@gmail.com

M. Elbakidze
Faculty of Geography, Lviv University of Ivan Franko,
Lviv, Ukraine
e-mail: marine.elbakidze@smsk.slu.se

O. Kaspruk • I. Solovyi
Ukrainian National Forestry University, Lviv, Ukraine
e-mail: ok@nepcon.net; soloviy@yahoo.co.uk

J.A. Parrotta and R.L. Trosper (eds.), *Traditional Forest-Related Knowledge:*
Sustaining Communities, Ecosystems and Biocultural Diversity,
World Forests 12, DOI 10.1007/978-94-007-2144-9_7,
© Springer Science+Business Media B.V. (outside the USA) 2012

selected, contrasting areas in the Russian Federation, Ukraine, the Caucasus, and Central Asia. Traditional uses of forest resources have survived in the huge area of barely modified ecosystems stretching from Scandinavia through the northern reaches of the Russian Plain, the Urals, and Siberia, and up to the Far East. Traditional uses of forest resources practised in the North, Siberia, and the Far East of Russia generally involve a combination of activities including reindeer breeding, hunting, fishing, and harvesting of non-timber forest products; traditional activities in northern villages also include the processing of yields from the above activities, breeding wild animals, honey, and hay production. Non-timber product gathering and hunting support the livelihoods of mountain dwellers in the Caucasus and Central Asia regions, where nomads have used forest resources for fuel and for making yurts and household utensils. Today, traditional forest-related knowledge throughout the region is at risk, complicated by the erosion of the role of families in the intergenerational transfer of this knowledge.

Keywords Biodiversity • Boreal forests • Eurasia • Forest-based livelihoods • Forest policy • Indigenous peoples • Local communities • Non-timber forest products • Traditional knowledge • Wildlife

7.1 Introduction

Eurasia is the world's largest continent, extending over a range of climatic zones from the tropics to the arctic. The ecological zones of northern Eurasia represent different natural landscapes including forests, steppes, semi-deserts, and deserts. Natural forests with a rich biodiversity have survived here in the remote northern and eastern areas of the Eurasian continent, bordering on landscapes modified by peoples living in western and southern regions. In this chapter, we consider the region which we refer to as 'northern Eurasia,' which includes countries that were formerly part of the Soviet Union, a vast territory featuring a wide variety of traditional uses of forest-related resources from very diverse natural landscapes.

There are significant differences in the development of traditional forest-related knowledge between the densely populated areas of Europe and the sparsely populated mountain and taiga forest landscapes of northern Eurasia. Within this region, we focus

H. Sayadyan
Armenian State Agrarian University, Yerevan, Republic of Armenia
e-mail: hovik_s@yahoo.com

E. Shukurov
"Aleyne" Ecological Movement of Kyrgyzstan, Bishkek, Kyrgyzstan
e-mail: shukurovemil@mail.ru

T. Urushadze
Ivane Javakhishvili Tbilisi State University, Tbilisi, Georgia
e-mail: t_urushadze@yahoo.com

Fig. 7.1 Forest and woodland cover in northern Eurasia (Source: Adapted from FAO (2001)). *Key: Dark green* closed forest, *light green* open or fragmented forest, *pale green* other wooded land, *yellow* other land

on these less densely populated forest areas of Russia; Ukraine; the Caucasus (Armenia, Azerbaijan, Georgia); and Central Asia (Kazakhstan, Kyrgyzstan, Tajikistan, Turkmenistan, and Uzbekistan) (Fig. 7.1). These 'case study' regions were chosen on the basis of their geographical and biotic significance (Bocharnikov 2005).

There are many geographical, ecological, cultural, and historical differences among these former Soviet republics, but their economic, cultural, and political relationships are still maintained. Adding to their shared history, these countries seek to preserve the common system of natural resource use and to share best practices for wildlife conservation. In these selected regions of northern Eurasia, the past and present characteristics and uses of, and future prospects for, traditional forest-related knowledge and practices of indigenous peoples and local communities will be discussed.

Traditional uses of forest-related resources in this region reflect the rich history of hundreds of culturally distinct peoples (Laletin 1999). Each group has developed its own ways of interacting with nature over the past 10,000–15,000 years of human settlement. Unique expressions of such interactions are found in the traditions, customs, folklore, and forest resource uses that have survived among contemporary communities in the area (Dmitriev and Dmitriev 1991; Fedorova 1986). As in other parts of the world, traditional knowledge and practices of the region's indigenous peoples exist in a much broader cultural context that includes spiritual dimensions. For example, traditional communities that rely on hunting for survival have typically developed rites that often include communication with spirits to ensure success in forest hunting.

The distinctive characteristics of the region's varied ecosystems have shaped the evolution of traditional knowledge and practices that sustain the cultures and livelihoods of indigenous peoples and local communities—guiding the hunting, fishing, and gathering activities of communities who are directly dependent on a diversity of natural resources to meet their daily needs. As indigenous peoples in different areas accumulated knowledge on efficient uses of local forest-related resources, communities across vast areas often developed similar hunting and gathering practices. This valuable knowledge was often shared among groups living considerable distances from each other. Archaeological studies have shown that biological resources—particularly species of fish and game animals, plants, and animals used in hunting—have also been shared among communities (Bocharnikov 1996).

The sustainable management of biodiversity by traditional communities has, for generations, provided foods, medicines, dyes, tanning agents, clothing fabrics, tools and utensils, construction materials, and means of transport. In the case of peoples who rely heavily on wild plants for food and medicine (activities practised mostly by women and their children), such knowledge includes a profound understanding of ecosystems and habitats, weather, and plant phenology in order to determine the timing and methods of collection, as well as storage and processing techniques for these products. Such knowledge, accumulated over hundreds of years of observation and experience and passed down through the generations, has enabled the survival of communities settling throughout the area to adapt to often harsh environmental conditions. For this reason, traditional forest-related knowledge is becoming increasingly recognized for its value in solving problems related to biodiversity conservation and sustainable use of natural landscapes.

The main objectives of this chapter are:

- To present an overview of the characteristics and management of forest ecosystems in the case study areas and of the role of traditional forest-related knowledge and practices in different environmental, ethno-cultural, and socioeconomic contexts;
- To discuss the modern role and application of traditional forest-related knowledge in education, research, and conservation activities; for securing livelihoods; and for strengthening national policy;
- To discuss and evaluate threats to traditional forest-related knowledge, and suggest possible ways to conserve and maintain this knowledge.

7.2 Overview of the Region's Forest Ecosystems and Their Management

7.2.1 Russia

Forests in Russia cover an estimated 809 million ha, or 49% of the country's total land area (FAO 2010). Russia's forests account for 20% of the world's total forest area and are of global interest for many reasons, including their importance to

Fig. 7.2 Mountain-taiga forest of southern Sikhote-Alin, Russian Far East (Photo: V. Bocharnikov)

indigenous peoples and local forest-dependent peoples. An estimated 26% of forest ecosystems in Russia, mainly in the northern and Asian regions of the country, remain in their pristine, undisturbed, state due to relatively low human population densities and consequently low anthropogenic pressures.

The mountain ecosystems of the Russian Far East and southern Siberia deserve special attention. Paleo-geographers believe that, during the Pleistocene epoch (2,588,000–12,000 years BP), forests disappeared from most of Siberia, except for these mountain refugia. Having survived glaciations of this period, the flora and fauna of the broadleaved and mixed forests in the Far East are distinguished by their age and abundance of relic species. Here subtropical species share their habitats with temperate and even arctic wildlife, which retreated from the north and northeast of the continent after the Quaternary glaciation and adapted to the local environment (Bocharnikov et al. 2004).

The Amur River basin is the only area where broadleaved and mixed forests can be found in the Far East. These forests are occupied by local oaks (e.g., Mongolian oak), maples, lindens, and other broadleaved species as well as conifers such as Korean pine, Manchurian fir, and Yezo spruce, among others. As a result of logging of conifers, today's secondary forests are dominated by deciduous tree species. At higher elevations, broadleaved forests give way to the mixed forest belt with a higher proportion of conifer species and, further up, to taiga forests and mountain pine shrubbery on the highest Manchurian peaks (Fig. 7.2). The Amur River basin ecosystems are among the most diverse in Russia. In contrast to the most southern Russian borders, here the Chinese floral and faunal biodiversity extends into the Far East. The mountain range stretching along the Russian border from the Altai Mountains to the Amur River divides Siberian taiga forests from the deserts and steppes of Central Asia. Because of their geographic location, these mountains are distinguished by a unique combination of alpine forests and alpine steppes.

During the Pleistocene glaciations, ancient peoples are believed to have crossed Siberia on their migration to, and eventual settlement in, other regions of Eurasia and beyond, including America and the South Pacific. The common traditional uses of forest-related resources in these areas provide support for this view. Modern ecosystems, however, vary in terms of their original integrity. Centuries of human activity considerably transformed natural landscapes, which are now replaced with their modified counterparts. By studying the traditional uses of forest-related resources it is possible to understand their origin and reference to human culture.

7.2.2 Ukraine

The geographical area now known as Ukraine, located in the centre of the great European plain, is the country through which the greatest number of European peoples approached their eventual homeland from the east (Davies 1997). A variety of forest ecosystems cover 57.9 million ha, or approximately 9.7% of the total land area (FAO 2010). These range from mixed and broadleaved forests in the north to forest-steppe and steppe in its central and southern regions. The Carpathian and Crimean Mountains represent an altitudinal transformation of ecosystems. Forest biodiversity in Ukraine can generally be described as limited to certain areas of the country by natural conditions, and modified by agricultural and industrial practices that decrease the forested area and species diversity. The Carpathian Mountains and Polissya (swamped woodlands) in the west and north of the country have the most forests and therefore are of most interest when studying traditional uses of forest-related resources.

Intensive forest exploitation in Eastern Europe began in the eighteenth century, and the forest industry (along with railway building) started to alter forest landscapes completely. The wood was exported to Germany, France, England, and Poland. Wood was also intensively used for potash production, iron and glass manufacturing, and sugar beet processing. The need for new agricultural lands caused a disastrous reduction of forest areas in the forest-steppe region. Today clear cutting is prohibited in more than half of Ukraine's forested area, of which 12% is under protection. Forest areas under protection are increasing, which is forcing some forestry units out of business, especially during the current economic recession (Synyakevych et al. 2009).

7.2.3 The Caucasus

The Caucasus—which includes the Republics of Georgia, Armenia, and Azerbaijan, and the Southern Federal District of Russia—is a region rich in biological diversity. It has a complex system of altitudinal belts and a varied humidity from subtropical landscape in the west to semi-desert in the east. Forests in this region are not only an economic resource but also have ecological, aesthetic, recreational, and cultural significance (Gulisashvili 1956). With more than 1,000 species of plants, its flora is the richest in northern Eurasia, especially in the eastern Caucasus, where conditions are more diverse and represent both alpine steppe and alpine forests. Countries of the Caucasus provide good examples of traditional uses of forest-related resources.

The Republic of Armenia, situated in the southern Caucasus, has a forest area of approximately 2.3–2.8 million ha, or between 8% and 10% of the total land area of 28,200 km^2 (FAO 2010; Hergnyan et al. 2007). The forest area of present-day Armenia has been significantly reduced during its long history, from ca. 10.5 million ha (35% of the land area) during the pre-Christian era (6,000–4,000 years BP), to ca. 5.3 million ha during the seventeenth to eighteenth centuries (Moreno-Sanchez and Sayadyan 2005; Makhatadze 1977; Maghakyan 1941). Extensive forest exploitation during the twentieth century led to further declines of natural forests in the country, and today's forests are unevenly distributed and fragmented.

Armenia's forests provide the population with a wide variety of plant resources. According to historical data, in Armenia about 2,000 plant species (60% of flora) have been traditionally used for a variety of purposes. The overwhelming majority of tree species in Armenia, particularly oaks, beech, ash, linden, and other hardwoods, provide valuable woods with diverse applications. According to the literature, at the beginning of the twentieth century the population used about 500 species of edible plants. Today the number of species used for food has been reduced to about 100 species, of which no more than 30–40 species are widely used. From more than 50 species of wild fruit and berry plants that are available, the population gathers only 14–15 species (Ministry of Nature Protection 1999).

The forests of Georgia, a predominantly mountainous country with average elevation of 1,691 m in the east and 1,314 m in the west, are divided by landscape conditions into mountain and plains forests. Total forest area is approximately 2.64 million ha, or 39.7% of the total land area (FAO 2010). Mountain forests constitute 98% of the total wooded area, and plains forests the remainder.

Forests have always played a significant role in Georgian life. In most parts of the country, forest cover is of secondary origin, with its species composition changed through human activities. The vegetative cover of Tbilisi's environs was exploited for fuel for many centuries. Centralized control and management of forests began in Georgia as early as the twelfth century, during the reign of Queen Tamar, with the establishment of the official position of a forest guard. Data on reforestation are found in many historical sources from the fifth to the tenth centuries. The expansion of permanent agriculture on the plains and on lower mountain slopes reduced forest cover over time; deforestation of steeper slopes resulted in soil erosion in many areas.

The southwest of the Caucasus is occupied by Azerbaijan and the Southern Federal District of Russia. According to FAO (2010) forest resource assessment (FRA) data, the total forest area of Azerbaijan is 936,000 ha, or 11.3% of the country's total land area (8.27 million ha). The Southern Federal District of Russia, which is twice as large as the territory of three Caucasus countries, has long been inhabited and has rich traditions of using forest-related resources.

7.2.4   Central Asia

The countries of Central Asia are dominated by dry habitats, with low precipitation and low winter temperatures hindering the growth and development of forest and woodland communities. Deserts and semi-deserts occupy the southern regions below latitudes 41–42° N. In the mountain systems of the eastern and southeastern regions—the Pamirs (in Tajikistan and Kyrgyzstan) and Tien-Shan ranges (including parts of Kazakhstan, Kyrgyzstan, and Uzbekistan)—moisture is absorbed from the upper atmosphere at elevations above the snow line, feeding the streams and rivers that flow into the Aral Sea and the Tarim basins, making agriculture possible over vast areas of the adjacent plains. The significant diversity of natural conditions, landscapes, and habitats within the region are reflected in the various cultures and natural resource use patterns of the peoples of Central Asia.

Forests and woodlands in the region are very limited in extent, placing severe constraints on the use of forest resources on local communities. Estimates of current forest and woodland cover in the region as a whole range from less than 2% to approximately 3% of the total land area (Shukurov 2008; FAO 2010). At the country level, according FAO (2010), forest and woodland areas (and percentage of total land area) are as follows: Kazakhstan–3.3 million ha (1.2% of the total land area); Uzbekistan–3.28 million ha (8.0%); Turkmenistan–4.13 million ha (8.8%); Tajikistan–410,000 ha (2.9%); and Kyrgyzstan–954,000 ha (4.5%).

Given the variability in climatic and landscape characteristics among the countries of Central Asia, forest types found in these countries are also highly variable. In Kyrgyzstan, for example, there are spruce, spruce-and-fir, juniper, walnut, pistachio, maple, poplar-and-willow, and birch forests (Golovkova 1957); while in Kazakhstan the forest areas consist of pine, birch, aspen, saxaul (*Haloxylon*), and other types of forests (Mirhashimov 2005). In the region as a whole, the major forest types include: semi-arid and sub-humid deciduous formations dominated by species such as *Pistacia, Amygdalus, Prunus,* and *Celtis*; open coniferous forests characterized by *Juniperus*; and sub-humid forests including species of *Juglans, Acer, Malus,* and *Haloxylon* (UNCCD 2003).

7.3   Overview of Traditional Forest-Related Knowledge and Its Application to the Livelihoods of Indigenous Peoples

Northern Eurasia, one of the world's largest polyethnic regions, is characterized by an enormous diversity of peoples, cultures, practices and conditions. This vast area is inhabited by dozens of indigenous peoples, each with their unique material and spiritual culture that is expressed in their traditional knowledge and uses of natural resources. The indigenous peoples of the region tend to perceive language, culture, nature, sustainability, and biodiversity as related to each other in the sense that they constitute co-dependent determinants (Posey 1999). Their knowledge about biodiversity and its sustainable utilization is vast because their lives and cultures depend on this knowledge.

7.3.1 Russia

Over thousands of years the indigenous peoples who inhabit the vast and undisturbed areas of the Russian North, Siberia, and the Far East have developed their particular cultures and livelihoods based on their traditional knowledge of their environment and efficient uses of the tundra-, steppe-, and forest-related resources to sustain their livelihoods and cultures. Among the region's indigenous peoples, environmental experiences have accumulated gradually, learned through a process of trial and error, and have been passed on from one generation to the next. Traditional hunting and fishing cultures are closely connected to religious beliefs and practices, including worship of particular animals (including the moose, bear, wild hog, tiger, and salmon).

Although Russia's vast forests are today managed by the Federal Forest Service of Russia (Gorbachev and Laletin 1998), traditional forest knowledge and practices are maintained by traditional communities, which include 40 'numerically small' indigenous peoples (Turaev et al. 2005). These cultures are distinguished by specific knowledge, beliefs, customs, norms, bans, and rules of natural resource use (Taksami and Kosyrev 1986; Kile 1997). However, hunting conditions in the mountain forests of southern Siberia are similar to those found in the mountain forest landscapes of Kazakhstan, Mongolia, and northwest China; consequently, similar traditional forest-related knowledge and practices for hunting and gathering have developed in these regions.

A survey of more than 500 respondents from Central Siberia (Vladyshevskiy et al. 2000) has shown that in the last years of the twentieth century the use of wild mushrooms and Siberian pine nuts increased two- to threefold (Fig. 7.3), the use of wild onion three- to fivefold, and berries one and a half to two times. Where forest industry enterprises have closed and rural jobs in the forest sector are lost, non-timber forest products (NTFPs) are often the main source of food and income for village populations, representing as much as 30–40% of family income. Different aspects of NTFP use by local population in Siberia were reviewed in publications by Vladyshevskiy and Laletin (2002) and Laletin et al. (2002).

Traditional fishing practices, found throughout Russia, vary across regions and play different roles in local economies. Both fishing and hunting activities have also been an important source of knowledge about the environment. Success in these activities by indigenous peoples, and therefore survival, has always depended on good general knowledge of the environment as well as specific knowledge of numerous ecological and other natural phenomena—such as the migration routes of animals, birds, and fish, natural disasters, etc.

Fig. 7.3 Korean pine nuts are an important traditional food source for people as well as wildlife in the taiga of Siberia and Russian Far East (Photo: V. Bocharnikov)

The seasonal changes of plants also helped to determine when it was time to start hunting and other activities. For example, by changes in larch the Udege and Oroch knew that it was time to start fur hunting. Dog rose blooming implied a start of salmon spawning, while a rich harvest of Siberian pine nuts was a sign of a good fur and hoofed animal hunting. The Udege people from the Bikin River basin are an example of an indigenous population that adapts its fishing and hunting activities based on an understanding of local ecological conditions; when salmon are abundant in rivers, for example, forest dwellers adjust by increasing their fishing activities and then drying and storing surplus catches for the winter months.

Indigenous hunters in the Russian Federation are well-adapted to life in the forests of Siberia and the Far East, and have long depended on hunting to provide not only food, but also materials for clothes, the home, and transport. Sustainable hunting and fishing play a key role in the economy of local and indigenous communities and have contributed to the conservation of the region's unique biodiversity. The indigenous communities of the Amur River area in particular have developed sound traditions of environmental conservation, which has played a key role in their forest-related resource use and livelihoods. These traditions had a noticeable impact on the conduct and culture of hunters, fishers, gatherers, dog handlers, and reindeer keepers. In these regions traditional rites are conducted to appeasing the spirits of animals and to ensure a successful hunt (Bocharnikov 2011).

Hunters in the region have traditionally used their knowledge of forest, rivers, and mountain wildlife and its behaviour to improve their hunting skills, a necessity for the survival of their families. The indigenous peoples of the Lower Amur River and Sakhalin Island are famous for their fur hunting skills, for example; hunting in taiga forests has always been the preferred activity of the Tungus-Manchurian ethnic group (Turaev et al. 2005); and local professional hunters in Siberia and

Fig. 7.4   Traditional winter hunting in Siberia (Photo: V. Bocharnikov)

the Russia's European North have long used large areas to provide for themselves and their families. Trade of hunting products is also common in the region; the Nanai and Udege peoples, for instance, exchange reindeer antlers and ginseng for other goods.

Traditional hunting practices, which vary with environmental conditions, are often shared and become common practices. Over time, these skills have expanded with new knowledge, such as finding the right place for a summer or winter camps (Fig. 7.4), building shelters in bad weather (Fig. 7.5), making and keeping fires, and knowledge of animal behaviour and the use of horses, dogs, hunting birds, traps, etc. The ability to find and make food in the wild is an equally important element of traditional hunting practices. There are also a wide variety of practices and means of transport used for hunting in the region. For example, the Orok people from the Far East use reindeer, while the Kety people from Central Siberia travel on skis, pulling hunting sledges behind them. Horses are popular in the south of Siberia, while dog sledges were used to hunt and move around the area in the North.

The indigenous peoples of the Far East treated forest-related resources carefully. The Nivkh and Oroch people harvested firewood not with an axe, but with a boat-hook, by breaking dried branches from standing trees. They used brushwood and fallen and sunken (submerged) trees for heating, felling trees with an axe only when urgently required (Taksami and Kosarev 1986). The Udege, Olchi, and Nanai peoples considered it a greatest sin to pollute water bodies; it was common for them

Fig. 7.5 Storage and sleeping place of Siberian hunters (Photo: V. Bocharnikov)

to remove windblown trees from berry fields and to burn the previous year's grass cover. Over the past few centuries some indigenous communities have started to use forest-related resources such as wild honey and to produce goods from tree bark and other non-timber forest products (Bocharnikov 2005).

Like those who hunted and fished, people who gathered non-timber forest products traditionally had a profound knowledge of vegetation and sustainable harvesting practices. Currently these traditional uses are practised only in some areas of the Urals and along the Volga River, where it is more of a heritage than a livelihood. Before the beginning of the nineteenth century, most of Russia east of the Ural mountains was covered by undisturbed taiga forests. Indigenous peoples of Siberia and the Russian Far East used their traditional practices. But during nineteenth and especially twentieth centuries, large territories in southern Siberia and the Far East were logged and occupied by agriculture. Modern forest-related livelihoods emerged at the time when agricultural and industrial trends were developing in the taiga forest area, shaped by peasants and forest loggers.

Throughout the indigenous communities of Russia and neighbouring countries, traditional forest-related knowledge and skills are being lost or are in sharp decline, along with the region's forest resources. In the indigenous territories of Siberia and the Far East, in particular, only a few traditional uses are still practised today, such as berry gathering, some hunting for meat and fur, fishing, and harvesting of a limited number of medicinal plant species.

7.3.2   Ukraine

In the Carpathian Mountains of Ukraine, the ethnic groups of the Boyko, Hutzul, and Lemko peoples have established a specific culture of forest landscape development. Their traditional culture of landscape zoning has survived time and today supports the traditional settlement system.

Traditional villages in the forested regions were divided into several functional spatial zones. From the centre to the periphery of a village these zones included: (1) a built-up area, (2) vegetable and fruit gardens, (3) fields, (4) meadows for hay, (5) pastures, and (6) forests, all of which satisfy the different needs of land users (Shuhevich 1898). For Hutzul villages these zones (except forests) were typical for each farm, and each zone was fenced. In all villages, the village centre was designated by a church or a local administration building.

Land distribution among different functional zones depended on the land use traditions of a particular ethnographic group. For example, in the Lemko region arable land occupied more space than meadows and pastures; in the Boyko region all zones had more or less the same size; and in the Hutzul region most of the area was occupied by meadows for hay and pastures. In Polissya there were additional zones with ponds and places for beaver hunting and beekeeping (Hrushevsky 1905).

Gardening was important for livelihoods. Fruit trees were planted near houses at a certain density that allowed for the gathering of hay. Ash trees were also planted near houses, because these trees consume a lot of water, and thus help drain the land of excess water; its hard wood was used for production of different tools and gadgets (Kobylnik 1936). Hutzuls had traditional knowledge that helped them to choose the right place for their houses. For example, they knew that spruce forests 'took all earth's juice,' and they did not place their vegetable gardens on such sites (Shuhevich 1898).

In Ukraine many villages in the forest landscapes were located along the main road, which typically followed a river, connecting households and neighbouring villages. The houses were built beyond a certain distance from a river to avoid flooding and accumulation of alluvium near their houses after flooding (Goshko 1983). In Hutzul regions houses were located far away from each other. It was widely known that 'Hutzuls need a space, and they live with space.'

Local people built their structures in places that satisfied the following criteria: (1) be dry, (2) be protected against wind and flooding of rivers, (3) be close to land with good soil for a vegetable garden, (4) have good climatic conditions during summer and winter, and (5) be situated along rivers. Building sites had to be 'free from ghosts.' To identify such favourable places, they observed the behaviour of domestic animals, and chose that place where the animals, especially cows, liked to rest. The history of a place also had a meaning. Places where accidents and murders had taken place or which were burial places were avoided. Villagers developed tricks in order to test a place. For example, in the Hutzul region, they put an upside-down clay pot together with a fleece in the evening, and if the fleece was dry in the morning it meant that the place was dry enough to build a house.

Villagers also developed knowledge of how to 'gain' land from forests and make it suitable for grain production. For example, in the fifteenth century in the Boyko region, a popular method for clearing forests for fields was to cut them down or burn them. The full shifting cultivation cycle lasted 15–20 years, including 1–2 years for cutting down the forests, 1–4 years for sowing, and 13–14 years for the forest fallow period. Forests were burnt in the fall and rarely in the spring. Entire forest sections were burnt in one session, during the course of 1 day or sometimes overnight, depending on the size of the section. Seeds were sown into ashes and then covered with soil and organic fertilizers (Goshko 1983).

However, many uncontrolled forest fires occurred as a result of these agricultural practices. Such fires led to a considerable decrease in coniferous species, such as Scots pine, Norway spruce, and silver fir, as well as in species with large and heavy seeds (oak and beech). Among the groups of plants that appeared in the locations of former fires, were tree species whose seeds could be carried far away by wind—such as hornbeam, maple, sycamore, birch, and aspen (Krynytskyj and Tretiak 1998).

One of the elements of traditional knowledge was to create 'natural shelter' for herders and shepherds. In the middle of pastures the person responsible for grazing animals would locate a young spruce growing in meadows and cut the top off the tree. Then the spruce stopped growing up; instead it grew in width. In several years the lower part of the tree would have many dry branches with lichens. Such a managed spruce tree had better resistance to wind storms and became a good shelter for sheep and cattle during storms and hot weather (Shuhevich 1898).

For centuries forests were important as a source of working places for many people in the forested regions of Ukraine. In the nineteenth century the income of villagers from their work in forests was much higher in comparison with their income from agricultural work (Kubiyovich 1938). The best time for logging trees was early spring or late autumn (when no leaves were on the trees), when forest stands with old trees were harvested. The forest owner hired a person who was responsible for all logging operations, including the hiring of people who would do the logging and count the trees. These forest workers built themselves a shelter for autumn and winter work.

There were many rituals associated with logging operations. The first tree had to be cut by the most successful forest worker; usually that was the person who had successfully done such work before (Shuhevich 1898). After cutting the tree it was de-barked and left on-site for 14 days to dry. After that they cut the tree into pieces. The villagers also had knowledge of how to dry wood for buildings. For this purpose, big trees were cut down between January and February. The branches were cut off, but the top was not touched, because the top would remove sap from the tree. The tree was left on-site until spring. In spring, bark was pulled off from one side of the tree, and the tree was left again until October. In October bark from another side of the tree was pulled off, and the top of the tree was cut off. The tree was then cut into pieces and boards were produced. The boards were kept in a dark place for 5–6 years to dry, and then used for building (Shuhevich 1898).

Before logging a road to a site had to be made. There were different types of roads: (1) from a logging place to a gathering place for wood from different sites, and (2) from this place to a river. Wood was transported during dry weather. Hutzul knowledge was widely used for wood transportation in the nineteenth century, when wood became important for trade in western Ukraine. There were three types of special 'routes' for wood transportation: ground, wooden, and water-wooden. A 'ground' route, adopted in the eighteenth century, was a gully with a width of 0.6–1 m and a depth of 0.4–1 m. 'Wooden' trails were built in the nineteenth century, consisting of a chute paved with wood; the bottom of the chute leant against the block dug into the ground. 'Water' chutes were used at the end of nineteenth century to the beginning of the twentieth century; the chute had the shape of a washtub and the bottom had a width of 30–35 cm. Water routes were made in the streams; to supply these routes with the needed amount of water, small dams were made in the upper parts of the stream (Goshko 1983; Shuhevich 1898). The wood that was transported by river had a special value for local people. For example, the Hutzuls thought that river water took all the 'juice' from wood; therefore, it would not be eaten by different insects (Shuhevich 1898).

7.3.3   Central Asia and the Caucasus

Traditional forest-related knowledge supports the cultures and livelihoods of local and indigenous people in Central Asia and the Caucasus, providing foods, medicines, dyes, and other non-timber products for local use and trade, particularly among mountain-dwelling communities. Non-timber product gathering and hunting, and dependence on forest ecosystems for livelihood needs, have also shaped attitudes and spiritual beliefs regarding forests and nature conservation within these societies (Attokurov 2011; Domashov 2011).

Traditional knowledge and practices associated with these activities vary among social groups. For example, gathering of non-timber forest products is typically an activity of women (assisted by their children), who have developed skills over generations for harvesting and processing wild plants; they use their knowledge of habitats, plant phenology, and techniques for storing and processing harvested products (Novikova and Yatimov 2011).

Local people in Central Asia have long used plants in traditional medicine. Among the most commonly used plants of medicinal value in this region are: dog rose (*Rosa canina*), oblong barberry (*Berberis oblonga*), felt burdock (*Arctium tomentosum*), karakol aconite (*Aconitum karakolcus*), jungar aconite (*Aconitum soongoricum*), and sea buckthorn (*Hippophae rhamnoides*). In southern Kyrgyzstan, where the world's largest natural walnut woodlands are found, local populations since ancient times have used both the fruits and leaves of the Circassian walnut (*Juglans regia*) for treatment and prevention of various diseases. The wood of this species is also highly valued for cooking and for the manufacture of traditional dishes and interior elements in houses.

Forest plants are widely used for food in this region, where unique methods exist for the drying and preserving fruits and berries. Many national dishes use forest plants, including spices and herbs such as Marshall's thyme (*Thymus marschallianus*), oregano (*Origanum vulgare*), and blue mint (*Ziziphora clinopodioides*).

The use of plants for preparation of dyes for wool and fabrics is widespread and has a long history in Central Asia. The plants most often used for dyes by local people are Circassian walnut, great nettle (*Urtica dioica*), juniper (*Juniperus*), and willow (*Salix*), among others. Some work has been done to preserve this traditional knowledge, including through programmes carried out by such organizations as Helvetas Kyrgyzstan, the Central Asian Mountain Partnership Program (CAMP), and others (Domashov 2011).

Nomadic communities in Central Asia use forest resources directly for provision of domestic heating and cooking fuels and for construction of yurts and household utensils. Within these societies, specific forest and woodland areas, sometimes named after famous or respected people, are typically controlled by communities and families, with permission required from the community to harvest trees. Within some sparsely populated permanent settlements in semiarid and desert areas, strict forest conservation measures are enforced by communities. Such locally enforced regulations restrict collection of fuelwood to dead stems and branches, and prohibit tree felling in forest areas 3–5 km around settlements. Violation of such tree-felling regulations can result in expulsion from the community (Anonymous 2006).

Traditional forest-related knowledge in Central Asia is widely used to control desertification and soil erosion in mountain areas. In Tajikistan, where the use of stepped terraces has a 1,000-year history, planted forests are widely used for stabilization of hill slopes (Anonymous 2006). Methods for slope terracing and cultivation of fruit and nut gardens, especially in the traditional system of land and water management known as *boghara*, has been known to inhabitants of mountains since ancient times. Throughout the region, traditional techniques including shelterbelts have been used to control windblown sands in the vicinity of settlements. Today the use of traditional knowledge in forest management in Central Asia is becoming rare and its role may disappear in the near future, primarily because of the diminishing role of families in the transfer of knowledge of sustainable nature management to younger generations.

The traditional forest-related knowledge of people in the mountainous Caucasian region is broadly similar to that of indigenous communities in southern Siberia and Central Asia. In the forest regions of Georgia, where pastures and haymaking resources are limited, local people use a traditional 'pasture turnover' system for regulated forest grazing. The creation of cultural pastures in open areas within forests increases animal productivity while preventing damage to sprouts and seedlings of valuable species (i.e., oak, ash, maple, beech) due to grazing in young, naturally regenerating forest stands. Once regenerating trees attain heights sufficient to prevent their damage by livestock, these forests are used on a temporary basis for grazing, while previously used pastures are managed to encourage restoration of forest cover and growth of valued tree species through natural regeneration (Eganov 1967).

Other innovative forest management practices developed by farmers to sustain their agricultural livelihoods have been documented. For example, during the late nineteenth century in the village of Mereti in western Georgia, one farmer developed orchards and vineyards in a forested valley susceptible to high winds and erosion by retaining alluvial plain forests (oak, ash, lime alder, and other hardwood species) in the windward side of his farm and planting a variety of useful tree species within these forests and among his cultivated areas as hedges (honey locust) and line plantings (poplars) to provide further protection. These practices were later reflected in the works of forest scientists of Georgia (Gulisashvili et al. 1975; Urushadze 2005; Kharaishvili 2008).

Throughout history a wide range of wild-growing fruit and edible plants have been collected and used by local communities. Many of these have been domesticated by these communities, and some are widely cultivated worldwide. More than 200 wild-growing plants in Armenia are used as food products, and many subspecies (varieties) are used as food in various forms: fresh, prepared, marinaded, or dried. Today, only 10 of the estimated 300 edible species of mushrooms are used by local residents on a regular basis (Ter-Ghazaryan et al. 1995). Approximately 10% of all plants in Armenia have some medicinal value and for many years have been used in traditional medicine. However, the excessive harvesting of some herbs, particularly those that are widely used for other purposes, such as species of thyme (*Thymus*) and the mints used for preparation of soft drinks, has diminished natural stocks of these plants to a critical level (Gabrileyan et al. 2004).

7.3.4   Forest-Related Cultural and Religious Traditions of Indigenous and Local Communities

Customs and religious beliefs regarding nature among indigenous communities of the Russian North, Siberia, and the Far East have shaped the evolution of their traditional forest-related knowledge and practices. Certain similarities exist among these customs and prohibitions and those of local and indigenous peoples living in Ukraine, Central Asia and the Caucasus.

Indigenous peoples of the northern Eurasian region clearly identify the connections between their cultures and the environment. Animistic traditions, including the belief that human beings as well as animals and plants are endowed with spirits, support the idea of mutual respect and esteem between humans and nature. Totems play an important role in nature conservation in these indigenous cultures. In the past, peoples of the Amur River area considered the moose, otter, wild hog, bear, tiger, and other animals as totems, and prohibited or strictly regulated their hunting. Similar prohibitions and restrictions existed in this region for certain birds, snakes, frogs, and turtles, which were considered sacred. Some trees, including the birch, larch, and oak, also represented protective spirits. Furthermore, the indigenous peoples' names for wildlife species show that they believed in an interaction of animals, plants, and people. For example, the people living along the Amur River and in Sakhalin Island believe that plants have a 'soul'; therefore, one cannot break trees, or tear grass without a household purpose. The knowledge of plants was also used in the aboriginal arts: hunters found plants that can be used for making traditional musical instruments, for example.

Traditional taiga forest hunting by the indigenous peoples of Siberia and the Far East, including those of the Finno-Ugric peoples inhabiting European Russia, were once wide-spread among the forest-dependent peoples in the forests of the temperate zone. In the harsh conditions of the northern areas, indigenous peoples and local communities were highly dependent on forest resources for their subsistence. In these communities strict hunting restrictions developed over centuries. These included ban on hunting female ungulates, and designation of 'protected areas' where no hunting was permitted. The Finno-Ugric peoples based these bans on ancient religious beliefs, while the Russian people developed restrictions through community consultations.

To the south, where elders are traditionally greatly respected, it was a common practice to name certain sites after famous people. Tracts of land and winter huts were commonly named after the man who settled there. Even certain trees were named, and these were protected from felling. In some walnut forests in Central Asia, local people still show with pride trees supposedly planted by Aleksander Makedonsky (Alexander the Great) or his soldiers, and springs created or visited by one saints. Such beliefs and practices, widespread in Central Asian countries, have helped to preserve many natural and cultural monuments, where activities such as tree felling and cattle were prohibited.

The conservation of sacred sites is the most ancient type of wilderness conservation, dating back to the prehistoric past of the Earth's peoples. Sacred sites are common throughout Russia and other countries of the northern Eurasian region where indigenous and long-established local communities still thrive. Such sites may range in size from small groves or even individual trees to extensive forested landscapes. Some areas are considered sacred because they provide major habitats for species with ritual or medicinal values; special ceremonies and rituals (bound to harvesting or hunting, for example) are conducted to honour the spirits of particular species or of special natural habitats or features such as lakes, springs, forests, rocks, etc., to ensure the survival of the community. The protection of such sites is extremely important for the health and spiritual well-being of a community. These sites also serve important functions for biodiversity conservation—serving as shelters, habitats, or protected areas for the conservation of certain species, to help ensure the survival and livelihoods of indigenous communities.

Protection of forest resources based on religious beliefs are characteristic of Central Asia, where the sacralisation of nature is expressed in cultural traditions and practices connected with particular species and sites. For example, one of the national traditions of the Kyrgyz people is reverence for holy sites. Here, the Abeliya bush (*Abelia corymbosa*), known as 'asa-musa' or 'Moses staff,' is considered sacred. Abeliya is strictly protected by traditional communities in the Tian-Shan mountain range. The shrub is imbued with religious significance, embodying the Source of Life, and pilgrims offer prayers and make special offerings by tying strips of fabric torn from their clothing to its branches. Forests surrounding holy places such as burial sites are also strictly protected, and tree planting in such areas is an important traditional activity. According to a study conducted in the Talas Oblast in Kyrgyzstan, 10% of 157 sacred sites surveyed were linked with forests (Aitpaeva et al. 2007).

From ancient times walnut forests have been used as a place for Anchorites. This practice still exists in the Jalalabad Oblast of Kyrgyzstan, notably in the Arkyt and Arslanbob villages. The names of the villages themselves are indicative of this tradition—Arkyt, for example, originates from the Sanskrit word 'Arhat' meaning 'Anchorite.' Many natural areas of religious significance, visited by pilgrims, are also protected in Kyrgyzstan, such as the lake and forests of willow and juniper found at Kulj-Ata, where the traditional epic hero Manas' war-horse was born.

7.4   Forest Loss, Degradation, and Erosion of Traditional Knowledge and Livelihoods

As discussed above, sustainable use of natural resources is one of the major components of traditional livelihoods in northern Eurasia. The region's indigenous peoples' philosophy has always viewed forests as part of their lives (Bocharnikov et al. 2003; Bocharnikov and Fedotenkov 2004). However, the integrative and disintegrative political, economic, and social processes in the countries of the region—resulting in continual mixing of people and their traditions and cultures related to utilization of forests and other natural resources—has had serious consequences for the forest ecosystems of the region as well as for traditional knowledge-based livelihoods and cultures of the local and indigenous communities who depend on forests.

The culture of woodcutters, arising in Western Europe, spread throughout the world, including Russia and the countries of Central Asia, without consideration of the specific features of forests growing there (Gorshkov and Makarieva 2006). In the 1920 and 1930s, the official view of the economic value of forests was summarized as follows: 'standing forest never could be considered as goods because it hasn't exchange value and also until it is cut down it will not have using value' (Kalinin 1932). This view shows that the real value of standing forests was not taken into account (in forest policies and practices). As in other parts of the world, forest management focused on the exploitation and replacement of mature forests with forests established by planting seedlings on their place. In Kyrgyzstan, for example, all old trees in walnut, juniper, and spruce forests were harvested, followed by clearance and replanting with young trees to create even-aged stands, which were considered to be of equal or greater value than the mature forests they replaced (Shukurov 2008).

At the beginning of the twentieth century, concerns about deforestation in the European North of Russia were voiced for the first time by foresters, even though it was a region of seemingly boundless taiga forests with only small patches of populated and intensively used land. These foresters were concerned that timber supplies of large and defect-free pines and spruces in northern Russian forests were declining. Unregulated selective cuttings during the nineteenth century, together with the extensive wildfires that followed, have gradually turned the largely intact, old-growth northern forests into less commercially valuable ones dominated by small-diameter and/or damaged trees of little use to the timber industry. This negatively affected both the forest industry and most of the northern populations who depend on supplies of valuable timber. Sawmills either were forced to use smaller and even defective trees to produce timber of lower value, or else they closed because of their unprofitability. There remain today decreasing areas of natural old-growth forests in more remote (but increasingly accessible) areas that are both highly valued by the forest industry, and last refuges for the biological diversity of the taiga forests.

Over the past two decades, the vast area of the Russian North has seen unfavourable economic and social conditions. As these conditions have worsened, fewer indigenous and other forest-dependent people have remained in this extensive and harsh environment. Today, the Russian Arctic region is populated by more than ten million people, the most numerous group being recent, first-generation, immigrants. Others, the 'old-timers' and indigenous peoples, have been forced away from their traditional lands by the recent immigrants involved in mining of non-ferrous metals, coal, oil, and gas, whose activities have had significant negative impacts on the environment (Bocharnikov 2005; Bocharnikov and Martynenko 2008). Environmental problems of the remaining intact taiga forests in European Russia are complex and diverse. The most urgent problem is that of forest decline in the European North of Russia, which has caused environmental, economic, and social concerns among foresters and forest managers for more than a century.

The problem of forest protection has become more urgent over the past century and in some northern regions of European Russia, Siberia, and the Far East. Intensive forest management by the state alongside extensive adverse processes—such as unfavourable after-fire dynamics of forests, drying of coniferous stands, and ecosystem failures to perform major ecological functions—have led to a considerable decline in composition, and a change in forest age and species structure, as well as a decrease in biodiversity over extensive areas.

7.5 Political, Legal, and Industrial Developments Affecting Traditional Forest-Related Knowledge and Practices

7.5.1 Changes Affecting Indigenous Communities and Traditional Natural Resource Use During the Soviet Era

In the 1930s the government of the Soviet Union first became actively involved with issues related to indigenous peoples and their traditional lands in Russia. While national autonomous districts were established during the early decades of the twentieth century, their administrative borders did not always match those of the traditional lands of indigenous peoples. During this period, the government made an official list of 26 'Indigenous Peoples of the North and Siberia.' These communities were entitled to government subsidies, but they were forced to adopt new types of livelihood practices that were alien to them and detrimental to their traditional livelihoods. Through the process of collectivization (establishing collective management units), for example, the state forced indigenous communities who practised reindeer herding to abandon their nomadic lifestyle to settle in permanent villages and engage in reindeer breeding. These communities maintained their common reindeer herds, however, with community members tending them on a rotating basis in the wild. Following a system of obligatory secondary education, children were taken from their parents to special boarding schools. During this period, many indigenous people, bearers of traditional knowledge in the region, were also killed (Bocharnikov 2005).

In the mid 1950s, industrial development in western Siberia began after discovery of oil and gas deposits. This led to occupation and exploitation of indigenous people' lands, loss of their traditional livelihoods in affected areas, and a number of ecological disasters in the following decades. While some autonomous districts were transformed into independent regions of the Russian Federation during the early 1990s, this political development has had little if any positive impact on the deterioration of living conditions for these regions' indigenous peoples.

7.5.2   Policies and Legislation Affecting Traditional Forest-Related Knowledge in the Post-Soviet Era in Russia and Ukraine

In recent years the indigenous peoples of Russia have been trying to restore their traditional livelihoods through legal efforts. Nature protection legislation of the Russian Federation enacted since the mid 1990s, takes into account the rights of indigenous communities to use nature resources in ways that are sustainable and consistent with their traditional livelihood practices.

The Russian Federation's Federal Law #49 of May 7, 2001,[1] 'On Territories of Traditional Nature Resource Use of the Indigenous Peoples of the North, Siberia, and the Far East of the Russian Federation,' includes specific provisions (in Article 15) on environmental protection within the territories of traditional nature resource use. According to these provisions, environmental protection is to be ensured by executive authorities of the Russian Federation and of the regions of the Russian Federation, local governments, indigenous people, and communities of indigenous peoples. The Russian legislation also ensures a new status for indigenous peoples by providing enabling conditions for traditional nature resource use within the so-called Territories of Traditional Nature Resource Use (TTNUs) for the indigenous peoples

---

[1] Available via http://base.garant.ru/12122856/. Cited 7 Mar 2011.

of the North, Siberia, and the Far East of the Russian Federation. These territories are designated so as to ensure environmental protection and to support indigenous livelihoods, religion, and culture. The legal norms for these TTNUs are related to the various natural resource uses (such as reindeer breeding, hunting, fishing, and non-traditional forest product collection) within different landscapes of the each territory, and they recognize the concept of sustainable resource use for indigenous peoples in complex natural ecosystems (Sulyandziga and Bocharnikov 2006).

According to another federal law, 'On Environment Protection' (2002), indigenous lands and traditional livelihoods are under special protection (Article 4). Under Article 15 of the federal law, 'On Specially Protected Natural Areas' (1995), traditional nature use areas are permitted in national parks situated in the territories populated by indigenous peoples. Under Article 24 of this same law, protection is given to traditional natural resource uses by indigenous communities within state nature reserves inhabited by such communities.

In the 1990s the Russian Federation also adopted many decrees related to the establishment of community-managed territories, although the government has yet to officially recognize these territories. In Russia today, community management has specific conditions for land tenure and use, which vary across different regions. For example, in 1992 in the Amur River and Khabarovsk regions, territories were established for nature management by traditional communities, and pre-existing leases were withdrawn from logging units in favour of the Udege, Nanai, and Oroch peoples. In the Primorye region, however, the community rights over these territories were ignored by authorities elected after their designation. In the Ternejsky district, for example, authorities approved a joint venture with Hyundai (the South Korean corporation) for logging of these forests, despite a rejection by a State Expertise Review. Hyundai was subsequently forced to abandon this venture when the Udege people took up arms to defend their community rights to manage these forested lands.

Similar events have recently taken place in Western Siberia, where a decree allocating community territories in Yamal-Nenets and Khanty-Mansiysk national districts has been adopted, although defining their boundaries has presented difficulties, particularly in cases when oil and gas exploration and exploitation are involved. Conflicts have arisen, as in the case of the Khanty people, who took direct actions to halt mining and oil and gas exploitation on their lands by moving their nomadic camp to areas to prevent mining, and organizing pickets in the district centre by placing nomadic tents by oilers' offices. On the Sob River, considered sacred by the Khanty, they blocked the river with their boats to stop a ship from dredging gravel out of the river. Unfortunately, indigenous peoples do not always succeed in their efforts to preserve their traditional livelihoods by efforts aimed at combating industrial exploitation and destruction of their traditional lands.

Issues of traditional uses of forest resources are closely related to forest ownership and tenure security. In Ukraine, while legislation defines state, municipal, and private forest ownership, in practice the state ownership predominates, with 66% of the country's forests managed by the departments of the State Forest and Game and Fish Committee. Although municipal ownership could be a basis for 'community forestry,' and restoration of traditional uses of forest-related resources, the public has little role in forest-related decision-making. (Soloviy and Cubbage 2007).

Unlike countries in Central Europe, property restitution was not considered in Ukraine (or in other countries of the former Soviet Union) during the process of reforming forestry in the years following the breakup of the USSR. This was due to various historical circumstances in the different regions of Ukraine and the public's fear that forest management would not be sustainable in privatized forests. This, combined with a lack of private forestry skills in the private sector, has limited private forest ownership and management to very limited areas in the country. There is a need in Ukraine to conduct forest research, train foresters, and raise public awareness about the values of traditional nature uses and forest protection.

Policy initiatives at the international level dealing directly or indirectly with the protection of rights of indigenous peoples and the preservation of traditional forest-related knowledge and associated management practices (discussed in Chap. 1 of this volume) play an important role in this regard. For example, in 2004 the fifth Conference of the Parties of the Convention on Biological Diversity developed and approved voluntary guidelines [known as Akwé: Kon Guidelines (SCBD 2004)] for the conduct of cultural, environmental, and social impact assessments for developments proposed to take place on, or likely to impact, sacred sites and on lands and waters traditionally occupied or used by indigenous and local communities. These guidelines are particularly valuable in the northern Eurasia region, where sacred sites (and more generally, indigenous lands and livelihoods) are increasingly affected by extractive industries (mining, oil and gas, industrial logging) and other industrial developments. If applied, these guidelines could be used to predict and therefore avoid or minimize adverse impacts of industrial projects on biological diversity, sacred sites, and more broadly on the well-being and safety of indigenous communities and ecosystems upon which they depend.

7.6 Traditional Forest-Related Knowledge in Rural Development, Research, Education, and Public Awareness-Raising

Indigenous societies accumulated their wealth of traditional forest-related knowledge based on their everyday experiences and connected with their primary activities—hunting, fishing, and gathering. Today there is increasing interest in the preservation and restoration of lost traditions for natural resource management, including forest management (Domashov 2011). In recent years, many scientists, politicians, and NGOs in Russia are considering alternative approaches to sustainable forest resource management because of the need to elaborate a national strategy for sustainable development in the face of social and ecological conditions that threaten the very survival of the indigenous peoples in some regions. This has included the development of innovative approaches combining traditional and formal scientific knowledge and practices for managing forests and other natural resources

(Bocharnikov 2011). In the area of forestry education, Ukraine provides an example of an education policy that involves traditional forest-related knowledge, including a comprehensive curriculum (Mountain School: Current State Problems and Prospects) developed by a research collective in the Ivano-Frankovskiy region.

In parts of Central Asia, for example, efforts are underway to integrate traditional knowledge and practices in local forest resource management through collaborative projects and programmes supported by the Netherlands, Switzerland,[2] and other countries, although research on traditional forest-related knowledge in Central Asia is currently very limited (Shukurov 2008). However, partial information on traditional sustainable land and water use in the countries of Central Asia has been gathered within the framework of the Central Asian Mountain Partnership Program (CAMP). This work has resulted in the organization of an exhibition of nature conservation technologies, including traditional approaches. Among the traditional technologies documented, several practices of forest exploitation and management in Kyrgyzstan and Tajikistan are notable—for example, the terracing of slopes for tree-planting; the use of poplar as a rhizofiltrator on saline lands and a technique for creating 'hanging gardens.' These gardens are used by mountain dwellers (who lack sufficient arable land or irrigation water) to create small oases with fertile soil on otherwise poor lands on which it is possible to plant trees that will survive and grow during the hot, dry summer season. This technology was widely used in the mountain villages of Tajikistan during the pre-Soviet period. The essence of the technology is the enrichment of very poor, stony, highly erosion-prone soils with fertile silt from riverbeds, which is mixed with dung and used to fill holes created for planting seedlings of fruit trees and agricultural crops. This technology allows for an increased fruit harvest and serves to decrease soil erosion and the amount of water needed for irrigation. Recent work has also been carried out in this region to document and share knowledge on traditional medicine, and on the traditional uses of wild plants for food and for extraction of dyes. For example, within the framework of Global Environment Facility (World Bank) projects in West Tien-Shan, traditional conservation and utilization practices of 40 species of plants used for dyeing wool and fabrics was described (Orolbaeva 2003).

In Azerbaijan for most of the twentieth century, forest specialists received their professional education and training at universities and colleges under All-Union (Soviet) programmes. While most students from outside the region knew little of local environmental and social conditions, many had local origins and had a good relationship with the local population, which helped them to combine traditional forest-related knowledge with modern forest management practices in their work. Experience in this region has shown that effective forest management requires not only formal (scientific) forest-related knowledge, but traditional knowledge as well. Effective integration of traditional and modern knowledge to resolve forest management conflicts will require effective dialogue among stakeholders and a common

---

[2]For information on projects in Kyrgyzstan, supported by the Swiss international development agency Intercooperation, see: http:// www.intercooperation.ch

understanding of forestry strategy in the country. In our view, it is necessary to document and assess available traditional forest-related knowledge and practices, identifying knowledge gaps that may be filled by formal (scientific) knowledge, as well as examples of sustainable management practices that combine both traditional and scientific approaches.

Traditional forest-related knowledge is gaining prominence in biodiversity research and conservation programmes. An example can be seen in the Tunguso-Manchzhury and Nivkh peoples, who have accumulated knowledge of local flora while searching for food and medicinal plants (Sem and Sem 1988; Podmaskin 2001; Kreinovich 1929). In so doing, they have categorized plant species as food, medicines, and useful for making utensils, etc. The Tunguso-Manchzhury and ancient Asian peoples were very familiar with the plants that were the basic forage for animals, birds, and fishes; such knowledge was necessary for successful hunting. Today special efforts are needed to ensure the transfer of such traditional forest-related knowledge within the extensive areas of Siberia, where all the indigenous peoples have similar knowledge of their local flora, having inheriting it from the ancient Tunguso-Manchzhury and Nivkh peoples. All of this knowledge is important for current education and for helping the public begin to better understand the role traditional forest-related knowledge can plan in the modern world.

7.7   Concluding Remarks and Recommendations

Posey (1999) has asserted that it is increasingly evident that the 'minority' and disenfranchised peoples of the earth speak for all humanity. Indigenous peoples are determined to preserve, develop, and transmit to future generations their ancestral territories, and their ethnic identity, as the basis of their continued existence as peoples, in accordance with their own cultural patterns, social institutions, and legal systems (Martínez Cobo1986/7). The efforts of indigenous peoples to use their traditional lands in a sustainable way and to protect these areas from over-use have yet to be adequately acknowledged in the spheres of natural resource policy and law (Posey 1999).

Indigenous peoples of the northeast Eurasian region view the biological diversity of the ecosystems in which they live as essential for the maintenance of their cultural identities and their physical survival. While biodiversity conservation is the indirect outcome of their traditional forest use practices (Berkes et al. 1995), the indigenous peoples of the region are among those groups that are most adversely affected by external efforts to protect their environment and other measures that disrupt their traditional activities. Unfortunately, their stakes and aspirations are not yet taken fully into account in the development of conservation and natural resource management policies (Jentoft et al. 2003).

The indigenous heritage for conservation and sustainable use of biodiversity and forest resources in Russia, Ukraine, the Caucasus, and Central Asia is being lost owing to a decline in traditional uses of forest-related resources and indigenous and community livelihoods, intensive exploitation of natural resources, and urgent social and economic problems faced by the societies in the states of the former USSR. Urgent action is needed in most countries of northern Eurasia to conserve and restore traditional sustainable use of forest-related resources, and to ensure harmonious relationships between human beings and the natural environment.

The preservation and development of traditional forest-related knowledge and practices in the region requires financial, economic, legal, and social actions to: (1) protect indigenous communities' forest ecosystems and their traditional uses of natural resources to ensure traditional livelihoods of the indigenous peoples of the region; (2) develop and improve traditional livelihoods of all indigenous peoples inhabiting the region; (3) improve the quality of life of the holders of traditional knowledge and experiences; (4) improve social conditions for village residents who practise traditional livelihoods and whose populations are currently declining; (5) ensure access to education facilities for indigenous peoples of the North who practise traditional livelihoods that reflect their cultural identities; and (6) assist in the development of communities and other self-government entities of the indigenous peoples of the North, Siberia, and the Far East of Russia and of the neighbouring states.

National governments have a major role to play, given the importance of the state in natural resource management in these countries. In particular, state forest management policies and practices need to give greater recognition to the importance of traditional forest-related knowledge and practices, and enable traditional uses of forests by local and indigenous communities.

References

Aitpaeva G, Egemberdieva A, Toktogulova M (2007) Mazar worship in Kyrgyzstan: rituals and practitioners in Talas. PH Maxprint. The Aigine Research Center, Bishkek

Anonymous (2006) Informacionnyi sbornik: tradicionnyie znaniya v oblasti zemlepolzovaniya i vodopolzovaniya (Information digest: traditional knowledge in the sphere of land and water utilization). Dushanbe, Tajikistan, NGO Fond podderzki grazhdanskogo obschestva, set RIOD. Available via http://www.undp.tj/files/pubs/booklet_rus.pdf. Cited 7 Mar 2011

Attokurov AT (2011) Assessment of human impacts on mountain landscapes of the northern slope of the Turkestan-Alai Range. In: Laletin A, Parrotta JA, Domashov I (eds.) Traditional forest-related knowledge, biodiversity conservation and sustainable forest management in Eastern Europe, Northern and Central Asia, vol 26, IUFRO World Series. International Union of Forest Research Organizations (IUFRO), Vienna, pp 63–66

Berkes F, Folke C, Gadgil M (1995) Traditional ecological knowledge, biodiversity, resilience and sustainability. In: Perrings CA, Maler KG, Folke C, Holling CS, Jansson BO (eds) Biodiversity conservation. Kluwer, Dordrecht, pp 281–300

Bocharnikov VN (1996) Forests and the Far East: a model for sustainable development and cultural survival in the Bikin River watershed. In: Proceedings of the workshop on trade and environment in Asia-Pacific: prospects for regional cooperation, Honolulu, 23–25 Sept 1994, pp 23–27

Bocharnikov VN (2005) Traditional forest-related knowledge: a case study of the Russian Federation. In: Newing H (ed) Our knowledge for our survival. vol. II: national case studies on traditional forest related knowledge and the implementation of related international commitments. International Alliance of Indigenous and Tribal Peoples of the Tropical Forest (IAITPTF) and Center for International Forestry Research (CIFOR), Chiang Mai, pp 364–394

Bocharnikov VN (2011) Traditional knowledge of indigenous peoples and its contribution to biodiversity conservation in Russia. In: Laletin A, Parrotta JA, Domashov I (eds.) Traditional forest-related knowledge, biodiversity conservation and sustainable forest management in Eastern Europe, Northern and Central Asia, vol 26, IUFRO World Series. International Union of Forest Research Organizations (IUFRO), Vienna, pp 13–20

Bocharnikov VN, Fedotenkov A (2004) Strengthening the participation of the indigenous peoples in the CBD. Biodiversity 5(3):40–43. Available via http://www.tc-biodiversity.org/biopreview5-3.htm. Cited 7 Mar 2011

Bocharnikov VN, Martynenko AB (2008) Ecologicheskoe raionirovanie dal'nego Vostoka (Ecological regionalization of the Far East). Izvestia Russian Academy of Sciences. Seria Geogr 2:76–85

Bocharnikov VN, Sulyandziga R, Peskov V (2003) Russia: World Bank group projects and the indigenous peoples of the North, Siberia and Far East of Russia. In: Caruso E, Colchester M, MacKay F, Hildyard N, Nettleton G (eds) Extracting promises: indigenous peoples, the extractive industries and the World Bank—synthesis report. Tebtebba Foundation, Baguio City, p 84

Bocharnikov VN, Martynenko AB, Gluschenko YN (2004) Bioraznoobrazie Dalnevostochnogo ecoregionalnogo komplexa (Biodiversity of the Far East ecoregional complex). Apelsin, Vladivostok

Davies N (1997) Europe: a history. Pimlico Random House, London

Dmitriev VA, Dmitriev SV (1991) Znaniya narodnye: svod etnograficheskikh ponyatii i terminov (Folk knowledge: list of ethnographic terms). Narodnye znaniya. Fol'klor. Narodnoe iskusstvo 4:45–47

Domashov I (2011) Overview of traditional knowledge for sustainable use of nature in Kyrgyzstan and Central Asia. In: Laletin A, Parrotta JA, Domashov I (eds.) Traditional forest-related knowledge, biodiversity conservation and sustainable forest management in Eastern Europe, Northern and Central Asia, vol 26, IUFRO World Series. International Union of Forest Research Organizations (IUFRO), Vienna, pp 21–26

Eganov KV (1967) Fodder value of grasses and meaning of pasturage of cattle in mountain forest of Georgia, vol XVI, Works of Tbilisi Forest Institute. Vasil Gulisashvili Forest Institute, Tbilisi

Fedorova EG (1986) Elementy tradizionnogo v sovremennyh hozjaistvennyh zanjatijah severnyh Mansi: kulturnye traditzii narodov Sibiri (Traditional elements in current practices of northern Mansi people: cultural traditions of Siberian peoples). Nauka, Leningrad

Food and Agriculture Organization (FAO) (2001) Global forest resource assessment, 2000. FAO forestry paper 140, FAO, Rome

Food and Agriculture Organization (FAO) (2010) Global forest resources assessment, 2010. FAO forestry paper 163, FAO, Rome. Available via http://www.fao.org/docrep/013/i1757e/i1757e00.htm

Gabrileyan E, Amroyan E, Zakaryan A, Grigoryan G (2004) Regulation of herbal medicinal products in Armenia. Drug Info J 3:273–281

Golovkova AG (1957) Rastitelnost Kirgizii (Vegetation of Kyrgyzstan). Ilim, Frunze

Gorbachev VN, Laletin AP (1998) Role of criteria and indicators for the sustainable management of Russian forests: the contribution of soil science to the development of and implementation of criteria and indicators of sustainable forest management. SSSA special publication no. 53, SSSA, Madison, pp 113–119

Gorshkov VG, Makarieva AM (2006) Biotic pump of atmospheric moisture, its links to global atmospheric circulation and implications for conservation of the terrestrial water cycle. Preprint no. 2655, Petersburg Nuclear Physics Institute, Gatchina

Goshko UG (ed) (1983) Hutzulzhyna: istoryko-etnografichne doslidzhennya (Hutzul land: historical and ethnographical study). Naukova dumka, Kiev

Gulisashvili VZ (1956) Gornoe lesovodstvo (Mountain forestry). Goslesbumizdat, Moscow

Gulisashvili VZ, Makhatadze LB, Prilipko LI (1975) Rastitelnost Kavkaza (Vegetation of Caucasus). Nauka, Moscow

Hergnyan M, Hovhannisyan S, Grigoryan S, Sayadyan H (2007) The economics of Armenia's forest industry, Economy and Values Research Center, Yerevan. Available via http://www. armeniatree.org/thethreat/resources/ev_forest_industry121007.pdf

Hrushevsky M (1905) Materialy do istoriji suspilno-politychnykh i ekonomichnykh vidnosyn Zakhidnoji Ukrajiny (Materials about the history of socio-political and economic relations in the Western Ukraine). Naukove Tovarystvo imeni Shevchenka, Lviv

Jentoft S, Minde H, Nilsen R (eds) (2003) Indigenous peoples: resource management and global rights. Eburon, Delft

Kalinin B (1932) Teoreticheskie osnovy 'lesnoy statiki' i teorii 'ocenki lesa' (Theoretical bases of 'forest static' and 'forest assessment' theory). Izvestiya Lesotehnicheskoy Akademii 2(40)

Kharaishvili GI (2008) Water regulation and soil protection functions of mountain forests and measures of its preservation, afforestation and intensification. Vasil Gulisashvili Forest Institute, Tbilisi

Kile NB (1997) Nanaitsy v mire prirody (Nanai people in the natural world). In: Etnos i prirodnaya sreda. Dal'nauka, Vladivostok, pp 34–44

Kobylnik V (1936) Materialna kultura sela Zhukotin, turchanskogo povity. Litopys Boykivzhyny 8:15–71

Kreinovich EA (1929) Ocherk kosmogonicheskikh predstavlenii gilyakov ostrova Sakhalin. Etnografiya 1:78–102

Krynytskyj H, Tretiak P (1998) Stan lisiv Ukrajinskykh Karpat, ekologichni problemy ta perspektyvy (Conditions of forests in the Ukrainian Carpathians, ecological problems and perspectives). Pratsi Naukovoho Tovarystva Shevchenka 9:54–65

Kubiyovich V (1938) Geographiya ukrainskih i sumizhnyh zemel (Geography of Ukraine and adjacent areas). Ukrainskiy Vydavnychy Istityt, Lviv

Laletin AP (1999) CIS (Commonwealth of Independent States). In: Verolme HJH, Moussa J (eds) Addressing the underlying causes of deforestation and forest degradation—case studies, analysis and policy recommendations. Biodiversity Action Network, Washington, DC, pp 63–70

Laletin AP, Vladyshevskiy DV, Vladyshevskiy AD (2002) Protected areas of the Central Siberian Arctic: history, status, and prospects. In: Watson AE, Alessa L, Sproull J (eds) Wilderness in the Circumpolar North: searching for compatibility in ecological, traditional, and ecotourism values. Proceedings RMRS-P-26. US Department of Agriculture, Forest Service, Rocky Mountain Research Station, Ogden, pp 15–19

Maghakyan AK (1941) The vegetation cover of Armenian SSR. Akademia Nauk SSSR: Otdel Armenii/Botanicheskiy Institute, Moscow/Leningrad

Makhatadze LB (1977) Dubravi Armenii (Oak forests of Armenia). Izdatelstvo Akademii Nauk Arm SSR (Publication of Academy of Sciences of Armenian SSR), Yerevan

Martínez Cobo, José (1986/1987) Study of the problem of discrimination against indigenous populations". UN Doc. E/CN.4/Sub.2/1986/7 and Add. 1–4. United Nations, New York. Available via http://www.un.org/esa/socdev/unpfii/en/spdaip.html. Cited 7 Mar 2011

Ministry of Nature Protection (1999) First national report to the convention on biological diversity. Ministry of Nature Protection of Armenia, Yerevan. Available via http://www.cbd.int/doc/world/am/am-nr-01-en.pdf. Cited 7 Mar 2011

Mirhashimov I (ed) (2005) Landshaftnoe i biologicheskoe raznoobrazie Respubliki Kazakhstan. (Landscape and biological diversity of Republic of Kazakhstan). Informacionno-Analiticheskiy Obzor Programmy Razvitiya OON. OO OST-XX Century, Almaty

Moreno-Sanchez R, Sayadyan H (2005) Evaluation of the forest cover in Armenia. Int For Rev 7(2):113–127

Novikova TM, Yatimov OR (2011) Biodiversity of mountain forest ecosystems supporting the socio-economic well-being of local communities in south-eastern Tajikistan. In: Laletin A, Parrotta JA, Domashov I (eds.) Traditional forest-related knowledge, biodiversity conservation and sustainable forest management in Eastern Europe, Northern and Central Asia, vol 26, IUFRO World Series. International Union of Forest Research Organizations (IUFRO), Vienna, pp 45–51

Orolbaeva L (2003) Gornye travy: lekarstva, krasiteli, specii (Mountain herbs: medicines, dyes and spices). Central Asian Mountain Partnership Program (CAMP), Bishkek

Podmaskin VV (2001) Ekologicheskie traditsii v kul'ture narodov Dal'nego Vostoka Rossii: Rossia I Kitai na dal'nevostochnykh rubezhakh. AGU, Blagoveshchensk, pp 72–76

Posey DA (ed) (1999) Cultural and spiritual values of biodiversity. United Nations Environment Programme, London

Secretariat of the Convention on Biological Diversity (SCBD) (2004) Akwé: Kon—voluntary guidelines for the conduct of cultural, environmental and social impact assessment regarding developments proposed to take place on, or which are likely to impact on, sacred sites and on lands and waters traditionally occupied or used by indigenous and local Communities. CBD Guidelines Series, Montreal. Available via http://www.cbd.int/doc/publications/akwe-bro-chure-en.pdf. Cited 1 Feb 2011

Sem JA, Sem LI (1988) Leksika, sviazannaya s rastitel'nym mirom, v nanaiskom yazyke (Lexics connected with vegetation in Nanai language): voprosy leksiki i sintaksisa yazykov narodov Krainego Severa. LPBI Publishing House, Leningrad

Shuhevich V (1898) Hutzulzhyna (Hutzul land), vol 1. Naukove Tovarystvo Shevchenko, Lviv

Shukurov ED (2008) Sochineniya (Works). Kyrgyzstan, Bishkek

Soloviy I, Cubbage F (2007) Forest policy in aroused society: Ukrainian post-orange revolution challenges. J For Policy Econ 10(1–2):60–69

Sulyandziga R, Bocharnikov VN (2006) Russia: despoiled lands, dislocated livelihoods. In: Colchester M, Caruso E (eds) Extracting promises: indigenous peoples, extractive industries and the World Bank, 2nd edn. Tebtebba-Forest Peoples Programme, Baguio City, pp 223–251

Synyakevych I, Soloviy I, Deyneka A, Keeton WS (eds) (2009) Ecological economics and sustainable forest management: developing a transdisciplinary approach for the Carpathian Mountains. Ukrainian National Forestry University Press-Liga-Press, Lviv

Taksami CM, Kosarev VD (1986) Ekologia i etnicheskie traditsii narodov Dal'nego Vostoka (Ecology and ethnic traditions peoples of the Far East). Priroda 12:28–32

Ter-Ghazaryan K, Karapetyan V, Barseghyan M (1995) Forest and forest product country profile: Republic of Armenia. Geneva Timber and Forest Study Papers, ECE/TIM/SP 8. UNECE Timber Committee and the FAO European Forestry Commission. Available via http://www.unece.org/timber/docs/sp/SP8-ARMENIA.pdf. Cited 7 Mar 2011

Turaev VA, Sulyandziga RV, Sulyandziga PV, Bocharnikov VN (2005) Encyclopedia korennyh malochislennyh narodov Severa, Sibiri i Dalnego Vostoka Rossiiskoi Federatzii (Encyclopedia of indigenous peoples of the North, Siberia and Far East of the Russian Federation). Biblioteka Korennyh Narodov Severa, Seria

UNCCD (2003) Subregionalnaya programma deystviy stran Centralnoy Azii po borbe s opustyini-vaniem v kontekste KBO (Subregional program of actions of countries of Central Asia on combating desertification in the context of Convention on Combating Desertification). Available via http://www.unccd.int/actionprogrammes/asia/subregional/2003/srapcd-rus.pdf. Cited 7 Mar 2011

Urushadze AT (2005) Alluvial soils of East Georgia. Pochvovedenie 1:31–39

Vladyshevskiy DV, Laletin AP (2002) Ecologicheskiy aspect ustoichivogo razvitiya na primere ispol'zovania lesov (Ecological aspect of sustainable development using forests as an example). KrasGU, Krasnoyarsk

Vladyshevskiy DV, Laletin AP, Vladyshevskiy AD (2000) Role of wildlife and other non-wood forest products in food security in central Siberia. Unasylva 202:46–52

Chapter 8
Northeast Asia

Youn Yeo-Chang, Liu Jinlong, Sakuma Daisuke, Kim Kiweon, Masahiro Ichikawa, Shin Joon-Hwan, and Yuan Juanwen

Abstract Northeast Asia including China has a vast land area that hosts many ethnic groups living in diverse environmental conditions. Its ecological diversity is matched with diverse cultural heritages, including forest-related cultures. There is a rich tradition of managing the villagers' common forests, with well-organized institutions in accordance with traditional religious customs of the local communities. People of village groves and community common forests such as the 'fengshui' forest in China, 'satoyama' in Japan, and 'maeulsoop' in Korea have kept their traditional

Youn Y.-C. (✉)
Department of Forest Sciences, Seoul National University, Seoul, Republic of Korea
e-mail: youn@snu.ac.kr

Liu J.
School of Agricultural Economics and Rural Development, Renmin University of China, Beijing, China
e-mail: liujinlong@ruc.edu.cn

S. Daisuke
Osaka Museum of Natural History, Osaka, Japan
e-mail: sakuma@mus-nh.city.osaka.jp

Kim K.
Department of Forest Resources, Kookmin University, Seoul, Republic of Korea
e-mail: kwkim@kookmin.ac.kr

M. Ichikawa
Faculty of Agriculture, Kochi University, Kochi, Japan
e-mail: ichikawam@kochi-u.ac.jp

Shin J.-H.
Korea Forest Research Institute, Seoul, Republic of Korea
e-mail: kecology@forest.go.kr

Yuan J.
Guizhou College of Finance and Economy, Guiyang, China
e-mail: yuanjuanwen@yahoo.com

knowledge about forest ecosystems that provide physical, cultural and spiritual amenities to local communities. People in this region have been protecting some of their forests as seed and water conservation reserves, while some forest areas have been used traditionally as sources of medicinal herbs and food. Agroforestry has long been a common practice for the production of food, timber, and fibre in the region and has gained vitality when integrated with modern technology, in particular in China. Many species of medicinal herbs as well as mushrooms continue be collected, and cultivated, in the forests of Northeast Asian forests. Traditional forest-related knowledge, however, has been eroding in Northeast Asia as countries in this region have become industrialized and influenced by economic globalization. This rapid loss has been exacerbated by government policies promoting infrastructure development and free trade. In order to protect traditional forest-related knowledge, it should be given proper recognition and an equitable role alongside scientific knowledge in social decision-making. Suitable regulations should also be implemented in order to protect the local peoples who have created and inherited traditional forest-related knowledge. It is imperative, therefore, that scientific knowledge be integrated with traditional forestry-related knowledge.

Keywords China • Fengshui • Forest management • Forest governance • Forest policy • Japan • Korea • Maeulsoop • Non-timber forest products • Satoyama • Traditional knowledge

8.1 Introduction

The Northeast Asia region—which includes China, Korea, Mongolia, and Japan—has a long history of civilization and cultural development that has included the evolution of rich traditional knowledge systems that influence many aspects of the peoples' lives. People in this region have had a deep association with forests in their daily lives, and have developed innovative ways of making use of forest resources and the environmental services that forests provide. The accumulation of knowledge about forest resources has enriched the cultures and livelihoods of people throughout the region.

Traditional forest-related knowledge (TFRK) in the Northeast Asian context includes knowledge relevant to production, livelihoods, customs, religion, and culture related to forests and the lives of the community people. It includes utilization and cultivation of plants, biodiversity conservation, integrated farming systems (agroforestry), forest management practices, and community regulations related to forest resource utilization.

Traditional forest-related knowledge in Northeast Asia is very rich, diverse, and complex, having evolved over many centuries under conditions of great ecological, geographical, social, economic, and cultural diversity. It has been developed and passed on over generations by farmers, and is closely linked to people's livelihoods, production systems, cultures, religious beliefs, and traditions. This chapter provides a broad overview of traditional forest-related knowledge in the Northeast Asia

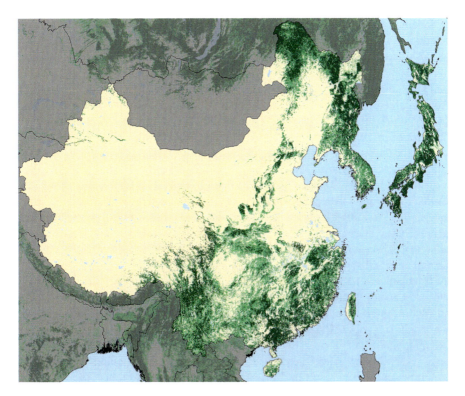

Fig. 8.1 Forest and woodland cover in Northeast Asia (Source: Adapted from FAO (2001)).
Key: *Dark green* closed forest, *light green* open or fragmented forest, *pale green* other wooded
land, *yellow* other land

region, the customary regulative framework and cultures that are closely related to
traditional forest-related knowledge, and the implication of this knowledge on the
ongoing policy debate regarding forest management in Northeast Asia. While it is
not possible to describe all existing traditional forest-related knowledge practices
and related environmental, economic, and social issues in Northeast Asia in this
chapter, specific practices and case studies have been highlighted, which will help
to illustrate the broader principles and issues considered. Cultural diversity of
Northeast Asia is due to its ecological diversity. Northeast Asia (from the Arctic to
tropical regions) covers an area crossing over 50° of latitude. It includes areas of
diverse temperature and humidity, with very versatile forests ranging from tropical
rainforest to taiga or boreal forest (Fig. 8.1). This region hosts about one-fourth of
the world's population and consumes 13% of the world's timber production It is
also the centre of genetic diversity for major crops such as rice and soybeans (Bailey
1998; Lee and Suh 2005; Ministry of Forestry of China 1994; UNEP 1995).

The large-scale patterns of phytogeography in Northeast Asia—covering the eastern part of China, Korea, Japan, and the eastern part of Russia—are strongly related

to latitude, which correlates with several climatic variables such as temperature. Northeast Asia has long been a focus of attention of botanists and biogeographers, partly because of its markedly high species diversity; relatively high proportion of Tertiary relicts of vascular plants; close floristic relationships with eastern North America; and a long, unbroken latitudinal gradient of forest vegetation, extending from the tree line at the Siberian Arctic southward to the tip of the Malay Peninsula (nearly at the equator). This is the longest latitudinal continuum (across nearly 60° latitude) of forest vegetation in the world, and it provides a unique opportunity to understand the origin and maintenance of many latitude-associated patterns in ecology and biogeography (Qian et al. 2003).

In this paper, we provide case studies from different regions, including Mongolia, China, Korea, and Japan. Mongolians have a proud history regarding nature conservation. The State of Mongolia, in ancient times, implemented a policy aimed at nature conservation and the proper use of natural resources. Mongolia has a severe continental climate and an ecologically vulnerable environment divided into six natural belts and zones: the alpine, mountain taiga, and mountain forest steppe belts; arid steppe; desert-steppe; and desert zones. These belts and zones differ from each other in terms of their soil quality, and plant and animal species, which in turn are adapted to the different habitats and climatic conditions characteristic of each of these belts or zones (MNEM 1998).

China has a vast territory with a complex climate, varied geomorphic types, a large river network, many lakes, and a long coastline. Such complicated natural conditions inevitably result in diversified habitats and ecosystems. The terrestrial ecosystem can be divided into several types: forest, shrubland, meadow, steppe, savanna, desert, tundra, and marsh, among others. It crosses frigid, temperate and tropical zones from north to south. Plateaus and high mountains occupy over 50% of the land. The fauna and flora are characterized by having many endemic and relic species because during the late Tertiary most regions had not been affected by glaciations. Thus, China is considered internationally as one of the mega-diversity countries in the world, where the number of species as a whole makes up about one tenth of the total number of species of the world (BCCAS 1992).

The climate of Korea is primarily continental, with a monsoon season in the summer. Korea is a mountainous peninsula with complex geology. The three major vegetation types of the Korean peninsula are warm temperate forest, cool temperate forest, and sub-boreal forest. The warm temperate forest covers the area south of 35°N, a part of the southern coastal regions, including many islands, where the annual mean temperature is above 14°. Evergreen broadleaved forests were once typical in the zone, but most of the natural forests were destroyed by overexploitation and fire. This area now supports deciduous broadleaved, pine, or mixed forests. Cool temperate forest accounts for 85% of the total forest area. Except for the mountainous highlands, the land between 35°N and 43°N belongs to this forest type. The annual temperature ranges from 6° to 13°. Deciduous broadleaved forests are found in the zone, but most of the forests have been disturbed by human activities and changed into pine forests. Today, deciduous broadleaved tree species can be found in the lower strata of most pine forests where there is no human-induced

Table 8.1 Biodiversity information related to countries in Northeast Asia

Country	NBI[a]	Ex situ collections[b]			Protected areas	
		NHM	Zoo	Bot	No.	Area (km²)
Mongolia	0.358	1	–	1	42	179,912
China	0.839	7	133	73	885	686,806
Korea, Republic	0.423	2	4	5	30	6,839
Japan	0.638	26	160	54	96	25,610

Source: CBD (2001)

[a]*NBI* National Biodiversity Index. This index is based on estimates of country richness and endemism in four terrestrial vertebrate classes and vascular plants

[b]*NHM* natural history, botanical and zoological museums, *Zoo* zoological gardens, *Bot* botanical gardens

disturbance. The sub-boreal forest covers the northern part of the peninsula and the mountainous highlands where the annual mean temperature is below 5°. Coniferous forests are representative of the zone (Shin 2002). These diverse habitats support more than 30,000 species, which is quite a high number for a small territory of 99,000 km² (ROK 2009).

The Japanese archipelago runs parallel to the eastern rim of the Eurasian continent, extending about 3,000 km, from 45°33′ to 20°25′, with climatic zones ranging widely from subarctic to subtropical. In general, however, the Japanese climate is temperate and humid, with marked monsoons and spectacular seasonal changes. With a geographical history of being alternately connected to and separated from the Eurasian continent, and composed of several thousand islands of various sizes, Japan hosts a unique and rich biota with many endemic and relict species. Nearly 40% of mammals and nearly 80% of amphibian species are endemic. The number of known species in Japan is around 90,000, and estimates place the total number at approximately 300,000. This rich biota is contained in a relatively small land area, about 378,000 km² (NCB 2008).

In this part of the world, because of the long history of civilization characterized by population growth, this inherently rich biological diversity has been exploited beyond its natural limit to regenerate. China, Japan, the Republic of Korea, and Mongolia have set aside some areas as national parks and other protected areas for the protection of biological diversity and have established natural history museums, and botanical and zoological museums, in order to preserve species that are under threat of extinction (Table 8.1).

Biogeographers have noted that human cultural diversity exhibits strong geographical patterns; thus, there exists a relationship between ecological conditions and human culture. According to Collard and Foley (2002), the density and diversity of human culture decline with latitude and increase with temperature and rainfall. According to a study on cultural diversity by Eearon (2003), which measured cultural diversity of countries based on an ethnic fractionalization index using languages as a proxy for cultural similarity, China was the most diversified in ethnic groups, followed by Mongolia. Japan and Korea were among the most homogeneous societies in terms of ethnic groups.

In addition to the local ecological conditions, religions have influenced the culture of the world's people. For more than a 1,000 years Buddhism and Confucianism have exerted a strong influence on the lives of people in Northeast Asia. These two advanced religions, with beliefs in natural spirits, have been imbedded in the conservation and management of natural resources, including forests. For example, the emperors and kings adopted the principle of royalty backed up by Confucianism and made use of symbolism on the nature of pine trees of unchanging colour. There is a strong cultural heritage of conserving pine trees in royal gardens and village groves. On the other hand, in villages remote from the power centres of nations in this region, like Yunnan Province of China, virtually a lion's share of the remaining natural forests and native trees have either a cultural or a spiritual purpose in the lives of people.

8.2   Traditional Forest Management Practices

Northeast Asia has 5,000 years of recorded history and civilization during which time its people had been living in close association with forests—practising hunting; gathering of wild food, fuel wood, medicinal plants, and other non-timber forest products; and forest-based agriculture. The people in this region have acknowledged the value of forest resources, as evidenced by their rich traditional forest management practices. The most common of these practices at the community level includes managing common forests or village groves: fengshui forests (in China), maeulsoop (in Korea), and satoyama (in Japan).

8.2.1   Village Groves/Community Forests: Fengshui Forest, Maeulsoop, and Satoyama

8.2.1.1   Fengshui Forests—Chinese Village Forests or Groves

In China, fengshui forests can be strictly protected, man-made or natural forests that have consequences for people's lives in terms of granting wishes for safety, longevity, family prospects, wealth, and promotion in jobs (Guan 2002a; Liu 2007b).

Fengshui forests are usually established at the village entrance, tomb area, rear hill to the village, backyard, and/or temple (Yang 1999; Park et al. 2006). Fengshui forests are categorized into three types according to location: village fengshui forests, grave fengshui forests, and temple fengshui forests (Guan 2002a, b). Maeulsoop has a variety of forms that are differentiated mainly by location and function (Kim and Jang 1994). Some of the fengshui forests and maeulsoop are protected by customary and regulatory laws or by-laws as nature reserves or natural monuments.

Fengshui forests or maeulsoop have important functions in the people's daily lives, including conservation of biodiversity, soil and water, and landscapes, and as

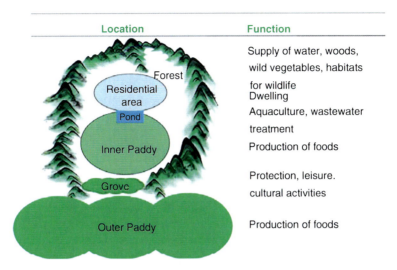

Location	Function
Forest	Supply of water, woods, wild vegetables, habitats for wildlife
Residential area	Dwelling
Pond	Aquaculture, wastewater treatment
Inner Paddy	Production of foods
Grove	Protection, leisure. cultural activities
Outer Paddy	Production of foods

Fig. 8.2 A typical location of a traditional village grove in Korea (Adapted from Lee and Park 2009)

religious places (Park et al. 2006; Zhu 2007; Liu 2007b; Zhong and Boris 2007; Lee 2008). It was not until the 1990s that scholars and administrators realized the value of maeulsoop and fengshui forests for forestry management in Korea and China. Liao et al. (2008) and Liu et al. (2007a, b) investigated the economic, social, and environmental values of fengshui forests in Guangdong, Yunnan, and Guizhou provinces of China, while Yuan and Liu (2009) conducted an investigation on fengshui forests of the Buyi minority at Guizhou Province.

8.2.1.2 Maeulsoop Woods in Korean Villages

Wishing to live in harmony with the land, the people of Korea have constructed village woods called maeulsoop or bibosoop. (Kim and Jang 1994; Lee 2008) This culture is somewhat related to the philosophy of fengshui, according to which a place surrounded by high and firm mountains to the north and relatively gentle but uninterrupted ranges to the west and east sides is a good land to live, so in the middle, people feel comfortable (Fig. 8.2). The ends of the ranges on both sides become the outlets of the village watershed. The outlets must be lower than the mountain, but should also have some height and be interconnected with each other to shield the village. In reality, however, these conditions often do not exist. If people thought there were some defects in the environment of their village, they complemented the defects by constructing village woods. The most common practice was to establish woodlands in the outlet of the village watershed. When the outlet was opened, they planted trees in the form of a grove to protect the village and to make it a good place to live in. The canopy of groves functions like a curtain. From the inner side of the village woods, people can look at the outer world while from the outer side, a passerby hardly sees

Fig. 8.3 Typical village grove located in Songmal-ri, Baeksa-myeon, Icheon City, Republic of Korea. The portion of woodlands marked by a dotted circle is village grove artificially established and maintained by the Yoo clan villagers

the village. It is also believed that the grove keeps the evil spirits from entering the outer world. People protect these village groves and traditionally use them for village meetings to discuss community affairs.

Village woods have various functions: for religious purposes as shrine woods; as windbreaks or shelter belts; for erosion control; and for aesthetic beauty. A typical well-maintained village woods is at Dundong village, Hwasoon County, Cheonnam Province. In Dundong village, the people used manpower to construct a bank along the river in front of the village to protect it against the water flow and then planted trees on the bank. The village woods became effective in protecting the village from strong winds and floods, and in keeping the village out of sight from strangers. The village woods also screen the people's eyes from ugly looking rocks. In spiritual terms, the woods help in the upbringing of youngsters in a manner that makes them see good things and makes their minds beautiful.

The distribution of and ecosystem services provided by maeulsoop woods in Korea were investigated by Lee et al. (2007). They found that the village woods established in front of a village like the one at Songmal-ri, Icheon City, can reduce wind speed by 10–40%, which can further reduce evaporation in agricultural fields protected by the village woods (Fig. 8.3). This confirms the fact that the main function of maeulsoop woods is protection of living space in the village, as observed in the case study of Jinan County's village groves by Park and Lee (2007). There are at least 51 village woods in the country, the main function of which is protection of the villages from winds, floods, and bad spirits. Most of village woods in the county are strongly related to the horse-ear, such as the

Fig. 8.4 House of land God located in the village grove of Seongnam-ri, Shinlim-myeon, Wonju City, South Korea

weird-looking peaks of Mai-san Mountain (Park 1999, 2000). It can be postulated that one of the functions of village woods in Jinan County is to screen the village from unwanted scenes such as those Mai-san Mountain peaks.

Many maeulsoop woods have been lost or degraded over the last century. The underlying causes of loss of village groves in Korea include change of land ownership, erosion of traditional culture and religion, and increased opportunity costs of keeping forests in places where the land is very expensive (Kim and Youn 2005; Youn et al. 2010). The privatization of forestland initiated by the First Forest Law in 1908 is considered to be one of the most important institutional changes that caused the loss of village woods in Korea. The fate of maeulsoop also depends on the existence of cultural capital such as traditional rituals based on natural worship and community organization in charge of woodland management. The maeulsoop of Seongnam-ri, Wonju City, Gangwon-do, includes two big worship trees and surrounding old-growth forests that have been well-protected by the community, including villagers and the city government. The villagers practise land god worship rituals today (Fig. 8.4).

In recent years there has been a growing interest among environmentalists in conserving the maeulsoop woods. A non-governmental organization called The Forest for Life has been working with the Korea Forest Service towards the restoration of traditional village groves, which have deteriorated because of mismanagement (Forest for Life 2008).

Fig. 8.5 An example of satoyama in Miyazaki Prefecture, Kyushu Island, Japan

8.2.1.3　Japanese Satoyama

Satoyama, meaning 'village mountain' in Japanese, is typically used by villagers who engage in paddy-field agriculture. It is a working landscape that surrounds one or more local agricultural communities. It is a land for complex and varied uses that may include a combination of forests, paddy fields, meadows, reservoirs, irrigation channels, gardens, and residential areas.

Satoyama landscapes (Fig. 8.5) are diverse in their structure and functions depending on the conditions of the locality including weather, geological history, topography, kinds of crops and farming activities, population, and distance to and size of the commodity markets. Satoyama forests are occupied by village commons, and the land is owned collectively by the villagers. Although trees are protected by village customary law, the utilization of satoyama by villagers has not been extensively restricted. Therefore, it is usual for village commons in satoyama to be overexploited, resulting in satoyama that is occupied by pine trees standing sparsely, or that is becoming nearly a grass land.

The meaning of satoyama is changing as the relationship of the people to the natural landscape changes. Satoyama is a rural landscape that was defined as 'forest utilized for agricultural activities (fuel and fertilizer), especially in relation with rice production' (Shidei 1973). Preservation of these cultural landscapes became a concern during Japan's rapid economic development period, which gave rise to a nature conservation movement against the conversion of forestland to golf courses, industrial areas, and other areas for economic development activities. As conservation interests have been focused on endangered species and the rural landscape of high biodiversity value, ecological research on satoyama was initiated and gave a new definition to satoyama's meaning not only forests, but the entire landscape that is necessary for rice cultivating agriculture (Tabata 1997; Sakuma 2008).

Japan's Ministry of the Environment and the United Nations University Institute of Advanced Studies (UNU-IAS) recently launched the Satoyama Initiative, which aims to promote and support socio-ecological production landscapes worldwide that have been shaped over the years by the interaction between people with nature. The Initiative was recognized in the 10th Conference of the Parties to the Convention on Biological Diversity held in Nagoya, Japan, in 2010.

8.2.2   Seed Reserves and Watershed Protection Forests

Seed reservoirs are a basis of sustainable cultivation in human civilization. Some forest reservoirs serve as a seed reservoir, which facilitates the regeneration of the original vegetation, most commonly evidenced in fallow fields of traditional shifting cultivation systems. The seed tree silvicultural system has its roots in the traditional knowledge of reserving old mother trees along the ridges above shifting cultivation fields (Chao et al. 1994).

People of the Jinuo minority group in China protect such forests and woodlands as permanent greenbelts marking their territory boundary between villages (Long et al. 1999). In a traditional Korean shifting cultivation system, patches of forest located in the valley were left for seeds so that the fallows could be established easily.

In the history of human civilization, conservation of water and soil has been the key to sustainable agriculture. Emperors and kings, even current governments in China and Korea, have been stressing the importance of water and soil conservation in ruling their people because without water and good soil they could not feed the people (Jiang 2008).

Water source forests exist quite often in Southwest China, where many minority ethnic groups live. Each ethnic village possesses an area of water source forest (Luo et al. 2001), usually located in a valley near the village. Sometimes it covers a little watershed. Most of the water used for human and livestock consumption is from water source forests.

In a remote minority region of Southwest China where shifting cultivation is still in practice, a fire protection strip that is a continuous greenbelt is usually established between two forest areas. The width varies from several meters to more than 10 m. The grass, shrubs, and ground litter in this 'fire proof' forest are cleared (Luo et al. 2001). In many regions, even where shifting cultivation has been eliminated for a decade, this fire protection forest still exists and now contributes to biodiversity conservation.

Community-based fuel wood forests remain traditionally managed in central and western parts of China, usually used by one or more villages for fuel-wood collection. These forests have been managed by an association whose members were elected by the villagers in a traditional way. In Liuyang County of Hunan Province, about 10 of the total forests are under this type of management (Liu and Zhao 2009). Management of the community-based fuel wood forest is allocated to individual households by virtue of a new forest policy initiative of the Chinese government, which gives priority to the economic returns of forest management.

8.2.3 Traditional Knowledge for Forest Plantations

Northeast Asia, especially China, has a long history of forest plantations, in particular, plantation of Chinese fir (*Cunninghamia lanceolata*), bamboo (*Phyllostachys pubescens*), and camellia oil trees (*Camellia* spp.). Bamboo is regarded as a virtue in East Asian culture and often selected as an ornamental plant for gardening (Liu 1996).

Chinese fir plantations have a history of about 400 years in west Hunan Province and southeast Guizhou Province. There is a tradition in these regions that when a girl is born, her parents plant Chinese fir trees. The newly established plantation is called 'Eighteen Year Chinese fir plantation,' meaning that when the girl is already 18 years old, these trees are ready to harvest and can be used as her marriage dowry. Even though Chinese fir was planted in the same soil after harvesting from generation to generation, the soil has never become degraded and its productivity has remained the same for over a 100 years (Liu 1996). It is worthwhile to note, though, that with the application of modern technology—including fertilization, gene improvement, and good spacing—the second generation of Chinese fir plantation often suffers from rapid yield decline. In some places, there is a yield reduction of about 20% in Chinese fir plantations, while that of the third generation drops even by 50% because of soil quality decline (Liu et al. 2007a, b).

The Dai minority people in Yunnan have been planting trees along the roadsides and farmland sites in order to supply fuel wood from pruning the trees. In the Huanghuaihai plain area, where there have been shortages of fuel wood for many decades, villagers have been planting *Salix* or *Acacia* species, which they prune for fuel wood production.

8.3 Non-timber Forest Products

8.3.1 Non-timber Forest Products in China

A large number of forest products are extracted from the forests for commercial and subsistence purposes. There are 6,000 multi-purpose tree species recognized in

China, but among them only 200 species have been developed as economic species. Except for the use of timber and forest products as fibre and energy sources, the potential values of many species have not yet been recognized. Other forest products include: (1) more than 100 species of known nut-producing trees; (2) more than 200 species of oil tree, including walnut, tea-oil tree, shiny-leafed yellow thorn (*Xanthoceras sorbifolium*), apricot, hazelnut, almond, and Chinese prickly ash; (3) bamboo species; (4) ornamental plants such as flowers and potted plants; (5) perfume and seasoning species such as anise; (6) medicinal plants such as eucommia (*Eucommia ulmoides*), cork tree (*Quercus suber*), and magnolia (*Magnolia officinalis*); (7) camellia oil; and (8) edible fungi and mushrooms, seabuckthorn (*Hippophare rhamnoides*), wild peach, and wild jojoba.

The traditional knowledge of making silk and paper, which originated from this region, consists of collecting the leaves and bark of mulberry trees (*Broussonetia kazinoki*) for raising silkworms and making paper. Farmers in this region have been planting mulberry trees for a long time period.

In China, bamboo has been incorporated into its history, art, handicrafts, music, religion, customs, architecture, and agricultural production, and has been used by various people for a variety of purposes. In Yunnan Province alone, there are 250 species of bamboo, which provide a source of income supporting local livelihood (Yang et al. 2004).

People of China have been using herbs for medical treatments for a long period of time. At least 12,807 plants have been documented as medical herbs in the Chinese medical resources inventory (Pei 2000). The variety of medicinal plants is evidenced by the rich list of species being traded as medicines in the local markets in mountainous regions, especially Southwest China (Pei 1996; Lee et al. 2007). The medicinal plant species collected by the people from forests tend to change when the primary forests are destroyed, as seen in Yunnan, China. Environmental changes such as forest degradation are accompanied by changes in people's lifestyles, shifting from hunting and gathering to sedentary cultivation, although people still depend on the forest and their folk medicinal knowledge, particularly those who are poor and elderly, who became more vulnerable to forest degradation. In addition, flowers have long been eaten by people in China. Pei (2000) reported that flowers of 59 wild plant species are used as food in Yunnan, China, including *Rhododendron decorum, R. pachypodum, and Ottelia accuminata*.

8.3.2 Non-timber Forest Products in Japan

In Japan, the villagers' rights to collect mushrooms and edible wild herbs from forests are governed by customary law. Usually, harvesting wild herbs is a common right for villagers, not restricted by land ownership. But the right to collect matsutake mushrooms is obtained for a fee from the village authorities.

The reason for this is that these mushrooms have been traded at a high value since the middle ages. Many written records are stored in temples and shrines showing that matsutake mushrooms were donated as a seasonal present from their adherent nobles. Tree branches were also used as commercial products in Japan, although their price was comparably low. Fallen leaves are also collected for private use as fuel. The collection of leaves by villagers is permitted according to certain rules, which are different among regions. In remote areas leaves are free for use, while in suburban areas collecting tree leaves is restricted for a few weeks in winter. Collecting tree leaves is not constrained by the ownership of the forest.

8.3.3   Non-timber Forest Products in Korea

Korea is highly populated, with limited arable lands available. Therefore Koreans have used forest resources intensively as a source of food. Various non-timber forest products are harvested for trade as well as subsistence. Acorns and nuts have been a basic food for Koreans since their arrival in the Korean peninsula. Discovery of acorns and grinding stones in the pre-historic settlements, dating back to at least 10,000 years, suggests the prehistoric Koreans cooked acorns and hazelnuts as foods (Kang 1990). For a long period of time, ginseng root (*Panax ginseng*) has been regarded as a medicine in Korea and is still in high demand. Natural ginseng grows only in pristine forests having high biological diversity and composed mainly of hardwood species. Forest mushrooms are also being collected as food by Korean people. Songyi (matsutake) mushrooms grow in pine forests and are collected mainly for commercial purposes. Collectors of ginseng and songyi mushrooms maintain important traditional knowledge about conditions in which ginseng and songyi mushrooms grow.

Collecting honey from natural forests is also a long tradition in Korea. Villagers in mountain areas have kept caves in the backyards of their homes for bees that originated from natural forests. The knowledge of keeping bees originated from their ancestors and is also learned from neighbours. The native honey bee keeping is based on traditional forest-related knowledge well-preserved by local people living near healthy natural forests and provides a good source of income for forest-dependent communities in Korea and China (Fig. 8.6).

The Korean diet is composed mainly of vegetables, with limited quantities of meat and poultry. Rural people collect natural vegetables from nearby forests as a part of their daily meals. Bamboo is grown by farmers in their backyards for domestic consumption and sale of the edible young shoots. Growing edible forest products is being targeted by small forest owners as a kind of forestry business, with support from the government.

Fig. 8.6   Practice of native beekeeping in Korea (Photo courtesy of Eurocreon, Korea)

8.4   Traditional Forest Management Practices

8.4.1   Farm-Based Agroforestry Practices

Agroforestry is a conventional farming system that combines agricultural and forest crop production and at the same time fulfils the objective of environmental conservation. It is still commonly practised in Northeast Asia, notably China (Lundgren 1990).

Farm-based agroforestry in the Northeast Asia region usually focuses on food rather than on wood production, with emphasis on the utilization of fruit and multi-purpose tree species (Liu 1996). Agroforestry is more prevalent in densely-populated rural areas and, in China it is commonly practised in the plateau and hilly land with slopes less than 15°.

Home gardens, where timber or fruit trees are grown in combination with animal husbandry and vegetable farming, are common in northern China and Korea. The main timber species used is poplar (*Populus* spp.), and the main fruit trees are walnut (*Juglans regia*), pear (*Pyrus* spp.), plum (*Prunus salicina*), apple (*Malus pumila*), grape (*Vitis* spp.), pomegranate (*Punica granatum*), persimmon (*Diospyros kaki*), peach (*Prunus persica*), and jujube (*Ziziphus jujube*). Different kinds of home gardens that produce various commodities are common in southern China, where the products of agroforestry farming include fish, crab, pigs, chicken, small animals, vegetables, mushrooms, flowers, and fruit trees (Liu 1996). The main species planted are peach, grape, pear, and persimmon.

Trees are often planted in crop fields in China and other parts of Northeast Asia. Tree species most frequently planted in crop fields include: fruit species such as walnut, pear, plum, peach, apricot (*Prunus armeniaca*), persimmon, pomegranate, and jujube; and multi-purpose trees and shrubs, including black locust (*Robinia pseudoacacia*), purple willow (*Salix purpurea*), pagoda tree (*Sophora japonica*), and Siberian elm (*Ulmus pumila*), among others. These provide timber and fuel wood, and also enhance soil fertility. Paulownia (*Paulownia elongate*) has been planted over a few million hectares of crop fields in an intercropping fashion and has become part of an important cultivation system in the plains of northern China and has recently become a source of raw materials for furniture.

In China, trees are maintained as shade in tea and coffee orchards (Liu 1996). Major tree species for shading tea shrubs are *Paulownia tomentosa, Pinus elliottii, P. taeda, P. serotina, P. banksiana, Phyllostachys pubescens, Acacia confuse, Leucaena leucocephala, Albizia kalkora*, and *Alnus cremastogyne*. Also, multipurpose trees such as Chinese waxmyrtles, orange, loquat, peach, plum, persimmon, pear, grape, tallow tree, laurel, rubber tree, chestnut and camphor are used to provide shade. Coffee is planted in combination with mango trees in the Yunnan Province of China, and with rubber trees in Henan Province.

Trees have been planted around agricultural fields in order to protect the crops from wind. Especially in China, people have developed shelter forests on farm land. These have been proven to create a favourable micro-climate for farming, to increase farming productivity, and to provide a regular source of fibre and energy materials to the rural populace. In China at least 123 tree species are planted in these shelter-forests (Liu 1996). In the warm temperate zones, Chinese white poplar (*Poplar tomentosa*), black Poplar (*P. nigra*), I-214 poplar (*P. xeuramericarca*), cottonwood (*P. deltoides*), willow (*Salix* sp.), black locust, dwarf elm (*Ulmus pumila*), paulownia, Chinese arbour-vitae (*Platyclalus orientalis*) and amorpha (*Amorpha fruticosa*) are planted. In the subtropical zones, pond cypress (*Taxodium ascendens*), bald cypress (*T. distichum*), and metasequoia (*Metaseqoia glyptostroboides*) are usually planted; while in the temperate zone, Lombardy poplar (*P. nigra*) and Chinese pine (*Pinus tabulae formis*) are regularly planted.

People in the mountain areas of Korea have recently planted maple trees, not only in the forest land but also in agricultural fields, in order to tap the sap, which is sold to tourists as a fresh drink in forest villages. This practice of tapping sap from maple trees was inherited from the ancestors, who considered maple sap to be healthy for fishermen. This traditional knowledge has been widely spread in recent years among urban residents who have high demands for uncontaminated natural products. In order to increase the production of maple sap, many forest villagers have started to plant maple trees in their marginal agricultural land, not in agricultural production. This is an example of using traditional forest-related knowledge by forest villagers for income generation (Youn 2009).

8.4.2   Forest-Based Agroforestry Practices

Many species of Chinese medicinal herbs such as cardamom (*Amomum villosum*), Chinese goldthread (*Coptis chinensis*), ginseng, and cacao (*Theobroma cacao*) have been cultivated in Northeast Asian forests. About ten centuries ago a Korean farmer invented the technology of ginseng culture by learning the ecology of wild ginseng in a forest environment. The technology has now spread to the rest of the world including North America, Europe, and Australia. As noted previously, oak mushrooms and pine mushrooms have been cultivated and collected in natural forests for centuries. Bees are raised in forests, and honey from hollow tree trunks.

Traditional taungya practices involving the intercropping of agricultural crops in timber plantations during the early years of tree stand development exist in many parts of China. In the tropical zones of China, teak (*Tectona grandis*), Simao pine (*Pinus kehaya*), Yunnan pine (*Pinus yunnanensis*), *Gmelina arborea*, mango (*Mangifera indica*), and rubber (*Hevea brasiliensis*) are planted; in the subtropical area, Chinese fir, alder, masson pine (*Pinus massoniana*), and bamboo; and in the temperate area, armand pine (*Pinus armandii*), with edible seeds, is the common intercropped tree in the early stages of this system (Liu 1996).

Shifting cultivation has been practised for a long period of time in Northeast Asia. Until recently it was practised in the mountains of southern China, but was banned by the Chinese government in recent decades (Liu et al. 2007a, b). Traditionally, farmers have had limited access to land, so they divided the available land into several pieces for the different stages of shifting cultivation. Upland rice and maize are cultivated on one piece of land for about 2 years after the forest is burned. Since soil productivity declines severely after this cultivation, the land is abandoned for a period of approximately 10 years, after which a new cultivation period of 2 years begins. This is a sustainable practice, but with the increase in population pressure, the area available for shifting cultivation becomes smaller and more food production is needed. In Yunnan, Sichuan, and Guizhou, Chinese mountain farmers replant with Nepal alder (*Alnus nepalensis*) to assist in soil recovery. Alder is a fast-growing, easily propagated nitrogen-fixing tree, providing small-diameter logs, fuel wood, and forage, while improving the soil properties with mulching and N-fixation.

In the mountainous area of southern China, it is common practice to raise geese, cattle, horses and pigs under the trees. Water-tolerant tree species like pond cypress and willow are planted in wetlands along the big watercourses. In the rainy season, these tree stands serve as a buffer to prevent further flooding. In the dry season farmers raise geese and ducks under these stands, whereas during the flood season, the areas serve as temporary fish ponds.

8.4.3   Forest Management for Timber and Energy Production

The use of charcoal has a long history in Northeast Asia. A document of the ancient Shilla Korean kingdom which dates back to the seventh century, records the use of charcoal for heating the houses in Gyeongju, the kingdom's capital. In Japan, commercial scale charcoal-making began in the fourteenth to sixteenth century in Japan. Timber production camps were already established in the eighth to tenth century around Nara and Kyoto in Japan. They were located in mountainous districts connected to Nara and Kyoto via river networks. Fuelwood production on a commercial scale started around the sixteenth century with the development of oak (*Quercus acutissima*) plantations. Until modern transportation systems were established timber and fuel wood production relied on river transportation.

Axes were used for cutting trees even during the prehistoric period. Especially, coppicing operations were done with axes to generate good sprouts. In Japan, during the fifteenth century, crosscut saws were used for processing boards. Previous to that time, only selected timber, such as the huge timbers of Japanese cedar and hemlock (*Tsuga diversifolia*), which can easily be split into straight planks, were used. Once saws came into use, many other wood species like pine could be used.

Cycles of forest operation vary depending on the place and species. In Japan, pine forests are usually cut once a 40–60-year cycle, and understory shrubs (e.g., *Rhododendron*) were cut annually. In oak woods, cutting intervals vary from 7 to 40 or more years. A villager in the Hokusetsu area north of Osaka developed a technique for making good charcoal, which is used for the tea ceremony. This charcoal has been given a brand name, Ikedazumi or Kikuzumi, and commands a high price. The charcoal maker uses pollard-like stools of oak cut at 8-year intervals. The oak woods are harvested in rotation and are a source of stable income for the charcoal maker.

8.4.4   Examples of Traditional Forest-Related Knowledge Applications in Forest Management

The Dong ethnic group of Hunan Province and the Guangxi Autonomous Region of China, in their long-term practice of utilization and management of mountainous land, has been applying good systems of forest management, including those known as rotating fallow system, fire break forest belt, and plantations (Liu 1996). They have also set up village rules for forest management. Fengshui forests and sacred mountains have served to protect natural forests. The Dong minority people living in remote, boundary mountainous areas of Hunan, Guangxi, and Guizhou Province in China rarely experience forest fires because they make fire barriers, ranging in width from several meters to more than 10 m, between their forest plots. These are used to block wind and fire during site preparation and burning. They also establish woods along river bank. Logging of the trees planted along both sides of the river is strictly forbidden. Usually, members of one village or several villages jointly negotiate to establish those village rules and regulations, and the members have to promise, taking oaths, to follow them.

The recent demand for non-timber production from forests could be met by the application of traditional forest-related knowledge in the commercialization of forest products. For example, maple sap was collected and disseminated to mountain villages in South Korea from a village Mount Baek-un in Gwangyang city. The people there developed this forest activity based on traditional knowledge of the maple sap's effect on the health of fishermen, and now maple sap collection is practised by mountain villagers throughout the country. Thanks to the traditional knowledge, production of maple sap and the value of natural forests have increased in Korea in recent years.

8.5 Traditional Governance and Management of Forest Commons

8.5.1 Categories of Forest Management

Some ethnic communities in China have their own way of categorizing their forests, for example in the case of a Dapingdi village in Yunnan Province. The local Yi minority people recognize six different forest categories:

Holy trees: There are three holy trees near the village, namely the dragon tree, the tree of the hunting god, and the God tree. They are situated in a closed forest on the hill above the hamlet. The God tree is considered to be the king of trees in the village. On the other hand, the tree of the hunting god is where the Yi people pray before and after hunting (Yang et al. 2006). The dragon tree is a pine tree with a diameter of more than 80 cm and height of 24 m, lying on a steep hill, with the upper end of the trunk directed to the top of hill, and the tree roots directed to the holy water pond.

Forbidden forest: A piece of forest located at the top of the hamlet (Fig. 8.7). It is a virgin forest of about 7 ha. The first inhabitants of the village decided to forbid human activities when they established the settlement. Thus, no extraction is allowed except that of edible wild mushrooms.

Temple forests: There are five stands of virgin temple forests covering the total area of 4.1 ha of land, and there is one temple located in each piece of temple forests. Temple forests are under collective management by customary regulation. These forests were attached with a sort of supernatural image, for instance, for healthy, safety.

Grave forests: There are five separate grave forests in Dapingdi village, covering a total of about 2.5 ha of land, containing buried ancestry of a family clan in each piece of forest. These forests are secondary natural forests under collective management. They have been managed by different families even before the forest policy of household responsibility system began in 1981.

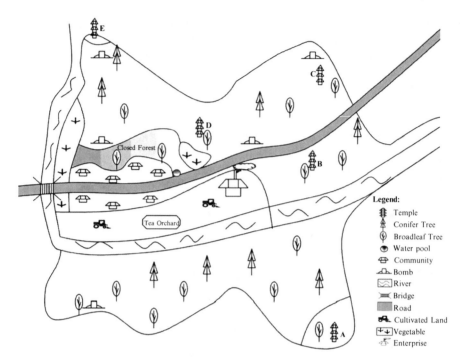

Fig. 8.7 Sketch of land use and forest management practices at Dapingdi village, Yunnan Province, China

Family contracted forests: 175.1 ha of forests, accounting for 92.8% of the total forestland, were contracted to 29 households in 1981. These forests were divided into 72 pieces that were equitably distributed in terms of location and quality of the forests, each household acquiring an average of 2.48 pieces.

Trees outside forestland, planted for agroforestry: There are many trees outside the main forests, but most of them are species of fruit trees including walnut, apple, tea, cherry, bayberry, and peach. These trees were usually planted in home yards, along contour lines and ridges of farm fields in a number of different forms of agroforestry. The trees are privately owned and covered by customary regulations.

8.5.2   Traditional Forest Regulations

Customary forest regulations have prevailed in many hilly and mountainous places of Northeast Asia. These are usually transmitted orally from generation to generation. The Buyi people of the Guntang village, Guizhou Province, China, are an example (Yuan and Liu 2009). These people observe community regulations such as no cutting of trees, no land reclamation, and no burning of the forest for ash. If a tree dies, nobody can make use of it, so that it decays naturally. Branches of the dead trees are not allowed to be collected even for fuel wood. This is to prevent people from intentionally killing the trees. Serious punishment is imposed for violation of these regulations. The rules also prescribe that trees in fengshui forests can be cut only for some collective purpose, such as for making special pole for a deceased person, but in this case, the pole must be kept in the village head's possession. Trees can be used for collective welfare on various special occasions. For example, several trees were cut and sold in 2003 for building the village road. Customary regulations are strongly and strictly enforced. Local villagers can easily identify any violation and once recognized steps are taken immediately to enforce the rule (Chen 1999; Weng et al. 2003). Weng et al. (2003) indicated that customary regulations contain forest management terms and facilitate self-governance in forest resource management. Whether it is in oral or written form, the villagers must strictly obey a regulation. But in some cases enforcement of local regulations for forest management still needs the support of the government, although the local government already recognizes and supports this kind of self-governance and self-management. As He and He (2000) pointed out, local regulations have been used for governing local forest resources by the central government and local governments, especially in the enforcement of local regulations. Actually, Guan (2002c) mentioned that the Song Dynasty also had the mandate to manage fengshui forests officially and issued official regulations to protect them.

The rules of the commons vary greatly depending on the place/location and time. In Japan until the twentieth century, the Nara-Osaka border had been a commons area for green manure, with open access for citizens of Osaka, Nara, and Kyoto. However, since the commons were open to citizens of a most populated and industrialized area, the landscape became almost bare rock mountains, and many species of grassland herbs and insects that lived there became endangered. Commons in Kizu at the south of Kyoto is an area of open access for the surrounding six villages, while at the same time some of the villages have their own individual village commons. Rules for collecting fuel wood are slightly different in each common. Sometimes villages located along river terrain have no commons of their own at all, so these villagers have to journey to open commons located in faraway places or have to buy green manure in the market. Rules on the utilization of forest litter and grasses vary in Japan from place to place. Until the eighteenth and early nineteenth centuries, many villages around Kyoto entitled the villagers to free access to private forests in the winter time to gather litter or fallen branches for fuel wood.

Prior to colonization by Japan in 1910, the forest policy of old Korean kingdoms maintained public ownership of forest and wetlands. Even though the entire forestland of Korea was proclaimed belonging to the king, some parts of forestlands were particularly distributed to royals and high officials. As commerce and the market economy developed in the seventeenth and eighteenth centuries, many of the forestlands were occupied by noble families and rich merchants through the establishment of private graveyards in the forests. As the private occupation became more and more apparent, less forestland was left from which poor villagers could collect fuel wood and timber for housing. The villagers formed a club, with representatives from every village household to manage a common forest, not privately occupied by powerful families. The community-based forest management club was called Song-gye, meaning 'club of pine forest,' which signifies the dominant tree species of the forestland at the time. Song-gye, which was based on a by-law of the community for managing the common forest, organized various activities such as controlling wildfires and illegal logging. It has played an important role in the formulation and implementation of forest policy for the rehabilitation of denuded forest landscapes in recent decades in South Korea (Chun and Tak 2009).

8.5.3   Traditional Cultural and Religious Practices Related to Forests

The psychological or spiritual effects of forests on human beings are reflected first in people's religious beliefs. Trees and forests are the foundations of various folk religions. By worshipping trees and forests, people attempted to draw comfort for their mental and physical weaknesses. 'Seo Nang Dang,' a Korean village shrine, is a place for the god protecting the village. It lies in the forest at the entrance to the village. It consists of a 'Dangzip,' where the god resides, and a holy tree called 'Seonang Namu,' the passageway for the god to descend from heaven. The god governs the villagers' fortune and prosperity, and for that reason each village holds a religious ritual at the start of the year, which is also a festival day for the village. Through these rituals, Seo Nang Dang worked to tie the villagers together spiritually. Such folk religions have faded away as they clashed with the modern world, and as new religions spread out and replaced them. Nonetheless, Seo Nang Dang is a definitive trace of a folk religion related to trees and forests. It represented forest cultures of the past, which played a critical role in determining the composition of the nature-cultural landscape of the countryside.

In Korea, 'Seongjushin' is the house guardian god who protects the house, and it is believed that he resides in a pine tree (*Pinus densiflora*). There is a legend saying that this god used to live in heaven, but after committing wrongful deeds he was sent to the world of humans. There, he found out that he had nowhere to live, so he prayed to the Heaven god, and the Heaven god allowed him to get seeds of a pine tree from a temple. He planted these seeds and then cut down the grown-up pine trees to build his house. That is why Seongjushin is believed to be living in pine trees. The Korean people perform a small ritual for Seongjushin by fixing in girders while building their houses. This legend signifies the importance of growing trees for the Korean people to be able to live in a comfortable house made of timber.

In traditional Chinese culture, the relationship between humans and nature has been described as a 'relationship between heaven and man' (Zhou 2006), requiring people to respect heaven and requiring harmony, integration, and unity to exist

Box 8.1 Animistic Traditions of the Yi Minority People Protect Forests

The Yi people usually regard forests and waters as the sole indicators of the wealth of their residences. They consider families who live in the barren hills to be poor and regard families that live in good forests to be rich. The typical house of a Yi family is made of wood and has wooden furniture. Utensils such as bowls and cups that are used on a daily basis are also made of wood. The objects used for local religious rituals such as Bimo, which is a typical religious ceremony of the Yi minority, are also made from wood. Traditionally, the Yi people collect a type of grass from the forest and use it for making a type of jacket. The Yi people in the Nanhua County of Yunnan Province still practise a local dance during which they wear the traditional jacket made of grasses.

The Yi nationality has a basic religion, which is concerned with nature (in particular mountains and forests), being close to nature, respecting the earth, and having a protective attitude towards the natural environment (Yang and Zhou 2007). The Yi people believe that every living thing has a god. They believe that there is a heaven god in heaven, an earth god on earth, a sun god in the sun, and a moon god in the moon. Similarly, mountains, trees, birds, wind, rain, and rivers are believed to have their own gods, which should be respected and worshipped. Based on the animism religion, the Yi people in Nanhua County have lived in harmony with nature for many generations; thus, the forests of the mountains have been sustainably managed by the Yi people.

God trees exist in each Yi ethnic village and are believed to protect their village. Most of these trees are hundreds of years old. The Yi people believe that these trees can provide safety, good fortune, and good harvests to their village. No one is allowed to cut down these trees and any person who does so will be punished. The Yi people place dead bodies in a cave near their villages, which is called an ancestor soul cave, which should be covered by well-protected forests. They believe these forests safeguard the soul of their ancestor. The Yi people also have a saying that a living man cannot be part of nature although he wishes to be. Only after passing will his soul become a part of nature.

between them. Academicians in recent decades documented the interrelationship among traditional culture, forest management, and biodiversity conservation. They disclosed that the local ethnic groups had developed very close interrelationships and interactions with local animals, plants, and forests, from which developed diversified indigenous knowledge systems and traditional cultural beliefs (Pei 2007). Due to the rich knowledge and practice of using and conserving animals, plants, and the ecosystems in which they dwell, these knowledge systems and traditional cultures have become the social foundations and technical support for mutual coordination and sustainable development of local inhabitants, local environments, and biodiversity (Xu and Liu 1995; Zhang 1995; Pei 1996; Rao 1996). Box 8.1 illustrates the beliefs of the Yi minority people related to forests. The Yi have many similarities with other ethnic groups who lived in the mountainous regions of Southwest China.

Among the long-enduring customs, there are many related to trees. For example, before New Year's Day (following the lunar calendar), people hang a branch of prickly castor-oil tree (*Kalopanax pictus*) in their main gate. The branch blocks harmful spirits from entering the house—mainly by its sharp thorns—and only lets good spirits in, so that on the New Year, only good fortune is admitted. There are other customs, such as throwing a pine branch on the roof before New Year's Day, burning a branch of juniper, and sticking pampas grass on a heap of ashes. All of these have the same interpretation: they bring plenty of good fortune over the New Year.

There was also a custom for prenatal care, where pregnant women expose themselves to the wind blowing through the pine forests. Another example relating to babies is planting a special tree whenever a new baby is born. When a male baby is born, they plant a pine tree, and when a female baby is born, they plant a paulownia. When the male baby grows up and finally gets married, they use the pine tree in building a new house. When he dies, they use the pine tree to build a coffin for him. In the case of the female baby, when she grows up and gets married, they use the paulownia for her wardrobe.

When a tree is cut, there is this other custom of having a ritual for the mountain god. This has two meanings: one is to protect people from any accidents by not making the god angry, and the other is to soothe the spirits of dead trees. These customs reflect valuable forest cultures and are still handed down to the present generation.

8.6   Loss of Traditional Forest-Related Knowledge

Traditional forest related knowledge has been eroding in Northeast Asia as the countries become industrialized and integrated with the rest of the world. Much evidence and research show that the loss of traditional forest-related knowledge has become serious (Youn 2009). In recent times, rapid economic growth geared towards privatization and globalization has hastened the loss of traditional forest-related knowledge (He and He 2000; Liu 2006).

8.6.1   Vulnerability of Traditional Forest-Related Knowledge to Rapid Social Changes

Traditional forest-related knowledge in Northeast Asia is in a state of transition as its practitioners are confronted with rapid economic and social changes and governmental policy interventions. In the case of China the vitality of traditional knowledge can be explained by two major factors. First is the tight and complex interrelationship between the livelihoods of the local people and the forests, cash revenue, self-subsistence, and integrated farming systems. Second is the synchronization of traditional forest-related knowledge into local beliefs, religion, and culture through religious activities, folklore, songs, temples, groves, and the like.

The traditional forest-related knowledge of Northeast Asia has become more vulnerable in recent years, because of two major factors. First, government policies and regulations, especially in China and Korea, neglect the importance of traditional forest-related knowledge. They even restrict access of local communities to forest resources in the hands of local people by nationalizing the forests and prohibiting traditional forest management practices by local people. Second, the expansion of globalizing market economies into previously self-sufficient rural areas, promoted by infrastructure development, and greater exposure to mass media, has led to a rapid erosion of traditional culture and traditional ecological knowledge and practices. This is particularly observed as the interests in traditional wisdom, knowledge, and lifestyles decline among younger generations.

8.6.2 Vulnerability to Outside Development Interventions

Biodiversity conservation in China and Korea has been designed based on: (1) the application of modern knowledge, technologies, and institutions; (2) establishment of new nature reserve administration; and (3) introduction of new crops and technologies for economic and social development of the nearby communities. Little emphasis has been given to the traditional knowledge, technologies, and grassroots institutional of the local ethnic groups, and their relevance to biodiversity protection and community development has often been ignored (Luo et al. 2009; Youn 2009).

To alleviate this problem, development efforts should be made to guide the local people to rediscover their traditional knowledge and biological resources in order to both develop the economy of the community and protect biodiversity. It is the local people who know their local resources best. Over time, they have developed a set of production systems that are adaptable to local natural conditions. They are able to make appropriate decisions suitable to their interests and situations. Therefore, we contend, the management of nature reserves should be based on the local knowledge of the community. The community should be endowed with the right of self determination, making decisions according to their recognition, values, and principles. Help from outside should never replace the local people in making decisions.

Baima Tibetans have no written word (He et al. 2004; Luo et al. 2009), so they pass their traditional knowledge from one generation to another, usually orally from the old people to the younger members of the village. In recent years, 85% of young generations of Baima Tibetans have left their villages to find jobs and live outside their communities. The number of people left behind is decreasing and many youngsters are losing even their own language while living outside their own home villages. Without being transmitted to the succeeding younger generations, the traditional knowledge of Baima Tibetans is vanishing.

8.7 Traditional Forest-Related Knowledge, Science, and Policy

8.7.1 Traditional Forest-Related Knowledge, Livelihoods of Forest-Dependent People, and Forest Resource Policy

There is an apparent lack of official recognition of traditional forest-related knowledge and traditional forest management in forest policy and legislations (Liu 2007a; Youn 2009). While governments give much emphasis to the protection of minority rights, and actively promote the expression of people's ethnic/cultural/religious identity, in actual policy formulation and implementation, minority people's traditional land-use practices are generally regarded as underdeveloped, primitive, and irreconcilable with the state's 'modern,' science-based bureaucratic and technological environment. An example can be seen in the Chinese forest policy directives contained in Article 22 of the Regulations for the Implementation of the Chinese Forest Law, enacted by the State Consul of the People Republic of China in 2000, which stipulates that 'Hillsides with a slope of over 25° shall be used for planting trees and grasses. Crop land on hillsides with a slope of over 25° shall, in light of the plan formulated by the local people's government, be withdrawn gradually from cultivation and be used for planting trees and grasses.' On the contrary, hillside farming on terraced fields has been practised by ethnic minorities for centuries without disastrous consequences, and can hardly be condemned per se as detrimental or unsustainable.

8.7.2 Integration of Traditional Forest-Related Knowledge with Formal Forest Science

Traditional knowledge has been given much recent attention in various fields, in particular among anthropologists, development experts, foresters, and local communities.

The academe has been working towards the integration of traditional forest-related knowledge with modern sciences, notably from ethno-botany, human ecology, development anthropology, and social forestry fields. The ethno-botanical knowledge held by minority people in Southwest China was investigated by Pei (2000) and his colleagues. Their research focused on: (1) traditional knowledge in botany; (2) qualitative evaluation of utilization and management of botanic resources; (3) laboratory testing of botanic resources used for commercial and subsistence purposes; and (4) exploring the commercial value of traditional knowledge of ecological and botanic resources. Pei (2000) and his colleagues have documented a great deal of knowledge on ethnic medicines, wild vegetables, edible plants, culture plants, ornamental plants, religion plants, and plants for daily subsistence.

In recent years, Korean and Chinese scholars have recognized the role of traditional cultures in reconstructing their society towards modernization (Song 1996; Zhou 2006; Chun and Tak 2009). The Chinese government is promoting the so-called ecological civilization in the new era of development. Song (1996) has zoned China's ecological culture based on indicators of ecological environments, local people's livelihood, and historic and state-of-art cultural elements. However, this classification is very limited in terms of forest management and development perspectives.

Anthropologists prefer to incorporate traditional knowledge within the scope of culture. Traditional forest-related knowledge has been viewed as knowledge that is practical, collective, and location-specific, integrating forest management practices with daily life of traditional community people. The burgeoning focus on indigenous knowledge is premised on the belief that many failures of development and underdevelopment are due to the priority being given to modern and scientific knowledge over local and traditional knowledge.

More and more evidence has shown that traditional forest-related knowledge is essential to maintaining the sustainability and complexity of natural ecosystems and providing tangible and intangible benefits for present and future generations. (Parrotta et al. 2009; Ramakrishnan et al. 2006; Ramakrishnan 2008; Youn 2009; Zuraida 2008; Nanjundaiah 2008; Camacho et al. 2011) Its roles in sustainable forest management, rural development, and continuation of local culture are irreplaceable. For this reason, research institutes and researchers have recommended learning from farmers and promoting equity of scientific technology and traditional knowledge. However, we are now facing another dilemma in which many poor farmers have been led to believe that shortages of advanced technology and capital are the main causes of their poverty. Farmers prefer to ask for assistance in terms of technology and capital and in the process tend to neglect the roles of their own traditional knowledge.

Since the early 1990s, community forestry or social forestry projects supported by foreign donors have been implemented in many places in China, in particular, in the southwest part of China where traditional forest-related knowledge is still relatively rich. Academics in this field regard traditional knowledge as a sort of people's knowledge that could play a significant role in the development of forests and in poverty alleviation. They have documented a large amount of traditional forest-related knowledge under their social forestry or community forestry programmes (Xu et al. 2004).

In China, governmental authorities and society believe that a lag of technology is prevailing and is seen as the main reason for less development. Thus the discourses of 'advanced technology' and 'modern science' dominate in contemporary forestry development, and these are embodied into the 'top-down' institutional regimes. Traditional forest-related knowledge has been marginalized in arenas of forestry norms, academe and research, and forest development practices on the ground (Liu 2006). In recent decades, Chinese academics have gathered and documented traditional forest-related knowledge using Western sciences and methodologies in order to disseminate localized traditional forest-related knowledge in a coherent and systematic way, which has been widely accepted as the so called 'scientific way.'

The rich pool of knowledge connected with forest resources and held in traditional practices and beliefs has been rather neglected by decision makers and scientists in Northeast Asia. It is hard to find any subject dealing with traditional ecological knowledge in the forest education curriculum. There is a need for further understanding of traditional forest-related knowledge by forest decision makers and scientists. It is imperative that incorporation of traditional knowledge with forestry education and research on forest-related problems be started.

8.8   Conclusion

8.8.1   Importance of Traditional Forest-Related Knowledge in Sustainable Forest Management

Many ethnic people in China and elsewhere in Northeast Asia have their own knowledge on forests, inherited from their ancestors and innovated over generations with continuous knowledge and policy intervention from outside (He and He 2000; Liu 2007a). This can be illustrated from every element of forest utilization, protection, and management, such as categorization of forests; tenure arrangements; practices of use rights for collecting fuel wood, mushrooms, and pine needles; and application of customary laws. Traditional knowledge, the roots of local knowledge systems and practices, provide the basis of efficiency in forest management and protection, and equity for resource access and benefits sharing. Traditional forest-related knowledge has demonstrated its significance in the protection of forest ecosystems at large and some tree species in particular, and in improving the value of forests through sustainable management. In many cases in China, Japan, and Korea they are still acknowledged by the people and applied in their daily life (Parrotta et al. 2009).

Forest-related cultures are a major consideration for the inclusion of traditional forest-related knowledge in sustainable forest management, since traditional forest-related knowledge is embedded in culture. Culture can be innovated, shifted, and even grafted onto or integrated with other cultures. Taking fengshui as an example, Yuan and Liu (2009), after conducting case studies at the Buyi minority of Guizhou province, China, concluded that the fengshui forests and their associated collective knowledge are an integral component of the rural life of the Buyi ethnic people. Their traditional forest-related knowledge is based on long historical experiences and their religion or beliefs in the dynamic interrelationship between people and nature. It has been passed on through generations in the form of legends, folklores, and festivals. Fengshui forests, maeulsoop, and satoyama are of special economic, ecological, cultural and social, and in some cases spiritual significance to local and ethnic communities and the individual household members. As in the case of Buyi people of China, the fengshui forests and the magic meaning attached to them could be considered as integral components in maintaining the distinctive cultural identities of the ethnic group. Fengshui forests, maeulsoop, and satoyama provide village as a whole and the individuals within it with a venue where they can exercise harmonization between human and nature, as well as among the members of the community. Moreover, fengshui forests, maeulsoop, and satoyama represent a symbolic meaning for the collective action of a community. It is noted that although there are similarities among the three common landscapes, the values (culture, spiritual, and religious) and ways to assess fengshui forests, maeulsoop, and satoyama differ with places and over time. Examination of fengshui forests and maeulsoop in China and Korea reveals that there are two common tendencies. With the intervention of scientific knowledge and Western culture in recent decades, fewer people are acknowledging the spiritual or religious functions of fengshui forests and maeulsoop in China and Korea. Instead, the functions of environmental protection, biodiversity conservation, recreational, or aesthetic services and culture heritage have been taking the place.

As the world recognized the role of satoyama in the conservation of biodiversity and the importance of traditional knowledge systems associated with it, the traditional knowledge related to forests like fengshui forests in China, maeulsoop in Korea, and satoyama in Japan could provide forestry profession with a sound basis for sustainable forest management. We believe that traditional and scientific knowledge systems can work hand-in-hand towards sustainable management of forests. In recent years the governments in this region started to recognize the potential roles of traditional forest land-use systems such as fengshui, maeulsoop, and sotoyama for watershed management, biological resources management, and ecotourism based regional development.

We can conclude that knowledge and perspectives of fengshui, maeulsoop, or satoyama forests can be very inclusive and can be merged with other religions and/or modern Western systematic knowledge. At the same time, these forests can be managed for diversified products and environmental services. It has been integrated into the Northeast Asian culture and as part of local culture. It is expected that fengshui, maeulsoop or satoyama forests will continue to be an important component of Northeast Asian rural landscape and culture.

8.8.2 Strategic Recommendations for Forestry Development

Firstly, recognizing the role of local community, especially ethnic groups, in conserving traditional forest-related knowledge, new initiatives in forestry development—such as community empowerment, capacity building, legislation, and financial incentives in support of local communities—need to be given more attention by academics and governments.

Secondly, only when an appreciation of the normative framework, evaluative standards, conceptual theory, and methodology for the protection and utilization of traditional knowledge has been established will traditional knowledge be in a negotiation position to share an equal and complementary role with scientific knowledge in development planning and decision-making. Suitable regulations should be stipulated to protect and acknowledge the local people who created and inherited traditional forest-related knowledge, to ensure that they participate in decision-making process and receive fair and equitable benefits for their contributions.

Thirdly, the integration of scientific and traditional forestry-related knowledge should be promoted. 'Transformation' and 'actor' are the core essential terminologies in development anthropology. We believe that culture can evolve and adapt. Weak cultures might be invaded by strong cultures. 'Localization' is always accompanied by 'modernization.' The history of scientific forestry in Northeast Asia is about a 100 years old, a short time compared to the long history of civilization in this region. We should not totally abandon the colourful, effective, and meaningful traditional forest-related knowledge, which had been practised for many generations, sometimes centuries or millennia. 'Grafting' and inheriting traditional knowledge with scientific knowledge is the optimal selection of forestry development in Northeast Asia.

References

Bailey RG (1998) Ecoregions: the ecosystem geography of the oceans and continents. Springer, New York

Biodiversity Committee of the Chinese Academy of Sciences [BCCAS] (1992) Biodiversity in China: status and conservation needs. Science Press, Beijing

Camacho LD, Combalicer MS, Youn Y-C, Combalicer EA, Carandang AP, Camacho SC, de Luna CC, Rebugio LL (2011) Traditional forest conservation knowledge/technologies in the Cordillera, Northern Philippines. Forest Policy and Economics. doi:10.1016/j.forpol.2010.06.001 (In press)

Chao G, Xue J et al (1994) Social forestry. Yunnan Science and Technology Publishing House, Kunming

Chen DS (1999) Research on the comprehension of community-based natural resources management in the mountainous areas of Guizhou. China Popul Resour Environ 9(4):22–26

Chun YW, Tak K-I (2009) Songye, a traditional knowledge system for sustainable forest management in Choson Dynasty of Korea. For Ecol Manag 257:2022–2026

Collard IF, Foley RA (2002) Latitudinal patterns and environmental determinants of recent human cultural diversity: do humans follow biogeographical rules? Evol Ecol Res 4(3):371–383

Convention on Biological Diversity [CBD] (2001) Global biodiversity outlook. Secretariat of the Convention on Biological Diversity, Montreal

Eearon JD (2003) Ethnic and cultural diversity by country. J Econ Growth 8:195–222

Food and Agriculture Organisation (FAO) (2001) Global forest resource assessment, 2000. FAO forestry paper 140, FAO, Rome

Forest for Life (2008) Ten years of forest movement by the Forest for Life. Seoul (In Korean)

Guan CY (2002a) Ancient fengshui forest. Agric Archaeol 3:28–33

Guan CY (2002b) Fengshui forest, afforestation and tree management in ancient China. J Wanxi Univ 1:65–68

Guan CY (2002c) Ancient Fengshui Forest and afforestation. Root Search 3:28–33, In Chinese

He PK, He J (2000) Local regulations and forest management in ethnic region. In: He PK et al (eds) Forests, trees and minorities. Yunnan Nationality Press, Kunming, pp 113–122, In Chinese

He P, He J, Wu X (2004) Practices and exploration of indigenous knowledge. Yunan Ethnic Publishing House, Kunming, pp 23–38, In Chinese

Jiang Z (ed) (2008) Modern forestry in China. China'sese Forestry Publishing House, Beijing, pp 48–52, In Chinese

Kang CK (1990) A historical study on fruits in Korea. Korean J Dietary Cult 5(3):301–312, In Korean

Kim HB, Jang D-S (1994) Maulsoop, the Korean village grove. Yeolhwadang Publishing Co., Seoul, In Korean

Kim IA, Youn Y-C (2005) Conditions for sustainable community forests: evidence from 17 village groves in Korea. In: Innes JI, Edwards IK, Wilford DJ (eds) Forests in the balance: linking tradition and technology; abstracts of the XXII IUFRO World Congress, Brisbanem, 8–13 Aug 2005. International Forestry Review 7(5):221

Lee D (2008) A Korean perspective of traditional knowledge forest-related knowledge. In: Proceedings of the 1st international conference on forest related traditional knowledge and culture in Asia, Korea Forest Research Institute, Oct 5–10, 2008. Korea Forest Research Institute, Seoul, pp 3–5

Lee D, Park C-R (2009) Modern implications of traditional ecological knowledge and practices in Korea: a sustainability perspective. In: Ministry of Culture, Sports and Tourism (ed) Aian eco-culture. Minsokwon, Seoul, pp 61–80

Lee DK, Suh SJ (2005) Forest restoration and rehabilitation in Republic of Korea. In: Stanturf JA, Madsen P (eds) Restoration of boreal and temperate forests. CRC Press, Boca Raton, pp 383–396

Lee D, Koh I, Park CR (2007) Ecosystem services of traditional village groves in Korea. Seoul National University Press, Seoul

Liao Y, Chen C, Chen H et al (2008) The community characters and plant diversity of fengshui woods in Liantang country, Guangzhou. Environ Ecol 17(2):812–817, In Chinese

Liu J (1996) China's agroforestry profile. In: Koppelman R, Lai CK, Durst PB, Naewboonien J (eds) Asia-Pacific agroforestry profiles, APAN field document no. 4. RAP Publication, Bangkok, pp 79–121

Liu J (2006) Forests in the mist. Dissertation, Wageningen University and Research Centre, Wageningen

Liu J (2007a) Contextualizing forestry discourse and normative framework of sustainable forest management in contemporary China. Int For Rev 9(2):653–660

Liu J (2007b) Traditional knowledge and its implication to development in China—development anthropological perspectives. J China Agric Univ 2:133–141

Liu J, Zhao L (2009) Have decentralization and privatization contributed to sustainable forestry management and poverty alleviation in China? FAO forestry policy and institutions working paper 23, Rome

Liu J, Zhang D, Zhang M (2007a) Community participation in forestry management. China's Forestry Publishing House, Beijing, pp 37–42, In Chinese

Liu SS, Lu HR, Ye YC et al (2007b) Green heritage—briefing introduction of fengshui forests in Gongguan Profecture of Guangdong Province. Guangdong Landscape Archit 29:17–18

Long CL, Abe T, Wang H, Li ML, Yan HM, Zhou YL (1999) Biodiversity management and utilization in the context of traditional culture of Jinuo society in S Yunnan, China. Acta Botanica Yunnanica 21(2):239. doi:CNKI:SUN:YOKE.0.1999-02-017, In Chinese

Lundgren BO (1990) ICRAF into 1990s. Agrofor Today 2(4):14–16

Luo P, Pei SJ, Xu JC (2001) Sacred sites and its implications in environmental and biodiversity conservation in Yunnan China. J Mt Sci 19(4):327–333, In Chinese

Luo Y, Liu J, Zhang D (2009) Role of traditional beliefs of Baima Tibetans in biodiversity conservation in China. For Ecol Manag 257:1995–2001

Ministry for Nature and the Environment of Mongolia [MNEM] (1998) Biological diversity in Mongolia. Ministry for Nature and the Environment of Mongolia, Ulaanbaatar

Ministry of Forestry of China (1994) Long-term research on China's forest ecosystems. Northeast Forestry University Press, Harbin

Nanjundaiah C (2008) The economic value of indigenous environmental knowledge in ensuring sustainable livelihood needs and protecting local ecological services: a case study of Nagarhole

National Park, India. In: Parrotta JA, Liu J, Sim H-C (eds) Sustainable forest management and poverty alleviation: roles of traditional forest-related knowledge, vol 21, IUFRO World Series. International Union of Forest Research Organizations (IUFRO), Vienna, pp 124–126

Nature Conservation Bureau [NCB] (2007) Our lives in the web of life: the third national biodiversity strategy of Japan 2007. Government of Japan, Ministry of the Environment, Nature Conservation Bureau, Tokyo, 23 p

Park J-C (1999) A study on the village groves in Chinan-Gun region, Korea. J Korean Soc Rural Plann 5(1):56–65

Park J-C (2000) Relationship among village, village grove of Jinan-Eub Region and prospected landscape about Mt. Mai. J Korean Plann Assoc 35(5):309–316

Park J-C, Lee S-H (2007) Maeulsoop in Jinan. Jinan Mun Hwa Won (Cultural Foundation of Jinan County). Jinan, Jeonbuk, In Korean

Park C-R, Shin JH, Lee D (2006) Biboosoop: a unique Korean biotope for cavity nesting birds. J Ecol Field Biol 29(2):75–84

Parrotta JA, Lim HF, Liu J, Ramakrishnan PS, Youn Y-C (2009) Traditional forest-related knowledge and sustainable forest management in Asia. For Ecol Manag 257:1987–1988

Pei SJ (1996) A tentative study on function of ethnic botany in development of botanic gardens (IV). Yunnan University, Kunming

Pei SJ (2000) Modernization of traditional medicines and continuation of ethnic medicines. J Chin Ethnic Med 1:1–3, In Chinese

Pei SJ (2007) Local people's traditional knowledge: an important tool for nature reserve. Man Biosph 5:65

Qian H, Song J-S, Krestov P, Guo Q, Wu Z, Shen X, Guo X (2003) Large-scale phytogeographical patterns in East Asia in relation to latitudinal and climatic gradients. J Biogeogr 30:129–141

Ramakrishnan PS (2008) Knowledge systems: the basis for sustainable forestry and linked food security in the Asian region. In: Parrotta JA, Liu J, Sim H-C (eds) Sustainable forest management and poverty alleviation: roles of traditional forest-related knowledge, vol 21, IUFRO World Series. International Union of Forest Research Organizations (IUFRO), Vienna, pp 150–152

Ramakrishnan PS, Saxena KG, Rao KS (2006) Shifting agriculture and sustainable development of north-eastern India: tradition in transition. Oxford and IBH Publishing Co., New Delhi

Rao RR (1996) Traditional knowledge and sustainable development: key role of ethnobiologists. Ethnobotany 8:14–24

Republic of Korea (ROK) (2009) Fourth national report to the UN convention on biological diversity. Government of the Republic of Korea, Seoul

Sakuma D (2008) Trace the historic elements of satoyama vegetation. Agric Hortic 83(1):183–189

Shidei T (1973) Value of forests. Kokin, Tokyo, In Japanese

Shin JH (2002) Ecosystem geography of Korea. In: Lee D (ed) Ecology of Korea. Bumwoo Publishing Company, Seoul, pp 19–46

Song S (1996) The relationship between anthropological study and ecological environments and culture in the minority region. J Central Univ Nationalities 4:62–67, In Chinese

Tabata H (ed) (1997) Nature in satoyama. Hoikusha, Osaka, In Japanese

United Nations Environment Programme [UNEP] (1995) Global biodiversity assessment. Cambridge University Press, Cambridge

Weng ZW, Zhu J, Wen QL (2003) Cunguimingyue and forest resource conservation—report from three Buyi villages in Duyun, Guizhou. For Soc 5:23–27

Xu ZF, Liu HM (1995) Relations of Dai Group's traditional knowledge of plants and sustainable botanic diversity. China Science and Technology Press, Beijing, pp 10–12

Xu JC, An D, Qian J (eds) (2004) Dynamics of community based resource management in Southwest China. Yunnan Science and Technology Publishing House, Kunming, pp 19–22, In Chinese

Yang GR (1999) Traditional forestry heritage—history of fengshui forest. For Econ 6:60–63

Yang F, Zhou H (2007) Forum of Chuxiong ethnic culture. Series no. 1, Yunnanz University Publishing House, Kunming, China, pp 1–3 (In Chinese)

Yang CM, Liang J, Pan GJ (2004) Traditional culture and sustainable forest management of Miao people. In: He PK, He J, Wu X (eds) Practice and exploration of indigenous knowledge 20–29. Yunnan Minority Press, Kunming, pp 18–24

Yang YH, Su KM, Wang ZH (2006) Discussion about strategies for management and protection of wild mushroom resources in Chuxiong Prefecture. J West China For Sci 4:154–158, In Chinese

Youn Y-C (2009) Use of forest resources, traditional forest-related knowledge and livelihood of forest dependent communities: cases in South Korea. For Ecol Manag 257:2027–2034

Youn Y-C, Lee E, Yun J, Choi S (2010) Ownership and management status of village groves in Korea: the case of five municipalities around the Jirisan Mountains. In: Proceedings of the 3rd international conference on forest related traditional knowledge and culture in Asia, Research Institute of Nature and Humanity, Kanazawa, 14–15 Dec 2010, pp 68–71

Yuan J, Liu J (2009) Fengshui forest management by the Buyi ethnic minority in China. For Ecol Manag 257:2002–2009

Zhang XS (1995) Views on biodiversity. In: Yingqian Q et al (eds) Progress on biodiversity study. China Science and Technology Press, Beijing, pp 129–138

Zhong ZQ, Boris DC (2007) Fengshui—a systematic research of vernacular sustainable development. In: Ancient China and its lessons for the future: proceedings, 7th annual general meeting, UK-Chinese Association of Resources and Environment (UK CARE), Greenwich, 15 Sept 2007

Zhou S (2006) Forest in China—historic transitions and industry developments. Thomson, Singapore, p 19

Zhu GW (2007) Managing village and fengshui forest to formulate a green and new village. J Chin Urban For 5(6):53–55

Zuraida (2008) Conservation and utilization of traditional forest-related knowledge for achieving sustainable livelihoods of local communities—a case study in Jambi Province, Indonesia. In: Parrotta JA, Liu J, Sim H-C (eds) Sustainable forest management and poverty alleviation: roles of traditional forest-related knowledge, vol 21, IUFRO World Series. International Union of Forest Research Organizations (IUFRO), Vienna, pp 210–212

Chapter 9
South Asia

P.S. Ramakrishnan, K.S. Rao, U.M. Chandrashekara, N. Chhetri, H.K. Gupta, S. Patnaik, K.G. Saxena, and E. Sharma

Abstract Forests of the South Asian region, including major 'hotspots' of biodiversity, have been sustainably managed for generations by ethnically and culturally diverse traditional societies. The rich traditional forest-related knowledge possessed by the traditional societies in the region is closely linked to cultural diversity as well as to biodiversity in all its scalar dimensions (i.e., genetic, species, ecosystem, and landscape diversity). This knowledge, generated through an experiential process, has ensured sustainability of diverse forested ecosystems as well as livelihoods of forest-dependent communities. In recent times this knowledge base has been severely eroded, due in large part to deforestation and associated land degradation, processes

P.S. Ramakrishnan (✉) • K.G. Saxena
School of Environmental Sciences, Jawaharlal Nehru University, New Delhi, India
e-mail: psr@mail.jnu.ac.ac.in; kgsaxena@mail.jnu.ac.in

K.S. Rao
Department of Botany, University of Delhi, Delhi, India
e-mail: srkottapalli@yahoo.com

U.M. Chandrashekara
Kerala Forest Research Institute, Nilambur Campus, Chandrakunnu,
Malappuram, Kerala, India
e-mail: umchandra@rediffmail.com

N. Chhetri • E. Sharma
International Centre for Integrated Mountain Development,
Kathmandu, Nepal
e-mail: chettrin@rediffmail.com; icimod@icimod.org.np

H.K. Gupta
Forest Survey of India North Zone, Shimla, India
e-mail: hemantgup@gmail.com

S. Patnaik
UNESCO Regional Office, New Delhi, India
e-mail: s.patnaik@unesco.org

J.A. Parrotta and R.L. Trosper (eds.), *Traditional Forest-Related Knowledge:*
Sustaining Communities, Ecosystems and Biocultural Diversity,
World Forests 12, DOI 10.1007/978-94-007-2144-9_9,
© Springer Science+Business Media B.V. (outside the USA) 2012

triggered by forces external to traditional socio-ecological systems. Successful efforts have been made towards conserving traditional forest-related knowledge and linking it with formal scientific forest knowledge to develop 'hybrid technologies' relevant to sustainable forest management. To facilitate this process, it has been helpful to elucidate broad, generalizable, principles of traditional forest-related knowledge, rather than viewing this knowledge stream as 'local knowledge.' One such key principle that has contributed towards community participation in sustainable forest management initiatives relates to socially valued species that typically have important ecological keystone values. The protected 'sacred groves' that are abundant in the region are important learning sites both for understanding ecosystem dynamics and as a resource base for sustainable forest management practices. This is the context in which emerging institutional arrangements in the South Asian region, such as community forestry, joint forest management, and forest user groups are to be seen.

Keywords Biodiversity • Collaborative forest management • Cultural landscapes • Forest history • India • Nepal • Non-timber forest products • Sacred groves • Shifting cultivation • Traditional knowledge

9.1   Introduction

South Asia—which includes Afghanistan, Bangladesh, Bhutan, India, Maldives, Nepal, Pakistan, and Sri Lanka—is the home of more than 20% of the world's population and is renowned for its rich cultural diversity. The region's high population growth, along with its rapidly expanding industrial economy, is placing increasing demands on an already stressed natural resource base, including its forests. World Bank (2005) estimates suggest that by 2050, South Asia's population is likely to exceed 2.2 billion; an estimated 600 million people live below the poverty line (on less than US$ 1.25 a day) in 2000, a substantial proportion of those being located in the forested areas of the region. Ongoing deforestation in the region, and the expected impacts of climate change on forest ecosystems and agroforest landscapes, will be felt most directly by the already impoverished sectors of society, particularly those in traditional communities living in the forested areas.

The population of forest dwellers in South Asian region is estimated to be between 120 and 150 million, of whom up to 90 million (World Bank 2005) constitute traditional societies generally referred to in the region as 'tribals.' In India alone are 427 distinct ethnic groups officially designated as 'scheduled tribes' comprising 7.8% of the country's total population according to 1981 national census data (Dube et al. 1998). Largely confined to hilly terrain and mountains, the population densities of forest-dwelling communities are generally very low, and particularly so in Bhutan. As an integral part of the forested landscape, they depend on local natural resources for their livelihood needs. In addition to these forest-dwelling communities, another 350–400 million people are directly dependent on forests for products and various ecological services (Poffenberger 2000).

The South Asia region is characterized by a rich cultural diversity. According to the Anthropological Survey of India there are 91 eco-cultural zones inhabited by 4,635 major communities, speaking more than 1,500 languages or dialects in India alone.[1] This includes the 645 officially recognized scheduled tribes mentioned above. Thousands of endogamous groups and sub-sects are structured around the Hindu caste system, which also contributes to the extraordinary cultural diversity of India that is indicative of the richness of the region as a whole (Dube et al. 1998). The majority of those who live in forested areas still hold and follow a variety of nature-based religious beliefs and practices, even if they are adherents of other religious belief systems such as Hinduism, Islam, Buddhism, Sikhism, and Christianity. The richness in cultural diversity in the forested areas is matched by an equally diverse range of subsistence-based natural resource use practices, including hunting, gathering, shifting cultivation, traditional settled farming, and nomadic herding. As a result of 'modernization' and many decades of external social, economic, and political influences, energy-subsidized and intensive agricultural practices such as plantation crops are now common, especially in areas closer to urban centres.

The region's cultural diversity extends throughout the broad range of eco-climatic zones found in the region, from the humid tropical evergreen rainforests, sub-tropical deciduous forests and dry desert scrub jungles at lower elevations, to sub-temperate, temperate forests, alpine scrub jungle, and meadows at higher elevations. Such an eco-culturally diverse environment has contributed towards the development of rich bodies of traditional forest-related knowledge, often referred to as 'local knowledge,' that forest dwellers continue to value. This knowledge base is now gaining more and more recognition and credibility as the basis for expanding community involvement in forest conservation and management in the region.

Given that policy planners and developmental agencies emphasize regional planning processes for conservation-linked management of natural resources for livelihood development in rural societies, one cannot afford to look at small location-specific socio-ecological systems as isolated entities. In other words, for a viable regional planning process towards conserving and indeed sustainably managing forests, it becomes necessary to arrive at generalized principles regarding traditional forest-related knowledge, so that this knowledge and associated practices may be effectively linked with those arising from formal scientific forest knowledge application on a regional scale. This is the context for the following discussion on the history and current status of forests and forest management in the South Asian region. In this discussion we will consider diverse pathways that may be used to link the two knowledge streams for sustainable management of forests and associated agricultural systems, with concern for the food security of all stakeholders and of the forest-dwelling communities in particular.

Linking knowledge systems as the basis for sustainable forest management is indeed complex, particularly since traditional forest-related knowledge, unlike

---

[1] The 1991 Indian census recognized 1,596 'mother tongues,' of which 114 had more than 10,000 speakers.

formal scientific forest knowledge, has a certain degree of location-specificity and includes socio-ecological and socio-cultural dimensions with tangible and intangible values. Linking knowledge systems at the socio-ecological process level poses additional challenges, although the South Asian region has had some experience with this since the early 1970s in the context of sustainable forest management in the shifting agricultural landscapes (Ramakrishnan 1992a). In this chapter, we will highlight both traditional forest-related knowledge (TFRK) and associated agricultural system management systems based on that knowledge, examples of generalized formal scientific principles that can be learned from these traditional practices, and selected examples of how traditional knowledge may be combined with formal scientific knowledge in 'hybrid technology' formulations for effective community-based sustainable forest management.

9.2 Forests of South Asia: The Socio-Ecological Context

9.2.1 Regional Overview—Forest Ecosystem Diversity and Extent

The countries of South Asia are endowed with rich, though shrinking, forests resources. According to FAO statistics, the total forest area in the region was estimated to be 79.2 million ha, or 2.0% of the world's total (FAO 2009) (Fig. 9.1). The region's remaining forests are concentrated in the Himalayan region from northern Pakistan and north-western India through Nepal and Bhutan to Arunachel Pradesh (in India), and in other higher elevation areas of northeastern India and southeastern Bangladesh, central and eastern regions of India, the mountainous Western Ghats of India from western Maharashtra to southern Kerala and Tamilnadu, and in Sri Lanka.

The complexity and diversity of the region's geologic history, soils, topography, and climate have given rise to a tremendous diversity of tropical, subtropical, temperate, and sub-alpine forest ecosystems of exceptionally high plant and animal species diversity (Champion and Seth 1968; Stainton 1972; de Rosayro 1950). A variety of tropical forest types (tropical wet evergreen, semi-evergreen, and moist deciduous) are found at varying elevations in Sri Lanka; in the Andaman Islands, and India's Western Ghats and northeastern states; and in Bangladesh. Extensive (though rapidly disappearing) mangrove forests are found in the coastal Sundarbans of Bangladesh and West Bengal (India) and in isolated patches on both the eastern and western coasts of India, and in Pakistan. Tropical dry deciduous, thorn, and dry evergreen forests occur in the sub-Himalayan regions of Pakistan and in central and southern peninsular India. Subtropical broadleaf, pine (coniferous), and dry evergreen forests occur at middle elevations in the western and central Himalayas from Pakistan eastwards through northern India, Nepal, and Bhutan; and in isolated locations in northeastern India, Rajasthan, and southeastern India. Wet, moist, and dry temperate broadleaved and evergreen (coniferous) forests are common at higher

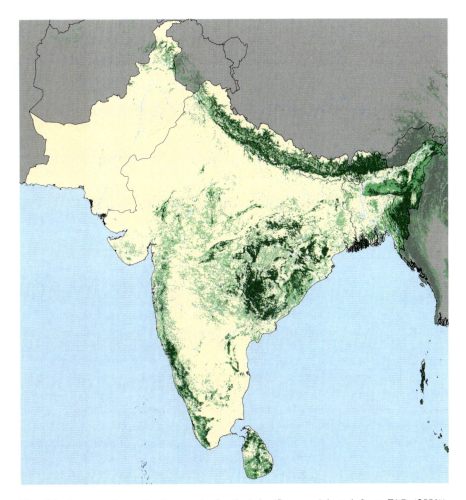

Fig. 9.1 Forest and woodland cover in South Asia (Source: Adapted from FAO (2001)). Key: *Dark green* closed forest, *light green* open or fragmented forest, *pale green* other wooded land, *yellow* = other land

elevations of the Himalayan region from northern Pakistan southeastwards through northern India, Nepal, and Bhutan to high elevation sites in Arunachel Pradesh (India). Finally, subalpine and alpine forests occur in the Himalayan region at elevations above the temperate forest zone.

According to FAO estimates for the period 1990–2000 (FAO 2001), the region as a whole has had a negative rate (0.13% per annum) of forest cover change. Forest cover in Bhutan and Maldives remained roughly the same during this period; it increased in Bangladesh and India but decreased in Nepal, Pakistan, and Sri Lanka. Plantation programmes contributed to the increase in forest cover within the region, the preferred species being teak *(Tectona grandis),* Indian rosewood *(Dalbergia sissoo),* and/or exotic Eucalyptus. While forest cover loss has generally slowed in many countries during the past 20 years in comparison with periods of very rapid deforestation during the nineteenth and particularly the twentieth centuries, the estimated annual losses remain high (1.3–1.9%) in some countries, notably Nepal, Pakistan, and Sri Lanka. In any case, the forest cover in the region, according to more recent assessments of FAO (2009), ranges from a low of 2.5 and 6.7% of the total land area in Pakistan and Bangladesh, respectively, to a high of 68% in the Himalayan kingdom of Bhutan, with India (22.8%), Nepal (25.4%) and Sri Lanka (29.9%) lying between these extremes. Per capita forest area in the region varies widely among countries, from 0.006 to 0.012 ha in Bangladesh and Pakistan to 4.93 ha in sparsely populated Bhutan (Table 9.1).

9.2.2   The Present Status of Forests

While most of the region's forests have been used and managed for a variety of purposes (such as wild foods and other non-timber forest products, shifting agriculture, forest grazing of livestock, and timber extraction) for centuries, if not millennia, the intensification of these uses during recent generations has resulted in extensive loss of forest cover (Chokkalingam et al. 2000), and expansion of degraded forests dominated by weedy species, including invasive alien species that now cover vast tracts of land (Ramakrishnan 1991).

Regardless of their condition, the region's forested landscapes are the home of most of the region's 'traditional societies.' In India alone these communities represent a substantial segment of the population, and include communities who are classified by the government as 'tribal,' as well as many other communities who are not included in this political category but nonetheless live as an integral part of the forested mountain ecosystems of the country. The near-term subsistence livelihood needs of these marginalized traditional societies are under constant threat given their direct or indirect dependence upon forest resources. Recognition of this reality has important implications for sustainable forest management in the region. Given the long predominance of secondary forest formations in the region, much of the forest-related knowledge possessed by traditional societies is largely derived from their long experience in the management of these secondary forests. A comparative analysis of shifting agricultural forested landscapes conducted in northeast Indian, where fallow cycles range from 5 to 60 years, is illustrative of this (Ramakrishnan 1992a).

9.2.3   The History of Forest Degradation and Its Consequences

The natural vegetation of South Asia has been subject to human disturbances of varying intensity during the thousands of years of recorded history in this region. From the Vedic period (ca. 1500 BC) through the until the early sixteenth century, human population densities and rates of natural resource exploitation appear to

Table 9.1 Forest cover change in South Asian region (1990–2000)

Country	Land area 000 ha	Forest area in 2000		Total forest			Forest area change 1990–2000	
		Natural forest 000 ha	Forest plantation 000 ha	000 ha	%	ha/ capita	000 ha/ year	%
Bangladesh	13,017	709	625	1,334	10.2	n.s.	+17	+1.3
Bhutan	4,701	2,995	21	3,016	64.2	1.5	n.s.	n.s.
India	297,319	31,535	32,578	64,113	21.6	0.1	+38	+0.1
Maldives	30	1	–	1	3.3	n.s.	n.s.	n.s.
Nepal	14,300	3,767	133	3,900	27.3	0.2	–78	–1.8
Pakistan	77,087	1,381	980	2,361	3.1	n.s.	–39	–1.5
Sri Lanka	6,463	1,625	316	1,940	30.0	0.1	–35	–1.6
Total	**412,917**	**42,013**	**34,652**	**76,665**	**18.6**	**0.1**	**–98**	**–0.1**

Source: FAO (2001). n.s. = not significant, indicates a very small value

have been relatively low compared to today (Erdosy 1998; Allchin 1998). While available information on the ecological history and particularly pre-colonial deforestation in South Asia is limited, studies examining the period from 500 BC to 1500, and particularly from 1600 to 1760, suggest that on the Indian subcontinent there were periods of rapid forest loss (Gadgil and Guha 1992). Extensive deforestation took place in the Indus and Ganges river basins and in semi-arid regions during very early periods, and further forest loss appears to have been taken place during periods of military expansion of pre-colonial kingdoms (Filiozat 1980; Lal 1985; Grove 1995).

Starting in the late eighteenth century, intensive commercial exploitation of teak, sal (*Shorea robusta*), and other valuable tropical timbers was initiated to supply European and, to a lesser extent, local markets for shipbuilding materials. The struggle for colonial domination of India between England and France in the late eighteenth century, which shifted from Bengal to the Malabar coast during this period, was spurred by competition for control of peninsular India's forest resources to meet the needs of these countries' shipbuilding industries following the decimation of their domestic forests and loss of their North American colonies (Grove 1995). In the case of the Terai forested region of Nepal, however, during this period, exploitation of forest wealth remained centred around a few ruling elites (Malla 2001; Gautam et al. 2004).

During the nineteenth century, the pace of deforestation accelerated first with the expanding influence of the East India Company over India's forest resources and later under direct British colonial rule (over present-day India, Pakistan, and Bangladesh), when administrative and land-use policies were established that usurped traditional land tenure arrangements and long-established forest utilization practices. These new policies resulted in centralized government control over the majority of India's forests to promote logging, develop roads and railroad systems, and expand India's permanent agricultural land base. Despite the introduction of modern (European) forest management practices intended to balance forest protection and commercial production in the late 1800s, India's forest area steadily declined during the first half of the twentieth century because of official disregard of timber cutting limits established for reserved and protection forests, illegal logging, and pressures from traditional forest users (Gadgil and Guha 1992). In Nepal, unlike India, a centralized forest service was not established until the 1950s. Consequently, local Nepalese communities lost control over the management of their forests for a much shorter period during the latter half of the twentieth century than was the case in India.

India's rapidly growing population and the demands of its agricultural and industrial development plans since 1947 have further eroded India's forest wealth. Expansion of permanent agriculture, often into marginally productive forest areas; construction of new roads and hydroelectric facilities; expansion of coal, iron, and bauxite mines; and the cumulative daily pressure of millions of rural communities that rely on a shrinking forest resource base for fuel, fodder, grazing land, and innumerable non-wood forest products have all contributed to reductions in both the extent and quality of the country's remaining forests.

In the development of forest policies in the South Asian region, local communities living within the forests have often been blamed by government agencies, usually unfairly, for mismanagement, over-exploitation, and degradation of forest resources. Consequently official perspectives on sustainable forest management are often based on incorrect premises. For example, extensive research in northeastern India showing that farmers practising shifting cultivation are not the primary causative agents for deforestation and land degradation, suggesting that a very different perspective is needed to evaluate sustainable forest management practices and policies in shifting agricultural landscapes (Ramakrishnan 1992a, 2008a).

During the colonial period, and subsequently, the dominant approach of state forestry authorities was to promote plantation forestry in areas where the natural forest cover had been destroyed. In the central Himalayan region, for example, large-scale wood extraction of timber from natural mixed oak forests began in the 1840s and 1850s, arising from increased demand for timber by government agencies for developmental works in the plains regions. With increasing state control, large-scale timber extraction from natural forests continued until the early 1920s (Dangwal 2005). Over time, extensive oak-dominated mixed forests in the central Himalayan region were converted into pine plantations and/or secondary forest formations (Sinha 2002). Deforestation in the Chittagong hill tracts of southeastern Bangladesh, as well as in as in the northeastern hill areas of India, has resulted in expansion of secondary successional bamboo forest formations in areas used for shifting agriculture (commonly referred to as *jhum* in this region (Ramakrishnan 1992a)).

The assertion that governmental policies and market pressures have been the primary drivers of deforestation and land degradation was reinforced by the results of a comprehensive set of Tri-Academy-sponsored studies involving India, China, and the United States (Wolman et al. 2001). The conclusions of these studies were supported by a global analysis done on land use and land cover change (Lambin et al. 2001). These studies strongly suggest that communities living in the forested areas are, at worst, proximal drivers of land use/land cover change only at the local level.

Human impacts on forested ecosystems of varying intensities and frequencies have occurred for centuries, and considerably longer, in most of the South Asia region. It is therefore not surprising that much of the traditional forest-related knowledge possessed by tribal and other communities who depend on forests for their livelihoods is connected with the extensive, diverse, secondary forest formations found in the region, since people in these communities have long managed such forests and their biodiversity for meeting their survival needs (Ramakrishnan 2008a, b). Much of the traditional knowledge that local communities possess tends to be largely centred around socially valued early-successional tree species, typically fast-growing species with narrow crown forms (Boojh and Ramakrishnan 1982a, b; Shukla and Ramakrishnan 1986) that allow good light penetration beneath their shade for sustaining traditional agriculture on which forest dwellers depend for their food security.

9.3 History of Traditional Forest-Related Knowledge

Traditional societies of South Asia practise a wide variety of natural resource (including forest) use systems that have evolved over centuries, adapting to diverse and often changing local environmental conditions. While the origins of traditional forest-related knowledge in South Asia certainly predate the first settled agricultural communities that appeared in the region about 10,000 years ago, the *documented* history of traditional forest-related knowledge in South Asia can be broadly considered in three phases: (i) the pre-colonial era from the Vedic period (ca. 1500–500 BC) until the end of the Muslim period in the mid-nineteenth century, (ii) the colonial era of the British Empire (1864–1947), and (iii) the contemporary era since the mid-twentieth century.

9.3.1 The Pre-Colonial Era (1500 BC to Mid-Nineteenth Century)

While there is evidence of human activity of Homo sapiens as long as 75,000 years ago, little is known about the forest utilization practices of peoples who inhabited the South Asia region prior to the Vedic period (ca. 1500–500 BC), ushered in by the mainly pastoralist Aryan tribes who migrated into the region from Central and Western Asia, presumably during the last centuries of the Indus Valley Civilization (3300–1300 BC). The Vedic Civilization, which eventually exerted control over much of the Gangetic plain, provided the earliest records of how people of the region viewed forests. According to *Vedas* (the earliest Sanskrit literature and oldest scriptures of Hinduism), forests were worshipped as Vanaspate, literally meaning 'Oh Lord of the Forest' (Vannucci 1993).

Throughout the pre-colonial phase, clear-cut ethnic boundaries remained in a state of flux, with geographical movement occurring over long distances, particularly among shifting agriculturists or communities practising nomadic pastoralism or transhumance (Sivaramakrishnan 2009). Even small endogamous groups generally did not occupy the same location for long periods of times and did not have exclusive delineated natural resource access and use rights (Guha 1999; Cederlof and Sivaramakrishnan 2006).

With the emergence of Buddhist and Jain cultural identities, and their political dominance from the sixth century BC onwards, rulers of kingdoms and empires during this early period enforced measures to protect animal life through protection of their forest habitats. Generally speaking, consistent with Buddhist and Hindu religious traditions, forests during this phase were commonly viewed as spiritual abodes, and special efforts were made to delineate specific forest areas to provide protection to wildlife (Karan 1963; Thapar 2001; Rangarajan 2002; Cederlof and Sivaramakrishnan 2006).

Between the late seventeenth and mid-nineteenth centuries, extensive forest areas under the control of Muslim rulers (Mughals) were protected as royal hunting preserves, which had implications for wildlife conservation. While villagers practising settled agriculture in forested areas during this period were granted usufruct rights, access to the forests was controlled by the community leadership. This situation existed throughout much of India as well as in Bangladesh, Nepal, Pakistan, Sri Lanka, and Bhutan.

In general, the rural landscape of the Indian sub-continental region prior to the eighteenth century included large areas of secondary forests with discrete and discontinuous patches of old-growth forests, and savannahs. The population density in India during most of this period is estimated to be relatively low (35 persons/ km^2), increasing to about 70 persons/ km^2 by 1881 (Guha 2000). On the basis of available descriptive information, Trivedi (1998) estimated that during the Mughal period (1526–1850), forest cover may have been somewhat more than 50% in the more populated Indo-Gangetic plains of the Indian sub-continent, and conjectures that in other parts of India and other regions of South Asia it may have exceeded 60% of the land area. Richards et al. (1985) suggests that in the last quarter of nineteenth century, about 33% of India was cultivated, 33% was forested, and about 20% of land was grassland or savannah.

Commercial agriculture did not emerge until the beginning of the nineteenth century (Ludden 2002; Rangarajan 2002), when intensified traditional agricultural practices were promoted, based on well-developed tank (reservoir) and canal irrigation networks that supported diversified farming systems for cereals, vegetables, oil seeds, and other crops (Ludden 2002), and a network of production centres and distribution networks. The increased concentration of political authority and land-based economic activities during this period suggests that vast tracts of land were temporarily depopulated or only sporadically used by nomadic groups, and that extensive farming resulted in net increases in patchy wooded areas well into the nineteenth century (Sivaramakrishnan 2009).

While most regions of South Asia followed the patterns discussed above, the case of Bhutan, though analogous in many respects, also differs from the perspective of land ownership and user rights. The unified kingdom of Bhutan came into existence only in the early seventeenth century, when the lama Ngawanag Namgyal defeated three Tibetan invasions, subjugated rival religious schools, and established himself as ruler (Shabdrung) over a unified ecclesiastical and civil administration system. Prior to this time the country was dominated by a number of 'valley'-based kingdoms ruled by hereditary kings, chiefs, or lamas (Karan 1963; Aris 1979). After Ngawanag Namgyal's death, divisions within the kingdom eroded the power of the shabdrung until 1885, when Ugyen Wangchuck once again consolidated power.

9.3.2 The Colonial Era

Early British colonial land management policies viewed forests as both a source of raw materials and areas for agricultural expansion. The advent of railways, specifically the increased demand for timber for their construction, dramatically changed forest utilization patterns in the region. Railway requirements for wood and selective harvesting of valuable timber like *Shorea robusta*, for example, degraded the entire northern hills and forested plains (Richards and Tucker 1989).

During the nineteenth century centralized forest management was seen by the British imperial rulers as a necessity, which led to the creation of the Imperial Forest Department in 1864. The British rulers during this period also asserted state control over remaining forested lands, and assigned ownership of all cultivated lands falling within the otherwise forested landscape. British control and management of forests from the mid-nineteenth century until 1947 in most parts of present-day India, Pakistan, and Bangladesh, marked a break from the pre-existing patterns of forest control and management. The creation of a separate forest service by the British colonial government, and related legal measures initiated by it during this period, conferred significant powers to the newly constituted forest departments. This led to the creation of separate 'reserved' and 'protected' categories of forests, supported by formal science-based silvicultural 'working plans' for each of the management units of state-owned forests.

Subsequent to the creation of a state forest service, forest dwellers were used as labour to clear the forest, and allowed to cultivate in the cleared land and take care of the introduced teak saplings, enabling the locals to cultivate the land from seedling growth till the canopy closed. They were then moved to another forested site to start the process all over—the taungya system of forest management, which could be viewed as a compromise approach towards shifting agriculture, prevalent at the time. The forest department reasserted control over and reforested the land with tree plantations of high timber value such as teak. In the context of state-sponsored forestry, the taungya system of community participation has to be seen as an important milestone in the recognition of community forest rights.

While the British are generally credited with initiation of the taungya system of community participatory forest management, it appears that this practice pre-dated their arrival. Local Karens from the Tonze forest of northern Burma had already been raising teak as part of their traditional fallow management practice, which perhaps was adopted by the British when they gained control of these lands (Blanford 1958; Gadgil and Guha 1992). This same taungya system of forestry is practised by the local ethnic minorities living in the high mountain areas of Yunnan (in China) that border Arunachal Pradesh State in India. In Yunnan, these practices have existed for at least three centuries before the British came into the region, suggesting that this knowledge is indeed, traditional (Menzies 1998). Unlike the British, who emphasized teak plantations, the Chinese ethnic minorities cultivate a multipurpose species preferred by the community, *Cunninghamia lanceolata*, for rehabilitating *Imperata cylindrica*-infested grasslands. This traditional wisdom of the local people was based on a combination of ecological, social, and economic considerations. What is implied in this traditional approach to taungya is that tree selection should be based on community values. More recent attempts made in the Darjeeling Himalaya by foresters to introduce tree species of high timber value, in an attempt to revive the taungya system, have not succeeded (Shankar et al. 1998).

The rural populations that had previously depended on forests and managed them in a more-or-less sustainable manner experienced an erosion of their traditional forest use rights, and contributed to the over-exploitation of forests now owned by the colonial government. This, combined with excessive harvesting of timber by the state, led to widespread forest degradation.

9.3.3 The Contemporary Era

The post-colonial governments (after 1947) maintained the forest policies created by the British. It is estimated that in more recent times, deforestation rates have been between 1% and 3% per year in Bangladesh, Pakistan, Nepal, and Sri Lanka. In India alone the rate of deforestation was estimated to be about one million ha per year between 1970s and 1980s (Poffenberger 2000).

These drastic changes in forest ownership, control, and management, begun by the British and continued by post-colonial governments in the region, also led to large-scale displacement of forest-dependant people and controversial conversions of natural forests to monocultural plantations of commercial tree species, as in the case of pine plantations in Bastar in central India. These policies and practices have sparked much public debate on state forest policies, particularly in India (Gadgil and Guha 1992).

With much of the natural forests of the region either converted to permanent agriculture or otherwise degraded to varying degrees, the 'biosphere reserve' concept—developed in the 1970s through UNESCO's Man and the Biosphere Programme (Batisse 1982)—has emerged as a useful model for integrating conservation of biological and cultural diversity while promoting sustainable economic and social development of cultural landscapes based on traditional values, local community efforts ,and conservation science, particularly in the 'buffer zones' of protected areas (Ramakrishnan et al. 2002). Government agencies responsible for forest conservation in all countries of the region are becoming increasingly aware that the forest management principles that are valid for 'buffer zone' management of nature reserves could be the basis for managing other forest areas that are not part of protected area networks. Leading this process were India's policy changes that resulted in 'joint forest management' initiatives (Gupta 2006), to be discussed further in Sect. 9.6.

There is also a growing realization in the region amongst all forest stakeholders, particularly environmental activists, that humans are an integral part of the cultural landscapes they have been created over generations, and that accepting this reality could effectively address sustainable conservation of what still remains of the region's biodiversity 'hotspots.' Building bridges between 'nature' and 'culture,' particularly in situations where traditional societies live, has emerged as one of the priorities arising from recent debates over conservation. Indeed, there is a rapidly developing environmental activist movement to conserve and/or recreate 'nature' around human settlements, even in urban centres (Ramakrishnan 2008a, b, Shutkin 2000). In these efforts, non-governmental environmental activists in the region, both national and international, are playing key roles in creating broader public environmental awareness. Of late, there has also been increasing efforts towards rediscovering and conserving the traditional institutional arrangements that have contributed, or could contribute, towards the integrity of forested cultural landscapes (Maurer and Höll 2003).

9.4   Conservation of Forested Cultural Landscapes Through Traditional Forest-Related Knowledge and Practices

Traditional local institutions that governed resource use in the pre-colonial and colonial eras can still be found in many locations. These include, for example, the Kipat system of forest management of the Rai and Limbu communities of Nepal, the Shingo Nua system of the Sherpas in the Everest Himalayan region of Nepal, and the Sokshing (woodlot for collection of leaf litter) system of the Bhutanese (Wangchuk 1998). All of these are participatory approaches for non-timber forest resource use by local communities, though governmental agencies retain decision-making powers regarding timber production and harvest. However, many non-governmental activist groups are working towards a better integration of non-timber forest product-related issues (which have been delegated to forest dwellers) with timber production-related issues that are pursued by the foresters. The following case studies are illustrative of the forest dwellers' viewpoints in which the concept of 'forested cultural landscape' bridges 'nature' and 'culture,' a perspective that is now serving as a basis for community participatory sustainable forest management, moving beyond the conservation vs development 'dilemma', and towards conservation linked sustainable development.

9.4.1   Conserving the Demojong Cultural Landscape in Sikkim Himalaya

The Demojong is a sacred landscape located in the west Sikkim Himalaya. Connected to the Buddhist Tibetan belief system, it has well-defined boundaries, and is supported by a traditional institutional system. It is a landscape rich in biodiversity, extending from the snow clad peak of Khangchendzonga through Alpine meadows interspersed with rhododendron scrub jungle, to conifer and mixed evergreen forests, and finally into sub-tropical rainforests (Ramakrishnan 2003). There are a number of glacial lakes in the alpine zone that feed the sacred river Rathong Chu that runs through the landscape (Fig. 9.2). The soil, the water, the biota, the visible water bodies, the river, and the less obvious notional lakes on the river bed, along with many monasteries and temples, are all held sacred by local communities of diverse cultural backgrounds who inhabit this watershed and manage a variety of traditional land use systems.

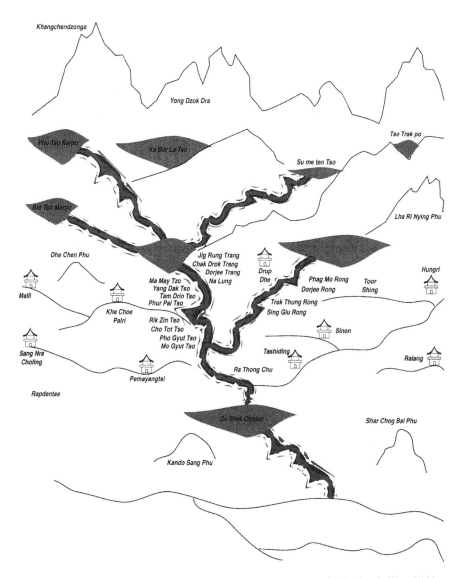

Fig. 9.2 Demojong, the land of hidden treasures. Pictorial depiction of holy sites in West Sikkim, Eastern Himalaya, with the sacred river system Rothong Chu running down the slope (*shaded thick line*); many sacred lakes (*shaded patches*) and monasteries are to be found scattered all over the landscape

The sacred nature of the landscapes is defined by Tibetan Buddhist philosophy (Box 9.1), which defines the sacred boundaries and identifies specific cultural and institutional norms and arrangements that often are codified to delimit human usage. The air, soil, water, and biota are all considered sacred. Alterations of the landscape

Box 9.1 Demojong, the Land of the Hidden Treasures! (Ramakrishnan 2003)

Padmasambhava, who is highly revered and worshipped by the Sikkimese Buddhists, is considered to have blessed Yoksum and the surrounding landscape in West Sikkim District. It is believed that a large number of hidden treasures (ter) were hidden by Lhabstsun Namkha Jigme in the Yoksum region, and that these sacred treasures are being discovered slowly and will be revealed only to enlightened Lamas, at appropriate times. It is said that the last such discovery was made by Terton Padma Lingpa 540 years ago. Conserving these treasures and protecting them from polluting influences is considered important for human welfare.

The area below Mount Khangchendzonga in West Sikkim, referred to as demojong, is the core of the sacred land of Sikkim. Yoksum is considered to be a lhakhang (altar) and mandala where offerings are made offerings to the protective deities. No meaningful performances of Buddhist rituals are possible if this land and water is desecrated. Any large-scale human-induced perturbation in the land of the holy Yoksum region would destroy the hidden treasures, the ters, in such a manner that the chances of recovering them sometimes in the future by a visionary will be diminished. Further, any major perturbation to the river system would disturb the ruling deities of the 109 hidden lakes of the river, leading to serious calamities, a past example of this being that of Lake Khecho-Palri, which is believed to have moved away from the river during a period of bloodshed.

Indeed, the very cultural fabric of Sikkimese society is dependent upon the conservation of the entire sacred landscape of interacting ecosystems. It is not merely a question of protecting a few physical structures or ruins. The uniqueness of this heritage site is that the value system of the people here is interpreted in a more holistic sense—the soil, the water, the biota, the visible water bodies, the river, and the less obvious hidden lakes on the river bed, are all to be protected along with the physical monuments.

are restricted, circumscribed by cultural norms, and the guiding principles for resource use are strictly enforced through social institutions. Broad community participation is ensured through a variety of rituals that involve the diverse communities living within the landscape boundaries, each of which have their own pre-determined rights for natural resource use. While these and many other sacred landscapes have use restrictions, traditional societies live as an integral part of it; they are involved in a variety of agriculture and animal husbandry practices and extraction of resources such as fuelwood, fodder, food, and medicinal species from the natural ecosystems, both fresh water and forest. In such landscapes, natural ecosystems are closely linked to the village ecosystem functions in a sustainable manner.

Social institutions do allow small ecosystem alterations, whilst larger destabilizing alterations are prohibited. For this reason there was a strong social reaction when government agencies attempted to initiate a hydro-electric project within the Demajong landscape region during the late 1990s. The project was abandoned within a few years after the developmental initiative was initiated, in response to community pressure, and the government agencies have taken steps to conserve this cultural landscape. Pant et al. (1996) studied land cover/land use changes in this area between 1960 and 1985 in this area using aerial photographs and Landsat TM false colour composite image (spectral bands 2, 3 and 4) of the respective years. The study revealed a total change of 27%, consisting of changes/conversion from one cover type/land use pattern to another out of the total area. It is significant to note, however, that between 1985 and 1998, adverse or negative changes were only 1.52%.

This case study is also indicative of the intensity with which traditional societies hold on to the intangible values that they cherish, even at the cost of their lives. Further, it shows that the human security angle of 'knowledge systems' in general and traditional knowledge in particular is very important at the local and regional levels (Ramakrishnan 2009), and raises important questions about the role that traditional forest-related knowledge could have in enhancing global human security under threat in this era of global environmental change (Brauch et al. 2009).

9.4.2 The Chipko Movement in the Central Himalayan Garhwal Region

Oak forests (dominated by *Quercus* spp.) in the Central Himalayan region are socially valued, including for their ecological keystone role in sustaining soil fertility management by providing high-quality leaf litter, and their dense root systems that help to conserve water in the soil profile (this dense root mat is often compared to the thick matted hair of the god Shiva of Hindu mythology, who is said to have used his matted hair to control the force of the mighty river Ganga (Ganges) as it descended from heaven!). Selective harvesting of these forests and their replacement with pine plantations in India's central Himalaya region inspired the grassroots movement in the 1970s known as Chipko, which gained global publicity (Box 9.2). More than 1,500 km to the east, in the northeast Indian State of Arunachal Pradesh, oaks are also socially valued. This parallelism that exists in social values centred around oaks in two geographically separated parts of the country is indicative of social selection operating independently and coming to similar conclusions, suggesting that what is socially valued invariably tends to have ecological value (Ramakrishnan et al. 1998). As will be discussed below in Sect. 9.7.1, this principle was important for initiating a community participatory conservation and developmental programme in Nagaland in the 1990s.

Box 9.2 The Story of the Chipko Movement

The 'Sarvodaya' (peoples' enlightenment) movement begun by the great Indian leader Mahatma Gandhi was the philosophical basis for the Chipko movement in the central Himalayan Garhwal region led by Gandhian workers Sundarlal Bahuguna and Chandi Prasad Bhatt, who were involved in community participatory social forestry activities in the early 1970s (Weber 1987).

For years tensions had been mounting between local communities and timber contractors coming into the region from outside, centred around harvest of timber from the natural oak-dominated forests. This conflict came into the open in the early 1970s when local women, led by Bahuguna and Bhatt, organized to prevent the impending harvest of oak trees by hugging them, the implication being that foresters would need to 'cut us down before cutting the trees.' This direct community action became known as the Chipko (meaning "to stick" in Hindi) movement, a protest against the exploitation of forest resources that had been ongoing in the region since the British colonial period that ended in 1947.

The communities living in the area perceive the oak species of their forests to be sacred. This value system is expressed by them through folk-music, documented in *Nanda Devi Prakriti Samrakshan Geet* ('folk-music linked with nature conservation'), a product of a UNESCO initiative (Ramakrishnan et al. 2000), and in numerous local poems and folk-tales also woven around oaks, locally called banjh.

In this region, oak litter is a highly valued resource that is used to maintain soil fertility in traditional settled agricultural systems; this organic litter from these keystone species has always been the mainstay for the sustainability of agriculture in the mid-elevations of Garhwal region. With extensive deforestation of mixed oak forests and their replacement by pine plantations, sustaining soil fertility and water in the soil profile became a major problem.

The general belief that was propagated when the Chipko movement got underway was that the local communities wanted to share in the benefits accruing to the timber merchants. However, the fundamental underlying reason for this movement was the erosion of traditional agricultural systems resulting from large-scale land degradation that has been going on for several decades in the region, arising from the conversion of biodiversity-rich mixed oak forests into pine plantations (Ramakrishnan 2008a; Ramakrishnan et al. 2000).

9.4.3 Sacred Khejri (Prosopis cineraria) Forests of the Bishnois in Rajasthan (India)

The Bishnois, who inhabit the desert region of Rajasthan in northwestern India, are a people whose cultural values include the belief in the absolute protection of all life. For the past 500 years they have managed their surrounding cultural (dry forest) landscapes consistent with these values. The Bishnoi sect was founded in 1486 by the spiritual leader Saint Jambeshwarji, whose name refers to the 29 ethical vows to which the Bishnois adhere, of which non-violence is an important one. Their concept of protecting nature arose in the year 1730, when 363 Bishnoi men, women, and children of the village Peepsar laid down their lives to protect their local forests when the then Prince of Jodhpur came to the area to hunt and to harvest trees to fire lime kilns. With the founding of this new sect under their spiritual leader, natural resource management practices, rules, and regulations that were consistent with their ethical principals were institutionalized.

The dominant forest species in this region is khejri (*Prosopis cinerarea*), a legume tree used for a variety of purposes. This species, revered by the community, is valued for its edible pods (for food), leaves (as livestock fodder and green manure), and branches (for fuelwood and construction materials). All parts of the tree are of medicinal value to the local people. A nitrogen-fixing species, it improves soil quality through its production of nutrient-rich leaf litter. It forms a deep taproot penetrating to more than 30 m depth; its extensive root mass helps to stabilize the sandy desert soil and shifting sand dunes, and as a windbreak it protects the rain-fed farmlands from strong desert winds. Khejri is also an eco-cultural keystone species; its presence endowes the landscape with a distinctive cultural identity and provides habitat for a diverse assemblage of associated plant and animal species that play important functional roles in the village ecosystem. Indeed, the Bishnoi villages stand out as islands of biodiversity in an otherwise over-exploited desert landscape. The sacred groves of the Bishnois, called orans, are widely known in the Indian region for their conservation value for khejri and its associated wildlife, particularly the blackbuck (*Antelope cervicapra*). The value system of the Bishnois, expressed in the management of khejri forests through social institutions that determine resource use, has resulted in the creation of these productive, species-rich cultural landscapes wherever a Bishnoi village, or cluster of villages, has been located.

In stark contrast to these small, sacred, islands of biodiversity created by the Bishnoi's traditional knowledge and wisdom, vegetation of the larger arid landscape is now dominated by the closely related, exotic, tree species *Prosopis juliflora,* which forms an extensive single-species stands over large tracts of land. This tree, known as mesquite in its native Central and South American range, was introduced to Rajasthan (and other dry locations in India and Pakistan) from Australia by foresters beginning in the late nineteenth century, primarily for rehabilitating the otherwise largely degraded dry and arid landscapes. This species has spread far beyond where it was originally planted, its movement and regeneration greatly enhanced by livestock, which relish the seed pods and thereby facilitate the spread and germination of seeds. For the traditional farmer, *P. cineraria* is an important agroforestry species, non-invasive, with great value for soil fertility management and food security, unlike the invasive exotic *P. juliflora*, which is generally regarded as a weed in the farmers' plots, which they attempt to control by uprooting and burning. Managing invasive alien tree species such as *P. juliflora* is a problem that often defies scientific solutions (Drake et al. 1988), and is major challenge for sustainable management of forested areas in many tropical areas (Ramakrishnan 1991).

9.4.4 Beyul Khumbu: The Sherpas in the Sagarmatha (Mount Everest) National Park, Nepal

The case study of the indigenous Sherpa people living in the buffer zone area of Sagarmatha (Mount Everest) illustrates how intangible values have contributed towards conservation and sustainable management of this national park in Nepal (Spoon and Sherpa 2008). The Sherpas consider this landscape—the beyul (the sacred hidden valley set aside by the progenitor of Tibetan Buddhism, Guru Rinpoche)—to be a refuge for people living in the area and for sustainable use in times of need. Refraining from negative actions that are inconsistent with Buddhist philosophy—which prohibits harming any living beings, plants or animals (including humans)—the Sherpas maintain their beliefs in numerous location-specific spiritual values and taboos that ensure sustainable practices within the beyul landscape. These include protector deities and spirits associated with natural features, such as mountains, trees, rocks, and water bodies. Since the area is now a protected area, rapidly increasing tourism has been a source of income for the Sherpas, and has promoted development of local infrastructure.

9.4.5 Traditional Agricultural Systems as an Integral Part of the Forested Landscape

Traditional societies living in forested landscape have developed a range of agricultural systems that are managed at low intensities relative to more 'conventional' permanent cropping systems. They view their traditional agricultural practices as integral parts of the cultural landscape that they treasure, reflected in the intangible values associated with their land use practices. Such values are intimately connected with the food security concerns that they must deal with in their forested environment. Uncertain food security within many such communities is exacerbated by the losing battle they have to face with developmental agencies' continual efforts to convert them to modern settled farming practices, despite their repeated rejection arising partly from the threat that such changes would pose to their very cultural identities (Ramakrishnan 2008b). These cultural values motivate their continued reliance on traditional agricultural practices. There are numerous examples of culture-linked practices and beliefs that are connected to the livelihood concerns of shifting agricultural and other traditional communities of the region, examples of which are highlighted in Box 9.3.

Learning from such examples of societies who maintain their livelihoods and cultural identities while working with nature (to provide tangible benefits linked with intangible values) is increasingly seen as an effective and viable basis for addressing forest sustainability concerns involving local communities. This desire to 'get back to nature' can be seen even in urban communities in the region (and elsewhere in the world), if only symbolically, through public interest and promotion of urban forestry and even urban agriculture (Shutkin 2000).

Box 9.3 Examples of Spiritual Beliefs and Practices Associated
with Shifting Cultivation and Conservation Practices in South Asia
and Himalayan Region

- The tradition of maintaining a sacred grove for each village, with a variety
  of religious ceremonies performed within the groves during the year to
  propitiate natural elements before initiating slash-and-burn agriculture, is
  indicative of the sacredness attached to them by different ethnic groups of
  northeast India. While many of these traditions are eroding because of
  modern influences on these societies, many remain well-protected, for
  cxample the Mawsmai sacred grove in Cherrapunji and Mawphlong grove
  nearer to Shillong in Meghalaya, India.
- In the Garo hills of Meghalaya (India), the first two Garos (the tribal group
  inhabiting this area) to initiate shifting agriculture, locally called jhum, are
  believed to be the spiritual couple Bone Nirepa and Jane Nitepa. Their bless-
  ings as well as that of their deity, Misipa, are sought for a good harvest.
- The Wanchos of Arunachal Pradesh (India), like many other ethnic groups
  in the region, traditionally sacrifice cocks, pigs, buffalos, and even the
  socially treasured domesticated gaur (*Bons frontalis*), known locally as
  mithun, to propitiate the spirits of nature on different occasions to sustain
  soil fertility and ensure good crop yields.
- The Baigas of Madhya Pradesh (India), who practise shifting cultivation, view
  the use of a plough to prepare their agricultural fields as tearing the breast of
  mother earth. They therefore prefer to directly sow seeds after clearing and
  burning secondary forest vegetation from their fields, without ploughing.
- The Buddhist Dai (T'ai) tribe of Xishuangbanna in Yunnan Province of
  China, bordering the northeastern Indian region, where shifting agricultural is also practised, have many holy hills, Nong Ban and Nong Meng,
  belonging to a village or village clusters. These holy sites are spread over
  a large area, and include hundreds of small or large forested reserves and/
  or designated sacred woodlands.
- Sri Pada (also known as Adams Peak) in Sri Lanka, a biodiversity-rich
  forested landscape, is considered by the Buddhists, Christians, Hindus,
  and Muslims as a place of worship and is protected (Wijesuriya 2001).
- For the Kanis, a hunter-gatherer society living in the Agasthyamalai hill
  region of the Western Ghat mountains in southern India, specific ecosystems, rock shelters, marshy swamps, and large trees with huge buttresses
  are considered abodes of their local spirits of worship, the mountain as a
  whole being the abode of the supreme God, Agasthyamuni, who is revered
  as an ancient sage of wisdom.
- The folk-music, festivals, and associated cultural and spiritual values of
  local communities living in the Nanda Devi cultural landscape (a UNESCO
  world heritage site located in the central Himalayan Garhwal mountain
  region in India) offers opportunities also for learning lessons on sustainable management of the larger Biosphere Reserve itself.

9.5   Integrating Traditional Forest-Related Knowledge with Formal Scientific Forest Knowledge

Conventional forestry in the region has often promoted monoculture plantation forestry as a replacement for natural forest management. In pursuing this, the most convenient and silviculturally well-known species have been promoted, with an exclusive focus on those valued for timber. In the central Himalayan region, for example, extensive plantations have been established of *Pinus roxburghii* at higher elevation sites, and various species of (non-native) *Eucalyptus* and *Acacia* at lower elevations. Replacement of natural mixed-species natural forests by plantations has been very often deeply resented by local communities, and has inevitably caused conflicts between forest managers and communities, such as those discussed earlier with respect to the Chipko movement. In recent times, fire events in pine plantations have been on the rise. These conflicts, in our view, strengthen the rationale for appropriate integration of traditional and formal scientific forest-related knowledge as a means to more effectively address the social, economic, and cultural dimensions of sustainable forest management (Ramakrishnan 1992b).

While governmental policies and market pressures since the colonial period have promoted exploitation of forest resources (Wolman et al. 2001), the blame for deforestation and land degradation has typically been placed on forest dwellers, in particular those practising shifting agriculture (Ramakrishnan 1992a; Ramakrishnan et al. 2006). Traditional forest dwellers have repeatedly rejected development pathways offered to them that are not based on a value system that they understand and appreciate. This has led to their marginalization and has created conflicts between them and government development agencies, which has had implications for peace at local and regional levels (Ramakrishnan 2009). In this context an examination of the linkages and potential synergies between traditional and formal scientific forest knowledge is particularly relevant.

9.5.1   *Ethnobotanical Knowledge: Traditional Medicine and Lesser-Known Species of Food Value*

The South Asian region is richly endowed both with extraordinarily rich biodiversity and the traditional forest-related knowledge related to its use to help meet the daily food and health security of people. Ethnobotany—the study of how people use plants for food, medicine, and in ritual contexts; how they view and understand them; and their cultural, symbolic, and spiritual roles—is an area of traditional forest-related knowledge that has received considerable attention by scholars and scientists in the South Asia region for the past 500 years, and increasing attention in the scientific literature over the past century. European studies of South Asia's flora, the source of the overwhelming majority of drugs used in the ancient traditional systems of medicine in the region—Ayurveda,

Siddha, Tibetan, and Unani (Van Alphen and Aris 1995)—can be traced to Garcia da Orta's 1563 *Coloquios dos simples e drogas he cousas medicinais da India* (Markham 1913) and van Rheede's 12-volume *Hortus malabaricus* in the late seventeenth century (Rheede tot Draakestein 1678–1703). Knowledge of traditional Indian medicine, and the service of its local practitioners, was critical to the survival of early European traders and colonists in India given the near total ignorance of tropical diseases and their treatment by European physicians (Parrotta 2001).

The documentation of traditional forest-related knowledge in the Indian subcontinent (particularly in present-day India, Pakistan, Bangladesh, and Sri Lanka) during the British colonial period was extensive, and built on the large pre-existing body of knowledge recorded in Sanskrit, Pali, Tamil, and other ancient South Asian languages. A great deal of information derived from or highly relevant to traditional forest-related knowledge was included in forest floristic surveys and forest management research conducted under the auspices of the British colonial government in the late nineteenth and early twentieth centuries (c.f. Hooker 1875–1897; Troup 1921). Such information was also found in works on economic and medicinal botany such as those by Ainslie (1826), Watt (1889–1896), Dymock et al. (1890), Nadkarni (1908), and Kirtikar and Basu (1935).

More recently, a large and expanding literature dealing with the traditional utilization of the region's flora by local and tribal communities has developed, including: major works by government organizations such as the Indian Council for Scientific and Industrial Research (cf. CSIR 1948–1992, 1986) and individual scientists (cf. Jain 1991, 1997; Saklani and Jain 1994; Maheshwari 1996; Manandhar 2002), as well as journal articles published in dozens of peer-review journals such as Ethnobotany, Economic Botany, Journal of Ethnobiology , and the Indian Journal of Traditional Knowledge, among others. In an effort to systematically document, and strengthen intellectual property protection of, traditional knowledge related to the use of plants in Indian systems of medicine (Ayurveda, Unani, Siddha, and Yoga), the Indian Government has created an extensive online database, the Traditional Knowledge Digital Library, available online since 2009. The database includes thousands of traditional drug formulations recorded in classical texts involving 308 plant species.[2] At the local level in India, 'people's biodiversity registers'—consisting of records of individual people's knowledge of biodiversity, its use, trade, and efforts for its conservation and sustainable utilization—have been established; these registers are recognized in the Indian Biological Diversity Bill of 2000 (Hansen and Van Vleet 2007). Throughout the region, numerous research and grassroots development programmes have been initiated in recent years to promote the cultivation and processing of locally (and in some cases internationally) valued medicinal plant species as well as their sustainable management in natural forests where their over-exploitation is a serious concern (Parrotta 2002).

---

[2] Available online at: http://www.tkdl.res.in/tkdl/langdefault/common/home.asp?GL=Eng

In light of the ongoing degradation and loss of forests (and their biodiversity) in the region, ethnobotanical aspects of traditional forest-related knowledge have gained recognition and importance in recent years. This interest is often focused on the conservation of species of medicinal value and 'lesser-known' wild food species from forests (CSIR 1948–1992; National Academy of Sciences 1975, 1979; Bodekar et al. 1997; Valiathan 1998; Parrotta 2001). In order to maximize the potential of the region's biodiversity and related traditional knowledge for human well-being, particularly for forest-dependent communities, in-situ conservation of these resources should be an essential element of sustainable forest management planning.

9.5.2   Socially Valued Tree Species of Ecological Keystone Value

Formal (scientific) silvicultural knowledge of individual tree species, while important, should not be the sole criterion in tree species selection for sustainable forest management activities. Rather, such knowledge needs to be complemented with knowledge of the social and religious values attached to tree species so as to enhance community support for and participation in sustainable forest management. As discussed earlier, socially valued species typically have ecological keystone roles. For example, in shifting agricultural landscapes of northeastern India, favoured nitrogen-fixing trees such as *Alnus nepalensis* and many bamboo species—such as *Dendrocalamus hamiltoni, Bambusa tulda,* and *B. khasiana*—have been found to selectively concentrate key elements such as N, P, and K in their biomass and litterfall; these processes contribute to accelerated forest succession, rapid soil nutrient accumulation, agricultural productivity (and food security), and enhanced of biodiversity at the landscape scale (Ramakrishnan et al. 1998). These insights regarding the ecological roles of socially valued species were used to formulate the Nagaland landscape redevelopment project (NEPED and IIRR 1999) through an 'incremental pathway' discussed elsewhere in this chapter (Sect. 9.7.1).

   People in forest-dependent communities are more often concerned less with the availability of large timber than with the supply of smaller trees, used primarily as fuelwood, and non-timber forest products used for a wide variety of purposes in their daily lives. There is growing public recognition of the importance of biodiversity conservation and sustainable utilization of forest resources and its relevance to the livelihoods and food security of local and tribal communities in the South Asia region. Perhaps less well-appreciated by the public and policy makers is the vital role that the holders and users of traditional forest-related knowledge play in biodiversity conservation, and the importance of such traditional knowledge, specifically that associated with their management of forest resources for non-timber products (medicinal plants, wild foods, etc.), and the role of socially valued species with ecological keystone value in biodiversity conservation and sustainable traditional agricultural in the region.

9.5.3   Sacred Groves as a Resource for Rehabilitation/Restoration of Degraded Sites

Religious beliefs and ceremonies associated with forests and individual plant species are very common across the cultural spectrum of India and elsewhere in South Asia. While these beliefs and practices are well-documented in Hindu and later Buddhist religious texts and mythologies, their roots are in many cases much deeper, arising from the ancient cultures, beliefs, and practices of diverse peoples of the subcontinent that pre-date the migration of the Indo-European (Aryan) peoples from the northwest, some of which survive in today's tribal communities (Jain 1997; Majapuria and Joshi 1997; Gupta 2001). Sacred groves, a reflection of the spiritual significance and cultural value of forests in traditional societies of the region, are excellent examples of the natural forests rich in biodiversity that once existed over more extensive areas in South Asia. These forest stands have been rigorously conserved for socio-cultural, religious, and economic reasons, typically for countless generations (Hughes and Chandran 1998). While much is known about the sacred groves of India (Ramakrishnan et al. 1998), the information available from other countries of the region is much more limited.

9.5.3.1   India

There are very early references to South Asian forests in Vedic texts dating from 1400 to 700 BC, in which forests were classified by their principal uses, such as Tapovana (for meditation; Vannucci 1993); Pashuvan (for conserving deer and other animals); Hastivana (for conserving elephants); Mrigvana (for a combination of wildlife conservation, hunting, and leisure activities); and Dravyavana (for production forestry; Rawat 1991). These detailed classifications indicate that use restrictions were imposed on the ancient peoples of the Indian subcontinent, with an important focus on wildlife conservation. In these very ancient Hindu scriptures, there also are references to: Mahavana, where Lord Shiva (a major Hindu deity) resides; Srivana, suggesting 'sacredness' in a general sense; and Devavana, referring to the sacred gardens of the gods, a forest of prosperity where presiding deities of villages are placed. The tradition of sacred groves is indeed very ancient and, amongst many traditional societies of the region, their protection remains important.

The precise number and area of sacred groves in the region is unknown. Malhotra et al. (2001) reported 4,415 sacred groves covering 42,278 ha, although this is probably an underestimate of their total number and extent. Sacred groves are found in a variety of forest ecosystem types, including coastal mangrove forests. Today, sacred groves in India survive primarily as isolated patches in otherwise degraded forest landscapes (Swamy et al. 2003), and most of these forests are degraded to varying degrees because of rapidly changing societal values that weaken the resolve of communities to protect them. Rao (1996) documented 13,720 such groves in India alone.

A survey of 79 of these, ranging in size from 0.01 to 900 ha with a total area of 10,511 ha, found that most were located in the catchment areas of major rivers and streams and that only a small fraction (totalling 138 ha) were totally undisturbed.

Sacred groves in India typically have a presiding deity and associated folklore, beliefs, taboos, and rituals (Swamy et al. 2003). Customary religious tradition often requires that permission be granted by the local priest for cutting a tree only after a sacrifice is offered. In other cases, protection of these forests is stricter, as in the Mawsmai sacred grove in the Cherrapunji region of Meghalaya in northeastern India, where the Khasi religious tradition prohibits the removal of even a fallen twig from the sacred grove (Ramakrishnan 1992a). In Meghalaya, each village is said to have had a sacred grove attached to it in ancient times, though many of these were said to have been destroyed, due to changing value systems associated with the spread of Christianity in this region. Despite impact of religious conversion and associated prohibition of earlier practices, traditional values still remain in many communities, integrated with newly acquired religions values. Given the diversity or cultures and religious practices throughout India where sacred groves are found and maintained by communities, it is not possible to generalize about offerings in the form of animal sacrifices associated with these sites, but these practices do occur, often in connection with festivals and religious ceremonies. Animistic values and organized religious beliefs often remain intermingled in India, as along the Kerala coastal areas and in the north-eastern hill regions.

9.5.3.2 Nepal

In Nepal there are a large number of 'sacred fields' or 'kshetras'—environmental complexes that may include religious edifices, shrines, sacred objects of the natural world (such caves, hot or cold springs, lake, rivers natural formations such as rocks and caves, lakes, rivers, springs, and of course protected forests, i.e., sacred groves) (Messerschmidt 1989; Ingles 1990; Hamilton 2002). Many of the plants protected within these forests are associated with Hindu religious mythology and ceremonies and are valued in traditional Ayurvedic medicine, a Nepali tradition shared with India (Singh et al. 1979; Majapuria and Joshi 1997; Parrotta 2001).

Among Nepal's traditional communities, many trees are viewed as sacred, among which species such as *Ficus religiosa, F. bengalensis,* and *Michelia kisopa* are more venerated than others (Manandhar 2002). These sacred forests are associated with protective deities of the Buddhist and/or Hindu pantheons. Authority over these community-managed forests is exercised by 'pujaris' or priests, well-versed in religious scriptures, who are selected by local communities on a rotational basis. Elsewhere in Nepal, Sherpas living in the Khumbu valley of Nepal maintain sacred groves surrounding Buddhist temples. With a rich cultural/religious traditions embedded in the society and with very many governmental initiatives taken towards conservation of sacred forests of Nepal, these community conserved forests seem likely to maintain their high conservation value in the future.

9.5.3.3 Sri Lanka

Buddhist tradition values forests as sites of unlimited benevolence, making no demands on anyone. Sacred groves in this country have an important role as sheltered environments for fostering peace and harmonious coexistence with nature, and as sites for revitalization of Sri Lanka's traditional culture of peace. In Sri Lanka, a land rich in Buddhist traditions, any geographical site may becomes a protected sanctuary if there is a consensus among the people who use that site that it be preserved for meditation and promoting peace. There are very many such protected groves in the Sri Lanka, arising from the Buddhist belief in protecting life in general and protecting forests in particular for their benevolence (Withange 1998). Many of the important sacred groves have now been given protected area status through various acts enacted by the Sri Lankan government. Some of the more well-known of these reserves are Domba Gas Kanda, Yagirala, Kalugala, and Viharakelle. There are larger forested landscapes such as Ritigala, and Samanola Kanda (Adam's Peak) that are also considered sacred, with many Buddhist traditions and mythological stories woven around them. In addition, smaller groves are to be found as 'temple forests' associated with monasteries and hermitages, protected and tended by Buddhist monks.

9.5.3.4 Bangladesh

Sacred groves are also found in Bangladesh, particularly in areas where Hindu and Islamic traditions coexist. These sometimes occur as small patches of trees attached to the shrines of Muslim saints (Islam et al. 1998). The sacred groves of the Chittagong hill tracts in southeastern Bangladesh (which has a close eco-cultural affinity with the northeastern hill region of Mizoram State in India) have been under constant threat, and little is known about the present status of the groves in this region.

9.5.3.5 Afghanistan

In Afghanistan, there are at present more than 150 recorded sacred groves, part of the historical cultural traditions that are now being revived and managed following local Islamic traditions and values (Zaman 1998). The sacred grove tradition most likely developed during the long period of history in which Hindu, Zoroastrian, Buddhist, and non-institutional religions were practised by the people of present-day Afghanistan prior to the influence of monotheistic religions (i.e., Judaism, Christianity, and Islam) in the region. Traditional Islamic communities in Afghanistan today continue to protect and manage these sacred groves. While often strictly protected, some tree-cutting may be permitted for specific reasons.

9.5.3.6   The Value of Sacred Groves for Biodiversity Conservation and Forest Ecosystem Restoration

With their often astonishing biodiversity, sacred groves have an important function as sites to learn about the composition, structure, and function of forest ecosystems otherwise absent from the deforested or degraded landscapes in which they occur. Such information can have important application value in forest rehabilitation and ecological restoration projects. This was illustrated in studies of the sacred grove in the village of Mawsmai, located near Cherrapunji in the northeastern Indian state of Meghalaya. This village grove stands out as an island of rich biodiversity, a small remnant of the forest ecosystem that once dominated what is now a highly degraded landscape (in spite of an annual average rainfall of 12 m). This grove is so well-protected that even removal of dead wood from it remains prohibited. Studies of ecosystem structure and function in some of the sacred groves and the surrounding degraded land led to identification of many socially valued keystone species in the groves, which have had meaningful implications for the rehabilitation action plan for this high-rainfall, hilly region (Ramakrishnan 1992a). As biodiversity conservation sites, this and other sacred groves are of great ecological value (Ramakrishnan et al. 1998; Swamy et al. 2003). Programmes to conserve them for their plant and animal biodiversity values have been initiated by governmental and non-governmental agencies in India, for example in the Khejri tree dominated sites of Rajasthan discussed earlier, and sacred groves in Kerala in southern India. These emerging efforts should be seen as major landmarks in the conservation and restoration of degraded systems in the South Asian region.

9.5.4   Applying Traditional Forest-Related Knowledge to Current Forest Management Challenges

Significant opportunities exist for better use of traditional forest-related knowledge to resolve the significant conflicts that exist between state forest management authorities and local and tribal communities in the South Asia region. As discussed above, socially valued tree species, which invariably have ecological keystone value, should be given a high priority in forest restoration initiatives. An emphasis on species and management practices that are valued by the traditional community greatly enhances prospects for their active support for and involvement in forest management activities aimed at benefitting multiple stakeholders. This requires, among other things, an effort to reconcile silvicultural attributes and timber values with local social values in selecting species for reforestation, watershed restoration, and/or forest enrichment plantings so as to ensure community involvement in forest landscape management.

Given the progressive degradation of forest landscapes and biodiversity loss over very extensive areas of this region, socially valued ecosystems such as sacred groves have a key role to play in forest landscape restoration and other conservation management initiatives, both as reference systems and sources of biodiversity for planting or natural recolonisation of restored forest sites. Thus, what is needed is a reconciliation of traditional forest-related knowledge and formal science-based knowledge to promote forest management approaches that aim to reconstruct natural cultural landscapes based on values that local communities understand and appreciate, and therefore participate in and benefit from.

9.6 Collaborative Forest Management Initiatives in South Asia

In an effort to develop an effective institutional framework and mechanisms for greater involvement of local communities in the management of forest resources, several participatory forest management approaches have emerged in different countries in South Asia. These include community forestry (CF), joint forest management (JFM), and forest user groups (FUG), which differ in their institutional, tenurial, decision-making, and benefit-sharing arrangements. While India and Nepal have pursued these approaches on national scale, other countries in the region have only begun making cautious moves towards community participatory forest management. These decentralized approaches have been integrated them into national policy frameworks in India and Nepal, whose experiences will be considered in below in Sect. 9.6.1.

According to Rasool and Karki (2007), while India, Nepal, Bangladesh, and Bhutan are all moving from centralized to participatory forest management through these new institutional approaches, the magnitude and pace of the movement is greater in Nepal and India. Aside from JFM, all participatory forest management mechanisms are supported by state legislation, although the degree of institutionalization may vary. While forest user groups in Nepal have full decision-making authority, community forest management groups in Bhutan and joint forest management committees in India have limited authority. Considerable variation also exists in the degree of participation of local people—in Nepal local participation is very high, in Bangladesh it is very low, and in Bhutan and India it is intermediate. Although all these initiatives emphasize regulated participation arrangements, effectiveness of the participatory processes efforts vary, and tend to depend on the quality and extent of the forest resources, as well as the quality of local-level leadership.

9.6.1 Joint Forest Management in India

Over the past two decades, the focus in forestry in India has shifted towards conservation forestry, with people's participation as part of the joint forest management initiative of the Government of India. In return for providing improved forest protection, communities receive better access to non-timber forest products for subsistence use as well as a share of net commercial timber revenues. The state retains most of the control and decision-making over forest management, regulation, monitoring, timber harvesting, and forest product marketing (Gupta 2006). Under JFM arrangements, forest governance issues are handled through a forest protection committee (FPC), or van sangrakshan samiti (VSS), along with an executive committee (EC). Typically, FPCs and VSSs having one member from each family of the specific village or a group of villages.

Since its initiation in the 1970s, JFM has spread throughout the country, and today involves around 85,000 village committees and 17.3 million ha of forest land. Since JFM as a model is weighted in favour of state forest department control over planning, investment, management, harvesting, and marketing, participating communities often fail to gain optimal benefit from this otherwise well-conceived initiative. They view it only as a means through which fuelwood, fodder, and non-timber forest products may be obtained to meet their subsistence needs. The more recent Scheduled Tribes and Other Traditional Forest Dwellers (Recognition of Forest Rights) Act of 2006 (applicable for the non-tribal population living in the forests) is an attempt to give the original forest dwellers legal forest use rights (to support their livelihoods and enhance food security) while addressing sustainability concerns in the management of community forest resources, specifically conservation of biodiversity and related sustainability concerns.

A number of shortcomings in implementation of JFMs have been reported. In many cases forest protection has not been successful because of poor monitoring of the forest conservation measures by relatively inactive FPCs, constraining meaningful interaction amongst the stakeholders. In other situations, decision-making power has remained vested with the state forest departments, which are often reluctant to accept genuine community-participatory management (Gupta 2006). Often, as in other largely male-dominated societies, women (the custodians of traditional forest-related knowledge) and their interests are not adequately represented in JFM decision-making processes (Sarin 1998). Although JFM represents a significant step towards ensuring greater community participation in forest management, and empowerment of forest-dependent communities, state forest departments for the most part continue to follow a top-down approach towards decision-making. Further, for reasons related to industrial exploitation of forest resources, community-based customary and informal institutions have also been undermined (Springate-Baginski and Blaikie 2007; Sarin et al. 2003). Gupta (2006) rightly points out that if JFM is to be an effective tool for community-participatory sustainable forest management in India, a change in mindset and organizational culture amongst a large section of the state foresters will be required.

9.6.2   Community Forestry in Nepal

Although organized state-linked forest management on a large scale began in Nepal around 1880, most of the forests of Nepal were not under state control, but rather under de facto community control prior to the mid-1950s (Nagendra 2002).

Traditional and indigenous forest management practices were prevalent in the Nepal hills during this period. Since populations were small, and forest resources relatively extensive, pressures on Nepal's forests in the past were far less than those which exist today. In response to the adverse environmental and socioeconomic impacts arising from nationalization of forests and widespread deforestation, since the early 1970s efforts have been made to engage local communities in forest management through community forestry, leasehold forestry, and park buffer zone projects. Subsequent to the National Forest Act of 1976, a community-oriented group of foresters working in the districts met to promote a new form of forestry based on their experience of working with local people in forest management.

The Forest Act of 1993 formally established five categories of national forests: (i) community forests that are entrusted to village-level user group for management and sustained utilization; (ii) leasehold forests on land leased by central or local authorities to individuals or groups; (iii) government-managed forests in which production forests units are managed by the central government; (iv) forests belonging to religious institutions; and (v) strictly protected forests. Being based on a legislative framework, the legal rights of the forest dwellers are assured in the Nepalese context, in contrast to the situation in India where the legal rights for forest resources are vested with the forest department, whose officials also have the power to transfer rights to non-timber forest product resources through administrative orders. By 2009 an estimated one-third of Nepal's population was participating in the direct management of approximately 25% of the country's forests (United Nations 2010).

As a consequence of community forestry, there has been a significant improvement in forest conditions and increased forest cover in the middle hills of Nepal (Nagendra et al. 2005; United Nations 2010). This has resulted in increased access to forest products and development of forest-based small-scale enterprises, resulting in a greater flow of economic benefits to local communities, which strengthened local institutions and contributed to improvements in provision of forest ecosystem services (United Nations 2010). These positive benefits are mostly confined to the middle hill regions, where local indigenous systems of management have been practised for decades or even centuries in many locations. While the participation of women and other poor or disadvantaged groups in community forest management appears to be improving in recent years, the equitable distribution of benefits within communities is a persistent challenge. For example, according to data from the Community Forestry Division of the Ministry of Forest and Soil Conservation, women's participation in forest user groups is only 24%, and their role in decision-making negligible in spite of the fact that their involvement in implementing community forestry activities is very high (Kanel 2004).

In the Terai region—the belt of marshy grasslands, savannas, and forests between the Himalayan foothills and the Indo-Gangetic Plain of the Ganges, Brahmaputra, and their tributaries—the results of community forest management have been poorer. There are a number of reasons for this, including the large geographical area covered, high ethnic heterogeneity within forest user groups, high timber value of these forests, and the relative weakness of local institutions in the region, which has limited participation of stakeholders in defining and implementing community forest management policies (Bampton et al. 2007). Pravat (2009) argues that over the past 3 centuries, exploitation of forests in the Terai gradually became institutionalized, creating significant barriers to changes in existing power structures necessary to create the conditions for equity and ecological sustainability in forest management. The state, he maintains, continues to have a major stake in the forests of the Terai given their significant revenue contribution to the government, which has favoured continued control over the region's valuable forests. If managed more efficiently, the forests of Terai have the potential to boost the local economy, while also generating significant revenue for the country as a whole, transforming the presently cost-intensive forestry sector to an income generating one. However, given years of political instability and the continued reluctance of the state to engage in democratic, participatory and inclusive governance, it has yet to be seen whether transparent, accountable, and sustainable forestry in the Terai can be achieved in the near future. The risk of retaining the leading role of the state, with its commercial interest in timber extraction and its reluctance to promote a governance framework that involves all stakeholders, is that Nepal will squander forest resources of the Terai with no long-term benefit to the country and its people. However, with the real involvement of people, sustainable forestry practices could be promoted, contributing to improved quality of life for local communities living in forest areas.

To conclude, while there have been significant recent developments in collaborative forest management in both India and Nepal forest areas outside of protected areas, participatory forest management efforts in other countries of the region are, at present, largely confined to buffer zone management in protected Biosphere Reserves (Ramakrishnan et al. 1998).

9.7   The Emerging Developmental Pathways for Forested Landscapes in South Asia

Forested landscapes of South Asia may be characterized by diverse typologies of socio-ecological systems that include: (i) traditional societies who are highly dependent on forest ecosystems for their livelihoods and food security; (ii) modernized rural communities living in highly degraded landscapes, with energy-intensive land-use practices; and (iii) those falling in between these two extremes. Although sustainable forest management in the region needs to focus on conservation and effective management of the region's existing forest resources, ecological restoration and/or rehabilitation of tree cover in extensive degraded landscapes is becoming an increasingly critical need. This consideration is the basis for the following discussion of three distinct developmental pathways, based on linking traditional and formal scientific knowledge to varying degrees, for designing appropriate landscape management strategies (Swift et al. 1996). The first of these, the 'incremental pathway' discussed below, is particularly relevant for traditional forest dwellers.

An important element of these land management pathways is the selection of appropriate tree species to be planted and managed. The fundamental principle involved in the linkage of traditional and formal knowledge systems is that socially valued species are ecologically significant keystone species. This principle is based on extensive research in northeastern India on the ecological, silvicultural and socio-cultural attributes of early successional tree and bamboo species (Ramakrishnan 2008a, b). These research findings have been confirmed by studies conducted elsewhere in India and other South Asian countries (Ramakrishnan et al. 2000, 2002), and through a global analysis (Ramakrishnan et al. 1998).

9.7.1 The Incremental Pathway: Improving Agroforest Landscape Management

Institutional arrangements for community participatory forest management can be effective, but only when forest management adequately addresses sustainable liveli-hood issues, which requires a perspective that considers sustainable land use more broadly (as opposed to just forests). Most people in traditional rural communities in the region are economically marginalized, and depend heavily on subsistence agri-culture. Sustainable forest management, therefore, needs to be connected with sus-tainable development of traditional agricultural practices, without making drastic departures from the traditional land use practices and cultural values (Ramakrishnan 2008b). By building upon their traditional forest-related knowledge in a step-by-step fashion to generate 'hybrid technologies' (the incremental pathway; Swift et al. 1996), both forest biodiversity conservation and local food security objectives can be met. The redeveloped land use systems resulting from this process can enhance both natural as well as human-managed agro-biodiversity.

For traditional societies that still possess a rich body of traditional forest-related knowledge, an appropriate strategy for restoration and/or rehabilitation of their cultural landscapes is to build upon this traditional knowledge, introducing formal science-based technologies only to the minimal extent needed to achieve desired outcomes. This developmental initiative, described below, is an outcome of an inten-sive trans-disciplinary study carried out by a large network of scientists in northeast-ern India, which led to the conclusion, discussed earlier, that socially valued species are invariably ecological keystone species (Ramakrishnan 1992a).

The principles outlined above formed the basis for a major developmental initiative in the Indian State of Nagaland, the Nagaland Environmental Protection and Economic Development (NEPED) project, based on the incremental pathway. This rural devel-opment initiative involved over a thousand villages organized into village develop-ment boards (VDBs). A primary objective of the initiative was to enhance the role of trees and forests in the landscape, which had declined over a period of time because of deforestation and land degradation (NEPED and IIRR 1999). It should be noted that the VDBs were also formed in a ways that blended the traditional modes of institution-building of each of the tribal groups involved, with modern elective pro-cesses already introduced and operationalized in the region. The objective of this blending of the 'traditional' and 'modern' in this process of local-level institution building (i.e., of the VDBs), was to make these institutions both participatory and more effective. Over a brief period of time, communities were able to derive higher economic returns by building, step by step, upon the traditional knowledge available with the local communities, without making drastic departures from traditional agricultural practices. The initiative emphasized participatory testing of about a dozen tree species in more than 200 test plots with the involvement of VDBs, rather than testing pre-selected species being transplanted into the field site by extension agents.

This participatory approach for sustainable landscape management has received wide acceptance in the region. Over a period of 10 years, this agroforestry-based approach for enhancing food security and improving soil, water, and forest biodiversity management was tested in more than 800 villages (covering a total area of about 33,000 ha), in replicated farmers' plots (approximately 5,500 ha).

The socially valued species used were fast growing, early-successional, tree species with narrow crown forms that allowed light penetration to the ground level for crop growth during the initial few years of tree growth. As forest cover increased, tree species with broader crown forms were planted as the late-successional trees to form biodiversity-rich late-successional forests. The previously available (formal scientific) knowledge regarding the tree growth strategies, architecture, and ecological characteristics of the species ultimately chosen for planting at different stages complemented the traditional forest-related knowledge of the local farmers (Ramakrishnan 1992a, b). By building upon the traditional knowledge available within the local communities in an incremental fashion, it was possible to accelerate fallow regeneration in the jhum-affected areas, and enhance biodiversity in the resultant secondary forests (Ramakrishnan 1992a). Inclusion of traditional forest-related knowledge as an integral part of this programme triggered community participation, too, since the forest enrichment and rehabilitation techniques developed were based on a value system that the diverse communities could relate to. Although different stakeholders involved in this initiative had varied interests and objectives, all were addressed: (i) the foresters' interest in forest management for biodiversity conservation; (ii) the local communities' interest in redevelopment of their agricultural systems based on improved fallow management principles; (iii) the objective shared by both foresters and farmers to improve soil fertility and productivity, reduce erosion, and the conservation of biodiversity across the agro-forest landscape; (iv) increased carbon sequestration, an important objective in India's climate change mitigation strategy; (v) conservation and further development of locally available traditional forest-related knowledge; and (vii) respect for and preservation of the cultural identity of diverse ethnic forest dwellers of the region, which has had implications for conflict resolution and peace (Ramakrishnan 2009), a critical issue in the region.

9.7.2   Other Development Pathways for Increasing Tree Cover in Highly Degraded Landscapes

Elsewhere in India, government-supported forest management objectives remain largely focused on increasing tree cover across the landscape based on applying agroforestry principles for sustainable agriculture; at the same time socially valued species are also used for social forestry and rehabilitation of forest cover, thus increasing tree cover in the landscape unit as a whole. To meet these objectives, 'rediscovered' traditional forest-related knowledge has been used to promote tree planting to restore forest cover in degraded landscapes, with species selection based on community values (i.e., use of socially valued tree species that have ecological keystone value). One such agroforestry initiative, carried out in tea gardens of the mountainous Kodagu (Coorg) region of southwestern Karnataka in southern India, involved the development and use of an in-situ soil fertility management technology known as bio-organic fertilization (fertilisation bio-organique, or FBO). This technology, developed by a group of Indian and French scientists in collaboration with tea garden managers as part of the international Tropical Soil Biology and Fertility programme (TSBF), is based on improved management of organic residues to improve the sustainability of tea cultivation, and uses keystone earthworm species in the soil subsystem as indicators of soil health (Senapati et al. 2002). The results of this work in the Kodagu tea-growing region have been increased tree cover at the landscape level, increased yields of high-quality of tea leaves with a substantial reduction (30 to 50%) in inorganic fertilizer use, enhanced biodiversity (above- and below-ground), and improved soil health in the tea plantations, (Senapati et al. 1999). The FBO technology was patented jointly by Indian and French scientists in 1997 and has since been adopted in China.

Ultimately, the design of land management pathways for sustainable livelihood development of local communities needs to be appropriate to, and compatible with, a diversity of socio-ecological conditions. Agroforestry models developed and promoted to improve food security of forest dwellers can contribute to improved tree cover and enriched biodiversity in deforested and/or degraded agroforest landscapes. While consideration of ecological economic principles is important in this process (Ramakrishnan 2008a), it is equally important that people in local communities—as well as their objectives, values, and traditional knowledge—be incorporated in the development of land use development practices (Ramakrishnan et al. 2005; Ramakrishnan et al 2006).

9.8 Traditional Forest-Related Knowledge in the South Asian Region: Where Do We Stand Today?

Countries in South Asia are moving through a historic reconsideration of their approaches to forest management, arising from the failure of government-controlled industrial management models to sustain the region's natural forests, which have only accelerated deforestation. At the same time, growing concerns of rural communities throughout the region over the deterioration of local ecosystems has led to the emergence of grassroots environmental movements oriented towards forest conservation. Nepal and India are taking leadership roles in formulating national policies for community participatory, decentralized, village-based approaches to forest management. Further, pressures are building upon political leaders, professional foresters, and the forest science community to develop more effective and more sustainable approaches to forest stewardship, as evidenced by experimentation in recent decades with joint forest management (JFM) initiatives in the region.

On the applied research front, much progress has been made in working towards a better understanding of traditional forest-related knowledge, and understanding its relationships with formal knowledge to elucidate generalized integrative principles that can be applied across socio-ecological systems. For such an effort to succeed at the community level, there is an increasing realization that two aspects of traditional knowledge need to be fully appreciated, namely: (i) the 'intangible' dimension that requires the respect and preservation of cultural values of people living in forested areas and allowing these communities to develop economically, socially, and culturally with minimal external interference; and (ii) the introduction of formal scientific forest knowledge to traditional communities to the extent required to facilitate adjustments that socio-ecological landscape systems need for sustainable management of forest and agricultural resources in the changing contemporary context, where external influences have already had significant impacts on communities and their agroforest landscapes.

Promotion of a more equitable relationship between traditional and formal scientific forest-related knowledge streams has made it possible to develop 'hybrid technologies.' While such efforts have begun to yield positive results in the region, much work remains to foster these developments.

In education, since the 1980s there has been an increasing emphasis on traditional knowledge-based conservation approaches in areas of South Asia where traditional societies predominate. The implications for sustainable livelihoods and other development issues are becoming more apparent. At the university level in India, for example, sustainability science with its emphasis on ecological economics is rapidly gaining importance. There is also an increasing realization within the academic and scientific communities that conserving cultural values is important for conserving biodiversity, for promoting natural resource sustainability, and for enhancing human well-being more generally. This is an area in which the Anthropological Survey of India is becoming increasingly involved in its evaluations of the socio-cultural dimensions of sustainable natural resource management (Mitra 2007).

Academic efforts in the field of sustainability science, at the interface between the natural and social sciences, has contributed towards promotion of traditional forest-related knowledge as a powerful tool in community participatory resource management efforts. It is an important part of the university teaching/research curriculum in northeast India, and is used in joint forest management training in this region (Malhotra et al. 2004), with potential applications throughout India (Gupta 2006). These efforts are gradually facilitating better interaction between natural and social scientists in forestry training and research, more so in India than in other countries of the region.

Within the South Asian region's extensive network of local, national, and international non-governmental organizations (NGOs) working for conservation-related sustainable development, traditional forest-related knowledge and technologies are increasingly viewed as powerful tools for enhancing community participation.

Many NGOs working in the region are already influencing changes in government policies in the region, particularly in India and Nepal, although progress is this direction is slow.

The challenges of emerging environmental uncertainties in forest resource management can no longer be faced from narrow perspectives, but require adoption of interdisciplinary sustainable forest management principles. The progressive marginalization of forest-dependent traditional communities, a historical process that has accelerated over the past century or more of excessive exploitation of forest resources, has led to increasing political and social conflict in South Asia, as in many other parts of the world. In this context, it is important to consider the varied dimensions of human security that Brauch et al. (2009) discusses at the global level, as well as those related to linking knowledge systems for socio-ecological system security of traditional forest dwellers, particularly in South Asia (Ramakrishnan 2009). In this regard the two major forest management decentralization initiatives of the South Asian region, i.e., Joint Forest Management (JFM) in India, and Forest User Groups (FUG) in Nepal, may be seen as positive development towards more sustainable forest management. The rapid empowerment of traditional forest-dependent communities, both through governmental and non-governmental efforts, is enabling traditional societies to make their voices heard in natural resource policy formulation and developmental action plans that are open to stakeholder participation. In this process, non-governmental voluntary organizations in general, and youth in particular, are playing a key role in the region.

References

Ainslie W (1826) Materia Indica; or, some account of those articles which are employed by the Hindoos and other eastern nations, in their medicine, arts, and agriculture. 2 vols. Longman, Rees, Orme, Brown and Green, London

Allchin B (1998) Early man and the environment in South Asia. In: Grove R, Damodaran V, Sangwan S (eds) Nature and the Orient. Oxford University Press, New Delhi, pp 29–50

Aris M (1979) Bhutan: the early history of a Himalayan kingdom. Aris and Phillips, Warminister

Bampton JFR, Ebregt A, Banjade MR (2007) Collaborative forest management Nepal's Terai: policy, practice and contestation. J For Livelihood 6:30–43

Batisse M (1982) The biosphere reserve: a tool for environmental conservation and management. Environ Conserv 9:101–111

Blanford HR (1958) Highlights of 100 years of forestry in Burma. Empire For Rev 37(1):33–42

Bodekar G, Bhat KKS, Burley J, Vantomme P (1997) Medicinal plants for forest conservation and health care, vol 11, Non-wood forest products series. Food and Agriculture Organisation (FAO), Rome

Boojh R, Ramakrishnan PS (1982a) Growth strategy of trees related to successional status. I. Architecture and extension growth. For Ecol Manag 4:355–374

Boojh R, Ramakrishnan PS (1982b) Growth strategy of trees related to successional status. II. Leaf dynamics. For Ecol Manag 4:375–386

Brauch HG, Grin J, Mesjasz C, Krummenacher H, Behera NC, Chourou B, Spring UO, Kameri-Mbote P (eds) (2009) Facing global environmental change: environmental, human, energy, food, health and water security concepts, vol 2. Springer, Berlin, pp 817–828

Cederlof G, Sivaramakrishnan K (eds) (2006) Ecological nationalism: nature, livelihoods and identities in South Asia. University of Washington Press, Seattle

Champion HG, Seth SK (1968) A revised survey of forest types of India. Government of India, New Delhi. Reprinted, 2005. Nataraj Publications, Dehra Dun

Chokkalingam U, Smith J, de Jong W, Sabogal C, Dotzauer H, Savenije H (2000) Towards sustainable management and development of tropical secondary forests in Asia: the Samarinda proposal for action. Indonesia, Center for International Forestry Research (CIFOR), Bogor

CSIR (1948-1992) and CSIR (1986): *See* Indian Council for Scientific and Industrial Research

Dangwal DD (2005) Commercialisation of forests, timber extraction and deforestation in Uttaranchal, 1815–1947. Conserv Soc 3:110–133

De Rosayro RA (1950) Ecological conceptions and vegetation types with special reference to Ceylon. Trop Agricult 56:108–131

Drake JK, Mooney HA, di Castri F, Groves RH, Kruger FJ, Rejmanek MK, Williamson M (1988) Biological invasion: a global perspective. SCOPE 37. Wiley, Chichester

Dube SC, Singh B, Mishra SN, Singh KS (eds) (1998) Antiquity to modernity in tribal India. Vols I–IV. Inter-India Publications, New Delhi

Dymock W, Warden CJH, Hooper D (1890) Pharmocographia Indica: a history of the principal drugs of vegetable origin met with in British India (reprinted in 1995), 3 vols. Low Price Publishers, New Delhi

Erdosy G (1998) Deforestation in pre- and protohistoric South Asia. In: Grove RH, Damodaran V, Sangwan S (eds) Nature and the Orient. Oxford University Press, New Delhi, pp 51–69

Filiozat J (1980) Ecologie historique en Inde du Sud: le pays des Kallar. Revue des Etudes d'Extrème Orient 2:22–46

Food and Agriculture Organization (FAO) (2001) Global forest resource assessment, 2000. FAO forestry paper 140. FAO, Rome

Food and Agriculture Organization (FAO) (2009) State of the world's forests, 2009. FAO, Rome

Gadgil M, Guha R (1992) This fissured land: an ecological history of India. Oxford University Press, New Delhi

Gautam AP, Shivakot GP, Webb EL (2004) A review of forest policies, institutions and changes in the resource condition in Nepal. Int For Rev 6:136–148

Grove RH (1995) Green imperialism: colonial expansion, imperial Island Edens and the origins of environmentalism, 1600–1800. Cambridge University Press, Cambridge

Guha S (1999) Environment and ethnicity in India, 1200–1991. Cambridge University Press, Cambridge

Guha S (2000) Health and population in South Asia from earliest times to the present. Permanent Black, New Delhi

Gupta SM (2001) Plant myths and traditions in India. Munshiram Manoharlal Publishers, New Delhi

Gupta K (2006) Joint forest management: policy, participation and practices in India. International Book Distributors, Dehradun

Hamilton LS (2002) Forest and tree conservation through metaphysical constraints. George Right Forum 19:57–78

Hansen SA, Van Vleet JW (2007) Issues and options for traditional knowledge holders in protecting their intellectual property. In: Krattiger A, Mahoney RT, Nelsen L, Thomson JA, Bennett AB, Satyanarayana K, Graff GD, Fernandez C, Kowalski SP (eds) Intellectual property management in health and agricultural innovation: a handbook of best practices. MIHR and PIPRA, Oxford and Davis, pp 1523–1538. http://www.ipHandbook.org. Cited 28 Jan 2011

Hooker JD (1875–1897) (Reprint 1990) The flora of British India. 7 vols. Reprinted 1990. Saujanya Books, Dehra Dun, India

Hughes JD, Chandran MDS (1998) Sacred groves around the earth: an overview. In: Ramakrishnan PS, Saxena KG, Chandrashekara UM (eds) Conserving the sacred for biodiversity management. UNESCO and Oxford and IBH, New Delhi, pp 69–86

Indian Council for Scientific and Industrial Research (CSIR) (1948–1992) The wealth of India: a dictionary of Indian raw materials and industrial products. Raw materials, vols. 1–11 (1948–1976); revised and enlarged vols. 1–3 (1985–1992). Publications and Information Directorate, New Delhi

Indian Council for Scientific and Industrial Research (CSIR) (1986) The useful plants of India. Publications and Information Directorate, New Delhi

Ingles AW (1990) The management of religious forests in Nepal, Department of Forestry Research Report. Australian National University, Canberra

Islam AKMN, Islam MA, Hoque AE (1998) Species composition of sacred groves, their diversity and conservation in Bangladesh. In: Ramakrishnan PS, Saxena KG, Chandrashekara UM (eds) Conserving the sacred for biodiversity management. UNESCO and Oxford and IBH, New Delhi, pp 163–165

Jain SK (1991) Dictionary of Indian folk medicine and ethnobotany. Deep Publications, New Delhi

Jain SK (ed) (1997) Contribution to Indian ethnobotany, 3rd edn. Scientific Publishers, Jodhpur

Kanel KR (2004) Twenty five years of community forestry: contribution to millennium development goals. In: Kanel KR, Mathema P, Kandel BR, Niraula DR, Sharma AR, Gautam M (eds) Twenty five years of community forestry: proceedings of the fourth national workshop on community forestry, Community Forestry Division, Department of Forest, Kathmandu, 4–6 Aug 2004, pp 1–18

Karan PP (1963) The Himalayan kingdoms: Bhutan, Sikkim and Nepal. van Nostrand, Princeton

Kirtikar KR, Basu BD (1935) Indian medicinal plants, reprint of 1933, 2nd edn. 4 vols. Bishen Singh Mahendra Pal Singh, Dehra Dun

Lal M (1985) Iron tools, forest clearance and urbanization in the Gangetic plains. Man Environ 10:83–90

Lambin EF, Turner BL II, Geist HJ, Agbola S, Angelsen A, Bruce JW, Coomes O, Dirzo R, Fischer G, Folke C, George PS, Homewood K, Imbernon J, Leemans R, Li X, Moran EF, Mortimore M, Ramakrishnan PS, Richards JF, Skånes H, Steffen W, Stone GD, Svedin U, Veldkamp T, Vogel C, Xu J (2001) The causes of land-use and land-cover change: moving beyond the myths. Glob Environ Chang 11:261–269

Ludden D (2002) Spectres of agrarian territory in Southern India. Indian Econ Soc Hist Rev 39:233–258

Maheshwari JK (1996) Ethnobotany in South Asia. Scientific Publishers, Jodhpur

Majapuria TC, Joshi DP (1997) Religious and useful plants of Nepal and India. M. Gupta, Lashkar

Malhotra KC, Gokhale Y, Chatterjee S, Srivastava S (2001) Cultural and ecological dimensions of sacred groves in India. Indian National Science Academy, New Delhi

Malhotra KC, Barik SK, Tiwari BK, Tripathi SK (2004) Joint forest management in north-east India: a trainers' resource book. North-Eastern Hill University, Shillong

Malla YB (2001) Changing policies and the persistence of patron-client relations in Nepal. Environ Hist 6:287–307

Manandhar NP (2002) Plants and people of Nepal. Timber Press, Portland

Markham C (1913) Colloquies on the simples and drugs of India. (Translation of: Coloquios dos simples e drogas he cousas medicinais da India compostos pello Doutor Garcia da Orta, originally published in 1563). Sotheran, London

Maurer M, Höll O (eds) (2003) "Natur" als politikum. RLI, Vienna

Menzies N (1998) Three hundred years of taungya: a sustainable system of forestry in south China. Hum Ecol 16:361–376

Messerschmidt DA (1989) The Hindu pilgrimage to Muktinath Nepal: natural and supernatural attributes of the sacred field. Mt Res Dev 9:105–118

Mitra KK (2007) Traditional knowledge in contemporary societies: challenges and opportunities. Pratibha Prakashan, New Delhi

Nadkarni KM (1908) Indian materia medica. 2 vols. Revised and enlarged edition, 2007. Popular Prakashan, Mumbai

Nagendra H (2002) Tenure and forest conditions: community forestry in the Nepal Terai. Environ Conserv 29:530–539

Nagendra H, Karna B, Karmacharya M (2005) Cutting across space and time: examining forest co-management in Nepal. Ecology and Society 10. Available via http://www.ecologyand society. org/vol10/iss1/art24/. Cited 28 Jan 2011

National Academy of Sciences (1975) Underexploited tropical plants with promising economic value. National Academy of Sciences, Washington, DC

National Academy of Sciences (1979) Tropical legumes: resources for the future. National Academy of Sciences, Washington, DC

NEPED and IIRR (1999) Building upon traditional agriculture in Nagaland. Nagaland Environmental Protection and Economic Development/International Institute of Rural Reconstruction, Nagaland/Silang

Pant DN, Das KK, Roy PS (1996) Digital mapping of forest fire in Garhwal Himalaya using Indian remote sensing satellite. Indian For 122(5):390–395

Parrotta JA (2001) Healing plants of peninsular India. CABI, Wallingford

Parrotta JA (2002) Conservation and sustainable use of medicinal plant resources—an international perspective. In: Kumar AB, Gangadharan GG, Kumar CS (eds) Invited papers presented at the world Ayurveda congress, Kochi, Kerala, World Ayurveda Congress Secretariat, Kochi, 1–4 Nov 2002, pp 52–63

Poffenberger M (ed) (2000) Communities and forest management in South Asia. A regional profile of the working group on community involvement in forest management, forests, people and policies. International Union for the Conservation of Nature (IUCN), Gland

Pravat SP (2009) Looking back to move forward: using history to understand the consensual forest management model in the Terai. Thesis, The Open University, Milton Keynes

Ramakrishnan PS (ed) (1991) Biological invasion in the tropics. National Institute of Ecology and International Scientific Publications, New Delhi

Ramakrishnan PS (1992a) Shifting agriculture and sustainable development: an interdisciplinary study from north-eastern India. Man and biosphere book series 10. UNESCO/Parthenon, Paris/Caernforth. Republished 1993 by Oxford University Press, New Delhi

Ramakrishnan PS (1992b) Tropical forests: exploitation, conservation and management. Impact (UNESCO) 42:149–162

Ramakrishnan PS (2003) Conserving the sacred: the protective impulse and the origins of modern protected areas. In: Harmon D, Putney AD (eds) The full value of parks: from economics to the intangible. Rowman and Littlefield, Lanham, pp 27–41

Ramakrishnan PS (2008a) Ecology and sustainable development: working with knowledge systems. National Book Trust of India, New Delhi

Ramakrishnan PS (2008b) The cultural cradle of biodiversity. National Book Trust of India, New Delhi

Ramakrishnan PS (2009) Linking knowledge systems for socio-ecological security. In: Brauch HG, Spring UO, Grin J, Mesjasz C, Kameri-Mbote P, Behera NC, Chourou B, Krummenacher H (eds) Facing global environmental change: environmental, human, energy, food, health and water security concepts, vol 2. Springer, Berlin, pp 817–828

Ramakrishnan PS, Saxena KG, Chandrasekhara U (eds) (1998) Conserving the sacred for biodiversity management. UNESCO and Oxford and IBH, New Delhi

Ramakrishnan PS, Chandrashekara UM, Elouard C, Guilmoto CZ, Maikhuri RK, Rao KS, Sankar S, Saxena KG (eds) (2000) Mountain biodiversity, land use dynamics and traditional ecological knowledge. UNESCO and Oxford and IBH, New Delhi

Ramakrishnan PS, Rai RK, Katwal RPS, Mehndiratta S (eds) (2002) Traditional ecological knowledge for managing biosphere reserves in South and Central Asia. UNESCO and Oxford and IBH, New Delhi

Ramakrishnan PS, Boojh R, Saxena KG, Chandrashekara UM, Depommier D, Patnaik S, Toky OP, Gangawar AK, Gangwar R (2005) One sun, two worlds: an ecological journey. UNESCO and Oxford and IBH, New Delhi

Ramakrishnan PS, Saxena KG, Rao KS (2006) Shifting agriculture and sustainable development of North-East India: tradition in Transition. UNESCO and Oxford and IBH, New Delhi

Rangarajan M (2002) Polity, ecology and landscape: new writings on South Asia's past. Stud Hist 18:135–147

Rao P (1996) Sacred groves and conservation. WWF-Indian Quart 7:4–8

Rasool G, Karki M (2007) Participatory forest management in South Asia: a comparative analysis of policies, institutions and approaches. International Centre for Integrated Mountain Development (ICIMOD), Kathmandu

Rawat AS (1991) History of forestry in India. Indus, Delhi

Richards JF, Tucker R (eds) (1989) World deforestation in the twentieth century. Duke University Press, Durham

Richards JF, Hagen JR, Haynes E (1985) Changing land use in Bihar, Panjab and Haryana, 1850–1970. Modern Asian Stud 19:699–732

Saklani A, Jain SK (1994) Cross-cultural ethnobotany of northeast India. Deep Publications, New Delhi

Sarin M (1998) From conflict to collaboration: institutional issues in community management. In: Poffenberger M, McGean B (eds) Village voices, forest choices: joint forest management in India. Oxford University Press, New Delhi, pp 165–209

Sarin M, Singh N, Sundar N, Bhogal R (2003) Devolution as a threat to democratic decision-making in forestry? Findings from three states in India. In: Edmunds D, Wollenburg E (eds) Local forest management: the impacts of devolution policies. Earthscan, London, pp 55–126

Senapati BK, Lavelle P, Giri S, Pashanasi B, Alegre J, Decaëns T, Jiménez JJ, Albrecht A, Blanchart E, Mahieux M, Rousseaux L, Thomas R, Panigrahi PPK, Venkatachalan M (1999) In-soil technologies for tropical ecosystems. In: Lavelle P, Brussaard L, Hendrix PF (eds) Earthworm management in tropical agroecosystems. CAB International, Wallingford, pp 199–237

Senapati BK, Naik S, Lavelle P, Ramakrishnan PS (2002) Earthworm based technology application for status assessment and management of traditional agroforestry system. In: Ramakrishnan PS, Rai RK, Katwal RPS, Mehndiratta S (eds) Traditional ecological knowledge for managing biosphere reserves in South and Central Asia. UNESCO and Oxford and IBH, New Delhi, pp 139–160

Shankar U, Lama SD, Bawa KS (1998) Ecosystem reconstruction through 'taungya' plantations following commercial logging of a dry, mixed deciduous forest in Darjeeling Himalaya. For Ecol Manag 102:131–142

Shukla RP, Ramakrishnan PS (1986) Architecture and growth strategies of tropical trees in relation to successional status. J Ecol 74:33–46

Shutkin WA (2000) The land that could be: environmentalism and democracy in the twenty-first century. MIT Press, Cambridge

Singh MP, Malla SB, Rajbhandari SB, Manandhar A (1979) Medicinal plants of Nepal: retrospects and prospects. Econ Bot 33:185–198

Sinha B (2002) Pines present in the Himalayas: past present and future. Energy Environ 13:873–881

Sivaramakrishnan K (2009) Forests and the environmental history of modern India. J Peasant Stud 36:299–329

Spoon J, Sherpa LN (2008) Beyul Khumbu: the Sherpa and Sagarmatha (Mount Everest) national park and buffer zone, Nepal. In: Mallarach J-M (ed) Protected landscapes and cultural and spiritual values: values of protected landscapes and seascapes, vol 2. International Union for the Conservation of Nature (IUCN), GTZ (Deutsche Gesellschaft für Technische Zusammenarbeit), and Obra Social de Caixa Catalunya, Kasparek Verlag, Heidelberg

Springate-Baginski O, Blaikie P (2007) The political ecology of reform in South Asia. Earthscan, London

Stainton JD (1972) Forests of Nepal. J. Murrey, London

Swamy PS, Kumar M, Sundarapandian SM (2003) Spirituality and ecology of sacred groves in Tamil Nadu, India. Unasylva 54(213):53–58

Swift MJ, Vandermeer J, Ramakrishnan PS, Anderson JM, Ong CK, Hawkins B (1996) Biodiversity and agroecosystem function. In: Mooney HA, Cushman JH, Medina E, Sala OE, Schulze E-D (eds) Functional roles of biodiversity: a global perspective, SCOPE Series. Wiley, Chichester, pp 261–298

Thapar R (2001) Perceiving the forest in early India. Stud Hist 17:1–16

Trivedi KK (1998) Estimating forests, wastes and fields c. 1600. Stud Hist 14:301–311

Troup RS (1921) The silviculture of Indian trees. 3 vols. Clarendon, Oxford

United Nations (UN) (2010) Discussion paper on the community forestry programme in Nepal: an example of excellence in community-based forest management. Economic and Social Council,

United Nations Forum on Forests, Ninth Session, Document E/CN.18/2011/9/Add.3, 10 p. Available via: http://daccess-dds-ny.un.org/doc/UNDOC/GEN/N10/601/84/PDF/N1060184. pdf?OpenElement. Cited 28 Jan 2011

Valiathan MS (1998) Healing plants. Curr Sci 75:1122–1126

Vannucci M (1993) Ecological readings in the Veda. D.K. Print World, New Delhi

Van Alphen J, Aris A (eds) (1995) Oriental medicine. Serindia Publications, London

Van Rheede Draakestein HA (1678–[1703]) Hortus Indicus Malabaricus. 12 vols. Fol. Joannis van Someren and Joannis van Dyck, Amsterdam

Wangchuk S (1998) Local perceptions and indigenous institutions as forms of social performance for sustainable forest management in Bhutan. Thesis, Department Wald-und Holzforschung, Swiss Federal Institute of Technology, Zurich

Watt G (1889–1896) A dictionary of the economic products of India. 6 vols. Allen, London

Weber T (1987) Hugging the trees: the story of the Chipko movement. Viking-Penguin, New Delhi

Wijesuriya G (2001) Protection of sacred mountains: towards a new paradigm. In: Proceedings of the UNESCO thematic expert meeting on Asia-Pacific sacred mountains. UNESCO, World Heritage Centre, Agency for Cultural Affairs of Japan, and Wakayama Prefectural Government, Wakayama City, pp 129–146

Withange H (1998) Role of sacred groves in conservation and management of biodiversity in Sri Lanka. In: Ramakrishnan PS, Saxena KG, Chandrashekara UM (eds) Conserving the sacred for biodiversity management. UNESCO and Oxford and IBH, New Delhi, pp 169–186

Wolman MG, Ramakrishnan PS, George PS, Kulkarni S, Vashishtha PS, Shidong Z, Qiguo Z, Yi Z, Long JF, Rosenzweig C, Solecki WD (eds) (2001) Growing populations, changing landscapes: studies from India, China and the United States. Indian National Science Academy, Chinese Academy of Sciences, U.S National Academy of Sciences, National Academy Press, Washington, DC

World Bank (2005) Climate change strategy for the South Asia Region (draft). http://siteresources. worldbank.org/SOUTHASIAEXT/Resources/Publications/448813-1231439344179/ 5726136-1232505590830/torSARCCS.pdf

Zaman M (1998) A note on the sacred groves in Afghanistan. In: Ramakrishnan PS, Saxena KG, Chandrashekara UM (eds) Conserving the sacred for biodiversity management. UNESCO and Oxford and IBH, New Delhi, pp 151–152

Chapter 10
Southeast Asia

Lim Hin Fui, Liang Luohui, Leni D. Camacho, Edwin A. Combalicer, and Savinder Kaur Kapar Singh

Abstract Rich in biological and cultural diversity important for human survival, tropical forests in Southeast Asia provide a major management challenge. Loss and degradation of forests in the region are driven by a complex interplay of economic, social, cultural, political, and demographic factors. In rural areas, local communities have used traditional knowledge in the management of forest resources for centuries. The arrival of Western colonial rulers and the introduction of scientific forest management gradually marginalized traditional local forest management. Modernization and economic development continue to erode cultural diversity and traditional knowledge. In some communities, villagers have adapted to externally driven socio-economic changes emphasizing commercial economic activities; conversion to institutionalized religions (such as Christianity, Buddhism, and Islam); and formal education. In other societies, local communities continue to manage local natural resources based on traditional knowledge and governance systems. Post-colonial governments from the 1940s to the 1970s maintained policies that gave the state legal control over all, or nearly all, forest lands. In most countries local, traditional management systems are not legally recognised or accepted by state forest management authorities, and their role

Lim H.F. (✉)
Forest Research Institute Malaysia, Kepong, Selangor, Malaysia
e-mail: limhf@frim.gov.my

Liang L.
Institute of Sustainability and Peace, United Nations University, Tokyo, Japan
e-mail: liang@hq.unu.edu

L.D. Camacho • E.A. Combalicer
University of the Philippines Los Baños, College, Laguna, Philippines
e-mail: camachold@yahoo.com.ph; eacombalicer@yahoo.com

S.K.K. Singh
Malaysian Environmental Consultants Sdn. Bhd, Ampang, Kuala Lumpur, Malaysia
e-mail: savinder.gill@gmail.com

J.A. Parrotta and R.L. Trosper (eds.), *Traditional Forest-Related Knowledge: Sustaining Communities, Ecosystems and Biocultural Diversity*, World Forests 12, DOI 10.1007/978-94-007-2144-9_10, © Springer Science+Business Media B.V. (outside the USA) 2012

in sustainable forest management has not been recognized. To achieve long-term forest sustainability in Southeast Asia, new approaches involving empowerment of local communities to manage natural forests, along with selective combinations of traditional and modern scientific management practices, may prove to be a way forward.

Keywords Biodiversity conservation • Forest governance • Forest history • Indigenous peoples • Non-timber forest products • Shifting cultivation • Southeast Asia • Sustainable forest management • Traditional knowledge

10.1   Introduction

Of the world's total estimated forest area (4,033 million ha in 2010), 592 million ha (15%) are in Asia, of which 34% (214 million ha) is in Southeast Asia (FAO 2010). Located nearly entirely within the humid tropics, the Southeast Asia region includes the second largest rainforest area in the world (Fig. 10.1). The region is divided between mainland Southeast Asia (Myanmar, Thailand, Lao PDR, Cambodia, and Vietnam) and insular Southeast Asia (Malaysia, Singapore, Brunei-Darussalam, Indonesia, Timor-Leste and the Philippines). The region includes uplands of greater bio-cultural diversity as well as lowlands in river deltas and coastal areas. Several countries in the region (namely Indonesia, Lao PDR, Malaysia, and Myanmar) still have relatively extensive areas of closed forests of between 40% and 50% (Table 10.1). The lush forests and rich biodiversity of the region provided a solid foundation for the early hunting, gathering, shifting cultivation and traditional agricultural societies before adoption of modern agriculture, and the early domestication of many tree and perennial crops such as coconuts, bananas, sugarcane, annual and root crops (rice, taro, etc.) and animals (chickens, pig, water buffalo, etc.) that were integrated into diverse agro-forestry systems.

The region is characterized by tremendous ecological diversity. Indonesia, Malaysia, and the Philippines are 3 of the world's 17 'megadiverse' countries. This outstanding biological diversity is manifested in the region's fauna and flora as well as at the landscape level. Malaysian forests are home to at least 15,000 species of flowering plants and trees, 600 species of birds, 286 species of mammals, 140 species of snakes, and 80 species of lizards (Malaysia 2004). Four biodiversity 'hotspots'—Indo-Burma, Philippines, Sundaland, and Wallacea—in Southeast Asia are among 25 of the Earth's biologically richest and most endangered ecoregions, featuring exceptional concentrations of endemic species and experiencing exceptional loss of habitat (Myers et al. 2000).

Southeast Asia deserves special attention for five main reasons: its high proportion of land under forest, high population, cultural diversity, economic growth rates, and rapid decline in forest area. At present about 48% of land in Southeast Asia is forested. The region has a total population of 575 million (Table 10.1), or about 9% of the world's population. All the nations in the region are composed of many ethnic groups with distinctive languages, customs, norms and values, religious beliefs, customary laws, and institutions. An indicator of its rich cultural diversity is the

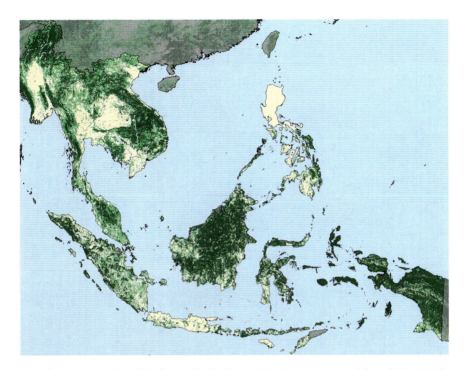

Fig. 10.1 Forest and woodland cover in Southeast Asia (Source: Adapted from FAO (2001)). *Key: Dark green* closed forest, *light green* open or fragmented forest, *pale green* other wooded land, *yellow* other land

large number of living languages, ranging from 17 in the small country of Brunei to 722 in the region's largest country, Indonesia. Of the total 1,519 living languages in Southeast Asia, 1,461 (96%) are languages of indigenous peoples.

A substantial percentage of the 557 million people of Southeast Asia are still dependent on the local forest ecosystems, of which humans are an integral part, for a wide array of goods and services needed to sustain their livelihoods. However, throughout the region, rapid population growth and economic development have resulted in widespread forest loss. According to FAO statistics, forest area in Southeast Asia declined at an average annual rate of 1%, from 247.1 million ha in 1990 to 214.1 million ha in 2010 (FAO 2010). In Thailand, for example, forest area has declined from 54% in 1961 (Poffenberger and McGean 1993) to 37% in 2010 (FAO 2010), while the forested area in Indonesia decreased from 62% in 1990 to 50% in 2010 (FAO 2010). The main causes of forest loss in the region are rapid timber exploitation; conversion of forests to agriculture associated primarily with population growth by poor migrants in search of farmland, forced migration, and resettlement of tribal communities (who are increasingly marginalized in their ancestral homelands); absence of effective forest management, which results in degradation of forest soils and vegetation; and conversion of forests to permanent agricultural or tree plantations (Poffenberger 1992). This situation of loss of forest land to oil palm plantation is illustrated in the case of Indonesia (Box 10.1).

Table 10.1 Selected land and forest characteristics of Southeast Asian countries

	Brunei-Darussalam	Cambodia	Indonesia	Lao PDR	Malaysia	Myanmar	Philippines	Singapore	Thailand	Timor-Leste	Vietnam	Total
Land area (million ha) (2010)	0.6	18.1	190.4	23.7	32.8	67.7	30.0	0.07	51.3	1.5	32.9	449.1
Forest area (million ha) (2010)	0.4	10.1	94.4	15.7	20.5	31.8	7.7	0.002	19.0	0.7	13.8	214.1
Percent forest area (2010)	63.3	55.8	49.6	66.2	62.5	47.0	25.7	2.9	37.0	49.3	41.9	47.7
Closed forest area (million ha)	n.a	6.70	91.7	11.9	16.0	26.8	4.3	n.a.	6.2	0.2	8.7	172.4
Percent closed forest area	n.a	37.0	48.2	50.2	48.9	39.5	14.2	n.a	12.1	11.3	26.5	38.4
Population (million) (2008)	0.4	14.6	227.3	6.2	27.0	49.6	90.3	4.6	67.3	1.1	87.1	575.5
Annual population growth rate (%) (2008)	1.8	1.7	1.2	1.9	1.7	0.9	1.8	2.9	0.6	3.2	1.1	1.7
Number of living languages	17	25	722	89	145	116	181	31	85	19	108	1,519
Number of indigenous languages	15	23	719	84	137	111	171	21	74	19	106	1,461
Indigenous population (million) (2007)	0.3	0.1	1.1	6.0	3.4	30	8.4	–	0.9	n.a.	7	57.2
Annual GDP growth (%) (2008)	–1.5	6.0	6.0	7.5	4.5	2.0	4.6	1.2	4.8	6.8	6.2	4.3

Sources: Land area, forest area, percent of forest area, population, population growth rate (FAO 2010); indigenous population and groups (Sobrevila 2008, www.vwam.com); living and indigenous languages (Lewis 2009); closed forest area (Guerra et al. 2006)

Box 10.1 Oil Palm Plantation Expansion and the Loss of Forest Resources in Indonesia

- Oil palm plantations have expanded rapidly in Indonesia in the past decade and currently cover seven million ha, managed by more than 600 companies.
- An additional forest area of 11 million ha was allocated to the oil palm industry but never planted; after cutting and selling the wood, the companies simply abandoned the lands.
- Over the next 10 years, local and provincial governments plan to issue licences for an additional 20 million ha for oil palm plantation. It is expected that most of the permits will be issued in forest areas, as the timber obtained from forest conversion can pay for plantation establishment costs.
- The government decentralization process, which started in 2000, affected the regulation of the plantation sector; the expansion of the oil palm plantations was no longer controlled nationally. The process of land acquisition for oil palm plantations is carried out locally, with political support given by the district government to oil palm plantation companies in exchange for financial support.

Source: Sirait (2009).

Very few large, ecologically intact natural forests, also known as the 'frontier forests' (World Resources Institute 1987), remain in Southeast Asia, largely confined to the islands of Borneo, Sumatra, Sulawesi, and Irian Jaya (Bryant et al. 1997, cited in Theilade et al. 2005). It is estimated that up to 42% of the total 200–300 fish species currently in Southeast Asia's peat swamp forests would disappear over the next century if habitat destruction continues at its present rate (Ng 2005).

In addition to conversion of forest lands for agricultural production, and unsustainable timber harvests, there have been uncontrolled, human-induced forest fires, particularly in recent decades, that have resulted both in forest loss and in serious impacts on human health. These fires, arising mainly from large-scale land clearing for industrial plantations and forestry practices that predispose forests to fire, have become an annual concern in many Southeast Asian countries, especially in Indonesia, where such large-scale fires occur almost every year. In 1982–1983, for example, forest fires destroyed 3.5 million ha of tropical forest in East Kalimantan (Boer 2002), of which 80% occurred in logged-over forest areas (Lennertz and Panzer 1983). The highly destructive peat forest fires in 1997 and 1998 affected five million ha of forest in East Kalimantan, Indonesia (Boer 2002). The resulting smoke from the peat swamp fires stretched over a million square kilometres, adversely affecting the health of up to 70 million people in the Southeast Asian region (Moore 2001). The haze between August and October 1997 caused by forest fires in Indonesia resulted in an estimated loss of US$300 million to Malaysia (Mohd Shahwahid and Jamal 1999) and up to US$9 billion in the Southeast Asian region as a whole between 1997 and 1998.

The impact of deforestation and forest fires on annual carbon release to the atmosphere is considerable. Based on an average carbon stock in live vegetation in Asian tropical moist and seasonal forests of 200 Mg C per ha (Houghton and Hackler 1995), an estimated 465 Tg C y^{-1} was released from Southeast Asia's forests between 1990 and 2000. This amounts to approximately 26% of net carbon release from tropical forests and 29% of world forests during this period. Available FAO statistics showed that in 2005 the carbon stock of Southeast Asian countries ranged from 50 Mt in Brunei to 6,725 Mt in Indonesia (FAO 2005). The region's terrestrial carbon stock will continue to decline if the loss of forest cover persists.

The manner in which forest resources are managed will have long-term impacts on Southeast Asia's biological and cultural diversity as well as global climate change. While modern scientific forest management practices are becoming more dominant in the region, the role of traditional knowledge in forest management is declining rapidly. In this chapter, we will discuss the role that these 'unscientific' traditional forest management practices have historically played in the sustainable management of forests in Southeast Asia, as well as the obstacles and opportunities that exist for the preservation and adaptation of these practices to meet the changing needs of their practitioners and society at large.

10.2   Traditional Forest-Related Knowledge in Forest Management

Traditional forest-dependent communities have developed a wide variety of forest and related agricultural management systems to sustain daily livelihoods and to ensure their food security and long-term survival. Studies have shown that in many situations swidden systems, or shifting cultivation, and associated forest management practices is a rational economic and environmental choice for farmers in the humid tropical uplands (Conklin 1957; Fox 2000; Mertz 2002). Nevertheless, more often than not these local management systems, and the associated local institutions and customary laws that support them, are not legally recognised or accepted by state forest management authorities, and their role in sustainable forest management has been ignored, if not forgotten.

10.2.1   Traditional Beliefs and Forest Management Practices of Forest-Dependent Communities

Over generations of living with forests, forest-dependent communities in Southeast Asia have developed their traditional forest-related knowledge (TFRK) to guide their use and management of forests. Many communities hold cultural and traditional beliefs conducive to forest conservation. These traditional beliefs and the customary institutions of local communities play an important role in regulating forest management.

For example, among the Kenyah of Long Uli in East Kalimantan, Indonesia, customary lands are designated and used for settlement and cultivation (ladang), protected forest (tanah ulen), fruit-tree groves (pulung bua), and fishing and hunting. Each village has regulations (hokum adat) for harvesting and distributing timber and non-timber forest products that determine when designated land can be opened for collection purposes and prohibit cultivation of this land. In addition, the community maintains two separate forest protection areas: one reserved for the village council, the products from which are to be used for village development, and another whose products are reserved for orphans and widows (Sirait et al. 1994).

Similarly, to the Dayak people in East Kalimantan, simpukng (a mixed fruit garden or forest garden) is established for environmental and economic benefits (vegetables, spices, honey, rattan, fruit trees, timber, construction materials, fuelwood, and medicinal plants) where rituals and spiritual beliefs influence its management and utilization (Mulyoutami et al. 2008). In Sumatra, Indonesia, the villagers in Jambi Province use the presence of ninik mamak (traditional leaders) and traditional taboos in conserving and using forest resources such as medicine plants, honey collection, rubber planting, fishing, hunting, rice, and vegetable farming with regulations that ensure the 'adat forest' (traditional forests) are managed on a sustainable basis (Zuraida 2008).

The Ikalahan people in the mountains of northern Philippines have a sophisticated system of forest management where forests are delineated for different functions, such as conservation, income-generation, and environmental service purposes, and their traditional farming practices are designed to conserve the soil and water (Dolinen 1997). For the aboriginal Semai community, 1 of 18 indigenous communities in Peninsular Malaysia, planted and inherited forest land species that have social, cultural and religious significance are either community-owned, family owned, or individual owned (Lim 1997).

Among forest-dependent traditional communities in Southeast Asia, certain forests or tree species are commonly believed to be associated with the tutelary spirits and deities of their villages and ancestors. These forests or tree species of cultural importance are locally protected with a perceived need to live in harmony with these spirits and deities. For example, the Khmu people in Laksip Village, Luang Prabang, Laos, believe that their tutelary deity lives in a high mountain to the northeast of the village. Forests there are well-preserved, and the mountain also serves as an important watershed for the village and others further downstream. The ethnic minority communities of Ratanakiri Province, Cambodia, believe that local spirits inhabit certain forests and forbid harvest of any forest products in those sacred areas (Poffenberger 1999). Sacred forest sites are important for forest and biodiversity conservation as well as local social-economic functions in Indonesia (Wadley and Colfer 2004; Soedjito 2006). Indigenous peoples in the Cordillera, Philippines believe many water-bearing tree species are associated with spirits (anito), so they conserve these trees (Carino 2004), which are important for watershed conservation in the region. In the northwestern mountain region of Vietnam, the local Thai ethnic community's beliefs in spiritual and magical power have been the basis of customary laws governing their management of natural resources (Pham and Trung 2004).

The relevance of local traditional forest-related knowledge in the sustainable management of forest resources is evidenced from various research findings. Even though much of the land in the mountainous regions of mainland Southeast Asia previously under shifting cultivation has been converted to permanent agriculture or plantations (for cash crops such as cabbage, fruit orchards, and plantations of rubber, tea and teak), the traditional knowledge and practices used in indigenous forest and land management remain (Rerkasem et al. 2009b). In this region, traditional forest management skills have been gradually adapted and incorporated into new cropping systems, and into forest management practices used for protection of headwater supplies for irrigation and domestic consumption, as well as for management of live fences and fuelwood forests (Rerkasem et al. 2009b).

The application of traditional forest management is clearly demonstrated in a study in Thailand conducted as part of the Highland Mapping Development and Biodiversity Management Project of the Inter-Mountain Peoples' Education and Culture in Thailand Association (IMPECT) in collaboration with the Forest Peoples Program (Charoenniyomphrai et al. 2006; Colchester et al. 2006). The study focused on the knowledge, customs, and traditions practiced by the Hmong and Karen (Pga k'nyau) peoples in the sustainable use and conservation of natural resources and biodiversity, and examined how these highland communities have adapted to the impacts of externally imposed laws, policies, and development processes (Box 10.2).

10.2.2 Technological Practices and Strategies

10.2.2.1 Traditional Agroforestry and Shifting Cultivation Systems

Agroforestry may be defined as 'land use based on planted trees, provid[ing] productive and protective (biological diversity, healthy ecosystems, protection of soil and water resources, terrestrial carbon storage) forest functions' (van Noordwijk et al. 2003). With their extensive knowledge of useful trees, plants, and animals, people in rural communities throughout Southeast Asia have developed a high diversity of agroforestry systems, many of which have been described in the literature (Christanty et al. 1986; Hong 1987; Soemarwoto and Conway 1992; van Noordwijk et al. 2003; Belcher 2001; Michon and Bouambrane 2004; Michon et al. 2007). The forests in which local agroforestry practices take place have been termed 'domestic forests' by Michon et al. (2007). People in traditional communities also rely heavily on traditional knowledge for predicting weather to plan and prepare their agroforestry activities and minimize impacts of extreme weather events, as described by Galacgac and Balisacan (2009) in Ilocos Norte Province in Northern Philippines

Agroforestry in Southeast Asia includes both simultaneous and sequential systems. Examples of simultaneous systems are the common forest gardens established by planting around houses, which persist even after people have

Box 10.2 Traditional Knowledge, and Customary and Sustainable
Use of Natural Resources Among the Hmong and Karen (Pga k'nyau)
in Thailand

The Hmong and Pga k'nyau communities in Thailand have a dual system of leadership made up of traditional leaders and elected village heads recognised by the government. The traditional leaders hold authority based on the history and structure of the community. They play the main role in building a relationship between people in the community and natural resources and biodiversity management, by guiding their customary use of soil, water, forest, animal, and plant resources. Indigenous knowledge about customary resource use is expressed as patterns of thought, production, beliefs, customs, traditions, and rituals. All of these tangible expressions result in a balance between the maintenance of life and dependence on nature as appropriate for each people. The process of transmitting this knowledge from one generation to another is incorporated into customs such as teachings, songs, legends, stories, rituals, and practical daily activities, especially those related to their agricultural and forest-dependent work.

There are similarities between the Hmong and Pga k'nyau in their categorization of, and beliefs about, the ownership of natural resources, in that the ultimate owners are supernatural entities such as the Lord of the Water, the Lord of the Forest, the Lord of the Mountain, or guardian spirits. Both communities believe that these spirits are the caretakers and guardians of natural resources. People using these resources ask permission to do so only in order to maintain their livelihoods. After having been granted permission by the spiritual owners, they are required to use the resources carefully and sensibly. These beliefs aid in the sustainable use and conservation of resources.

The work of various government and non-governmental agencies has caused adaptations to occur or has been met with community resistance in the past. For example, Mae Pon Nai village has struggled against the thinking of government and religious organizations, and has not accepted a great deal of outside cultural influence. Khun Ya village has not adopted monoculture cash crops because they believe it destroys their indigenous agricultural methods. Community members have joined together at the community, watershed network, and national people's network levels in order to find appropriate solutions to their problems.

Source: Charoenniyomphrai et al. (2006).

moved away. These forest gardens are rich in edible fruits, oil seeds, rattans, medicinal plants, resins, and other useful species meeting the needs of the Dayak people in Borneo (Brookfield et al. 1995). In the case of Sarawak in Malaysia, the local communities maintain forest land cleared for padi farming (temuda),

Fig. 10.2 Agroforest in Chiang Mai Province, Northern Thailand (Photo: Liang Luohui)

and pulau, forest areas set aside for production of essential items such as timber for house construction, boat building, jungle vegetables, rattan and other non-timber forest products, hunting, and protection of water catchment areas (Majid-Cooke 2005).

Sequential systems include a variety of rotational agroforestry and shifting cultivation practices, all of which include a short phase of cropping (usually lasting from one to a few years) followed by a relatively long phase of forest fallow. The cropping phase for producing mainly annual food crops involves clearing of secondary growth, burning, dibbling the seeds of annual crops, weeding and harvest. The cropping phase typically involves diverse mixtures of crops (cereals, root crops, vegetables, etc.) that support a balanced diet for shifting cultivators, with minimal disturbance of soil structure, soil seed banks, and root-mats because of zero tillage, while preserving some mature trees for subsequent forest regeneration (Figs. 10.2 and 10.3). The fallow phase is essential to help restore soil fertility lost during the preceding cropping phase. The forest is an integral and essential part of the shifting cultivation system (Conklin 1957; Colfer et al. 1997).

Fig. 10.3 A local farmer in Chiang Mai Province, Northern Thailand explaining his techniques for managing this agroforest for producing a special type of bamboo to make traditional pipes (Photo: Liang Luohui)

The duration and management of the fallow phase differ across cultures, space, and time. The duration of the forest fallow phase is normally above 10 years—sufficient for recovery of soil fertility—with a general preference for the forest re-growth of between 10 and 20 years (Brookfield et al. 1995). As forests re-establish through natural successional processes during the fallow phase, products of these natural forest fallows are collected by the local people. Inter-planting of useful trees with food crops during the cropping phase and enrichment planting during the forest fallow phase are practices widely used by indigenous communities to produce forest products and to accelerate soil regeneration (Burgers et al. 2005; Michon et al. 2007).

10.2.2.2 Management of Non-timber Forest Products

The indigenous peoples of Southeast Asia have been using and managing non-timber forest products for centuries, if not millennia (Fig. 10.4). Swidden farming, collection of forest fruits, traditional medicines, building and handicraft materials, and hunting and fishing continue to be practiced by these forest-dependent communities (Plant 2002). The aromatic resin, gaharu is a valuable non-timber forest product produced by trees of the genus *Aquilaria* found in tropical forests South and Southeast Asia (Lim et al. 2008). Until recently the process of the resin formation was poorly understood by forest scientists, but

Fig. 10.4 Edible pupae of forest insects are a traditional non-timber forest product in Luang Prabang Province, Northern Lao PDR (Photo: Liang Luohui)

Penan collectors of Indonesian Borneo have long recognized the complex ecology of resin formation involving site conditions, soils, associated trees, microclimate, fungi, and insects (Donovan and Puri 2004). In the Philippines, NATRIPAL (Nagkakaisang Tribu ng Palwan), an association of indigenous people in Pelawan Province, was organized in 1989 to advocate for recognition of ancestral tenurial rights and to enhance indigenous people's organizational capacity in sustainable management, trading, and marketing of non-timber forest products such as wild honey and almaciga (*Agathis philippinensis* Warb) resin (Ella 2008).

Traditionally, fruit trees, bamboo, and rattan are inter-planted with annual crops during the cropping phase in rotational agroforestry systems in Indonesia (Christanty et al. 1986; Belcher 2001). Rubber has been integrated by local farmers into their rotational agroforestry based on local knowledge to create a biodiversity-rich jungle rubber system (Penot 2004). Benzoin production from *Styrax tonkinensis* is integrated with hill rice under shifting cultivation in Laos (Kashio and Johnson 2001; Takeda 2006). In Sumatra (Indonesia), the establishment of damar (*Shorea javanica*) forest gardens with fruit trees such as durian (*Durio ziebethinus*), nangka (*Artocarpus heterophyllus*), manggis (*Garcinia mangostana*), petai (*Parkia speciosa*), and duku (*Aglaia dookkoo*) generates cash income and has biodiversity values (Poffenberger 2006).

In their traditional management systems, local people in northern Laos enrich forest fallow with paper mulberry (Burgers et al. 2005). Extensive plantations of teak (*Tectona grandis*) were established in this way during the nineteenth and early twentieth centuries in Myanmar, Thailand, Indonesia (Ball et al. 1999), and currently in Laos. Apart from wild gathering, local people have long cultivated useful tree species for non-timber products in Southeast Asia, such as damar (*Shorea javanica*) to produce resin and timber in agroforestry farms (Michon et al. 2007), and rattan (*Calamus caesius*) in swidden fields, to earn income and mark land-holdings (Belcher 2001).

10.2.2.3   Landscape Management

In many Southeast Asian countries, a variety of local ecosystem or land use types form complex landscape mosaics (Forman 1995). Local communities sustain themselves with the natural resources available from these landscapes, of which forests are an integral part. Their traditional wisdom and experience enable traditional communities to make the full use of the landscape through classification of the landscape into appropriate land use units that are managed under different customary regulations. Land units important to watershed and social-cultural development are often designated as sacred sites for protection, as discussed earlier (Sect. 10.2.1). The Kenyah people in East Kalimantan, Indonesia, allocate the land and forests in the village landscape to different land uses such as settlement and cultivation (lading), forest protection (tanah tulen), fruit-tree groves (pulung bua), fishing, hunting, and harvesting of timber and non-timber forest products (Sirait et al. 1994). In western Java, Indonesia, the indigenous Kasepuhan community in the Halimun Mountains uses the surrounding land and forests for various uses by adopting different land use models including sawah (paddy field), huma (swidden cultivation), talun (agroforestry garden), kebon (garden), leuweung titipan (entrusted forest), leuweung tutupan (closed forest) and leuweung bukaan (open forest) for watershed conservation and limited non-timber forest product collection (Hendarti and Youn 2008). Elsewhere in Indonesia the traditional coastal landscape management integrates paddy fields for rice-fish culture, tambak ponds for polyculture, and mangrove forests for coastal fisheries enriched by waters flowing from the paddy fields (Davidson-Hunt and Berkes 2001). Indigenous farming communities in the Cordillera region of the Philippines harmonize muyong system (meaning forest or woodlot), swidden farms, and rice terraces on the landscape (Camacho et al. 2011).

10.3   Emergence of 'Scientific' Forestry and Conservation Management

10.3.1   Colonialism and Scientific Forestry

The history of Southeast Asian countries is marked by waves of Western commercial exploitation followed by political domination over the past few hundred years. Apart from Thailand, all other countries in the region were under some form of European colonial rule during their recent history, in addition to their occupation by Japanese forces during World War II (1941–1945): Brunei (1888–1984), Malaya (1824–1957), Singapore (1824–1963), and Myanmar (1824–1948) under the British colonial rule; Cambodia (1863–1953), Lao PDR (1894–1954), and Vietnam (1885–1954) under French rule; Indonesia (1800–1945) under the Dutch; and the Philippines under the Spanish (1521–1898) and the United States (1899–1941). Colonialism transformed nearly all aspects of forest resource management and utilization, as well as the forest-based livelihoods of local communities. For example, during the French colonial period, extensive rubber plantations were developed in forest areas in Vietnam and Cambodia, with forest-dependent indigenous peoples employed to work in the plantations (Lang 2001).

Colonialism in Southeast Asia resulted in dramatic changes in forest control, use, and management by the state. The erosion of customary forest management systems and the rise of state agencies and private companies as forest managers coincided with the rapid loss of forests in the twentieth century (Poffenberger 2006). In the Philippines, for example, highly regulated, centrally controlled (and industry-biased) forest policies and management approaches were applied from the colonial period until the early 1980s (Pulhin et al. 2008). Post-colonial governments in Southeast Asia from 1940 to the 1970s maintained centralized policies stipulating that all lands in the public domain, which included all classified forest lands, belonged to the state. Indigenous peoples were deprived of their customary rights to their land, including forests, and the benefits arising from the utilization of forest resources.

10.3.2   Protected Areas and Biodiversity Conservation

One impact of Western colonialism can be seen in current approaches to protected area management. The Western-based conservation system that predominates in Southeast Asia and much of the rest of the world—seen by many as 'eco-colonialism' (Cox 2000)—imposes Western conservation paradigms and power structures on indigenous forest-dwellers (Nicholas 2005). In contrast to traditional (i.e., indigenous) approaches, including customary law and governance systems, Western conservation ethics tend to emphasize resource protection, legal use, intellectual property rights, and wise use of the protected areas. As a result, the presence of indigenous peoples in and around protected areas has and continues to be seen as a liability rather than an asset for effective protected area management.

There is a growing appreciation of the fact that protected area establishment has very often been detrimental to indigenous peoples' interests, most obviously when their access to protected areas is denied. Traditional knowledge, institutions, and practices have usually been and continue to be sustainable in terms of meeting livelihood needs without jeopardizing long-term biodiversity conservation objectives and forest ecosystem functions. In the Philippines, the 'muyong' system of the Ifugaos contributes to forest conservation and regeneration, soil conservation, watershed rehabilitation, sustainable farming (Camacho et al. 2011), and meeting livelihood needs such as food, fuelwood, construction materials, and medicine (Butic and Ngidlo 2003). The muyong are typically rich in plant biodiversity. A study of 67 muyong woodlots plots (each 625 m²; total sampling area 4.2 ha) reported a total of 264 plant species belonging to 71 plant families in (Rondolo 2001). The average number of plant species per woodlot was 30, and ranged from 13 to 47. Nearly 90% of these species (234 of 264) were considered useful (Ngidlo 1998).

10.3.3 Industrial Forestry

The rise of Western colonial powers in Southeast Asian countries, nationalization of forests in the twentieth century, the continued emphasis on timber production in the post-colonial period, economic globalization, and rising demand for timber by emerging economic giants such as China and India have resulted in both the development and expansion of the commercial timber industry and concentration of ownership and control of forest lands, principally by states. By 2005, FAO (2010) shows forest ownerships in Southeast Asia is mainly with the states—Brunei, Cambodia, Lao PDR, Myanmar, and Singapore (100%); Malaysia (98%); Indonesia (91%); Thailand (88%); Philippines (85%); and Vietnam (72%). The process witnessed the expansion of timber industry operations throughout Southeast Asia where vast areas of forests were deforested. As noted earlier, the rise of state forest agencies and private companies as forest managers has coincided with an accelerating loss of natural forests the region during the post-World War II era. In Southeast Asia, tropical rainforests receded from 250 million ha in 1990 to 60 million ha in 1989 (Poffenberger 2006), a decline of 76%. In the meantime, indigenous systems of management were displaced, erosion of customary forest management occurred, and this further accelerated deterioration of forests. In short, the state gained legal and administrative control over most if not all forests at the expense of local and indigenous communities (invalidating their customary laws and other aspects), favouring industrial interests (individuals and firms) by granting concessions for, or giving away, large forest areas.

By promoting 'scientific forestry,' the colonial and later modern states gradually took control of the forests and imposed centralized technocratic management from the nineteenth and twentieth centuries until the present. In Vietnam, until the 1980s, almost all lands and forests were administered directly by the state and its subsidiary bodies. In the 1980s, the state began to experiment with means of devolving rights over land to households and individuals and to the private sector. Such land reforms resulted in increases in agricultural production, and since the 1990s similar reforms have begun to be undertaken in the forests. About 18% of the country's 12.6 million ha of natural forests were leased to individuals and firms to manage, and only 4.4% to local communities, indicating priority given to individuals and firms. Government policies that encourage capital investment and allow joint ventures and corporations to control forest and agricultural lands are expanding in scope, and state agencies continue to pursue policies aimed at curbing swidden farming

Fig. 10.5 Conversion of traditional shifting cultivation to teak plantation in Luang Prabang Province, Lao PDR (Photo: Liang Luohui)

(Colchester and Fay 2007). Similarly, the Indonesian government controls all forests; of Indonesia's estimated 86 million ha of forest, about 27% was allocated to private companies for logging and plantation, while about 0.2% was given limited use rights as 'community forest' (Colchester and Fay 2007). This eventually led to state management of forests mainly for the sake of industrial forestry interests. The state hands the rights to harvest timber to private companies, thus excluding local communities from the forests (WRM 2002b). Consequently, swidden agriculture, once the dominant form of land use throughout the uplands and much of the lowlands of Southeast Asia, is being replaced by other land uses (Rerkasem et al. 2009a)(Fig. 10.5).

10.3.4 Modern Agriculture Development

Many innovative practices have been developed and used by local communities to improve the productivity and sustainability of shifting cultivation systems in Southeast Asia (Cairns and Garrity 1999; Rerkasem 2003; Burgers et al. 2005; Rerkasem et al. 2009a). In spite of the availability of these innovations, government policies in the region have generally aimed to curtail what are perceived to be 'backward' traditional shifting cultivation practices and to encourage conversion to sedentary agriculture. Such policies have often led to unforeseen and counter-productive consequences. For example, a land allocation program initiated in Laos in the early 1990s to stop shifting cultivation sought to increase land tenure security, enable farmers to invest in land development for sedentary agriculture, and preserve previously fallow land for forest conservation. However, the allocation of extensive fallow lands for forest conservation led to a shortening of the fallow phase in the remaining lands to less than 3 years (rather than a conversion to sedentary agriculture), which resulted in land degradation and yield declines (Rerkasem et al. 2009b). The shortening of the fallow phase has also made it impossible for these farmers to continue benzoin production from *S. tonkinensis,* as these trees take 7 years to mature (Takeda 2006).

In summary, the overall impact of Western colonialism, development of modern states and the adoption of scientific forestry emphasizing timber production for international markets, and agriculture development, has been the reduction in the forest areas in Southeast Asian countries. Infrastructure development, improved communication and transportation networks, and intrusion of external markets have transformed and marginalised the traditional forest communities, resulting in denial of their customary rights, as well as erosion and loss of traditional knowledge, practices, and institutions.

10.4   Current Challenges in Conservation and Sustainable Utilization of Traditional Forest-Related Knowledge

10.4.1   Governance Issues and the Legal Rights of Local Communities

In the Southeast Asia region, despite the existence of laws that consider the rights of local communities (Table 10.2), competing interests in forest land use have often led to serious local problems such as increases in rural poverty and land conflicts, as illustrated below in the following examples from, Cambodia, Thailand, and Vietnam (Colchester and Fay 2007).

In Cambodia, the Community Forestry Sub-Decree (CFSD) was passed by the Royal Government in 2003 (Oberndorf 2006). However, the weakness of government implementation of forest laws and policies, and competing interests in forestlands from the state and the private sector, have prevented effective enforcement of existing laws protecting the rights of local communities. Delineation of indigenous lands is at an experimental stage; individual land titling has focused on urban areas, and little attention is given to agricultural and forest lands of local and indigenous communities (Colchester and Fay 2007). Community rights regarding forests are largely ignored by forest concessionaires in Cambodia, leading to serious land conflicts. While nearly one million ha of forest concessions have been granted, only 20% of farmlands have been registered and only 200 small pilot community forestry areas have been established (Colchester and Fay 2007). In addition to timber concessions, controversial land concessions for agro-industrial crops like cassava, sugar

Table 10.2 Specific laws related to land and forest ownership, customary use and management rights, and traditional forest-related knowledge in selected Southeast Asian countries

Country	Law	Remarks
Cambodia	Land Law (2001)	Grants collective land ownership rights to indigenous communities. This enables the indigenous peoples to self-determine development (NGO Forum on Cambodia 2006)
	Forest Law (2002)	Leases areas of production forest and permanent forest reserve to local communities to manage and benefit from the resources
Indonesia	Constitution	Accords the state a controlling power over land and natural resources and recognize customary law communities (Colchester and Fay 2007)
	Basic Agrarian Law (BAL) (1960)	Recognizes the collective rights in land of customary law communities but treats these as weak usufructs on state lands subordinate to state plans and interests (Colchester and Fay 2007)
	Forestry Law (1999)	Limited use rights in forests can be accorded as long-term 'customary forests,' 'special purpose areas,' and 'village forests,' and as short-term 'community forests'(Colchester and Fay 2007)
Lao PDR	Constitution	All forest land lands belong to the 'national community represented by the State.' The law allows the government to allocate rights of 'rational usage' of forest areas to villages and individuals, while the customary use of natural resources within village boundaries is also explicitly recognized (Colchester and Fay 2007)
Philippines	Indigenous Peoples' Rights Act (1997)	Protects indigenous communities' rights in general, including their rights to traditional knowledge (Disini 2003)
Malaysia	Aboriginal Peoples Act (1954)	The customary rights of aborigines' lands within the national legal framework are stated to include aboriginal reserves, aboriginal areas, rights of occupancy, compensation for fruit or rubber trees, and compensation for use of aboriginal areas and aboriginal reserves
	Sabah's Land Ordinance (Cap 68)	Native customary rights to land are outlined
	Sarawak Land Code (Cap 81)	Determines land tenure and administration in the state, and provides for native land and native customary land. (Majid-Cooke 1999)
Thailand	Thai Constitution	Provides recognition of customary natural resource management by tribal and indigenous peoples (Anonymous 2006)
	Community Forestry Act (1996)	Enables the rural community to have a legal right to manage the land and forest in the village (Rerkasem and Yimyam 2011)
Vietnam	Constitution	Vests all lands and forests in the state (Colchester and Fay 2007)
	Land Law (1993)	Allocates forest land to other users such as forest companies, communities, households, and individuals (Sam and Trung 2001)
	Law on Forest Protection and Development (1991)	Identified three categories: protection (watershed, sandy and sea wave); special use forest (parks, conservation, historical and cultural parks); and production forest (Sam and Trung 2001)

cane, rubber, pulpwood, and palm oil have brought about local land use conflict and resistance from local villagers (WRM 2006). Although national data on the status and poverty of forest peoples is lacking, hardship and increasing conflict have resulted from the loss of forest resources and encroachment of land concessions (Colchester and Fay 2007).

In Thailand, under Royal patronage and through various foreign assistance programs, efforts have been made to promote alternative upland economies in forested regions, including substituting flower, fruit, and vegetable production for swidden farming and opium cultivation. Co-management of protected areas is also being tested, with mixed results (Colchester and Fay 2007). While there have been some successes in protected area co-management, these have not been extended to other communities owing to lack of an enabling legal framework.

However, the spontaneous spread of 'community forestry' in Thailand, where village committees assume management responsibilities over forests and woodlots, is notable. According to Asia Forest Network,[1] area under community forest management totalled 328,000 ha in 1998 or 1% of the total 30 million ha of forest land in Thailand. Studies by the University of Chiang Mai (Colchester and Fay 2007) suggest these community forests extend over some 1.3 million ha, a figure that does not include the wider areas under customary land management systems. Approximately 0.2 million ha of community forests have been 'permitted' by the Royal Forest Department, although the legal basis for this is unclear. On the other hand, the Thai government continues to pursue an intermittent policy of exclusion and resettlement of upland communities, while encouraging tree plantations and tea estates in uplands forests (Colchester and Fay 2007).

In Vietnam, while existing laws allow communities to hold forest and associated agricultural lands, the tenure afforded them is weaker than that available to individuals and households. Under the 1993 Land Law and a decree issued in 1994 (Poffenberger 1999), forest land is allocated on long-term, transferable leases to individuals, households, and enterprises, providing them with strong rights to control and manage resources. The law is less clear about transfers to communities. Moreover the Civil Code does not recognize communities as having legal standing, and only 4.4% of forests in Vietnam are currently allocated to communities as such (Colchester and Fay 2007).

10.4.2 Forest Resource Degradation and Intensification of Agricultural and Tree Crop Production

While it is now recognized that the principal causes of forest destruction and degradation in the Southeast Asia region are conversion of forests to permanent agriculture and large-scale plantations for wood, fibre, and commodities such as oil palm (Climate Change Monitoring and Information Network 2009), the practice of

---

[1] www.asiaforestnetwork.org/tha.htm

shifting cultivation and those who practise it continue to be unfairly blamed for destruction of forests (Cairns 2008).

Of particular concern is the major threat to traditional forest-related knowledge in the region posed by the sustained drive to replace shifting cultivation by permanent commercial agriculture (especially oil palm), accompanied by efforts to deprive the forest dwellers and forest-dependent communities of their decision-making rights. By 2009, oil palm plantation development in Indonesia covered more than seven million ha managed by 600 companies and one million small landholders. There are plans to develop another 20 million ha of oil palm plantation, mainly in forest areas (Sirait 2009).

External influences such as deforestation and forest degradation, development of agricultural crops in the surrounding areas, and depletion of forest resources in the vicinity of forest-dependent communities have resulted in changes in local ways of living. Such changes give rise to new needs and values among local populations that often result in the erosion and loss of traditional knowledge. In the case of Indonesia, oil palm plantation development has already resulted in detachment of indigenous peoples from their customs and cultures because of individualization of ancestral lands, descendant group lands, and household lands. The land conversion process also creates conflicts that damage community solidarity and local institutions (Sirait 2009).

The case of the forests associated with upland rice production in the Philippines further illustrates the threats, issues, and concerns about traditional forest-related knowledge. The owners of 'muyong' forests (Box 10.3) are now practising enrichment planting to enhance depleted muyong areas. Some muyong owners are choosing to use non-native, often fast-growing, tree species for assisted natural regeneration or enrichment plantings to take advantage of their short rotation periods and/or potential future income potential. These include fast-growing exotic species such as *Gmelina arborea,* and *Cassia spectabilis,* or high value species like *Swietenia macrophylla* (planted as in investment for retirement or college education expenses of children and grandchildren). Inclusion of these exotic species may pose threats to muyong biodiversity (due to their high potential for regeneration in disturbed habitats). There have been cases where muyong owners were tempted to clear portions of their woodlots for replacement with such exotic tree species. In some cases, local people cleared muyong forests for tambo or tiger grass (*Thysanolaena maxima*) production, or to implement sloping agricultural land technology (SALT[2]) practices that are completely alien to what the people have been doing in the area for many generations. Consequently, over the past 50 years, the size of the cultivated terraces had significantly shrunk, from 15,000 ha to just about 5,000 ha today. In a 2003 survey, local officials mentioned that one third of the rice terraces in the Barangay of Bangaan in Banaue are already damaged (Cagoco 2006).

---

[2]SALT is a technology package of soil conservation and food production that integrates several soil conservation measures to minimize soil erosion and maintain soil fertility, and involves planting field crops (such as legumes, cereals, and vegetables) and perennial crops (such as cacao, coffee, banana, citrus and fruit trees) in bands 3–5 m wide between double rows of nitrogen-fixing shrubs and trees planted along the contour (MBRLC 1988; Tacio 1993; Evans 1992).

Box 10.3 Agroforestry Practices and Local Forest Management of the Ifugao in the Philippines

The Banaue Rice Terraces of the Ifugao, a UNESCO World Heritage Site in Ifugao Province in the Cordillera Administrative Region in north central Luzon, have been supported for 2,000 years by indigenous knowledge management of muyong, locally owned agroforests that cap each terrace cluster. These communally managed forests areas are rich in biodiversity (Conklin et al. 1980). The terraces serve as a rainwater and filtration system and are saturated with irrigation water year-round. The Ifugao rice terrace paddy farming system allows protection and conservation of significant and important agricultural biodiversity and associated landscapes, while promoting tourism based on their aesthetic value. All these are made possible by local agroforestry and forest management practices.

Ifugao culture and laws. Ifugao culture revolves around rice, which is considered a prestige crop. There is an elaborate and complex array of rice culture feasts inextricably linked with taboos and intricate agricultural rites, from rice cultivation to rice consumption. Partaking of the rice beer (bayah), rice cakes, and betel nut is an indelible practice during the festivities and ritual activities.

Ifugao's muyong system of management. Muyongs are privately owned forests or woodlots typically up to 5 ha, which are inherited from generation to generation. Muyongs are managed through collective effort under traditional tribal practices and play an important role within the tribal economy. They are the primary source of fuel wood, construction materials, wood carving materials, food, and medicines. The muyong system involves a forest conservation strategy, watershed rehabilitation techniques, a farming systems, and assisted natural regeneration (ANR), all of which are living proof of the Ifugao's knowledge of silviculture, agroforestry, horticulture, soil and water conservation, enrichment planting and protection, and efficient silvicultural, harvesting, and good wood utilization practices.

Agroforestry and multiple cropping. Traditionally, the Ifugaos adopted agroforestry in woodlots and multiple cropping in swiddens as economic insurance in case of crop failure in the terraces. The integration of value-added tree crops and herbs in natural muyong vegetation and swiddens has been found to be highly compatible. Species preferred for enrichment of natural vegetation are mostly rattan, coffee, santol (*Sandoricum koetjape),* and citrus, while bananas, taro, and cadios (*Cajanus cajan*) are integrated into swidden farms (Butic and Ngidlo 2003). Rondolo (2001) found that almost all woodlots contained commercial plantings of coffee (88%), bananas (66%), and citrus (49%). Edible rattan (*Calamus manillensis*) is also included in almost all woodlots. Rattan is integrated in woodlots for its edible fruits and poles/canes

(continued)

Box 10.3 (continued)

for handicraft. Betel palm (*Areca catechu*) and ikmo (*Piper* spp.) are also cultivated in the woodlots for betle nuts, ritual purposes, and medicinal values.

The 'ala-a' system way of forest sustainability. Besides the privately owned muyong, traditional communal forest (ala-a) management is also important. The ala-a are generally located on lands not cultivated as swidden, lands too far from villages to be covered by a private claim, or lands not identified as hunting grounds. The ala-a are communally managed for fuelwood collection, construction materials, food, medicine, and other products that may be used in homes or farms. The use of the ala-a is controlled by consensus that the resource is to be shared, following two basic rules—no burning, and no gathering beyond what is needed for personal use. The ala-a are not used as sources of wood for sale outside the village.

Sources: Conklin et al. (1980), Ngidlo (1998), Dacawi (1982), Klock and Tindungan (1995), Rondolo (2001), Butic and Ngidlo (2003), Elazegui and Combalicer (2004), Camacho et al. (2011).

10.4.3   *Erosion and Loss of Traditional Forest-Related Knowledge*

The widespread land alienation that accompanied agricultural crop development in Southeast Asia by colonial rulers, nationalization of forests, and expansion of timber industry in the twentieth century have generally brought about the progressive erosion and loss of traditional forest-related knowledge throughout Southeast Asia (Poffenberger 2006). Land alienation and the granting of land concessions by the governments have resulted in removal of extensive native forest areas and forest resources from local management and use.

The colonial period witnessed the creation of 'dual societies' (Boeke 1966), namely a capital-intensive growth sector involving extractive industries, manufacturing, and estate agriculture on the one hand and an 'underdeveloped' subsistence sector on the other. The subsequent expansion of modern commercial agriculture into the subsistence sector saw the erosion and even loss of traditional knowledge. In rural areas, swidden agriculture was maintained in some areas for social and cultural reasons, amidst cash income generation from commercial crop cultivation. Padoch et al. (2007) noted that 'swidden is gradually disappearing in most parts of Southeast Asia.'

Hence, despite their importance in sustaining rural livelihoods, traditional forest-related knowledge and practices are fast disappearing in most Asian countries (Parrotta et al. 2009). The indigenous communities of Peninsular Malaysia of about

150,000 people are also rapidly losing their traditional knowledge as a result of increasing loss of forest areas, integration into mainstream society, urbanization, changing lifestyle, and lesser interests among the younger generation to learn traditional knowledge (Lee et al. 2009; Chai et al. 2008). Consequently, most indigenous communities at present are facing serious difficulties maintaining their traditional cultures, natural resource utilization practices, and traditional forest-related knowledge because of varying degrees of influence from outside economic, social, and cultural forces (Molintas 2004). Traditional shifting cultivation, hunting, collecting forest products, and fishing for livelihood among the aborigines of Peninsular Malaysia has now diversified to include non-traditional activities such as planting of cash crops (corn, fruit trees, vegetables, and rubber) and earning income from nearby oil palm estates (Fadzil and Hamzah 2008).

10.4.4   Modernization and Its Impacts

Colonialism also brought about the spread of formal education to local indigenous communities. While formal education increased literacy in rural communities, it also transformed rural livelihoods as formal education tends to discourage the use of indigenous knowledge and the practice of local traditions. In the Philippines, the Ifugaos are inclined to set aside their indigenous knowledge systems as they gain more education (Save the Ifugao Terraces Movement 2008).

Together with modern education, institutionalized religions have been introduced relatively recently in many areas, displacing traditional religious beliefs and practices, which has changed the manner in which forest resources were perceived, understood, and used at the local level. Formal education erodes traditional knowledge when youth are taught that their parents' religious beliefs, cultural practices, and traditions are 'superstitious,' 'backwards,' or otherwise lacking or inferior, and that their traditional practices are not scientific. Institutionalized religion generally erodes traditional knowledge by denying the existence of the spiritual world and perspective on human relationships with nature that often underlie traditional knowledge and practices. Institutionalized religion (evangelical and Catholic) forbidding traditional practices such as animal offerings among the Ifugaos in the Philippines have resulted in the loss and extinction of rituals (Cagoco 2006). The conversion to modern religion has changed body of knowledge, belief, and customs of affected communities, thus resulting in the neglect and loss of traditional knowledge, values, norms, and practices.

The development and increased availability of modern medical services has reduced the dependence of many rural communities on traditional medicine, as shown among indigenous peoples in Malaysia (Lim 1997; Baharuddin et al. 2009). Expertise in traditional medicine is transferred from one generation to the next through the practices of traditional knowledge, often associated with traditional belief systems. The preference for modern (allopathic) treatments and medications, especially among the young, contributes to the decline of traditional medical practices.

Changing economic conditions erode traditional practices as youth are more interested in generating cash income from modern agricultural farming and being engaged in wage-earning opportunities in the commercial agricultural and industrial sectors. In general, rural youth tend to migrate to the urban areas seeking employment and income-generating activities. In Thailand, for example, the main factors that play a role in cultural erosion and changes in social and economic activities taking place in upland communities relate to the influences of the external development system that emphasizes commercial economic concepts, religion, and formal education (Charoenniyomphrai et al. 2006).

Contacts between indigenous peoples and the West during colonialism and the successive independence of local societies have intensified the process by which local communities have become less attached to and seemingly less dependent on nature and specifically on forests compared to the past. Many peoples in the developing Southeast Asian countries have felt that the progress to be achieved is measured by the degree of technological achievement imported from the West, the change from a subsistence economy to a market-oriented economy, and the change from the traditional mode of production to the modern mode of production. Many tribes in Indonesia are undergoing a degradation of their practices because of intense contact with outsiders (Boedhihartono 1997).

10.5   The Role of Traditional Forest-Related Knowledge in Solving Contemporary Forest and Associated Agricultural Resource Management Challenges

Despite the varying consequences of external social, economic, and environmental circumstances on local communities, the practice of traditional knowledge persists and has a great potential to continue to contribute to sustainable management of forests and complex agroforest landscapes. The integration of modern development and local forest management based on traditional forest-related knowledge has taken place in a variety of ways in various contexts since the colonial era. It is important to note that many traditional farming practices in Southeast Asia evolved to minimize soil erosion and fertility loss, and to ensure healthy and rapid forest succession. The case of the Ifugao peoples in the mountainous region of the Philippines, discussed earlier (Box 10.3), is an excellent example of the integration of traditional agricultural and forest knowledge and management practices to sustain livelihoods, forest biodiversity, and environmental services over countless generations. Their traditional management systems emphasize environmental preservation and protection against landslides and other forms of erosion, as well as management of woodlots for the supply of needed fuel, construction materials, and irrigation (Conklin et al. 1980).

10.5.1   Agrodiversity Practices

An important means of integrating modern and traditional practices by local communities is through agrodiversity, which is defined as the dynamic variation in cropping systems, outputs, and management practices in agroecosystems. Agrodiversity is characterized by bio-physical differences and the changing ways in which farmers manage diverse genetic resources, natural variability, and their agricultural practices in dynamic social and economic contexts (Brookfield 2001). The success of agrodiversity has been demonstrated in the case of mountains farmers in northern Thailand, who manage agroforest edges to produce medicinal plants, building materials, food plants, and timber in addition to cultivation of modern crops such as cabbage, pepper, carrot, lettuce, tomato, and potato (Rerkasem et al. 2002).

10.5.2   Community/Social Forestry

Over the past few decades, some Southeast Asian governments have initiated forest tenure reforms in favour of customary tenure and management, in recognition of the land claims of indigenous and other local communities, the limits of state forest management, and the growing evidence of the potential of indigenous knowledge and customary management for sustainable forest management. In the Philippines, the previously centralized forest policy has been reformed through the democratization of access to forest land and resources (Pulhin et al. 2008). The Community-Based Forest Management Program (CBFM) was implemented, particularly after the end of Marcos' dictatorial rule in 1986, to tap the potential of people to sustainably manage forest resources and to tackle the problems of inequitable access to forest resources and massive forest depletion. Under the CBFM program, organized upland communities (People's Organizations) receive titles to manage, and benefit from, forest lands for a period of 25 years, renewable for another 25 years subject to certain conditions. Nationwide, thousands of these People's Organizations have received titles for a total of approximately 5.7 million ha through the implementation of this program.

10.5.3   Traditional Forest Protection Strategies

Local forest protection strategies have been applied by local communities in the Philippines. Such strategies, known locally as the lapat (literally, 'to prohibit') system, are implemented by the Isneg and Tingguian people of Abra Province. The lapat is an age-old traditional system of regulating the use of natural resources among the upland tribal communities (Parades 2005) and is passed on from generation to generation. The system involves imposition of taboos within a designated area (which may be as large as a whole mountain) over a period of time, which prohibits the exploitation of forest resources such as rattan vines, lumber, fruits, animals, fish, and wild vegetables. The lapat enables natural recovery of forests from earlier anthropogenic disturbances, allowing trees and other plants to regenerate and wildlife to reproduce for the benefit of current and future generations. The system is based on customary and local laws whose enforcement is overseen by an organization of the elders (elected or informally recognized) called lapat panglakayen. The lapat penglakayen is authorized to issue permits to cut trees for community use; when timber is transported outside the area or sold for commercial purposes, the Department of Environment and Natural Resources (DENR) issues the permits (Parades 2005). All in all, the practice of lapat and the policies imposed by the DENR blend harmoniously in this region, facilitating implementation of government policies on forest protection.

Such local forest protection strategies are also practised among the Tai and Mong communities in northwestern Vietnam (Poffenberger 2006), where land is held under communal tenure. In these communities, villagers classify forests according to their function, such as old-growth protected areas (Pa Dong), younger secondary forests that are part of long rotation swiddens (Pa Kai), early regenerating forests (Pa Loa), and bamboo forests (Pa). Similarly, in Kompong Phluk Village on the northeastern shores of Tonle Sap (Great Lake) in Cambodia, the community has since 1948 protected the flood forest, which provides habitat for more than 200 species of fish in the lake (Evans et al. 2004).

10.5.4   Co-management of Natural Forests

Application of forest co-management in Southeast Asia is not new. In Eastern Indonesia, the Dutch colonial government and post-colonial Indonesian government promoted kemiri (*Aleurites moluccana*) planting by local communities in reforesting abandoned swiddens and degraded forest areas (Koji 2002). The Southeast Asia Sustainable Forest Management Network Secretariat, based at the Centre for Southeast Asian Studies at the University of California at Berkeley, has initiated a research program on forest co-management, applying community and 'scientific' forest management systems. In Thailand, rural communities and the Royal Forest Department have worked together to regulate access and regenerate access degraded natural forests, as described in Box 10.4 (Poffenberger and McGean 1993). In the Philippines, with the passing of the Indigenous People's Act in 1997, the area of forests under Community-Based Forest Management (CBFM) increased from less than 200,000 ha in 1986 to about six million ha in 2005 (Pulhin et al. 2005), or about 38% of the country's total forest land involving more than 690,000 households (Pulhin et al. 2008). In 2005, community holders of management rights of public forests amounted to 47% in the Philippines, 100% in Timor-Leste, and 2% in Cambodia (FAO 2010).

Box 10.4 Local Community/Forestry Department Cooperation in Forest
Management in Northern Thailand

Case 1. The Dong Yai's 12 communities in northeast Thailand are heavily
dependent on non-timber forest products to sustain daily livelihoods. The forest resources in the vicinity were under persistent threat as the forest had been
periodically cleared by villagers for agriculture during the past 100 years.
This motivated villages to take action to ensure sustainability of forest benefits. With the assistance, support, and leadership provided by Tambon Council,
the Royal Forest Department's regional office, and researchers at Kasetsart
University, local communities led by village elders were empowered to organize local forest protection committees and establish their own use rules and
responsibilities. Consequently, the former kenaf (hemp: *Hibiscus cannabi-
nus*) fields established since the late 1950s in Dong Yai are now under community protection and have regenerated into the largest remaining lowland
stand of dry dipterocarp forest in the region.

Case 2. In the northern sub-watershed of Nam Sa in northern Thailand, unsustainable land-use practices led to conflicts over land resources use between
midland and upland indigenous communities, resulting in rapid forest and
environmental deterioration. To reduce social conflicts, micro-watershed
land-use committees and resident community group networks were organized.
Ecological resource and land-use mapping tools were used to enhance villagers' understanding of the importance of upstream–downstream watershed
linkages. The Royal Forest Department and the midland indigenous peoples
(Karen, Hmong, and Lisu) worked together to replace their steep-slope swidden practices with upland forest protection and lowland irrigated paddy cultivation. The application of decentralized controls over clearly defined
micro-watershed areas by the local villages has reduced the threat of fire,
illegal logging, and upland erosion, and has also resulted in impressive natural
forest regeneration.

Source: Poffenberger and McGean (1993).

10.5.5   Co-management of Protected Areas Established
for Biodiversity Conservation

To complement the modern protected areas, recognition of the indigenous sacred
natural sites, even in some official protected areas, could enhance and restore social
norms for protection of natural forests important to the spiritual well-being of local
communities. There is thus a need to encourage and strengthen the positive contribution of indigenous peoples to the conservation of biodiversity and protected areas
(Borrini-Feyerabend et al. 2004a).

The efforts and experiences of the Dayak in East Kalimantan province, Indonesia, provide an excellent example of the integration of traditional and modern forest management. The local communities formed the Alliance of the Indigenous People of Kayan Mentarang Park (FoMMA) in 2000. The park, situated in the interior of East Kalimantan, Indonesia, at the border between Indonesia and Malaysia, is central to the World Wildlife Fund (WWF) Heart of Borneo initiative; the initiative aims to protect the transboundary highlands of Borneo, which straddle the three Southeast Asian nations of Indonesia, Malaysia, and Brunei Darussalam. FoMMA represents the concerned indigenous people in the Policy Board (Dewan Penentu Kebijakan) of Kayan Mentarang National Park, which was set up in 2002 under a Decree of the Ministry of Forestry to preside over the park's management, and includes representatives from the central government (the agency for Forest Protection and Nature Conservation), as well as the provincial and district governments and FoMMA. The operating principles of the Policy Board emphasize coordination, competence, shared responsibilities, and equal partnership among all stakeholders (Borrini-Feyerabend et al. 2004a; Borrini-Feyerabend et al. 2004b). The 2002 Decree dictated that the park was to be managed through collaborative management, indicating the Indonesian government's recognition of customary law in national park management. Similarly, experiments in co-management of national parks are underway in Thailand at the Kuiburi National Park through the establishment of two working groups, namely a core management team comprising park personnel and a park management board working group that includes local people and other stakeholders (Parr et al. 2008).

10.5.6 Climate Change Mitigation and Adaptation

In Southeast Asia, the main direct relevance of traditional forest-related knowledge to climate change lies in its potential to enhance or augment scientific knowledge of forest fire and its management. In the case of Indonesia, recent extensive forest fires have reduced not only forest cover but also the quality of forests and their environmental functions such as carbon sequestration and climate change regulation (Boer 2002). The smoke and haze from these fires have affected the health of people, not only in Indonesia but also in neighbouring countries.

There is a growing appreciation that forest-dependent communities have relied and continue to rely on traditional knowledge and skills to protect forests against fire. In Thailand, for example, practices using traditional knowledge that are widely used in fire management include creating fire lines around homes and temples, using backfires to stop approaching fires, and controlling the spread of fire to minimize destruction of community properties (Makarabhirom et al. 2002).

Local forest management practices to reduce the incidence of forest fires was illustrated in the case of Vietnam, where the government made the decision to improve forest management and protection in 1993 by allocating forestlands to organizations, households, and individuals for long-term sustainable forest uses.

Villagers' efforts to prevent logging, grazing, and land clearance by burning and other damaging activities in forested areas have benefited communities tremendously. They actively reduced fuel loads and practise sustainable agriculture. Consequently, forest cover increased with regeneration and the incidence of forest fires has decreased (Linh 2002).

However, the application of local traditional forest-related knowledge to prevent and control forest fires in East Kalimantan, Indonesia, has had limited impact as governmental and private company interests generally take priority over community needs. In the long run, it is necessary to apply traditional knowledge in combination with scientific knowledge of fire behaviour and fire management technologies (Boer 2002).

While traditional forest-related knowledge may be very important to climate change mitigation and adaptation, a key concern, or question, is how best to create synergies for the benefit of forest-dependent communities between the use of their traditional knowledge on forest management and the evolving international efforts such as REDD (Reducing Emissions from Deforestation and Forest Degradation). Despite the potential carbon and biodiversity conservation benefits of these emerging opportunities, current formulations of REDD programs are viewed by many as having severe potential negative impacts on indigenous peoples. For example, since only forests that are immediately threatened to be destroyed or degraded would be considered under REDD, this could actually encourage deforestation and subsequently the erosion and loss of land rights, livelihoods, and traditional forest-related knowledge of local and indigenous communities (Climate Change Monitoring and Information Network 2009; Rai 2009). The challenge then is to achieve a win-win situation where local and indigenous communities can play a specific role in mitigating climate change without losing their rights, livelihoods, and traditional forest management practices.

10.5.7 Use of Expert Farmers to Promote Indigenous Knowledge and Forest Conservation

Promotion of indigenous knowledge and technical practices must follow a 'bottom-up' approach, rather than conventional extension approaches to introduce technologies. Farmers typically possess superb understanding of and good skills in management of natural resources and are often successful in responding to natural and social changes (Brookfield 2001). A good example of collaborative research on farmers' management of natural resources is the United Nations University's recent People, Land Management and Environmental Change (PLEC) project. Implemented from 1999 to 2002, the project has created a national farmer-to-farmer training program, known as the Farmers' Field School project, for tribal communities in northern Thailand (Rerkasem and Fuentes 2002).

Scientific evaluation of these farmers' practices has found that they are both profitable and environmentally friendly (Rerkasem 2003). Therefore, it is best to enlist these expert farmers to teach others about local knowledge and practices. The teaching by expert farmers does not mean simple extension or replication, but instead aims to explain local knowledge in both social as well as technical respects and to show how local knowledge facilitates the expert farmers' rational decisions in resource management. Such 'learning from expert farmers' can occur through informal exchanges or formal meetings, and the organization of learning activities can be facilitated directly by outsider researchers or through farmers' associations. Through the PLEC project discussed above, many researchers, technicians, and officials who had previously looked down on farmers' practices have come to respect these farmers' knowledge and innovations. The learning process from farmers has also raised farmers' self esteem and enabled them to feel more responsible for resource conservation.

10.6 Conclusions and Recommendations

Rich in both biological and cultural diversity, the tropical forests in Southeast Asia and the socio-economic and ecological goods and services they provide are important for human survival at local, national, and international levels. Over the past few decades, traditional forest-related knowledge and practices in this region have been changing rapidly in response to myriad external economic, social, cultural, political, and demographic changes. Such changes are also threatening traditional practices and sustainable forest management.

Despite these changes, the experience of Southeast Asian countries in involving local communities in forest management may provide some insights and hope for long-term forest management. Indigenous forest management systems could provide important insights and tools for promoting forest sustainability. By involving local people in forest management activities, villagers can become local custodians of forest genetic resources, as shown through the experience of a number of effective partnerships between local communities and government agencies for the protection, conservation, and management of forests in the region.

The future of sustainable forest management in Southeast Asia lies in the partnership between 'scientific' forest managers and local forest-dependent communities. These communities are in an appropriate position to protect their forest resources, because the villagers depend on these resources to meet their social, economic, cultural, and religious needs. Based on these examples from Southeast Asia, achievement of sustainable forest management should give serious consideration to community-based development strategies. However, if local communities are to play a greater role in managing forests, a greater degree of authority over forest management and utilization needs to be shifted back to them.

Governments worldwide that are signatories to international agreements such as the Convention on Biological Diversity (CBD) and UN Convention to Combat Desertification (UNCCD) have recognized and agreed that forests need to be managed sustainably to ensure planet Earth's health and human survival. It is essential to incorporate local traditional knowledge in forest management in view of changing forest management paradigms (from an emphasis on timber yield to sustainable multipurpose forest management) and market demands for implementation of forest certification that requires respect for traditional use rights of local communities and protection of their sites of economic, cultural, and religious significance (Lim 2008). Forest policies based on the notion that local communities are ignorant and destructive need to be re-examined and changed (WRM 2002a). Forest-dependent communities who still possess their rich traditional knowledge and wisdom regarding ecosystem function and sustainable management practices clearly have a role to play in sustainable forest management in Southeast Asia.

The first step to be taken by governments in Southeast Asian countries is to mainstream traditional forest-related knowledge in policy formulation with regards to sustainable forest management. While most governments in the region have laws that respect and recognize the rights of indigenous peoples, the unfortunate reality is that their management of forest resources is mainly confined to customary lands outside of protected forest areas and forest reserves. It is time for governments to adjust their policies for sustainable forest management, and to develop and implement appropriate strategies to combine the application of traditional forest-related knowledge with scientific forestry to ensure better management of forests for present and future generations.

References

Anonymous (2006) Indigenous knowledge, customary use of natural resources and sustainable biodiversity management: case study of Hmong and Karen communities in Thailand. Highland mapping development and biodiversity management project, Inter-Mountain Peoples' Education and Culture in Thailand Association (IMPECT) in collaboration with Forest Peoples Programme. Available via http://www.forestpeoples.org/sites/fpp/files/publication/2010/08/10 cthailandimpectjun06eng.pdf. Cited 9 Mar 2011

Baharuddin IN, Yusof N, Fui LH (2009) A case study of the knowledge and use of medicinal and aromatic plants by Semelai community in RPS Iskandar, Bera, Pahang, Malaysia. In: UfoRIA UiTM. Examining contemporary Malaysia: critical knowledge from research, vol 2, Universiti Teknologi MARA, Unit for Research and Intellect Application UfoRIA, Seri Iskandar, pp 11–23

Ball JB, Pandey D, Hirai S (1999) Global overview of teak plantations. Paper presented to the regional seminar on site, technology and productivity of teak lpantations, Chiang Mai, 26–29 Jan 1999

Belcher B (2001) Rattan cultivation and livelihoods: the changing scenario in Kalimantan. Unasylva 205:27–34

Boedhihartono (1997) Local religion and traditional healing practice: the indigenous minority groups in Indonesia. In: Sarasswati B (ed.) Integration of endogenous cultural dimension into development. D. K. Printworld Private Limited, New Delhi

Boeke JH (1966) Indonesian economics. W. van Hoeve, The Hague

Boer C (2002) Forest and fire suppression in East Kalimantan, Indonesia. In: Moore P, Ganz D, Tan LC, Enters T, Durst PB (eds.) Communities in flames. Proceedings of an international conference on community involvement in fire management. Food and Agriculture Organization of the United Nations (FAO), Regional Office for Asia and the Pacific, Bangkok, 69–71. Available via http://www.myfirecommunity.net/discussionimages/NPost5126Attach1.pdf. Cited 1 Mar 2011

Borrini-Feyerabend G, Kothari A, Oviedo G (2004a) Indigenous and local communities and protected areas. World Commission on Protected Areas (WCPA), vol 11, Best Practice Protected Area Guidelines. World Conservation Union (IUCN), Gland

Borrini-Feyerabend G, Pimbert M, Farvar MT, Kothari A, Renard Y (2004b) Sharing power: learning-by-doing in co-management of natural resources throughout the world. International Institute for Environment and Development (IIED) and the World Conservation Union (IUCN), Centre for Sustainable Development and the Environment (Cenesta), Tehran

Brookfield H (2001) Exploring agrodiversity. Columbia University Press, New York

Brookfield H, Potter HL, Byron Y (1995) In place of the forest: environmental and socio-economic transformation in Borneo and the Eastern Malay Peninsula. United Nations University Press, Tokyo

Burgers P, Ketterings QM, Garrity DP (2005) Fallow management strategies and issues in Southeast Asia. Agric Ecosyst Environ 110:1–13

Butic M, Ngidlo R (2003) Muyong forest of Ifugao: assisted natural regeneration in traditional forest management. In: Patrick DC et al (eds.) Advancing assisted natural regeneration (ANR) in Asia and the Pacific. RAP Publication 2003/19. Bangkok, Food and Agriculture Organization of the United Nations (FAO), Regional Office for Asia and the Pacific. Available via http://www.fao.org/docrep/004/AD466E/ad466e03.htm#. Cited 1 Mar 2011

Cagoco JL (2006) Saving Ifugaos' endangered rituals. Available via http://www.truthforce.info/?q=node/view/1732 (7 June 2007). Cited 1 Mar 2011

Cairns M (2008) Voices from the forest: integrating indigenous knowledge into sustainable upland farming. Nat Resour Forum 32:84–85

Cairns M, Garrity DP (1999) Improving shifting cultivation in Southeast Asia by building on indigenous fallow management strategies. Agrofor Syst 47(1):37–48

Camacho LD, Combalicer MS, Youn Y-C, Combalicer E, Carandang AP, Camacho SC, de Luna CC, Rebugio LL (2011) Traditional forest conservation knowledge/technologies in the Cordillera, Northern Philippines. Forest Policy and Economics. doi:10.1016/j.forpol.2010.06.001(in press)

Carino J (2004) An assessment of the implementation of the Philippine Government's international commitments on traditional forest-related knowledge from the perspective of indigenous peoples. Case study presented at the expert meeting on traditional forest-related knowledge (TFRK), International Alliance of Indigenous and Tribal Peoples of Tropical Forests, San Jose, Costa Rica. Available via http://www.international-alliance.org/tfrk_expert_meeting.htm. Cited 1 Mar 2011

Chai PPK, Tipot E, Henry J (2008) Traditional knowledge on utilization of natural resources among the Penan communities in northern Sarawak, Malaysia. In: Parrotta JA, Liu J, Sim H-C (eds.) Sustainable forest management and poverty alleviation: roles of traditional forest-related knowledge, vol 21, IUFRO World Series. International Union of Forest Research Organizations (IUFRO), Vienna, pp 16–18

Charoenniyomphrai U, Phichetkulsamphan C, Tharawodome W (eds.) (2006) Indigenous knowledge, customary use of natural resources and sustainable biodiversity management: case study of Hmong and Karen communities in Thailand. Inter Mountain Peoples Education and Cultures in Thailand Association (IMPECT), Chiang Mai. Available via http://www.forestpeoples.org/sites/fpp/files/publication/2010/08/10cthailandimpectjun06eng.pdf. Cited 1 Mar 2011

Christanty L, Abdoellah OS, Marten GG, Iskandar J (1986) Traditional agroforestry in West Java: the pekarangan (homegarden) and kebun talun (annual perennial rotation) cropping systems. In: Marten GG (ed.) Traditional agriculture in Southeast Asia: a human ecology perspective. Westview Press, Boulder

Climate Change Monitoring and Information Network (2009) REDD and indigenous peoples. Available via http://ccmin.aippnet.org/index.php?option=com_content&review=article&id=14 &Itemid=27. Cited 1 Mar 2011

Colchester M, Fay C (2007) Land, forest and people: facing the challenges in Southeast Asia. Rights and resources initiative, listening, learning and sharing. (Asia Final Report RP-0235-08). World Agroforestry Centre, Nairobi

Colchester et al (2006) Forest peoples, customary use and state forests: the case for reform. Paper for 11th biennial congress of the international association for the study of common property, Bali, Indonesia, 19–23 June 2006. Available via http://www.forestpeoples.org/sites/fpp/files/publication/2010/08/10coverviewiascpjun06eng.pdf. Cited 9 Mar 2011

Colfer CJP, Peluso N, Shin SC (1997) Beyond slash and burn: building on indigenous management of Borneo's tropical rainforests, vol 11, Advances in Economic Botany. New York Botanical Garden Press, New York

Conklin HC (1957) Hanunoo agriculture. A report on an integral system of shifting cultivation on the Philippines. Food and Agriculture Organization of the United Nations (FAO), Rome

Conklin HC, Lupaih P, Pinther M (1980) Ethnographic atlas of Ifugao: a study of environment, culture and society in northern Luzon. Yale University Press, New Haven

Cox PC (2000) A tale of two villages: culture, conservation and ecocolonialism in Samoa. In: Nerner C (ed.) People, plants and justice: the politics of nature conservation. Columbia University Press, New York, pp 330–334

Dacawi R (1982) The Ifugao way of forest conservation. Philippine Upland World 1(2):14–15

Davidson-Hunt IJ, Berkes F (2001) Traditional ecological knowledge and changing resource management paradigms. In: Duchesne LC, Zasada JC, Davidson-Hunt IJ (eds.) Forest communities in the third millennium: linking research, business and policy toward a sustainable non-timber forest product sector. (General Technical Report NC-217). U.S. Department of Agriculture, Forest Service, North Central Research Station, St. Paul, pp 78–92. Available via http://nrs.fs.fed.us/pubs/828. Cited 1 Mar 2011

Disini JJ (2003) Survey on laws on traditional knowledge in Southeast Asia. Available via http://cyber.law.harvard.edu/openeconomies/okn/asiatk.html. Cited 7 Mar 2011

Dolinen L (1997) Enriching upland development through indigenous knowledge systems: the case of Kalahan, Nueva Vizcaya. In: Castro CP, Bugayong, LA (eds.) Application of indigenous knowledge systems in sustainable upland development. General Technical Report Series 1. The Forestry Development Center (FDC), UPLB College of Forestry, in cooperation with the Department of Environment and Natural Resources (DENR), Quezon City, pp 71–86

Donovan D, Puri R (2004) Learning from traditional knowledge of non-timber forest products: Penan Benalui and the autecology of Aquilaria in Indonesian Borneo. Ecol Soc 9(3):3. Available via http://www.ecologyandsociety.org/vol9/iss3/art3/. Cited 1 Mar 2011

Elazegui DD, Combalicer EA (2004) Realities of the watershed management approach: the Magat watershed experience. Discussion paper series no. 2004–21. Philippine Institute for Development Studies, Makati City. Available via http://dirp4.pids.gov.ph/ris/dps/pidsdps0421.pdf. Cited 1 Mar 2011

Ella AB (2008) Indigenous collection and marketing of wild honey and almaciga resin in Palawan: the Natripal experience. In: Parrotta JA, Liu J, Sim H-C (eds.) Sustainable forest management and poverty alleviation: roles of traditional forest-related knowledge, vol 21, IUFRO World Series. International Union of Forest Research Organizations (IUFRO), Vienna, pp 49–51

Evans J (1992) Plantation forestry in the tropics: tree planting for industrial, social, environmental, and agroforestry purposes, 2nd edn. Clarendon, Oxford

Evans P, Marshke M, Paudyal K (2004) Flood forests, fish, and fishing villages: Tole Sap, Cambodia. Food and Agriculture Organisation (FAO) and Asia Forest Network, Bohol

Fadzil KS, Hamzah KA (2008) Challenges in applying traditional forest related knowledge in sustainable forest management and poverty alleviation in Malaysia: a case study of a fish farming project with the Jakun community in the Southeast Pahang peat swamp project. In: Parrotta JA, Liu J, Sim H-C (eds.) Sustainable forest management and poverty alleviation: roles of traditional forest-related knowledge, vol 21, IUFRO World Series. International Union of Forest Research Organizations (IUFRO), Vienna, pp 82–84

Food and Agriculture Organisation (FAO) (2001) Global forest resources assessment 2000, main report FAO forestry paper 140. Food and Agriculture Organization of the United Nations, Rome

Food and Agriculture Organisation (FAO) (2005) Global forest resources assessment 2005, progress towards sustainable forest management. FAO forestry paper 147. Food and Agriculture Organization of the United Nations, Rome

Food and Agriculture Organisation (FAO) (2010) Global forest resources assessment 2010, main report. FAO forestry paper 163. Food and Agriculture Organization of the United Nations, Rome

Forman RTT (1995) Some general principles of landscape and regional ecology. Landscape Ecol 10(3):133–142

Fox J (2000) How blaming 'slash and burn' farmers is deforesting mainland Southeast Asia. Asia Pacific Issues 47:1–8

Galacgac ES, Balisacan CM (2009) Traditional weather forecasting for sustainable agroforestry practices in Ilocos Norte Province, Philippines. For Ecol Manag 257:2044–2053

Guerra CA, Snow RW, Hay SI (2006) A global assessment of closed forests, deforestation and malaria risk. Ann Trop Med Parasitol 100(3):189–204

Hendarti L, Youn Y-C (2008) Traditional knowledge of forest resource management by Kasepuhan people in the upland area of West Java, Indonesia. In: Parrotta JA, Liu J, Sim H-C (eds.) Sustainable forest management and poverty alleviation: roles of traditional forest-related knowledge, vol 21, IUFRO World Series. International Union of Forest Research Organizations (IUFRO), Vienna, pp 73–76

Hong E (1987) Natives of Sarawak: survival in Borneo's vanishing forests. Malaysia Institut Masyarakat, Pulau Pinang

Houghton RA, Hackler JL (1995) Continental scale estimates of the biotic carbon flux from land cover change: 1850–1980. ORNL/CDIAC-79, NDP-050. Oak Ridge National Laboratory, Oak Ridge

Kashio M, Johnson DV (eds.) (2001) Monograph on benzoin (Balsamic resin from *Styrax* species). RAP Publication: 2001/21. Food and Agriculture Organization (FAO) of the United Nations, Regional Office for Asia and the Pacific, Bangkok

Klock J, Tindungan M (1995) The past and the present. A meeting of forces for a sustainable future. FTP program network. Forest, trees and people newsletter no. 29. Food and Agriculture Organization (FAO) of the United Nations, Regional Office for Asia and the Pacific, Bangkok. Available via http://www.fao.org/docrep/004/ad466e/ad466e06.htm

Koji T (2002) Kemeri (*Aleurites moluccana*) and forest resource management in Eastern Indonesia: an eco-historical perspective. Available via http://www.asafas.kyoto-u.ac.jp/publication/pdf/no_02/p005_023.pdf. Cited 1 Mar 2011

Lang C (2001) Deforestation in Vietnam, Laos and Cambodia. In: Vajpeyi DK (ed.) Deforestation, environment and sustainable development: a comparative analysis. Praeger, Westport, pp 111–137

Lee SS, Chang YS, Noraswati MNR (2009) Utilization of macrofungi by some indigenous communities for food and medicine in Peninsula Malaysia. For Ecol Manag 257:2062–2065

Lennertz R, Panzer KF (1983) Preliminary assessment of the drought and forest fire damage in East Kalimantan. Transmigration Areas Development project (TAD). German Agency for Technical Cooperation, Eschborrn

Lewis MP (ed.) (2009) Ethnologue: languages of the world, sixteenth edition., SIL International, Dallas. Available via http://www.ethnologue.com/. Cited 1 Mar 2011

Lim HF (1997) Orang Asli, forest and development. Malayan forest records no. 43. Forest Research Institute Malaysia, Kepong

Lim HF (2008) Enhancing the value of forest traditional knowledge through forest certification. In: Parrotta JA, Liu J, Sim H-C (eds.) Sustainable forest management and poverty alleviation: roles of traditional forest-related knowledge, vol 21, IUFRO World Series. International Union of Forest Research Organizations (IUFRO), Vienna, pp 101–103

Lim HF, Mohd Parid M, Chang YS (2008) Local gaharu trade and its contribution to house hold economy of harvester. FRIM report no. 88. Forest Research Institute Malaysia, Kepong

Linh HT (2002) Experiences in community forest fire control in Vietnam. In: Moore P, Ganz D, Tan LC, Enters T, Durst PB (eds.) Communities in flames. Proceedings of an international conference on community involvement in fire management. Food and Agriculture Organization of the United Nations (FAO), Regional Office for Asia and the Pacific, Bangkok, pp 66–68. Available via http://www.myfirecommunity.net/discussionimages/NPost5126Attach1.pdf. Cited 1 Mar 2011

Majid-Cooke F (1999) The challenge of sustainable forests: forest policy in Malaysia, 1970–1995. Allen and Unwin, Sydney

Majid-Cooke F (2005) State, communities and forests in contemporary Borneo. Australian National University Press, Canberra. Available via http://epress.anu.edu.au/apem/borneo/mobile devices. Cited 1 Mar 2011

Makarabhirom P, Ganz D, Onprom S (2002) Community involvement in fire management: cases and recommendations for community-based fire management in Thailand. In: Moore P, Ganz D, Tan LC, Enters T, Durst PB (eds.) Communities in flames. Proceedings of an international conference on community involvement in fire management. Food and Agriculture Organization of the United Nations (FAO), Regional Office for Asia and the Pacific, Bangkok, pp 10–15. Available via http://www.myfirecommunity.net/discussionimages/NPost5126Attach1.pdf. Cited 1 Mar 2011

Malaysia (Government of) (2004) Malaysian rainforests: national heritage, our treasure. Ministry of Primary Industries, Kuala Lumpur

Mertz O (2002) The relationship between fallow length and crop yields in shifting cultivation: a rethinking. Agrofor Syst 55:149–159. doi:101023/A:1020507631848

Michon G, Bouambrane M (2004) Can deforestation help rebuild forests? The Indonesian agroforests. In: Babin D (ed.) Beyond tropical deforestation. United Nations Educational, Scientific, Cultural Organization (UNESCO) and Centre de Cooperation International en Recherche Agronomique pour le Développement (CIRAD), Paris, pp 123–134

Michon G, de Foresta H, Levang P, Verdeaux F (2007) Domestic forests: a new paradigm for integrating local communities' forestry into tropical forest science. Ecol Soc 12(2):1. Available via http://www.ecologyandsociety.org/vol12/iss2/art1/. Cited 1 Mar 2011

Mindanao Baptist Rural Life Centre (MBRLC) (1988) A manual on how to farm your hilly land without losing your soil. Mindanao Baptist Rural Life Center, Davao del Sur

Mohd Shahwahid HO, Jamal R (1999) Cost based evaluation approach: an illustration on the damage cost of the Indonesian forest fires and transboundary haze on neighbouring Malaysia. In: Mohd Shahwahid HO (ed.) Manual on economic valuation of environmental goods and services of peat swamp forest. Malaysian-DANCED Project on Sustainable Forest Management of Peat Swamp Forest, Peninsular Malaysia. Forestry Department of Peninsular Malaysia, Kuala Lumpur, pp 85–93

Molintas JM (2004) The Philippine indigenous peoples' struggle for land and life: challenging legal texts. Arizona J Int Comp Law 21(1):265–306

Moore PF (2001) Forest fires in ASEAN: data, definitions and disaster? Asean Biodiversity, July–Sept 2001, pp 22–27. Available via http://www.arcbc.org.ph/arcbcweb/pdf/vol1no3/22-27_special_reports.pdf. Cited 1 Mar 2011

Mulyoutami E, Rismawan R, Joshi L (2008) Knowledge and use of local plants from 'simpukng' or forest gardens among the Dayak community in East Kalimantan. In: Parrotta JA, Liu J, Sim H-C (eds.) Sustainable forest management and poverty alleviation: roles of traditional forest-related knowledge, vol 21, IUFRO World Series. International Union of Forest Research Organizations (IUFRO), Vienna, pp 121–123

Myers N, Mittermeier RA, Mittermeier CG, da Fonseca GAB, Kent J (2000) Biodiversity hotspots for conservation priorities. Nature 403:853–858

Ng P (2005) Swamped in richness. Malaysian Nat 58(4):8–9

Ngidlo RT (1998) Conserving biodiversity: the case of the Ifugao farming system. Philippine Council for Agriculture, Forestry and Natural Resources Research and Development, Los Baños

NGO Forum in Cambodia (2006) Indigenous peoples in Cambodia. Available via http://www.ngoforum.org.kh/Land/Docs/Indigenous/INDIGENOUS%20PEOPLES%20IN%20CAMBODIA_final(3).pdf. Cited 1 Mar 2011

Nicholas C (2005) Who can protect the forest better? Pitching Orang Asli against professionals in protected area management in Peninsular Malaysia. International Symposium on Eco-Human Interactions in Tropical Forests, Kyoto University, Japan, Society of Tropical Ecology, Kyoto.

Available via http://www.coac.org.my/codenavia/portals/coacv2/code/main/main_art.php?par
entID=11400226426398&artID=11374532336034. Cited 1 Mar 2011

Oberndorf RJD (2006) Legal analysis of forest and land laws in Cambodia. Community Forestry
International, Santa Barbara. Available via http://www.communityforestryinternational.org/
publications/research_reports/LEGALA.PDF

Padoch C, Coffey K, Mertz O, Leisz SJ, Fox J, Wadley RL (2007) The demise of swidden in
Southeast Asia? Local realities and regional ambiguities. Geografisk Tidsskrift-Danish J Geogr
107(1):29–41

Parades AK (2005) How Tingguians protect forests. Inquirer News Service. Available via http://
www.truthforce.info/?q=node/view/834. Cited 10 July 2010

Parr JWK, Jitvijak S, Saranet S, Buathong S (2008) Exploratory co-management interventions in
Kuiburi National Park, central Thailand, including human-elephant conflict mitigation. Int J
Environ Sustain Dev 7(3):293–310

Parrotta JA, Lim HF, Liu J, Ramakrishnan PS, Youn Y-C (2009) Preface: traditional forest-related
knowledge and sustainable forest management in Asia. For Ecol Manag 257(10):1987–1988

Penot E (2004) From shifting agriculture to sustainable complex rubber agroforestry systems (jun-
gle rubber). In: Babin D (ed.) Beyond tropical deforestation. United Nations Educational,
Scientific and Cultural Organization (UNESCO), Centre de Cooperation International en
Recherche Agronomique pour le Développement (CIRAD), Paris, pp 221–249

Pham, T-V, Trung PC (2004) Indigenous knowledge and customary-based regulations in managing
community forest by the Thai ethnic group in Vietnam's northern mountain region. Centre for
Natural Resources and Environmental Studies (CRES), Vietnam, and Institute for Global
Environmental Strategies (IGES). Asia-Pacific Forum for Environment and Development,
Hanoi

Plant R (2002) Indigenous peoples/ethnic minorities and poverty reduction: Cambodia. Asian
Development Bank, Environment and Social Safeguard Division, Regional and Sustainable
Development Department, Manila

Poffenberger M (1992) Sustaining Southeast Asia's forests. Southeast Asia sustainable forest management network, research network report no. 1. University of California, Center for Southeast
Asia Studies, Berkeley

Poffenberger M (ed.) (1999) Communities and forest management in Southeast Asia: a regional
profile of the working group on community involvement in forest management. Asian Forest
Network and IUCN-World Conservation Union, Berkely and Gland. Available via http://www.
asiaforestnetwork.org/pub/pub30.htm. Cited 9 Mar 2011

Poffenberger M (2006) People in the forest: community forestry experiences from Southeast Asia.
Int J Environ Sustain Dev 5(1):57–69

Poffenberger M, McGean B (1993) Community allies: forest co-management in Thailand.
Southeast Asia sustainable forest management network, research network report no. 2.
University of California, Center for Southeast Asia Studies, Berkeley

Pulhin JM, Amaro MC, Bacalla D (2005) Philippines' community-based forest management 2005:
a country report. In: O'Brien N, Mathews S, Nurse M (eds.) Regional community forestry
forum: regulatory frameworks for community forestry in Asia. Proceedings of regional forum,
Bangkok, 24–25 Aug 2005, pp 85–100

Pulhin JM, Dizon JT, Cruz, RVO (2008) Tenure reform and its impact in the Philippine forest land.
Paper presented at 12th biennial conference of the international association for the study of
commons (IASC), Cheltenham, 24–18 July 2008. Available via http://dlc.dlib.indiana.edu/dlc/
bitstream/handle/10535/1995/pulhin_1233.pdf?sequence=1. Cited 28 Feb 2011

Rai M (2009) REDD and the rights of indigenous peoples ensuring equity and participation in
World Bank funds. Bretton woods update no. 65 (17 April 2009). Available via http://www.
brettonwoodsproject.org/art-564322. Cited 1 Mar 2011

Rerkasem B, Fuentes E (2002) People, land management and environmental change: Thailand
final report 1999–2002. Part 2: People, Land Management and Environmental Change (PLEC)
Report. Chiang Mai University, Chiang Mai

Rerkasem K (2003) Thailand. In: Brookfield H (ed.) Agrodiversity: learning from farmers across
the world. Chapter 14. United Nations University Press, Tokyo, pp 293–315

Rerkasem K, Yimyam N (2011) Sustainable land management and community empowerment in northern Thailand. In: Saxena KG, Liang L, Tanaka K (eds.) Land management in marginal mountain regions: adaptation and vulnerability to global change. Bishen Singh Mahendra Pal Singh, Dehra Dun, In press

Rerkasem K, Yimyam N, Korsamphan C, Thong-Ngam C, Rerkasem B (2002) Agrodiversity lessons in mountain land management. Mt Res Dev 22(1):4–9

Rerkasem K, Lawrence D, Padoch C, Schmidt-Vogt D, Ziegler AD, Bruun TB (2009a) Consequences of swidden transitions for crop and fallow biodiversity in Southeast Asia. Hum Ecol 37:347–360

Rerkasem K, Yimyam N, Rerkasem B (2009b) Land use transformation in the mountainous Southeast Asian region and the role of indigenous knowledge and skills in forest management. For Ecol Manag 257:2035–2043

Rondolo MT (2001) Fellowship report. Yokohama, Japan, International Tropical Timber Organization. Trop Forest Update 11(4):22–23

Sam DD, Trung LQ (2001) Forest policy trends in Vietnam. Policy Trend Rep 2001:69–73

Save the Ifugao Terraces Movement (2008) The effects of tourism on culture and the environment in Asia and the Pacific: sustainable tourism and the preservation of the world heritage site of the Ifugao rice terraces. United Nations Educational, Scientific and Cultural Organization (UNESCO), Bangkok

Sirait M (2009) Indigenous peoples and oil palm plantation expansion in West Kalimantan, Indonesia. A report commission by Amsterdam University Law Faculty, The Hague

Sirait M, Prasodjo S, Podge N, Flavelle A, Fox J (1994) Mapping customary land in East Kalimantan, Indonesia: a tool for forest management. Ambio 23(7):411–417

Sobrevila C (2008) The role of indigenous people in biodiversity conservation. Discussion paper. The International Bank for Reconstruction and Development/World Bank, Washington, DC. Available at http://siteresources.worldbank.org/INTBIODIVERSITY/Resources/RoleofIndigenous PeoplesinBiodiversityConservation.pdf. Cited 1 Mar 2011

Soedjito H (2006) Biodiversity and cultural heritage in sacred sites of West Timor, Indonesia. In: Schaaf T, Lee C (eds.) Conserving cultural and biological diversity: the role of sacred natural sites and cultural landscapes. United Nations Educational, Scientific and Cultural Organization (UNESCO), Paris, pp 64–67

Soemarwoto O, Conway GR (1992) The Javanese homegarden. J Farm Syst Res Ext 2(3):95–118

Tacio HD (1993) Sloping agricultural land technology (SALT): a sustainable agroforestry scheme for the uplands. Agrofor Syst 22:145–152

Takeda S (2006) Land allocation program in Lao PDR: the impacts on non-timber forest products (NTFPs) and livelihoods in marginal mountainous areas. In: Saxena KG, Liang L, Kono Y, Miyata S (eds.) Small-scale livelihoods and natural resources management in marginal areas: case studies in Monsoon Asia. Bishen Singh Mahendra Pal Singh, Dehra Dun

Theilade I, Luoma-aho T, Rimbawanto A, Nguyen HN, Greijmans M, Nashatul ZNA, Sloth A, Thea S, Burgess S (2005). An overview of the conservation status of potential plantation and restoration species in Southeast Asia. Paper presented at the symposium on tropical rainforest and restoration: existing knowledge and future directions, Kota Kinabalu, 26–28 July 2005. http://www.apforgen.org/pdf_files/FGR-Status-SoutheastAsia.pdf. Cited 1 Mar 2011

Van Noordwijk M, Roshetko JM, Murnianti, Angeles MD, Suyanto, Fay C, Tomich TP (2003) Agroforestry is a form of sustainable forest management: lessons from South East Asia. UNFF intersessional expert meeting on the role of planted forests in sustainable forest management conference, 24–28 Mar 2003, Wellington

Wadley RL, Colfer CJP (2004) Sacred forest, hunting, and conservation in West Kalimantan, Indonesia. Hum Ecol 32(3):313–338

World Rainforest Movement (WRM) (2002a) Community forest management: a feasible and necessary alternative. Bulletin no. 61, Montevideo. Available via http://www.wrm.org.uy/bulletin/61/viewpoint.html

World Rainforest Movement (WRM) (2002b) Community forest management: forests for the peoples who sustain the forests. Bulletin no. 63, Montevideo. Available via http://www.wrm.org.uy/bulletin/63/viewpoint.html

World Rainforest Movement (WRM) (2006) Cambodia: sustainable use of the forest by villagers. WRM bulletin no. 114, Montevideo. Available via http://www.wrm.org.uy/bulletin/104/viewpoint.html

World Resources Institute (1987) Frontier regions. Washington, DC. Available via http://multimedia.wri.org/frontier_forest_maps/

Zuraida (2008) Conservation and utilization of traditional forest-related knowledge for achieving sustainable livelihoods of local communities: a case study in Jambi province, Indonesia. In: Parrotta JA, Liu J, Sim H-C (eds.) Sustainable forest management and poverty alleviation: roles of traditional forest-related knowledge, vol 21, IUFRO World Series. International Union of Forest Research Organizations (IUFRO), Vienna, pp 210–212

Chapter 11
Western Pacific

Suzanne A. Feary, David Eastburn, Nalish Sam, and Jean Kennedy

Abstract The forests of the Western Pacific range from tropical in Oceania to cool temperate in the Australian state of Tasmania, and all have been manipulated by humans for thousands of years. Indigenous communities across the Western Pacific used forest resources for food, medicine, and raw materials, based on an intimate knowledge of local ecologies, understood though a cosmological lens. Differing colonial histories have influenced the degree to which traditional knowledge has been retained and valued. New Zealand Maori and Aboriginal Australians lost their land and much associated knowledge, whereas customary forms of land tenure are largely intact across the oceanic Pacific, where traditional knowledge continues to underpin integrated systems of subsistence agriculture and forest use. Traditional forest-related knowledge is threatened by modernity across the Western Pacific, and its diminution has been linked with deforestation in the Pacific Islands, with calls by non-governmental organisations (NGOs) and local people to replace large-scale commercial logging with more sustainable systems

S.A. Feary (✉)
Conservation Management, Vincentia, NSW, Australia
e-mail: suefeary@hotkey.net.au

D. Eastburn
Landmark Communications, Chapman, ACT, Australia
e-mail: Eastburn.landmark@gmail.com

N. Sam
Fenner School of Environment and Society, The Australian National University,
Canberra, Australia
e-mail: Nalish.Sam@anu.edu.au

J. Kennedy
Department of Archaeology and Natural History, School of Culture, History and Language,
The Australian National University, Canberra, Australia
e-mail: Jean.Kennedy@anu.edu.au

J.A. Parrotta and R.L. Trosper (eds.), *Traditional Forest-Related Knowledge:*
Sustaining Communities, Ecosystems and Biocultural Diversity,
World Forests 12, DOI 10.1007/978-94-007-2144-9_11,
© Springer Science+Business Media B.V. (outside the USA) 2012

that give more credence to traditional knowledge. In Australia and New Zealand, indigenous people are partnering with government agencies to ensure their cultural values are adequately recognised and protected in publicly owned forests.

Keywords Australia • Cultural expression • Forest history • Forest management • Indigenous peoples • New Zealand • Non-timber forest products • Oceania • Traditional agriculture • Traditional knowledge

11.1   Introduction

There is an emerging body of populist and scientific literature urging modern society to look to indigenous societies and their traditional knowledge for guidance on the sustainable use of natural resources (e.g., Knudtson and Suzuki 1992; Sveiby and Skuthorpe 2006; Jansen and Tutua 2001). It has been suggested that sustainable forest management in a modern world would benefit from increased recognition and application of the knowledge and traditions of indigenous peoples (Colfer et al. 2005).

This chapter explores the state of traditional forest-related knowledge (TFRK) and its potential contribution to sustainable forest management (SFM) in the Western Pacific, an area comprising many Oceanic islands and the larger land masses of Australia, New Zealand, and New Guinea. Indigenous peoples of the Western Pacific are culturally diverse, reflecting different origins and geographical locations, as well as the varying effects of colonial history, particularly on customary land ownership. Equally diverse are the forests of the Western Pacific region, being mainly tropical, with the vegetation of New Zealand and Australia largely temperate. Many indigenous populations in the Western Pacific rely on forest products for their livelihoods, and across the region, forests continue to have spiritual and economic values that are deeply embedded in cultural traditions.

Contestations over forests are a political and social reality in many Western Pacific countries and many are grappling with social, environmental, and economic problems that are having direct and indirect impacts on forests. Global concern continues to be expressed over unsustainable logging and land conversion in the larger Melanesian countries of the Solomon Islands and Papua New Guinea (Siwatibau 2009; Kanowski et al. 2005). Unstable governments, internal conflict, contested land tenures, poverty, and economic constraints of small islands all contribute to unsustainable forest management. But solutions must go beyond forest policies or codes of logging practice, to address the underlying social causes, many of which are embedded in an intercultural space where modern and traditional practices are not fully reconciled.

Due to space limitations, this chapter is not able to provide comprehensive coverage of the state of traditional forest-related knowledge across the entire Western Pacific. For example, little attention has been paid to the islands of Micronesia. Rather, the links between traditional forest-related knowledge and sustainable forest management are explored through specific themes and geographical locations.

This chapter recognises that indigenous societies inhabiting small Oceanic islands have both differences and similarities with those of New Zealand, Australia and New Guinea in regard to culture, governance and history. While it has been a challenge to treat the Western Pacific as a single unit for the purposes of the book, it hopefully enables a deeper understanding of how traditional forest-related knowledge has evolved across this vast area.

In this chapter, the word forest refers generally to natural forests, whilst recognising that most have been subjected to human influences over many millennia, particularly in Australia. Agroforestry describes the traditional process of Pacific agricultural peoples in habitual, deliberate manipulation of certain native tree species by transplanting or encouraging their growth for a desired product.

11.2   Overview of the Western Pacific

The Western Pacific can be defined geographically as the area between latitudes 10° and 50° S and longitudes 120° and 130° W. It straddles the Equator and the international dateline and falls eastward of Wallace's Line[1] (Fig. 11.1). The thousands of islands, comprising 22 countries and territories, are scattered across 180 million km^2 of the Pacific Ocean, with the larger landmasses of New Zealand, New Guinea, and Australia delineating the southern and western/northwestern edges respectively.

The region is often referred to in terms of its cultural sub-regions Australasia, Melanesia, Micronesia, and Polynesia, although the terms 'Near' and 'Remote' Oceania are also used, based on linguistics, degrees of insular isolation, and human adaptation to living on small islands (Merlin 2000). The human history of the Western Pacific is immensely long, with settlement beginning around 40,000 years ago in the Pleistocene continent of Sahul (Australia and New Guinea), through to settlement of New Zealand, around 1,000 years ago. The past 3,000 years of history are characterised by ocean navigation over vast distances, dispersing cultural traditions, people, their material culture, and their domesticated plants and animals throughout the Pacific islands.

Today, there are significant differences between Australia/New Zealand and the rest of the Western Pacific in relations between indigenous and non-indigenous populations, reflecting a complex mix of historical, political, and socioeconomic factors. These political and economic circumstances often determine how forests are valued and used at national and local scales, which in turn are influential in the retention, recognition, and application of traditional forest-related knowledge.

---

[1] Wallace's Line is the boundary between two major biogeograhical provinces.

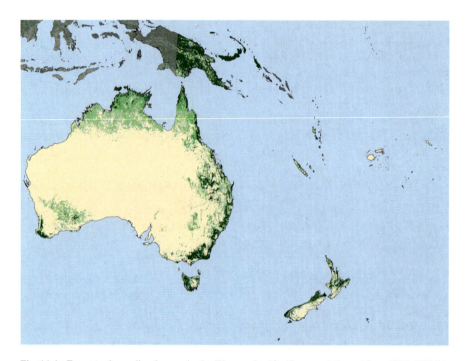

Fig. 11.1 Forest and woodland cover in the Western Pacific (Source: Adapted from FAO (2001)). Key: *Dark green* = closed forest; *light green* = open or fragmented forest; *pale green* = other wooded land; *yellow* = other land

11.2.1 New Zealand and Australia

The nation states of New Zealand and Australia have advanced economies, stable democracies, and rich resources (especially Australia). However, their indigenous peoples are minority populations who occupy a 'fourth world'[2] position in society. In both countries, usurping of land by British colonisers in the eighteenth century jeopardised the existence of customary land tenures and the traditional systems that underpinned them.

In New Zealand in 1840, the British Government and 50 Maori chiefs signed the Treaty of Waitangi, which guaranteed Maori in possession of land, forests, fisheries, and other property, in return for ceding Maori sovereignty to the Queen (Sinclair 2000). Many of the promises of the Treaty were never honoured and in 1975, the Waitangi Tribunal was established for assessing claims brought by Maori relating to actions or omissions of the Crown that breach promises made in the Treaty of Waitangi, including the return of forest land.

---

[2] A term describing minority and often marginalised indigenous peoples encompassed within modern nation states (Manuel and Polsuns 1974).

In Australia, traditional indigenous knowledge has been profoundly compromised by colonisation and successive government policies of exclusion and assimilation of indigenous people. A declaration by the landed British in 1788, that Australia was 'terra nullius'[3] was followed by a mostly brutal colonial history. Absence of a treaty and forced movements of people from their lands into institutions where speaking native languages was forbidden have all contributed to this loss. Social reforms since the late 1960s, passing of state and territory land rights legislation, and belated passing of the Native Title Act in 1993 have gone some way to addressing the impacts of colonisation.

Today, Maori constitute approximately 15% of the New Zealand population, with approximately 6% of land owned by Maori under communal title. In Australia, Aboriginal people constitute 2.5% of the total population and communally own approximately 20% of the continent, mainly in remote areas of the Northern Territory (Pollack 2001). Both Maori and Aboriginal populations fare poorly according to national indicators of economic and social well-being.

11.2.2   Oceania

The island states of Oceania comprise thousands of small islands, many uninhabited. Most are defined as Small Island Developing States (SIDS), recognised since 1991 with the establishment of the Alliance of Small Island States (Wilkie et al. 2002). Many Oceanic SIDS are well-endowed with forests (e.g. Cook Islands, Palau, Vanuatu, and the Solomon Islands), whereas others have less than 10% cover (e.g., Tonga). Generally, indigenous people constitute the majority of the SIDS populations.[4]

Sixteen Pacific Island countries are self-governing states identified as the Pacific Island Nations, while others are territories of other countries, including French Polynesia, New Caledonia, and Wallis and Futuna (France); American Samoa (United States); and Irian Jaya/West Papua (Indonesia).[5] Some have stable governance systems while others have been torn by internal conflict and ethnic tensions.[6]

---

[3] Literally 'belonging to no-one,' i.e., indigenous systems of ownership of land and resources were not thought to exist and were not recognised.

[4] Fiji's population is approximately 44% Indian, descended from 60,000 indentured labourers brought over to work in the sugar cane plantations. Melanesians constitute less than half the population of New Caledonia, owing to the presence of a French penal colony in 1864 (Crocombe 1989). French Polynesia is characterised by a mixing of ethnic groups, with Polynesians constituting around 66% of the population.

[5] See: http://www.forumsec.org/pages.cfm/about-us/

[6] In the Solomon Islands, the conflict of 1999–2003 led to a failed state and regional partners, including Australia, established the Regional Assistance Mission to Solomon Islands (RAMSI) to restore law and order and address the serious fiscal situation. Political unrest has occurred in PNG, and ethnic and political tensions are current in Fiji.

Despite colonisation and missionaries, indigenous populations have persisted as the dominant ethnic groups and, importantly, most land is still owned under systems of customary land tenure. These are inherited through traditional kinship systems and rely on traditional knowledge to identify boundaries and proscribe rights.

The mostly rural populations depend on subsistence farming of root crops— primarily taro, cassava, and yams—and continue to apply traditional systems of agroforestry to supplement the diet and provide medicines and raw materials. Many SIDS have low export capacity, with increasing reliance on cash crops such as copra and cocoa. Timber production is a major industry in Fiji, PNG, Samoa, Solomon Islands, and Vanuatu, much of which is at unsustainable levels (Wilkie et al. 2002).

11.2.3   Wood and Cultural Expression

No contextual section on the Western Pacific would be complete without reference to its diverse and beautifully crafted wooden material culture. Some examples are more utilitarian, such as the unique Australian boomerang and the magnificent ocean-going canoes of Oceania (Fig. 11.2a). Others, such as wooden carvings and personal adornments (Fig. 11.2b), are deeply symbolic and highly valued components of ceremonial life (Moore 1995).

The manufacture of wooden and other forest-based items of material culture for an expanding tourism market is becoming increasingly important as an economic base for many impoverished communities across the Western Pacific. Knowledge about these ancient crafts is privileged through customary law, but its application in the market economy is both a positive and a negative. Traditional knowledge can be kept alive and relevant, but economic imperatives can override tradition, leading to unsustainable or inappropriate use of the forest resource.

The 'high' art of Aboriginal people from northern Australia, stemming from traditional techniques of painting with ochre on bark, is perhaps the best known

a

Fig. 11.2 (a) Maori canoe prow (Source: Wikimedia Commons)

Fig. 11.2 (**b**) Face mask,
Torres Strait islands. Rietberg
Museum, Zürich (Source:
Wikimedia Commons)
(**c**) Ochre on bark painting
of a traditional story from
Blue Mud Bay, Northern
Territory Australia. Private
Collection

example of the commodification of culture. Many modern artists are merging
traditional stories with contemporary expression, in perhaps one of the most powerful demonstrations of cultural response to modernity (Fig. 11.2c).

11.3 Environmental and Social History of the Western Pacific

The term 'traditional knowledge'[7] has come into popular use, where its meaning
has become blurred and its cultural and historical contexts frequently missing. It is
not possible to fully understand traditional knowledge without understanding how
it has evolved and continues to evolve. Hence, this section begins by journeying
back to the time when indigenous peoples and natural forests co-existed prior to
white settlement.

Broadly speaking, archaeological, linguistic, and anthropological research points
to a history of human occupation of the Western Pacific from around 40,000 years
ago in the Pleistocene-aged Sahul landmass of a joined Australia and New Guinea
(White and O'Connell 1982). Much later and separately, it moved eastwards into
the islands of the Pacific Ocean culminating in settlement of New Zealand around
1,000 years ago (Bellwood 1979).

The nations that make up the Western Pacific have both shaped and been shaped
by their land and seascapes over many millennia. A distinctive difference exists

---

[7]This term is used interchangeably with 'indigenous knowledge,' 'traditional ecological knowledge,'
and 'traditional forest-related knowledge.'

between the larger land masses of New Zealand, New Guinea, and Australia, and the increasingly smaller and more isolated Oceanic islands moving eastwards to island Polynesia and northwards to Micronesia. Table 11.1 contains basic data on people and forests in the Western Pacific, and gives an indication of the relative status of forests in each country.

11.3.1   Biogeography

The region is geologically varied and spans a great period of time, with modern vegetation communities in New Zealand, Australia, and New Guinea owing much to their Gondwanic origins (Paijams 1976). Nutrient-poor lateritic soils of Gondwana evolved a diversity of major vegetation communities, from cool temperate beech forests to tropical rainforests (Specht and Specht 2005).

Eastward from New Guinea, curving to the south, tectonic activity formed the rest of Melanesia in a broad arc of islands composed of continental rocks of volcanic, sedimentary, and metamorphic origin (Howells 1973). This area lies within a zone of seismic instability, and volcanic eruptions are common (Bellwood 1979). Volcanic eruptions can destroy forests and gardens but they also produce the rich volcanic soils that benefit subsistence agriculture. Further eastward, the Polynesian islands (except New Zealand) are small and widely separated, the result of volcanic activity and subsequent erosion (Bellwood 1979).

Continental and island origins have produced different soil types, which have, in turn, been a major determinant of the nature and extent of original forest cover and of subsequent agroforestry and agricultural development by their human inhabitants. The rich volcanic soils of the high islands and continents contrast with the coral-based infertile soils and lack of potable water on atolls (Wilkie et al. 2002). The isolation and limited size of Oceanic islands have had a dominant impact, not only on humans, but on the entire island biota (Kirch 1979).

11.3.2   Human Colonisation

The Pleistocene landmass of greater Australia (or Sahul)—comprising New Guinea, Australia, and Tasmania—was colonised by humans from Southeast Asia around 40,000 years ago, during a period of low sea level (White and O'Connell 1982). There is also evidence for human settlement in island Melanesia (New Ireland and Solomon Islands) soon after (Mountain 1993). The first inhabitants were hunter-gatherers, but pollen and archaeological evidence suggest that techniques of plant management through forest burning and small-scale clearing were being practised and developed in the New Guinea highlands long before full agriculture commenced around 10,000 years ago (Mountain 1993). Indigenous societies practising early forms of agriculture would have still relied on wild resources to supplement the products of untested food production techniques. While some of

Table 11.1 Key statistics about forests and people in Oceania

Sub-region and country	Land area (sq. km)	No. of islands	Population 2002 (1000)	Population density people/sq.km	Forest area per capita (ha)	Forest area (1000 ha)	Forest area % of land area
Australasia							
Australia	7,687,000	1 mainland + 1 L	19,547	2.5	8.3	164,000	21
New Zealand	268,680	2 L	3,910	15	2.1	8,200	30
Melanesia							
Fiji	18,380	2 L + 300	856	47	1	935[a]	51
New Caledonia	18,575	1 L + 11	208	11	1.8	370[a]	20
Papua New Guinea	462,243	Mainland + 7 L + 593	5,172	11	6.5	36,000[a]	78
West Papua	421,981	Mainland + 12	c.2,200		–	33,000[a]	81
Solomon Islands	28,370	7 L + 985	495	17	5.9	2,200[a]	88
Vanuatu	12,190	80+	196	16	2.4	914[a]	75
Micronesia							
Federated States of Micronesia	702	607	136	702	0.1	[a]	–
Guam	541	1	161	293	–	[a]	–
Marshall Is	181	34	77	181	–	[a]	–
Nauru	21	1	12	587	–	[a]	–
Northern Mariana Is	471	17	77	162	0.2	[a]	–
Belau	488	200	19	42	1.8	35[a]	76
Polynesia							
American Samoa	200	5 in 2 groups	69	199	0.2	12	60
Samoa	2,935	8	179	61	0.6	105	37

(continued)

Table 11.1 (continued)

Sub-region and country	Land area (sq. km)	No. of islands	Population 2002 (1000)	Population density people/sq.km	Forest area per capita (ha)	Forest area (1000 ha)	Forest area % of land area
Tuvalu	26	9	11	429	–	[a]	(77% coco-nuts)
Kiribati	810	33 in 3 groups	96	119	0.3	28[a]	70
Cook Islands	240	15 in 2 groups	20	87	1.2	[a]	–
French Polynesia	3,521	118 in 5 groups	258	62	0.5	–	–
Niue	258	1	2	8	3	6	25
Tokelau	12	3	1.5	125	–	[a]	–
Tonga	649	170	106	142	–	4[a]	6
Wallis & Futuna	255	3	16	57	–	–	–

Source: Kanowski et al. (2005)

Notes: Data correct for 2005

L large island

[a] Forests either largely agroforestry systems (AFS) or combined AFS and natural forests

the food resources—for example, wild yams and bananas—may have been known to the colonisers; others would have required trial and error in the domestication process (Powell 1976).

Later arrivals, the Austronesians, brought already domesticated crops, and these were incorporated into established systems situated between horticultural and hunter/gatherer socioeconomic cultures.[8] The production of food crops and manipulation of forests to increase food supply foods and other products would have brought a new dimension to the corpus of traditional knowledge centred on the natural environment and its resources (see next section).

Farther south in what is now the Australian mainland, archaeological evidence indicates that Aboriginal people used the resources of all forest and woodland ecosystems from at least 40,000 years ago (Mulvaney and Kamminga 1999; Feary 2005), although intensive utilisation of tropical rainforests may have been more recent. One of the features of the pre-contact history of Aboriginal society is a continued reliance on hunting and gathering of wild resources, while the majority of the world's cultures became sedentary agriculturalists. There are various explanations for this phenomenon, one being the absence in Australia of fauna and flora suitable for domestication and/or a climate of unreliable rainfall, combined with generally poor soils. Alternatively, the presence of rich and varied foodstuffs supplied by nature throughout the year on a seasonal basis may have nullified the need to develop reliable sources of food through agriculture (Sahlins 1972).

The islands of remote Oceania are eastward of a major biogeographical boundary, Wallace's Line, where many naturally occurring edible plant foods do not occur. This has had important consequences for humans, who needed to bring in their own domesticated crop plants (and associated traditional knowledge) in order to colonise new islands (Kirch 1979). Production of food crops and development of shifting/swidden agriculture of root crops characterises all countries of the Western Pacific except Australia. Agroforestry systems for managing and manipulating tree crops and the forests surrounding the gardens were and are integrated with management of root crop staples (Kennedy and Clarke 2004). (See next section for discussion of traditional agroforestry).

The islands of Polynesia were the last places on earth to be settled by humans. Pottery and linguistic and other evidence point to a culture that grew out of the earliest settlements in island Melanesia around 3,000 years ago. Settlement occurred first in western Polynesia and spread to the tiny remote islands of eastern Polynesia (Gathercole 1977).

Campbell's rendition of the discovery of the islands of Tonga epitomises the early history of Oceania:

> An adventurous, sea-faring people capable of undertaking long sea voyages and of transporting colonising groups that had all the necessitates for survival in an exotic and

---

[8] Austronesians are people with a shared ancestry belonging and belong to a widespread family of languages with a possible origin in Taiwan around 5,000 years ago (Bellwood et al. 1995).

impoverished environment. In other words, colonists of both sexes made up the parties; there were probably older people and children, as well as men and women in their prime; they took with them seeds and roots of important plants that would supply not only food but also building materials, clothing, dyes, cosmetics and medicines. Seedlings of tree species that would enable the future construction or more sea going canoes must also have been included..... The food plants they brought with them to Tonga were the same as were used elsewhere in the pacific—the coconut, talo [taro], breadfruit, yam and banana (Campbell 1992, p. 18).

New Zealand was the last island group to be colonised by Polynesians, and Maori demonstrate a remarkable level of 'cultural localisation' (Gathercole 1977). Many economically important plants found in the rest of Polynesia such as coconut and pandanus did not exist in New Zealand but could not be grown because it was too cold. Maori adapted weaving traditions to make use of the locally available flax plant comprising several genera and species, producing the distinctive storytelling patterns of Maori weaving (Riley 2005).

This overview demonstrates that traditional forest-related knowledge of the people of the Western Pacific is not only about the resources and cultural values of natural forests but also about their modification in the context of horticulture and agroforestry. Nevertheless, natural forests remain an important source of food, raw materials, and medicines. Polynesians are described as horticulturalists who continued to appreciate the nutritive and economic value of many wild plants and animals, particularly when yields from domesticates were unavailable or low (Gathercole 1977). The languages of Polynesia demonstrate the relative importance of crop plants and forest foods. Kai (food) comprises two elements; 'staple starches', such as the cultivated taro and yams, and the 'relishes.' The latter includes opportunistically harvested nuts and fruits from the forest, and while considered to be highly desirable, they do not constitute a meal (Kirch 1979).

11.3.3 Human Interactions with the Environment

Human settlement of the scale and longevity described above could not have happened without environmental consequences. All the forests of the Western Pacific have been altered by human settlement and there is sound palynological and geomorphological evidence that swidden agriculture and agroforestry in the New Guinea Highlands has resulted in major changes to the forested environment over the past 10,000 years (White and O'Connell 1982; Paijams 1976; Golson 1977). Recently, new paradigms have emerged, advocating a greater role for climate change in modifying forests in Micronesia, which has implications for interpretation of landscape change elsewhere in the Pacific (Hunter-Anderson 2009).

In New Zealand, a third of the original forest was cleared for gardens prior to the arrival of the British (Roche 1990; Guild and Dudfield 2009). The slash-and-burn techniques the Polynesians brought with them from the tropics were unsuitable in the temperate climate, and the cleared forests did not regenerate, resulting in major deforestation in the North Island (Metcalf 2006).

Vegetation communities in most of islands of the Pacific are now dominated by secondary regrowth in abandoned gardens and highly altered forests manipulated over millennia to favour certain species. Primary forests contain shade-tolerant, long-lived trees, but thousands of years of agroforestry have replaced many of these with secondary forests of shade-intolerant, short-lived tree species. Most modern forests are a mixture of native species, early introductions, and later introductions (Mueller-Dombois 2008).

Traditionally, indigenous people were part of a functioning socioeconomic system in which the natural environment was managed and its resources used through a cosmological lens. Application of traditional law ensured that a particular resource was not overexploited. However, it is clear that early agricultural systems did lead to unbalanced and degraded ecosystems. It has been argued that Pacific Islanders 'prospered by disturbing the natural order' (Sauer, cited in Clarke and Thaman 1993). Newly arrived swidden agriculturalists cleared forest patches to plant the root crop plants they had brought with them, and established tree crops by selective planting and manipulation. Clarke and Thaman (1993) argue that modification of the closed forests rendered the islands more productive of food and more congenial to human occupation. Similarly, Geertz has argued that mixed planted crops eventually resemble the bush they have replaced; a 'natural forest is transformed into a harvestable forest' (Geertz, cited in Kirch 1979).

The extent to which Aboriginal people have altered Australia's vegetation is a contentious area of debate in both the scientific and populist literature. On the one hand it has been argued that modern Australian landscapes are a direct reflection of thousands of years of farming by fire and digging stick, for example (Gammage 2003; Rolls 2005). Others have taken a more precautionary approach by recognising that Aboriginal burning may have changed the floristics and structure of some forest types, but it is difficult to disentangle climatic from anthropogenic effects, especially prior to 10,000 BP (Head 1993).

As hunter-gatherers looking to maximise returns from the natural environment, Aboriginal people manipulated certain species and excluded use of others through religious taboos. Evidence for replanting the tops of harvested native yams and protecting the seedlings of certain species germinating around campsites in the Northern Territory is well-documented (Hynes and Chase 1982; Berndt and Berndt 1981). Aboriginal people living in the fertile and productive Murray Darling basin were observed harvesting *Panicum* grasses to enhance dispersal of seeds (Allen 1974), and there is some evidence to suggest that the economically important cycads were transplanted into different forests (Boutland 1988).

Following the late Rhys Jones' seminal paper on fire stick farming (Jones 1969), management of the land and its resources through judicious use of fire has become a central premise of reconstruction of traditional Aboriginal life (e.g., Hallam 1979; Hill 2003). In fact, an increased frequency of high-intensity wildfires has been attributed by some Aboriginal people and researchers to the cessation of traditional burning practices, which kept fuel loads at low levels (Langton 1998). Ethnohistorical and ethnographic sources referring to Aboriginal burning have been used to justify proposals for frequent, broad-scale burning to reduce fuel loads, particularly following catastrophic bushfires, but this has come under criticism (Benson and Redpath 1997). Nevertheless, palynological research has demonstrated major changes in Quaternary vegetation that can be reasonably attributed to Aboriginal burning (Singh et al. 1981) and this, together with early records, suggests that fire was a major management tool, used according to customary laws of land ownership and kinship, for a wide range of reasons. Oral traditions and fire management by traditional owners in northern Australia are testimony to a long history of land management by fire. However, the same cannot be said of much of southern Australia, where historical legacies have denied Aboriginal people the fire ecology knowledge that could have been passed on to subsequent generations.

Throughout the Western Pacific, traditional ecological knowledge bonded human societies to these ancient settings to form an integrated, holistic system. The forests and the manipulation of them by fire, clearing, and planting contributed to provision of food and raw materials that sustained traditional socioeconomic systems. The next section explores the cultural filters—the traditional knowledge—that determined how societies connected with these settings.

11.4   Understanding Traditional Forest-Related Knowledge

The traditional forest-related knowledge of indigenous peoples is embedded in their oral traditions, much of which is still extant in the Western Pacific, particularly in Oceania and Northern Australia, where people still speak their languages and practice customary forms of land management. Indigenous knowledge is also documented in the early records of non-indigenous explorers, settlers, and government officials, many of whom kept detailed accounts of their encounters with indigenous peoples. While both have inherent limitations—reliance on a flawed memory on one hand, and the selective, value-judged observations of the colonial imperialists on the other—together they provide a valuable insight into traditional indigenous cultures.

Traditional forest-related knowledge in the Western Pacific can be understood as a spectrum. Beginning with the western edge of the region, the ancient, primarily hunter-gatherer knowledge of Aboriginal Australians grades eastward through Oceanic cultures' integrated knowledge of natural and modified environments. Successful settlement, especially the depauperate islands of eastern Polynesia, relied on application and modification of traditional knowledge to new environments for successful propagation of root and tree crops and a concomitant decrease in use of natural forests.

11.4.1   Cosmologies

As with traditional knowledge systems worldwide, those of the indigenous people of the Western Pacific are embedded in complex and ancient spiritual connections with the natural world. The material world is suffused with spiritual forces which must be respected for continuation of human survival and well-being (Reid 1995).

Cosmologies of the Western Pacific's indigenous peoples include stories relating to the origin and creation of the world and its human inhabitants by gods or spirit beings. They link the present to the past through the actions of ancestral beings and impose responsibilities on humans in their relations with the natural world. In Australia, the term dreaming is ascribed to a time before, but which still exists, when creation beings made and named all the features of the natural world and all the plants and animals, including humans. These ancestral beings are everlasting and remain as part of nature and people. Anthropologist Debbie Rose comments that 'country' is the manifestation of creation, so everything that happens has creation as its precondition. Knowledge—local, detailed, tested through time—is the basis for being in country (Rose 1996).

Polynesian societies ascribe similar connections between people and the environment. In Tongan society, 'fonua' invokes a cosmology in which the environment is regarded as an extension of human society (Francis 2006). As a result, human agency is integral to a physical landscape that includes the land, the ocean, and the sky. The concept of fonua 'people of/and place' described a local territorial entity that incorporated the land and natural surrounds associated with a chiefly title holding, and the people residing on that land. Fonua is also a descriptive term for the soil that grips the roots of plants when pulled from the earth. The old Tongan word for placenta was fonus, a reference to the practice of burying the placenta after birth of a child (Francis 2006). Similarly in New Zealand Maori cosmology, the word for land, whenua, is the same as that for placenta (Walker 2004).

Cosmologies include improvements to the land for sustaining humans. At specified times of the year, Aboriginal Australians conducted increase ceremonies, asking the creation beings to maintain and replenish wild resources (Berndt and Berndt 1981). Similarly, the yam planting cycle in Oceania is intricately tied to human health and well-being. Yams are one of the most important crops in Oceania and New Guinea, with great utilitarian as well as symbolic significance. Historical accounts of traditional Kanak life in New Caledonia state that people perceived themselves by analogy with objects of nature such as the yam, whose cycle symbolised the cycle of life (Dahl 1989).

Forests, trees, and plants feature strongly in the cosmology of Western Pacific indigenous cultures. In Vanuatu, the Banyan tree is the most prominent arboreal symbol of place (Patterson 2002). Banyan trees are also important in Aboriginal cosmology, as expressed by Yolngu leader Galarrwuy Yunupingu:

> That tree is a special place, inside it are important things. Its like the heart of the country, our beliefs about our land reside in that tree and at the site of the tree, they reside in the rocks, in the waters and in our minds. We know these things to be true (Sculthorpe 2005, p. 172).

A powerful creation story in Maori lore concerns the kauri *Agathis australis*. This majestic tree was not only the god of the forest but also the creator of the first humans (Fig. 11.3).

Fig. 11.3 The New Zealand Kauri (*Agathis australis*), Tāne mahuta (Lord of the Forest), North Island, New Zealand. This tree is at least 600 years old Waipoua Forest (Source: Wikimedia Commons)

Tāne created the forests when he separated his parents, Ranginui (the sky father) and Papatūānuku (the earth mother), and let light into the world. As Tāne Mahuta he is god of the forest, presiding over its plants and birds. As Tānenui-a-rangi he is creator of the first human. Respect for Tāne's forest was shown by performing certain tikanga (customs). Their importance is reflected in the story of Rātā. Rātā went into the forest, cut down a tree, and began to carve it into a canoe. When he returned the next day to continue his task, the tree was miraculously standing in its original position. He felled it again and set to work, but the same thing happened the following day, and the next. Finally, Rātā hid behind a bush and saw the hakuturi (forest guardians in the form of birds, insects and other life) replanting the tree. When he confronted them, they told him he had failed to perform the appropriate rites. He then did so, and the hakuturi released the tree.

The great trees of Tāne, god of the forest, were called Ngā Tokotoko-o-te-rangi (the posts that hold the heavens aloft) because they held Ranginui (the sky father) above Papatūānuku (the earth mother) (Te Ara–the Encyclopedia of New Zealand 2009).

Taboos or tapu acted as a sustainability measure in traditional forest use and are usually imposed by village chiefs to prohibit use of a particular resource. The prohibitions generally related to resources in decline, with strict punishments for ignoring the bans (Government of Samoa 1998; Nalail 1996). In Australia, the sacred prohibitions on exploitation of certain species could be due to totemic associations with an individual, or exclusion of certain forest areas from burning or exploitation (Rose 1996).

Sacred trees and magic plants are also part of the cosmos. For Nyungar people living in the forests of southwest Western Australia, the karri (*Eucalyptus diversicolor*) and jarrah (*E. marginata*) trees were part of women's dreamings, while the marri was identified as male (Crawford and Crawford 2003). The changing nature of vegetation is an important symbolic representation of continuity and change in Aboriginal society, as are unusual ecological contexts such as species on the edge of their range or relict vegetation communities (Cooper 2000). Planting of *Cordyline fruticosa* in specific areas of Melanesian gardens was done to secure protection by the ancestors (Manner 2005).

The tall eucalyptus forests of southeastern New South Wales in Australia are home to the Doolagarl, a supernatural being who acts as a guardian of the forest—taking care of people who belong there and harming people who do not (Rose 1996). The Doolagarl is believed to be represented in some of the rock art paintings in the area and his presence continues to be a powerful one for local Aboriginal people (Feary 2007).

11.4.2 Traditional Forest-Related Knowledge of Aboriginal Australians

As hunter-gatherer peoples, Aboriginal Australians were intimately concerned with the growth cycles of the plants and animals on which they depended, and a detailed and precise knowledge of the local natural environment was essential for survival (Berndt and Berndt 1981). The life cycles of plants and animals and best times of year for harvest or hunting were learned by young people through observation and as part of gendered storytelling and ritualised passing on of information. Reading the landscape ensured that forest products were hunted or harvested in a sustainable manner. The finely detailed knowledge of relationships between living things—the seasonal calendar—informed knowledgeable humans of when a particular resource was ready to be exploited; for example, middle Victoria River crocodiles are laying their eggs when the jargala tree (*Sesbania formosa*) is flowering (Rose 1996).

Aboriginal people classified the landscape according to its biophysical and cosmological values. The Yankunytjatjara people of the Western Desert recognise five major ecosystems, and specialised techniques were used for the seasonal procurement and processing of their resources (Table 11.2)

As well as remembered oral traditions, archival records of the early colonial era often contain great detail on Aboriginal use of plants. Although relatively late, Donald Thomson's seminal work on seasonality in Aboriginal communities in far north Queensland (Thomson 1939), is a sympathetic and detailed evaluation of traditional life that showcases the critical role of indigenous knowledge of the local environment in maintaining a balance between resource availability and use. Thomson makes many references to the use of forests by Aboriginal people and provides detailed information on the different wood types used in manufacturing implements, such as the use of lancewood (*Acacia rothii*) for making hunting and fishing spears.

Table 11.2 Traditional ecosystem classification of desert Aboriginal communities

APU	Karu	Tali	Puti	Pila/Uril
Rocky outcrops, hills, ranges	Watercourses (gullies, creeks, rivers)	Sandhills, sand dunes, sandhill country	Woodlands, shrublands	*Spinifex* grasslands, plains/open country, grasslands, plains
Sparse vegetation with *Spinifex* and wattles (*Acacia* spp.)	Wide sandy riverbeds, adjacent floodplains, Seasonal wet/dry	Open, sparse shrubland, *Spinifex* dominant groundcover	Collectively the most important sandplain vegetation communities, mulga the dominant species	

Source: Institute for Aboriginal Development 1985

Bush foods and bush medicines and their ethnotaxonomy are a familiar manifestation of traditional forest-related knowledge in Australia, and there are literally hundreds of books on the subject by Aboriginal people (e.g., Thancoupie 2007), botanists (e.g., Maiden 1889), and for a general audience (Cribb and Cribb 1974; Isaacs 1987). The most insightful are collaborative, where a non-Aboriginal person converses in the local language or dialect, as this picks up the subtleties of the relations between resource use and cultural practices, such as participatory research with the Yankunytjatjara people in central Australia (Institute for Aboriginal Development 1985).

Specific records of Aboriginal forest use are more elusive than those for other ecosystems such as savannahs or the coast. This is probably due to several reasons. Forests occur in the high rainfall belt on the edge of the continent, where colonisation had the most and earliest impacts, leaving little time for detailed observation of traditional lifestyles. Furthermore, much forest-based activity was carried out by women, whose activities may not have been noticed by the predominantly male early settlers and anthropologists. Evidence from the rainforests of New South Wales in the nineteenth century comes from explorers who observed groups of Aborigines in the rainforests, although their camps appear to be outside or on the edge of the rainforest. They left detailed descriptions of organised game hunts by Aboriginal groups of men and women. Aboriginal women were observed twisting the large leaves of the Bangalow palm to make a waterproof container. The bark of the stinging tree *Laportea gigas* was processed and woven or knotted into dilly bags, or made into fishing and hunting nets (Byrne 1984). Sullivan's analysis of ethnographic records from northern New South Wales demonstrates the considerable economic and social importance of subtropical rainforests in secular and sacred life (Sullivan 1978), and Campbell (1978) discusses the highly specialised techniques required for processing many rainforest plants.

Ethnoecological research by David Harris demonstrated that Aboriginal people of the rainforests of northeastern Queensland were morphologically and culturally distinct from other Aboriginal groups. In addition to the ability to processes toxic cycads, rainforest people constructed dome-shaped thatched huts in forest clearings, made bark cloth from hammering the inner bark of fig trees, and wove sieve bags from lawyer cane and rush for leaching toxic tubers (Harris 1978). Specialised nut-cracking stones were used for opening the hard but highly nutritious nuts of rainforest trees, whose prolific production during the fruiting season supported large numbers of people gathering for ceremonial activities including the spectacular bunya festivals (Huth 2001). Although the colonisers stopped the festivals, the asso-ciated traditional knowledge has been kept alive (Haebich 2005).

Specific uses have been recorded for parts of a wide range of *Eucalyptus* species (Macpherson 1939), and a database of timbers used traditionally by Aboriginal people was put together in 2002 (Kamminga 2002).[9] It currently contains around 300 entries, listing the botanical name, the ethnographic source of the information, and the observed use of the timber. While compiled primarily for the purposes of demonstrating the relationship between woodworking properties and wooden imple-ments, it has much broader application.

11.4.3 Traditional Forest-Related Knowledge of Agricultural Societies

Agricultural peoples of the Western Pacific have worldviews that maintain strong connections between their living space and the surrounding forests. Mosko (2006) describes the traditional classificatory patterning of 'inside' and 'outside' spaces with reference to the North Mekeo people of the central province of Papua New Guinea. Mekeo daily life is divided between the village and the bush, with con-stant movements and transformations between the two. The village is conceptua-lised as the outside space, which was initially 'cleared out' of the inside space of the bush. This means that the outside village contains much that originated in the inside bush and is a metaphor for transformations between the human body and the outside world.

Within this broader cosmos, traditional forest-related knowledge of agriculturalists focuses on cultural relationships with the biota of native forests and with 'trans-ported landscapes' of agroforests, irrigated swamps, dry field agriculture, and a suite of consciously and accidentally introduced organisms. Merlin (2000) provides an excellent analysis of ethnobotanical research in remote Oceania, noting that observations of traditional plant use and agroforestry have been occurring since the voyages of Captain Cook in the late eighteenth Century.

For Papua New Guinea, Powell provides an excellent synthesis of previous eth-nobotanical research. Based on this and his direct observation, he gives comprehen-sive species lists of plant foods, distinguishing six main categories of levels of plant domestication and use by local indigenous people. Powell's analysis indicates that staple foods are primarily starchy underground roots and tubers, whereas supple-mentary foods are a very wide range of fruits, nuts, and greens. No wild foods were cultivated as staples, although wild forms of root crops such as taro and yam exist.

---

[9] Held by AIATSIS in their Aboriginal studies electronic data archive, but it is not currently active.

Most forest foods were identified as supplementary foods and many wild foods are transplanted and cultivated as crops within forests or as productive trees around villages (Paijams 1976).

Farther south, in New Zealand, the large forest tree, totara (*Podocarpus totara*) was used by Maori for fire-making, manufacturing canoes, and carving. A new totara tree had to be planted each time one was felled, to appease Tane, the god of the forests, for removing one of his children. This practice also ensured sustainable practices for use of valuable trees. The tall, straight kahikatea (*Dacryridium dacryoides*) bears sweet fruits that were highly prized by Maori. Fruits of the Miro (*Prumnopitys ferruginea*), are very attractive to the New Zealand pigeon, and Maori used to place snared water troughs in miro trees so the pigeons could be caught while drinking (Metcalf 2006).

Letters from John Deans to his father in 1845 contained observations of traditional Maori life in the Riccarton-Christchurch region of the South Island, where a sophisticated socioeconomic system produced surplus goods for trade and supported powerful tribal chiefs:

> Effective techniques for obtaining the different foods, while at the same time conserving the resource, had been perfected, and a sophisticated social system had been developed to do the required work…. Each whanau (extended family) had its allotted rights to take its requirements within the rules laid down, and its allotted part to play in producing a surplus for the tribal headquarters, and for the use of groups in other parts of the tribal territory, who would supply something else in return. Thus a Maori community living at Putaringamotu would specialise in products from the local forests—preserved pigeons, carved totara and canoes…. (Molloy 1995, pp. 3–4).

11.4.4   Cultivated Landscapes of the Western Pacific

Today approximately six million Pacific Islanders rely on traditional agriculture for their subsistence needs (Manner 2005). Most traditional agricultural systems depend on connections with a range of natural ecosystems, and the original Pacific Islanders were geographic opportunists and ecological practitioners who recognised the capabilities and developed different methods for their exploitation and maintenance (Manner 2005). Many terms have been used to describe land use practices in the Pacific; for example, agroforestry, agroecosystems, arboriculture (Kennedy and Clarke 2004; Thaman 1989), and classificatory systems have been developed to describe the various methods of cultivation and land use (Manner 2005). The term cultivated landscapes is used here, after Kennedy and Clarke (2004), who use archaeological, geographical, and anthropological sources to demonstrate the antiquity of forest transference and manipulation within a broader socioeconomic system of landscape management.

All traditional agroforestry systems across the Pacific exhibit common characteristics of high species diversity, incorporating cultivated and protected native and introduced tree species (up to 300 on the larger Melanesian islands). These include tree crops such as coconuts, breadfruit, and bananas as well as a wide range of fruit and nut species, either deliberately planted, encouraged, and protected in the regeneration of regrowth, or spared when clearing new garden plots (Thaman 1989). Within a given species of tree or root crops there are many locally differentiable cultivars and varieties, with variable yields and seasonalities, spreading distribution of food more evenly across the annual cycle. Different cultivars exhibit differential adaptations to ecological conditions such as salt spray, pest and disease resistance, soil type, and shade tolerance, and each was selected for a specific use (Thaman 1989).

Siwatibau (1984) discusses traditional environmental practices in Fiji to demonstrate that Fijians had developed methods of shifting cultivation that optimised resource utilisation without depleting the resource base. The optimum location for a new garden was carefully chosen and the garden was established by slashing the primary or secondary forest, often leaving the large trees. Ground disturbance was minimised (and effort reduced) by leaving tree stumps to rot. The slashed vegetation was burnt and weeded, and if slopes were steep the rubble was placed along contours to reduce soil erosion. New crops were then planted using methods appropriate to the plant species.

Manner (2005) also notes that in swidden preparation, not all trees and shrubs are killed; some are cut off and allowed to sprout, to enhance reforestation. Nitrogen-fixing plants such as *Causarina oligodon* were planted to enrich the soil. Thaman et al. (nd) note that Pacific Island agricultural and land use systems were built on a foundation of protecting and planting trees, and were developed and managed for both human need and ecological services.

Traditional land management in Tonga involves a complex mix of trees, shrubs, and short-term ground crops incorporated into short-term shifting agriculture on small plots of land (Fig. 11.4). Cleared vegetation is allowed to dry, and is then burnt. Large, important food trees (cultivated or native) are left. Some are pruned to

Fig. 11.4 A typical Polynesian garden within a forest. Ha'apai, Tonga (Photo: S. Feary)

allow light in and to produce leaves, which fall to the ground and act as mulch. Larger branches are used as trellises for the yams, and then as firewood after yam harvest. Trees such as *Pandanus* or *Hibiscus* are planted along garden edges to act as boundaries and windbreaks and to provide food and other products.

Living fences could consist of important timber trees such as *Casuarina* spp. The root crops are planted and harvested sequentially, and the cycle can be extended by planting kava or paper mulberry for making tapa cloth. The garden returns to fallow over 4–10 years, and the existing trees continue to provide food, medicines, and other products.

The vignettes presented above demonstrate that traditional land use systems at the time of European contact were productive and efficient. Such systems could not operate without detailed local knowledge of the natural and cultural environment. Thaman et al. (nd) aptly describe these systems as 'long term investment of time, knowledge and effort in a living, growing bank account.'

However, non-indigenous people often do not see these systems as being productive and have endeavoured to change them, as described in a narrative from Papua New Guinea (Box 11.1 and Fig. 11.6).

11.5   Contemporary Forest Management in the Western Pacific

Indigenous peoples are often closely associated with forests; forests provide habitat and are important to them for economic, social and cultural reasons. Concerns about conserving and managing forests often coincide with concerns about the survival and integrity of the cultures and knowledge of Indigenous peoples (Ruis 2001, p. 8).

International forest debates on sustainable forest management have been important for bringing indigenous rights and forest management into the same discourse (Feary et al. 2010). The Convention on Biodiversity has also influenced the global dialogue on forests by supporting recognition of forest-related knowledge of indigenous and forest-dependent people and raising issues of intellectual property rights (Ruis 2001).[10]

Sustainable management of forests is an important global issue. Current discourse on climate change has identified Western Pacific countries with significant areas of forests, such as Papua New Guinea and the Solomon Islands, can potentially contribute to amelioration of greenhouse gas emissions (von Strokirch 2008). Furthermore, many of the low-lying atolls of Polynesia are the most vulnerable places on earth to even minor rises in sea level.

The 'developed' nations of New Zealand and Australia have both undergone major reforms in forest management over the past 30 years, resulting in a greater recognition of indigenous interests and rights over forests, although many would argue that there is still a long way to go.

---

[10] See http://www.biodiv.org/convention/articles.asp

Box 11.1   Felling Trees on Top of the Garden—Revealing 'Monocultures of the Western Mind' by D. Eastburn

When an Australian patrol officer first conducted a survey and census of the people living in the rainforests of the Great Papuan Plateau in southwestern Papua New Guinea, he was confronted with an unusual agricultural practice of planting crops under the forest canopy, and when they begin to grow, felling trees on top of them. To his 'Western eyes' that process appeared outrageous, 'inferior even by primitive standards,' and he estimated that 'as much as 40% of the crop was destroyed' (Schieffelin 1975, p. 31).

The unusual method of cultivation while mechanistically apparently outrageous is in fact ecologically brilliant. It is uniquely suited to maintaining the resilience of the high rainfall lowland rainforest environment by minimising soil disturbance and enabling the area to be quickly reclaimed after the garden is abandoned. No fire is used in the preparation of the garden. The undergrowth beneath the forest canopy is first cleared, and planting material such as taro tops and banana suckers, is planted into the minimally disturbed humus on the forest floor with only a few centimetres showing above the surface. The canopy protects the new garden from rain and filters the sun in its first weeks. Each tree in the garden area then has a scarf cut into one side to direct its fall, and after a few weeks when the crop has 'taken root' a large tree is cut, creating a 'domino effect' resulting in all of the trees falling 'slowly' together in a tangled mass. The fall is also softened by the branches breaking and dissipating energy like the safety 'crumple zone' of a car. Anthropologists, who have studied the process, have found that the impact on the crops ranges from 'virtually no damage' to less than 5% (Schieffelin 1975, p. 31), and the food plants can easily grow through the tangle and quickly form a structure like a microcosm of the surrounding forest.

Other advantages of this method of agriculture include the protection of the soil from leaching and baking by the initial mulch of leaves and twigs, the release of nutrients as the mulch decomposes, and breaking up heavy raindrops by the tangle of fallen trees preventing erosion from runoff. The original condition of the forest floor is maintained enabling a quick reversion to rainforest after the garden is abandoned. Additionally, it is a low energy method of clearing in a sparsely populated area, and the fallen tangle of trees form an excellent defence for houses built in the centre of the garden area, as well as a ready supply of firewood and fencing materials.

In the early 1970s an agricultural officer (Didiman–Melanesian Pigin) trained in Western science cleared bare a rectangular patch of rainforest as a 'demonstration plot' to teach the local villagers, who use the same methods as described above, how to make a 'proper' neat garden. Unfortunately the timing,

(continued)

Box 11.1 (continued)

during the heaviest rainfall period, was not good. The exposed soil was leached and its surface packed by the impact of the heavy rain, and the sun baked it hard. The officer attempted to recover the situation by ordering a large amount of artificial fertiliser that was delivered by government-chartered aircraft. When I visited the area around a decade later, the 'Didiman's Demonstration Plot' remained like an oversized abandoned tennis court in the forest.

Fig. 11.5 Felling trees on top of the garden, Papua New Guinea (Photo: D. Eastburn)

11.5.1 New Zealand

Since the citizen debates over logging in the 1980s, most native forests in New Zealand are now in Crown-owned conservation reserves. However, Maori own 29% of the privately owned native forests, the majority occurring in the North Island (Hammond 2001). Some Maori landowners wish to derive economic benefits by harvesting and selling the timber from their forests, but some such as the *Nothofagus* (Beech) forests in the South Island, originally granted to Southern Maori under the South Island Landless Natives Act 1906, have been identified as having high conservation value (Wheen 2002).

New Zealand's Indigenous Forest Policy has introduced Nga Whenua Rahui covenants as a way of preserving significant natural ecosystems on Maori land. Similar covenants exist in relation to non-Maori land, but Nga Whenua Rahui recognises the unique situation of Maori, who can seek compensation for loss of income from the merchantable timber (New Zealand Department of Conservation 2006).

Forestry in New Zealand is heavily dependent on plantations for commercial timber, particularly the exotic species *Pinus radiata*. Maori have extended their cultural connections to native forests to embrace exotic commercial forestry as an 'adopted son' who protects remaining lands and provides employment and economic benefits (Miller et al. 2007, p. 15).

Maori now own 14% of the planted forest estate although most of this is managed by the Crown or private forestry on behalf of Maori owners. In recent years, however, Maori interest in more active involvement in forestry operations has increased, as a way of providing employment and training opportunities to young Maori (Thorp 2003). For example Ngati Porou Whanui Forests, Ltd. is a Maori-owned company currently managing 10,000 ha of forest in a joint venture arrangement with a Korean company (Miller et al. 2007).

11.5.2   *Australia*

After decades of citizen unrest over logging practices in public forests, the Commonwealth Government formulated a National Forest Policy Statement (NFPS) in 1992, which is an agreement between the Commonwealth and state governments about objectives and policies for the future of Australia's public and private forests (Commonwealth of Australia 1995). The NFPS committed to protecting Aboriginal values of forests and linked a potential enhanced employment in forest industries with an increased capacity to utilise traditional knowledge (Commonwealth of Australia 1995). State based regional Forest Agreements (RFAs) have been criticised for not giving sufficient recognition to native title rights (Dargavel 1998; Lane 1997) and Aboriginal people in most states have expressed some level of dissatisfaction (Lloyd et al. 2005; Engel 2000). Other researchers considered that Aboriginal interests had been defined too narrowly (Rangan and Lane 2001) and that there was a tendency for the RFAs to look backwards into Aboriginal culture instead of forwards (Lane 1997).

In 2005, the Australian government released a National Indigenous Forestry Strategy (NIFS) with the aim of establishing a framework for facilitating greater involvement by Indigenous people in the forestry industry, with the ultimate goal of alleviating economic and social disadvantage (Commonwealth of Australia 2005). Improved forest management is an objective of the strategy, but its narrow focus on employment in mainstream forest industries limits the strategy's ability to incorporate Aboriginal values and knowledge into forest management (Feary 2007).

Approximately 14% of Australia's forest and woodland area is under Aboriginal communal ownership (Fig. 11.5). This includes timber production forests, plantations, and forest ecosystems of high conservation value (Montreal Process Implementation Group 2008).

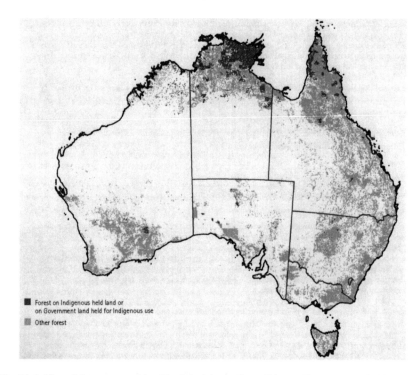

Fig. 11.6 Map of forests owned by Aboriginal Australians (Source: Commonwealth of Australia (2005). National Indigenous Commonwealth of Australia (2005). The Australian Bureau of Agricultural and Resource Economics and Sciences (ABARES) is an independent research agency of the Australian Government)

11.5.3 Pacific Countries

Sustainable forest management (SFM) is currently an unachieved goal in the majority of Pacific countries, and the prognosis for SFM continues to be a negative one (Siwatibau 2009). Over the past decade or so, reviews by international organisations such as FAO have expressed concern over forest management in most Pacific Island countries, especially the larger Melanesian islands of Papua New Guinea and the Solomon Islands (e.g., Brown 1997; Hammond 1997; FAO 2000). The Pacific is often combined with Asia in international reporting, and detail on specific Pacific countries can be lacking (e.g., Brown and Durst 2003).

Recent reviews (Bond 2006; von Strokirch 2008) have used deforestation statistics to paint a gloomy picture for the Pacific Islands, and concluded that the major primary natural and accessible forests of the Pacific will be logged out by 2020. Bond states that impacts of commercial exploitation of forests can be indirectly linked to impacts on human health; for example, HIV/AIDS incidence is relatively higher around remote logging camps (Bond 2006).

The two principal causes of unsustainable forest management have been identified as high levels of forest clearing for subsistence agriculture and cash cropping, and poor logging and silvicultural practices (von Strokirch 2008). Increasingly, damage from cyclones is also contributing to deforestation. While the diversity of situations across the vast area of the Pacific eludes generalisation, the reasons behind unsustainable forest management are closely tied to a range of social, economic, and political factors affecting both the informal, subsistence economy and the formal market economy. They are a direct result of population increases and efforts to achieve economic development goals and move into the mainstream market economy.

Addressing unsustainable forest management it is proving difficult for a number of reasons. These include poor quality forest data and unreliable forest inventories, under resourced and ineffective government departments, and lure of alleged benefits and royalties to local land owners from large-scale logging. Business ignorance of landowners and disputed land tenures are contributing factors (Wilkie et al. 2002).

Development of sound forest policy which is effectively implemented and monitored is a critical step towards sustainable forest management, and several Pacific countries have or are developing forestry and land use policies. Two examples are discussed below—one each from Melanesia and Polynesia.

11.5.3.1   Samoa

The independent country of Samoa is the only Polynesian island with significant forest cover, primarily tropical rainforest, and is notable for its unique forest biodiversity (FAO 2000). The Village Fono Act of 1989 recognises traditional management systems for control of village resources and village management (Government of Samoa 1998). Despite the existence of customary systems, the forests of Samoa suffered extensive deforestation from land conversion and logging operations between 1977 and 1990 (Government of Samoa 1998). Logging on the main island of Upolu was halted in 1989 and it was predicted that all merchantable forests would be gone within the next 5–6 years at the current rate of clearance (Groome Poyry 1993; Brown 1997). Since this time, Samoa has undertaken major reforestation, with the German Government Aid agency (GTZ) running a project on natural forest management of logged-over forests on the island of Savaii.

Samoa ratified the Convention on Biodiversity in 1993 and formulated a National Forest Policy in 1994 (Government of Samoa 1998). A revised policy in 2007 included two objectives relevant to traditional forest-related knowledge—economic and social benefits arising from pharmaceuticals based on native plant species, and ecotourism development. The first draws attention to the need for protection of the intellectual property rights of Samoans.

Samoa has a strong commitment to conservation and participatory approaches to forest management, and a number of community forest conservation projects have been established (FAO 2000), which are based on traditional rainforest preserves (Cox and Elmqvist 1994). In American Samoa, the national park concept fits well with the traditional Samoan way of life, the fa'asamoa. In keeping with the meaning of the word Samoa—'sacred earth'—the park helps protect fa'asamoa, the customs, beliefs, and traditions of Samoan culture.

11.5.3.2 Vanuatu

The republic of Vanuatu is one of several Melanesian islands where forestry is important to rural communities, and is one of their main sources of cash income (Gerrand and Bartlett 2001). As with all Pacific nations, forests are also important in the subsistence economy of most 'ni-Vanuatu' and have been managed as part of customary socioeconomic systems. Today, most land is under inalienable customary land tenure and traditional laws governing use of land and its resources have been enshrined in the constitution since independence in 1980,[11] although policy debate around land reform has been limited (Manning and Hughes 2008).

The 1998 National Forest Policy recognised that forests are vital to the country's cultural heritage, with resource owners not only beneficiaries of the forests, but also managers and developers of this resource (Tamla 2002). The policy's vision was for the government to achieve sustainable forest management through working cooperatively with landowners, customary chiefs, and forest industries (Gerrand and Bartlett 2001). Several components of the policy are related to traditional forest-related knowledge, including respecting traditional tapus, recording traditional knowledge, and benefitting holders of that knowledge via profits arising from its commodification.

A review of Vanuatu's Forest Policy implementation noted that many of the problems identified during policy development had been addressed, due primarily to the inclusive approach adopted with landowners; clear recognition that responsibility for policy implementation and sustainable forest management is shared across the key stakeholder groups; and effective consultation (Gerrand and Bartlett 2001). A more targeted review is required to address the issues specifically related to traditional forest knowledge; although there are many improvements, it is difficult to determine to what extent traditional forest-related knowledge has been a contributing factor.

11.6 Traditional Forest-Related Knowledge in a Modern World

History has shown that the Western Pacific's indigenous cultures have been remarkably resilient to the impacts of outside cultural influences. Changes in language, material culture, and cultural behaviours attest to a shaping by external forces, to which local cultures responded by ignoring or by adapting them to existing cultural systems

---

[11] Vanuatu was once known as the New Hebrides and managed jointly by the UK and France from 1880s until independence movements resulted in Vanuatu becoming a republic in 1980.

(Sahlins 2005), despite efforts to overwhelm them by 'civilising forces,' especially Christianity. More than 200 years of cross-cultural interaction have resulted in hybrid cultural systems, emergence of a neo-traditionalism, or in some cases, a complete loss of traditional knowledge. Studies of cultural change in Australia suggest that modern Aboriginal culture reflects a shared history (Wolfe 1994), where the 'traditional past' is not replaced by a 'changed modern' version of culture (Suchet 1996), but contains both a continuity of cultural traditions and deliberate cultural reconstruction.

11.6.1 Threats to Traditional Forest-Related Knowledge

Despite this resilience, traditional forest-related knowledge and the socio-cultural systems in which it is situated are under threat across the globe. Pacific Island countries face additional pressure from their reliance on international donor agencies advocating increased economic development that can potentially undermines traditional land management (Thaman 1989).

Generally, as indigenous cultures modernise and become part of the market economy, traditional systems and knowledge become less relevant, particularly for younger generations. For example, in Vanuatu, the Vanuatu Cultural Centre is concerned that traditional systems of sustainable development could disappear as a result of alienation of youth from cultural traditions and the increasing demand for a Western education (Nalail 1996).[12] Siwatibau (2009) refers to trained foresters conducting inventories without recourse to the knowledge held by local villagers. In regard to the application of traditional knowledge to forest conservation in Papua New Guinea, local villagers articulated the tensions between modern education and traditional knowledge in the following way:

> We cannot use the same practices used by our ancestors nowadays, because educated people do not respect and listen to the village elders. Educated people are proud of themselves, they think they have been to school and are more knowledgeable than the village elders. Therefore, if village elders make rules to conserve a certain area, people that have some form of western education will not adhere to those rules (Ellis 1997, p. 57, cited in Filer 2000, p. 14).

Traditional systems based on localised laws and cosmology can be undermined by pressure from the market economy. In Pacific Island societies, chiefs play a vital role in regulating use of certain resources through applying taboos on their exploitation. The power of chiefs and elders is compromised if international aid agencies or government bureaucracies do not heed their position in indigenous societies. The May 2000 Country report for Vanuatu noted that economic development was leading to a loss of purpose for traditional taboos and a reduction of their effect on conserving resources, including trees (Republic of Vanuatu 2000). Furthermore,

---

[12]The Vanuatu Cultural Centre is a national statutory cultural heritage management body comprising all the major cultural heritage institutions apart from the National Archives. See http://www.vanuatuculture.org/.

even traditional societies are not immune to greed, treachery, and disloyalty (Hooper 2005). The erosion of the ability of traditional forces to hold these in check has resulted in numerous examples of chiefs doing deals with big logging companies to further themselves and their families rather than for the greater good of the community.

There are other, more insidious threats to the retention of traditional forest-related knowledge. Villagers in Melanesia may be reducing their exploitation of local forest resources in order to disassociate themselves from the hallmarks of tradition and their associated stigma of 'primitivism' (Filer 2000). In Australia's Gulf Country, some Aboriginal people place culture in a past they perceived to be largely irrelevant to the present (Trigger 1997).

Indigenous people are therefore placed in a difficult situation of not wanting to be seen to be opposing development or 'living in the past' but are also fearful that modernisation will render traditional ways redundant (Horowitz 2002).

Exposure of many Pacific countries, especially Melanesian countries, to the global market economy has resulted, in many instances, to the dominant values of forests being equated with large-scale timber extraction (Lindberg et al. 1997). Growth of large-scale logging is a major economic shift away from traditional forest uses (Cashore et al. 2006), and although forest owners have strong bargaining power through ownership, agreements tend to work in favour of big logging companies and a few individuals, which destabilises and undermines customary practices (Wairiu 2006, p. 143).

Severing connections between indigenous people and their land is a significant factor in loss of traditional knowledge. Colonial dispossession and forced removal of half-caste children from their families had a devastating effect on traditional Aboriginal life in much of Australia. Elsewhere in the Pacific, a lack of under- standing by donor agencies and other external parties, of the collective and com- munal nature of indigenous land ownership has been a major cause of historical unrest and conflict. Traditional ideas about land and territorial entitlements within the region have had to be constantly renegotiated (Reuter 2006a, b). Today, some customary land tenure systems are subject to new and rapidly changing forces, severely testing their ability to adapt (Commonwealth of Australia 2008). Reuter's volume of case studies explores social change relating to traditional land rights in Pacific countries and identifies sources of conflict and tension, including state appropriation of customary land for timber companies (Reuter 2006a, b).

It follows that conflicts over land property and resource rights have the potential to erode the knowledge systems that codify the landscape. Furthermore, modern Western economic practice favours the individual or corporate over the communal and there are few tools for accommodating communalism. Communal land owner- ship is therefore often identified as a barrier to economic development, and some international aid agencies call for an abolition of collective land tenure in the Pacific. Neo-liberal advocates in Australia have also attempted to bring in individual land ownership but with little success thus far.

The Australian Agency for International Development (AusAID), in its compre- hensive investigation of land policy reform in the Pacific, has recommended improvements in tenure security, which would result in fewer disputes over land, access to finance for new businesses or housing, and greater investment by government in social services and infrastructure. AusAID advocates establishing better links between customary and formal land institutions through the formal registration of customary land title, which is effectively a documentation of the oral traditions that give credence to an individual's connection to place (Australian Government 2008). This can contain detail of reciprocal obligations; for example, a family or individual may be granted rights to harvest timber for subsistence use on the condition that they help the group to defend the forest (Vegter 2005).

11.6.2   Linking Traditional Forest-Related Knowledge with Sustainable Forest Management

Loss of traditional forest-related knowledge has consequences for both people and forests in Western Pacific countries. The more altered the natural landscape, the more people become disconnected from it, with a concomitant reduction of an appreciation of its social, environmental, and economic values (Filer 2000). Deforestation and forest clearing for cash cropping in the Pacific islands has gone hand in hand with erosion of people's knowledge of nature, together with decreasing respect for their relationships with it. Loss of forest species has also led to people becoming less reliant on forests and thus losing their appreciation and knowledge of forests (Siwatibau 2003, 2009). A study in the Solomon Islands showed that instead of the 87 plants harvested from forests less than 50 years ago, there are now fewer than 10 still being utilised. Thus, forests are no longer perceived as a culturally valuable, multi-functioning phenomenon, capable of sustaining humans, but as more of a commodity (Jansen and Tutua 2001).

Basu (2000) states that large-scale logging and other extractive activities are creating environmental imbalances that affect the lifestyle of traditional people and create a cultural vacuum. The push for economic development has side effects that have posed serious threats to the natural environment and traditional life and culture, with youth resorting to violent crime because of the loss of their traditional roots. Lindberg et al. (1997) argue that forest destruction undermines the capacities of forest-dependent people to survive, economically, culturally, and spiritually. Most Pacific cultures still practice traditional swidden agriculture, but increasing population numbers, especially on the small atolls, have resulted in more forest clearance for gardens, much shorter fallow cycles, and subsequent degradation of forest biodiversity and soil quality. Expanding human populations are also placing much greater demands on the forest for fuelwood, which is an essential ingredient for cooking, light, and warmth. Because of a shortage of forest-based fuelwood, communities are using the waste from gardening and food processing, such as coconut husks, instead of putting them on fallow gardens as mulch and to replenish soil nutrients. With this practice no longer occurring, soil quality is diminishing, resulting in long-term consequences for livelihoods and health (Clarke and Thaman 1993).

Thaman (1989) coined the term 'agrodeforestation' to describe the fact that present generations of Pacific Islanders are planting fewer trees around their villages; trees that traditionally would have been used for a wide range of purposes. This reduction in tree number and diversity is a major issue on small atolls where there is little native forest to provide supplementary foods and raw materials for building and fuelwood. Thaman (1989) notes the absence of any strong policies for retaining and enhancing traditional agroforestry systems, which inevitably leads to loss of traditional forest-related knowledge and associated ideologies and worldviews about the trees and their products.

11.7 Making a Difference with Traditional Forest-Related Knowledge

The 'healthy people healthy country' paradigm has considerable resonance in Australia, where links have been made between the health and well-being of indigenous peoples and their ability to learn, retain, and apply traditional knowledge systems through being 'on country,' including being involved in forest management (Baker et al. 2001). Recent studies in Australia have provided conclusive evidence that being involved in keeping the country healthy assists greatly in improving the health and well-being of Aboriginal people (Petty et al. 2006).

Acknowledging the existence of traditional forest-related knowledge as a valid knowledge system is one thing; applying it in a modern world is quite another. Much traditional knowledge has been lost or is fragmented because of the impacts of time on generations of orally transmitted information, colonial history, and modernisation. Many aspects of traditional knowledge are inappropriate or irrelevant in a contemporary context, and it is unrealistic to believe they offer a panacea for overcoming environmental or social problems (Howitt et al. 1996).

At the other end of the scale, economists tend to ignore the informal economy and its associated traditional knowledge systems. Economic development for indigenous people in Pacific countries needs to focus on a broad and more complex conceptualisation, made in socioeconomic, rather than economic categories (Hooper 2005). In the Pacific, the informal economy of customary landowners is not inconsequential, as subsistence agriculture continues to be practiced by the majority of the population. A series of papers on culture and sustainable development by practitioners working in the Pacific, while not idealising cultural traditions, recognise that 'culture' is a powerful political and social force (Hooper 2005). Similarly in Australia, the Centre for Aboriginal Economic Policy Research (CAEPR) undertakes extensive research in remote Aboriginal communities, exploring ways to achieve economic independence while retaining customary knowledge, beliefs, and traditions. Among remote Australian Aboriginal communities, a hybrid economy has been identified that includes the customary sector (hunting and gathering); monetary input and capacity building by the state (training, grants, etc.); and employment provided by the private sector (usually resource companies) (Altman 2001a, b).

11.7.1 Alternative Economic Activities to Large-Scale Logging: Mobilising Traditional Knowledge for Economic and Social Development

The development of commercial alternatives to large-scale logging and mining in Pacific countries is being increasingly sought by indigenous communities desirous of conserving their forests and their culture (Wilkie et al. 2002). Apart from concerns about the sociological and environmental effects of large-scale resource extraction, there is growing domestic and international concern over the effects of logging on non-timber forest products (NTFP), such as many of the indigenous nuts found in the region. This has hastened the need to attach an economic value to NTFPs by developing them commercially, both to provide an alternative to logging and to support their conservation (Stevens et al. 1996).

11.7.1.1 Non-timber Forest Products: Traditional Forest-Related Knowledge for Commercial Benefit

Thousands of years of plant domestication and a wealth of ethnobotanical knowledge have the potential to return financial benefits to indigenous communities in parallel with continuing subsistence and cultural uses and retaining the integrity of forests and agro-forests. The *State of the World's Forests 2009* report (FAO 2009) identified commercialisation of NTFPs as an alternative to large-scale logging and a key area of FAO's forestry programme.[13] The successful commercialisation of NTFPs relies on intact traditional forest-related knowledge and intact forests; thus there are links between development of these industries and conservation of forested areas, although development and conservation goals do not always align (Arnold and Ruiz-Perez 2005).

In 1994 a workshop was held to examine the commercial potential of a range of South Pacific nut species (Stevens et al. 1996). Commercial harvesting and processing of species such as *Canarium indicum* and *C. harveyi* (ngali nut) in the Solomon Islands are based on traditional knowledge, as are the same two species (nangai) in Vanuatu. Development of export markets for ngali nut products would help preserve the rainforest by providing an income for producers, and as the tree is a protected species, its presence helps to restrict destruction of the rainforest, particularly at the forest edge (Roposi 1994).

Food products based on traditional bushfood are enjoying increasing popularity in Australia. Figure 11.7, illustrates a product made from wild bush tomatoes marketed under the Outback Spirit brand by Robins Foods Pty Ltd. This is an Australian owned company and family business comprising the Robins and the company's Indigenous suppliers. The Outback Spirit Foundation also funds sustainable agricultural programmes in remote Aboriginal communities, to achieve economic independence and maintain cultural traditions.[14]

---

[13] See http://www.fao.org/forestry/nwfp/en/

[14] http://outbackspirit.com.au/

Fig. 11.7 Outback tomato
chutney. One of the many
bushfood products from the
Outback Spirit range (Image
courtesy of Robins Foods PL)

One of the advantages of bushfood businesses is that they enable Aboriginal
people to remain on their land when there is otherwise no other source of employment
or income. However, distance from markets, lack of business training, and tensions
between cultural and economic values can be problematical.

Although there are links between conserving forest areas and non-traditional
forest product industries, in the absence of land use planning, extensive cash cropping of NTFPs can lead to deforestation. Commercialisation of the kava/sakan
plant, *Piper methysticum,* on the Micronesian island of Pohnpei is a case in point.
This plant has been used for centuries in remote Oceania to produce a psychoac-
tive drink prepared traditionally from chewing and pounding the roots of the tree.
Kava is a sacred plant to many Oceanic peoples and its use was controlled through
customary law, being restricted to chiefs, religious people, and for medicinal
purposes. The non-traditional use of kava has caused both social and environmental
problems in Pohnpei, particularly destruction of the montane forests (Merlin and
Raynor 2005).

However, a partnership between a non-governmental organisation (NGO) and
local community groups has enabled an integration of cash crops, including kava
back into the traditional agroforests, rather than have farmers neglect their subsistence activities to concentrate on cash cropping (Merlin and Raynor 2005).

11.7.1.2   Ecotourism/Cultural Tourism: Traditional Forest-Related
Knowledge for Education

The *State of the World's Forests 2009* report (FAO 2009) identifies ecotourism as an
increasingly important industry in the Asia-Pacific, and a way for local communities
to benefit from protected areas.

A review of ecotourism across the Asia-Pacific identified that community-based
social forestry, with its focus on sale of non-timber products for the tourist market,

can raise local revenue and may be a viable alternative to such activities as poaching wildlife in protected areas (Lindberg et al. 1997).

Although the report does not make specific reference to connections between ecotourism and traditional forest-related knowledge, it implies that the visitor experience is greatly enhanced when traditional peoples articulate traditional beliefs, values, and knowledge.

Ecotourism is a significant factor in the economic growth of some Pacific Island countries, although it is has been argued that it would never be as financially lucrative as, for example, large-scale logging, because it is ecologically rather than economically driven (United Nations 2003). The United Nations report emphasises that culturally responsible ecotourism should allow local people to set the acceptable levels of impact on cultural traditions and to decide how to interlink traditions and ecotourism activities. The report's review of ecotourism in ten Pacific countries demonstrates the great variability across the region. In American Samoa, ecotourism includes homestays in villages, learning about traditional and medicinal plants, and traditional fishing with local indigenes. The key selling point for the tiny, isolated Federated States of Micronesia is the 'pristine' natural environment and an intact traditional culture.

Tourism based on Aboriginal culture is an important component of tourism in Australia and there is a growing interest in 'Aboriginality' by overseas visitors. Indigenous Tourism Australia was established in 2005, and in 2007 the website listed 282 indigenous tourism operators. This list demonstrates that there is a demand for Aboriginal cultural tourism, and that Aboriginal organisations are active participants in the tourism industry (Feary 2007).

A consequence of cultural tourism across the whole region has been a revival of interest in cultural traditions and a realisation that traditional knowledge about the natural environment and utilisation of its resources can attract revenue. A review of ecotourism ventures in Fiji concluded that it strengthened social cohesion and helped local people to recognise and value the preservation of their culture and heritage. The review noted that sustainable ecotourism fits more appropriately into the traditional economy than does large-scale resource extraction such as mining and logging, and 'helped to save indigenous knowledge' (United Nations 2003, p. 66). Commodification of culture in this way also has its disadvantages, and tourism is itself an agent of social and cultural change, with traditional dance, material culture, etc., being repackaged to suit the market. Although tourism has a role to play in the preservation of cultural knowledge, it can also lead to loss of authenticity and ultimately a bad experience for the ecotourist, although it is difficult to draw the line between authenticity and cultural change (Dallen and Prideaux 2004).

In New Zealand, Kaupapa Maori development is a unique values-based Maori approach whereby sustainable economic development is based on a distinctly Maori epistemology, with the aim of protecting and developing Maori social and cultural capital. Kaupapa Maori has been applied to Maori tourism to counteract misrepresentation of Maori culture in tourism. Nine interrelated cultural values were identified as essential for Maori-centred tourism, which recognise the desire of Maori to protect and develop Maori cultural and intellectual property (McIntosh et al. 2004).

11.7.2 Land and Resource Management

11.7.2.1 Pacific

Application of traditional forest-related knowledge in the Pacific occurs primarily on customary lands, where subsistence farming competes for space with cash cropping and commercial timber production. Thaman and others argue that integration of traditional agroforestry practices into appropriate cash cropping systems is essential for well-being and sustainable development of Pacific societies (e.g., Thaman 1989; Kennedy and Clarke 2004). They argue that in order to maintain species diversity there needs to be a balance between monocropping of commercial export crops and subsistence crops, and between modern agroforestry and preservation of traditional polycultural agroforestry. Thaman is a strong advocate for a more traditional, less capital-intensive, and less monocultural approach to modern agricultural economies in the Pacific, through legislation and systems for protecting and promoting important or endangered tree species as part of agricultural programmes. This would involve going back to traditional holistic approaches rather than government or overseas aid-driven, compartmentalised approaches (Thaman 1989, 2002).

Replanting with traditional species and other reforestation programmes can be effectively developed through forestry and agricultural extension programmes. Many local farmers are keen to plant in current or old garden sites, especially if they are provided with seedlings free of charge. These programmes are also an effective way to retain and pass on knowledge about traditional agroforestry. Local farmers want to grow cash crops but need adequate planning and training to ensure that forests are not cut out as a consequence. Donor support will continue to be required but capacity is being built; there is also a need to involve women forest owners, who are often excluded from negotiations.

The Traditional Tree Initiative[15] is a practically based programme aimed at restoring and applying traditional knowledge about local tree species; their products and uses; and their application in sustainable economic development, resource conservation, and food security. Landholders seeking information on tree species to use for a wide range of purposes are encouraged to use local and native trees rather than introduced ones, which may pose biological threats. The project arose from recognition that although public interest in agroforestry and tree crops was increasing, information on local tree species for the region was scarce, difficult to access, and failed to provide the detail required to make informed decisions about effective integration of local tree species. One of the many benefits of expanding the planting and conservation of native and traditional trees across the landscape is a strengthening of traditional tree-based land use practices, which can in turn promote sustainable development and protect the culture and ecology of the region.

The Traditional Tree Initiative has produced fact sheets covering 50 of the most important species in the region. There is an electronic information resource for

---

[15] See http://traditionaltree.org/ for more information.

reforestation, conservation, and agroforestry, *The Overstory*, which has subscribers in more than 170 countries.

In Vanuatu, the Cultural Centre has entered into several programmes aimed at restoring traditional knowledge into modern agricultural practice. The Vanuatu Environment Unit's Landholders' Conservation Initiatives project aims to establish a database to improve access to information relating to traditional biodiversity knowledge and management practices. This database will be an important mechanism in making appropriate knowledge accessible for policy, planning, educational, and community needs.

The Cultural centre is also developing a joint project with the Ministry of Education, the Environment Unit, and UNESCO/LINKS titled 'Strengthening Indigenous Knowledge and Traditional Resource Management through Schools.' This project will involve a partnership among local communities, teachers, resource managers, and culture specialists, which aims to implement the incorporation of traditional resource management knowledge into the formal school curriculum.

The Okari Enterprise in the Managalese plateau of Papua New Guinea used participatory processes to incorporate indigenous knowledge into the design and implementation of the project. Types of knowledge contributed by local communities included: a preliminary record of local customary practices of tenure and rights of access; baseline data on land use patterns; current slash-and-burn agricultural practices; management of sacred forest plots; current forest harvesting practices; locally appropriate nut-cracking technology; and locally appropriate packaging and transport from the forest to roads and airstrips (Olsson 1994).

Programmes such as the Babatana forest project, implemented through the Kastom Garden and the National Herbarium Ethnobotany programmes in the Solomon Islands, is an example of small-scale, practical, village-based natural resource management initiatives that strengthen indigenous knowledge. The programme uses participatory approaches with local communities that strengthens and improves traditional life rather than undermining it (Jansen and Tutua 2001).

11.7.2.2 Settler Societies

In the settler societies of Australia and New Zealand, indigenous knowledge has been supplanted by Western science in modern forest management. Movements for land rights and social justice over the past 30 years have fostered a reassertion and renegotiation of traditional forest-related knowledge in modern management, which has occurred principally through participation in natural resource management and protected area management, rather than in commercial forestry operations (Feary 2007). Additionally, numerous studies have demonstrated that there are significant environmental benefits to involving Aboriginal people in natural resource management programmes 'on country' (Gillespie et al. 1998). Access to land is seen as a critical factor in addressing social and economic disadvantage in symbolic and practical ways. It can be shown that being 'on country' contributes strongly to improved health and well-being (Petty et al. 2006; Kingsley et al. 2009). Indigenous people talk constantly about land in relation to their health and feelings of well-being (Rose 1996).

Co-management of national parks in Australia is aimed at finding a balance between the interests and rights of traditional owners and the need to conserve natural systems; it involves a partnership between protected area managers, local Aboriginal communities, and traditional owners (Smyth 2001). Joint management arrangements encourage integration of modern scientific knowledge with traditional local knowledge; for example, in far southeastern New South Wales (NSW), local Aboriginal land councils are working with the NSW Department of Environment, Climate Change and Water[16] in undertaking field surveys to detect koala and other wildlife species (Kilham 2004). In the jointly managed Kakadu National Park, scientists are using knowledge from traditional owners in seeking the reasons for declines of several ground-dwelling marsupial species (Woinarski et al. 2007).

Outside of jointly managed parks, a large proportion of the Aboriginal-owned estate in Australia is of high conservation value and makes a substantive contribution to the conservation of biodiversity (Altman et al. 2007). A proportion of this land, 23 million ha, comprises the Indigenous Protected Areas (IPA) network, which is a national system of land management and tenure where Aboriginal-owned land of high nature conservation value is incorporated into the national reserve system (Commonwealth of Australia 2009). While most IPAs are over deserts in Central Australia, many incorporate woodlands, and those in southeastern Australia include significant tracts of forest. Cultural values are equal to the natural values, and Aboriginal owners have a major say with regard to land and resource management (Smyth and Sutherland 1996). As with joint management, one of the goals of the IPA system is to integrate scientific land management with local Indigenous knowledge.

Numerous funding programmes exist at national and state levels to encourage Aboriginal people to undertake natural resource management, including management of forests on crown land and on their own land. Any programmes aimed at keeping Aboriginal people on their traditional country and assisting in its sustainable management automatically fosters revival and revitalisation of traditional knowledge, and assists in building self esteem and cultural identity as well as developing capacity in modern land management (see papers in Baker et al. 2001).

11.7.3   Certification

Forest certification arose from realisations by scientists and citizens in the 1960s that forests were under stress from uncontrolled exploitation across the globe. Because of the comparatively greater impact of unsustainable forest practices on indigenous and forest-dependent people, their calls for recognition of their rights and livelihoods, supported mainly by non-governmental organisations, have been a driving force behind some certification schemes, such as the Forest Stewardship Council (FSC)

---

[16]This department is responsible for administering the National Parks and Wildlife Act 1974, which protects Aboriginal heritage in New South Wales.

(Kanowski et al. 1999; Fanzeres and Vogt 2000). Forest certification demonstrates that a forest is being sustainably managed, and since indigenous participation is one of the platforms of sustainable forest management, it follows that certified forests are more likely to reflect an indigenous perspective in their management.

11.7.3.1 Australia and New Zealand

Two forest certification schemes operate in Australia—the international Forest Stewardship Council (FSC)[17] and the Australian Forestry Standard (AFS).[18] Principle No. 3 of the FSC identifies the 'legal and customary rights of indigenous peoples to own, use and manage their lands, [and] territories,' and that 'resources shall be recognized and respected.' Recent research in Australia has identified that FSC certification held by the large plantation company, Integrated Tree Cropping, was a driver for employing local Aboriginal people in plantation management (Feary 2007).

Benchmarking of the AFS showed that it comprehensively recognised 'the traditional uses of forests as well as the socioeconomic benefits that forests and timber processing may have to local communities' (Oy 2003). The AFS also compared favourably against Canada's National Sustainable Forest Management Standard in regard to 'respect for Aboriginal rights, traditional knowledge and values' (Abusow International Ltd and Canadian Standards Association 2004).

There are currently no certified Aboriginal-owned forestry enterprises in Australia. By contrast, in New Zealand, the Lake Taupo Forest achieved Forest Stewardship Council (FSC) Certification in 2002. The Lake Taupo Forest comprises around 22,000 ha of planted *Pinus radiata* harvested at a level of 480,000 m^3 per annum, on a 30-year rotation. The forest is administered by The Lake Taupo Forest Trust (LTFT), a Maori authority representing the interests of almost 10,000 owners of 65 separate Maori land titles located on the eastern shores of Lake Taupo, in the middle of the North Island. The Lake Taupo Forest Trust aims are: 'to protect the integrity and ownership of *Nga Taonga tuku iho* (core asset of land and resources)' administered by the Trust on behalf of the beneficial owners, and 'to strive for optimal and sustainable asset growth and financial returns through development of the Trust assets to assist the long-term social, cultural and economic development of the beneficial owners' (http://www.ltft.co.nz/default.asp?cid=1). The forest has no traditional associations for Maori (although the land it is on may have) and is managed under contract to a forest management agency. Therefore, the benefits of certification for retaining traditional forest-related knowledge may be limited in this instance, although it has the capacity to generate wealth for Maori to pursue cultural activities (see Sect. 11.5).

---

[17] See http://www.fsc.org/ for information on the FSC.

[18] See http://www.forestrystandard.org.au/ for information on the AFS.

11.7.3.2   Pacific

Forest certification has advanced little in the Pacific. Cashore et al. (2006) have identified several constraints to certification, including political instability, lack of community skills in managing forests for commercial production, disputes over forest tenure and resource allocation, and weak and/or corrupt governments that cannot or will not enforce logging codes and that condone illegal logging.

These countries have no national certification schemes, and foreign logging companies are generally not interested in becoming certified. Most forest is under customary land tenure, and forest owners and non-governmental organisations (NGOs) are working together to develop viable, more culturally appropriate alternatives to reliance on large-scale destructive logging, which destroys forests, marginalises customary resource owners (especially women), and often does not deliver on the promised financial returns. NGO certification programmes are designed to address illegal logging and to prevent more community forestland from being granted as concessions through unsustainable logging agreements with landowners. Certification of these small-scale, community or village based operations is seen as a tool for addressing unsustainable and illegal logging. In Papua New Guinea and the Solomon Islands, ecoforestry certification has been developed through collaboration between Greenpeace New Zealand and the International Tropical Timber group, tailored for community-owned and run forestry operations. This simplified and less costly certification scheme is aimed at assisting communities to build capital and skills to proceed to full FSC certification (McDermott et al. 2006).

A recent assessment of forest certification in the Solomon Islands is indicative of the situation more widely in the Pacific. Wairiu (2006) has shown that certification has had little effect on the forestry industry in the Solomon Islands. Reasons given are lack of demand for certified timber by the market, close relations between the government and foreign-owned timber companies, negative consequences of adopting sustainable forest management (e.g., loss of jobs), and lack of government support. Over the past 15 years, a number of landowners with assistance from NGOs have developed small-scale operations that involve all tribal members, in an effort to attain maximum community benefit from forest exploitation. If certified, these operations can earn more than large-scale logging (Wairiu 2006). NGOs such as Greenpeace and World Wildlife Fund (WWF) are working with landowners in raising awareness of sustainable forest management, small-scale forest enterprises, and forest certification. In 1998, an eco-certification called Solcert was set up and adapted FSC principles to local conditions to improve the chances of take-up by local communities. However, response has been poor, due primarily to socio-cultural factors. One argument against certification is its requirement for continuity of a labour force to meet the market demand. This takes men away from their customary work in the gardens, leaving women to shoulder the additional burden. The quantity and regularity of timer demand does also not fit well with needs of landowners who may only need to extra cash at certain times of the year.

Although prospects are not good for FSC-type certification in the Asia-Pacific generally (Cashore et al. 2006), some social benefits were identified in the Solomon

Islands study, especially in the area of community capacity and skill building. Some communities were able to halt commercial logging in their forest areas through awareness training in certification standards. The earnings shared from sale of certified timbers also reinforce traditional social networks of wealth redistribution by chiefs (Wairiu 2006).

11.7.4   Timber Production

11.7.4.1   Pacific

Numerous projects in the Pacific are aimed at improving the sustainability of forest management on customary land. The most successful use participatory processes and involve capacity building and two-way knowledge exchange. The SPC/GTZ Pacific-German Regional Forestry project is an example of an internationally funded project that has used participatory land use planning to assist landowners to develop a sustainable forestry enterprise in Fiji. An essential component was reaching agreement among landowners over forest areas to be excluded from exploitation (SPC/GTZ 2005). In the Solomon Islands, the Australian Centre for International Agricultural Research (ACIAR) is funding a project to interplant commercially valuable teak with Pacific agroforestry species. This recognises community reluctance to thin teak by encouraging wide planting and interplanting of traditional agroforestry species that can be commercially harvested early. It also provides training and capacity building and is of a scale appropriate for smallholders.[19]

The portable sawmilling industry is one seemingly well-suited to traditional sociocultural systems and offers an economic alternative to large-scale logging. Returns from small mills have also been shown to be greater and more equitable than from logging royalties. Portable mills can potentially lead to improved livelihoods through sustainable use of the forest resource, while retaining land, forests, and traditions. Using portable mills keeps people on their land, is low-impact, small-scale, locally controlled, and not highly mechanised. Not surprisingly, the industry has expanded rapidly in the Pacific over the past 20 years, with the support of NGOs that envisage that more sustainable forest management will arise from the increased local participation.

A recent review of the use of portable sawmills in Papua New Guinea and the Solomon Islands identified around 7,000 portable sawmills (Holzknecht and James 2009). Importantly, the review noted that performance could not be measured solely on commercial objectives, as many mills did not have commercial success as their primary function. Often a family or village operated a mill only to meet short-term requirements such as school fees or purchase of material goods. The existence of coherent and functioning customary social units such as clan or family groups operating on their own lands was found to be critical to the success of the operation (Holzknecht and James 2009).

---

[19] http://aciar.gov.au/project/FST/2007/020

Fig. 11.8 Kgwan community members establishing a portable sawmill, near Gembogl, Papua New Guinea Highlands (Photo courtesy Julian Fox)

Nevertheless, lack of knowledge by villagers about silvicultural practices and commercial forest management has lead to unsustainable forestry practices. Various ecoforestry projects in Papua New Guinea are demonstrating that these can be overcome by appropriate training and participatory learning. The KGWan Eco-Habitat Project[20], in the Chimbu Province of the Eastern Highlands, is run by Village Development Trust (VDT), an eco-forestry based local NGO. VDT has assisted setting up and operating a small-scale sawmill and is facilitating the sale of locally milled *Nothofagus grandis* as Community Based Fair Trade product to an Australian timber merchant (Fig. 11.8).

The Solomon Islands Forestry Management Project is encouraging purchase of timber from landowners cutting trees on their own land and producing sawn timber using chainsaws or portable sawmills. They maintain a directory of suppliers and the timbers they can supply and potential uses; for example, rosewood (*Pterocarpus indicus*) is good for furniture and boatbuilding.

11.7.4.2   Australia

Portable mills are also being used in Australia by some remote Aboriginal communities, to produce timber for the domestic and overseas market. Although fell-

---

[20]http://www.forestscience.unimelb.edu.au/research_projects/ACIAR_Projects/PNG_Project/Kgwan_Project.html

ing whole trees is not a traditional Aboriginal activity, research into the economic potential of Aboriginal-owned forests on Cape York Peninsula in far North Queensland has demonstrated that local communities support logging as a culturally appropriate activity generating employment and reducing welfare dependency (Venn 2004). The participatory processes used in assessing the economical potential demonstrated a complex mix of economic and socio-cultural values placed on forests by Aboriginal people that defied conventional economic approaches. Venn was told by elders that they were to be consulted over the intensity of logging operations to determine whether they were culturally appropriate for the forest being harvested (Venn 2004).

Culture is evident in the management of sawmilling operations elsewhere on Cape York, which are based on customary social boundaries and a desire to build social capital rather than maximise profit. Until recently, Darwin stringybark (*Eucalyptus tetrodonta*) forests in the lease area of the mining company Comalco were cleared and burnt to allow access to the bauxite layer beneath (Annandale and Taylor 2007). In the early 1990s, an Aboriginal-owned business was established to run a portable sawmill, at the Aboriginal settlement of Napranum on the western side of Cape York, to harvest timber prior to mining. Nanum Tawap, Ltd. is set up along traditional lines of land tenure and power relations, with management by the five clan groups whose traditional country is covered by Comalco's mining lease.[21]

A recent evaluation of the timber resource indicated that the sawmill's capacity would need to be substantially increased to maximise economic returns from timber salvage operations (Annandale and Taylor 2007). The Nanum Tawap Steering Committee supported a business model that would maximise local employment by having three small sawmills, one in each community, rather than a single large mill requiring fewer people with high levels of technical skills. This decision is significant for demonstrating that building social capital in local communities can take precedence over maximising economic returns (Feary 2007).

It has been argued that it is not always necessary to separate business from other community activities in order to achieve success (Altman 2001a); however, like any other small business, Nanum Tawap's main objective is to make a profit, and the approach taken has been to separate the business from the activities of the rest of wider Aboriginal community. Mark Annandale, then project manager with Queensland's Department of State Development and prime mover behind the project, attributes its success to local Aboriginal control and ownership, with the latter recognising customary boundaries and systems. He is a strong supporter of a separation between business and social customs:

> One of the things that was actively worked on and identified by community leaders was that the sawmill was business… that they don't want to mix all other elements of the social

---

[21]Tawap is an acronym formed from the names of the five clan groups: **T**hanikwithi People, **A**nathangayth People, **W**athayn People, **A**lngith People and **P**eppan People.

environment with that because that's what always happened in the past and the business failed...it's been very conscious, there's a big line there. *Leave it at the gate for the operational* [author's emphasis]. (Feary 2007: 168)

Mark Annandale summed it up:

I guess it's more accurate to say that the operational side of Nanum Tawap is removed as best they can from that [social and cultural life], but the broad policy direction is not (Feary 2007: 169).

11.7.4.3   New Zealand

Maori are major actors in New Zealand's timber production, owning 29% of private native forests and 14% of the planted forest estate (Miller et al. 2007). From the 1960s a significant proportion of Maori-owned land was committed to pine plantations under long-term leases, with the idea of guarding ownership of Maori land while fostering their development and use by forestry organisations (Nuttall 1981). Although royalties were paid to Maori landowners, the forestry company lessee was the major beneficiary of the timber, and promised employment opportunities for Maori in rural areas did not always eventuate (Nuttall 1981). The Waitangi Tribunal has had impacts on the forest sector in relation to benefits flowing to Maori land-owners (Caddie 2003).

Plantation forestry on Maori land continues to be distinctively different from that on non-Maori land (Thorp 2006). One of the major differences is recognition by most owners that they have limited time as kaitiaki (custodians) and are obliged to leave the asset in good condition for future generations. A study of the Lake Taupo Forest Trust (see Sect. 11.4) showed that traditional and recreational use, especially pig hunting, was considered as one of the main benefits of owning forests. Recreational use maintains connections to the land, and many elders are comforted by younger Maori also using the lands. The Lake Taupo Forest Trust lands also contain many sites with strong associative historical value such as old pa sites (traditional fortified villages, often located on terraced hillsides), graves, and pathways. Forest management ensures that these sites remain unplanted or are blessed before any work occurs on them (Thorp 2006).

Maori response to Western systems of forest management in New Zealand is a holistic decision-making framework called the 'Mauri' model. This sustainability model, based on that used by indigenous people in Canada, allows for preservation of indigenous lands, sovereignty, and culture, while supporting economic development, capacity building, and technological advance. 'Mauri' is the binding force between the physical and the spiritual and is a common attribute of all things, reflected in the traditional stories relating to forest use. These stories demonstrate the importance of knowing the rites, reciprocity, and knowledge associated with taking natural resources. Felling trees without appropriate ritual, sawmilling, and large-scale forest clearing damage the life forces of the timber (Morgan 2007).

11.8 Conclusion

With the use of specific examples, this chapter has explored the state of indigenous forest-related knowledge across the Western Pacific and concluded that it continues to have great relevance to the cultural identity, behaviours, and belief systems of indigenous people. Pacific Islanders rely on traditional forest-related knowledge in subsistence agriculture and use the surrounding forests for many foods, medicines, and raw materials. Aboriginal Australians and New Zealand Maori also retain a significant amount of traditional forest-related knowledge, despite having lost ownership of most of their forests as a result of colonisation.

Traditionally, forests were managed as part of complex socioeconomic systems, where spiritual beliefs about the forests played a significant role in their use. Thus, from a historical perspective, traditional forest-related knowledge is intrinsically valuable as a repository of information about a particular type of relationship between nature and culture that existed for millennia before Western civilisation came to the Western Pacific. However, the relevance of traditional knowledge across the Western Pacific is being severely threatened by the market economy and modernity. Outside of New Zealand and Australia, a striking feature of forest management in the Pacific countries is the power of international donor agencies, whose economic development imperative drives the processes that undermine traditional knowledge, including extensive cash cropping and encouraging large-scale timber extraction to be seen as the major forest value.

Traditional forest-related knowledge relies on transmission of oral information across generations, but it is the youth who most question its relevance. Poor health of community elders, increasing urbanisation (especially in New Zealand and Australia), loss of traditional social structures that enable knowledge transference, and education of Pacific children in distant cities exacerbate this situation. Both indigenous and non-indigenous people have expressed alarm at loss of traditional forest-related knowledge and argue that it is directly connected with both loss of social well-being and loss of forests, with a concomitant view that both can be restored through revitalisation of traditional knowledge in forest management.

There is evidence to suggest that traditional forest-related knowledge is more likely to remain relevant if it can be tied directly to economic development and the cash economy, to the extent that modernity becomes 'indigenised,' rather than the other way round. The latter sections in this chapter have described a wide range of programmes for recognising traditional knowledge as a valid knowledge system that can be used for economic development rather than be overwhelmed by it. The way forward embraces participatory processes that enable a mingling of traditional knowledge with scientific approaches to forest management. These include such activities as ecotourism, replanting of useful trees outside forests, using local species in agroforestry, mobile sawmilling, and protected area management. And, while traditional forest-related knowledge per se may appear be of limited relevance to some activities, the critical importance of customary processes for resolving disputes over land and payments in logging areas must not be forgotten.

Like any other knowledge, traditional knowledge is fluid and dynamic and capable of adapting to new situations, challenging paradigms of what is or is not 'traditional,' as was seen in the example of modern Aboriginal art. The challenge across the Western Pacific is to establish mechanisms for balancing protection of forests for their social, cultural, and environmental values against the need for economic development. Finding such a balance first requires a respect for and recognition of the applicability of traditional systems by both resource owners and governments. Respecting the non-utilitarian component of the forest may challenge its economic viability; provision must be made for consultation and adherence to cultural protocols in the use of forest resources. Modern economics will need to adapt to ensure the survival of traditional forest-related knowledge.

References

Abusow International Ltd, Canadian Standards Association (2004) Australian and Canadian sustainable forest management standards. A comparative analysis of AS 4708 (Int)-2003 and CAN/CSA Z809-02. Annex E, Australian Forestry Standard, Yarralumla

Allen H (1974) The Bagundji of the Darling basin: cereal gatherers in an uncertain environment. World Archaeol 5(3):309–322

Altman J (2001a) Indigenous communities and business: three perspectives, 1998–2000.Working paper no. 9/2001, Australian National University, Centre for Aboriginal Economic Policy Research, Canberra

Altman J (2001b) Sustainable development options on aboriginal land: the hybrid economy in the twenty-first century. Discussion paper no. 226/2001, Australian National University, Centre for Aboriginal Economic Policy Research, Canberra

Altman J, Buchanan G, Larsen L (2007) The environmental significance of the indigenous estate: natural resource management as economic development in remote Australia. CAEPR discussion paper no. 286/2007, Australian National University, Centre for Aboriginal Economic Policy Research, Canberra

Annandale M, Taylor D (2007) Forest futures—indigenous timber and forestry enterprises on Cape York. In: Feary S (ed) Forestry for indigenous peoples. Proceedings of technical session 130, XXII IUFRO world congress, 8–13 Aug 2005, Brisbane, Australia. Australian National University, Fenner School of Environment and Society, Canberra, pp 49–59

Arnold J, Ruiz-Perez M (2005) Can non-timber forest products match tropical forest conservation and development objectives? In: Sayer J (ed) Forestry and development. Earthscan, London, pp 130–144

Baker R, Davies J, Young E (2001) Working on country: contemporary indigenous management of Australia's land and coastal regions. Oxford University Press, South Melbourne

Basu PK (2000) Conflicts and paradoxes in economic development tourism in Papua New Guinea. Int J Soc Econ 27(7–10):907–916

Bellwood P (1979) The Oceanic context. In: Jennings J (ed) The prehistory of Polynesia. Australia National University Press, Canberra, pp 6–26

Bellwood P, Fox J, Tyron D (eds) (1995) The Austronesians: historical and comparative perspectives. ANU E Press, Canberra

Benson J, Redpath P (1997) The nature of pre-European native vegetation in south-eastern Australia: a critique of Ryan, D.G., Ryan, J.R. and Starr, B. J (1995) The Australian landscape—observations of explorers and early settlers. Cunninghamia 5(2):285–328

Berndt R, Berndt C (1981) The world of the first Australians. Lansdowne Press, Sydney

Bond A (2006) Pacific 2020. Background paper: forestry. Commonwealth of Australia, Canberra

Boutland A (1988) Review paper: forests and aboriginal society. In: Frawley K, Semple N (eds) Australia's ever changing forests: the first national conference on Australian forest history, Canberra. Australian Defence Force Academy (ADFA), Canberra, pp 143–168

Brown C (1997) Asia-Pacific forestry sector outlook study. regional study—the South Pacific. Working paper no. APFSOS/WESTERN PACIFIC/01, Food and Agriculture Organisation of the United Nations (FAO), Rome

Brown C, Durst P (2003) State of forestry in Asia and the Pacific, 2003—status, changes and trends. Food and Agriculture Organisation of the United Nations (FAO), Bangkok

Byrne D (1984) Archaeological and aboriginal significance of the New South Wales rainforests. National Parks and Wildlife Service (NPWS), Sydney

Caddie A (2003) The Treaty of Waitangi— a forestry overview. N Z J Forest 48(3):23–28

Campbell V (1978) Ethnohistorical evidence on the diet and economy of the Aborigines of the Macleay River Valley. In: McBryde I (ed) Records of times past. Australian Institute of Aboriginal and Torres Strait Islander Studies, Canberra, pp 82–100

Campbell I (1992) Island kingdom: Tonga ancient and modern. Canterbury University Press, Christchurch

Cashore B, Gale F, Meidinger E, Newsome D (eds) (2006) Confronting sustainability: forest certification in developing and transitioning countries. Yale School of Forestry and Environmental Studies, report no. 8, Yale University, New Haven

Clarke WC, Thaman R (1993) Agroforestry in the Pacific Islands: systems for sustainability. United Nations University Press, Tokyo

Colfer C, Colchester M, Laxman J, Rajindra P, Nygren A, Lopez C (2005) Traditional knowledge and human well-being in the 21st century. In: Mery G, Alfaro R, Kanninen M, Lobovikov M (eds) Forests in the global balance—changing paradigms, vol 17, IUFRO world series. International Union of Forest Research Organizations (IUFRO), Helsinki, pp 173–182

Commonwealth of Australia (1995) National forest policy statement. Commonwealth of Australia, Canberra

Commonwealth of Australia (2005) The national indigenous forestry strategy. Commonwealth of Australia, Canberra

Commonwealth of Australia (2008) Making land work: reconciling customary land and development in the Pacific, vol 1. AusAID, Canberra

Commonwealth of Australia (2009) Indigenous protected areas. http://www.environment.gov.au/indigenous/ipa/index.html. Cited 21 Nov 2009

Cooper D (2000) An unequal coexistence: from 'station blacks' to 'Aboriginal custodians' in the Victoria River District of Northern Australia. Dissertation, Australian National University, School of Resource and Environmental Management, Canberra

Cox P, Elmqvist T (1994) Ecocolonialism and indigenous knowledge systems: village controlled rainforest preserves in Samoa. Pac Conserv Biol 1:6–12

Crawford P, Crawford I (2003) Contested country: a history of the Northcliffe area, Western Australia. University of Western Australia Press, Crawley

Cribb A, Cribb J (1974) Wild food in Australia. Fontana Books, Sydney

Crocrombe R (1989) The South Pacific. An Introduction. Fifth revised edition. University of the South Pacific, Suva, Fiji

Dahl A (1989) Traditional environmental knowledge and resource management in New Caledonia. In: Johannes RE (ed) Traditional ecological knowledge: a collection of essays. International Union for the Conservation of Nature (IUCN), Gland, pp 45–53

Dallen T, Prideaux B (2004) Issues in heritage and culture in the Asia-Pacific region. Asia-Pacific J Tour Res 9(3):213–223

Dargavel J (1998) Politics, policy and process in the forests. Aust J Environ Manage 5(1):25–30

Engel A (2000) Dilemmas of participation: a case study from the southern regional forest agreement (RFA) process. Thesis, Australian National University, Department of Forestry, Canberra

Fanzeres A, Vogt K (2000) Roots of forest certification: its developmental history, types of approaches, and statistics. In: Vogt K, Larson B, Gordon J, Vogt D, Fanzeres A (eds) Forest certification: roots, issues, challenges, and benefits. CRC Press, Boca Raton, pp 11–54

Feary SA (2005) Indigenous Australians and forests. In: Dargavel J (ed) Australia and New Zealand forest histories: short overviews. Australian Forest History Society, Canberra, pp 9–16

Feary SA (2007) Chainsaw dreaming. Indigenous Australians and the forest sector. Dissertation, Australian National University, Canberra

Feary S, Kanowski P, Altman J, Baker R (2010) Managing forest country: Aboriginal Australians and the forest sector. Aust For 73(2):126–134

Filer C (2000) How can Western conservationists talk to Melanesian landowners about indigenous knowledge? RMAP working paper no. 27, Australian National University, Canberra

Food and Agriculture Organisation (FAO) (2000) Asia and the Pacific National Forest Programmes. Update no. 34, FAO, Rome, pp 89–194

Food and Agriculture Organization (FAO) (2001) Global forest resource assessment, 2000. FAO forestry paper 140, FAO, Rome

Food and Agriculture Organisation (FAO) (2009) State of the world's forests 2009, FAO, Rome. http://www.fao.org/docrep/011/i0350e/i0350e00.HTM. Cited 10 Mar 2009

Francis S (2006) People and place in Tonga: the social construction of fonua in Oceania. In: Reuter T (ed) Sharing the earth, dividing the land. ANU E Press, Canberra, pp 345–364

Gammage B (2003) Australia under aboriginal management. Barrie Andrews memorial lecture 2002. Australian Defence Force Academy, Canberra

Gathercole P (1977) Man and environment in Polynesia. In: Megaw JVS (ed) Hunters, gatherers and first farmers beyond Europe. Leicester University Press, London, pp 189–198

Gerrand A, Bartlett T (2001) Managing change: lessons learned from the development and implementation of Vanuatu's national forest policy. Paper presented at Forests in a changing landscape, Commonwealth Forestry Association Conference, Fremantle, 18–25 Apr 2001

Gillespie D, Cooke P, Taylor J (1998) Improving the capacity of indigenous people to contribute to the conservation of biodiversity in Australia. Report to Environment Australia for the Biological Diversity Advisory Council, Canberra

Golson J (1977) No room at the top: agricultural intensification in the New Guinea Highlands. In: Allen J, Golson J, Jones R (eds) Sunda and Sahul: prehistoric studies in Southeast Asia, Melanesia and Australia. Academic, London, pp 601–638

Government of Samoa (1998) National report to the convention on biological diversity. Division of Environment and Conservation, Apia

Groome Poyry Ltd. (1993) Western Samoa forestry policy review. Draft report No. 135/1993. Groome Poyry Ltd., Auckland

Guild D, Dudfield M (2009) A history of fire in the forest and rural landscape in New Zealand: part 1, pre-Maori and pre-European influences. N Z J For 54(1):34–38

Haebich A (2005) Assimilating the bunya forests. Aust N Z Forest Hist 2:27–33

Hallam S (1979) Fire and hearth, a study of aboriginal usage and European usurpation in south-western Australia. Australian Institute of Aboriginal Studies, Canberra

Hammond D (1997) Commentary on forest policy in the Asia-Pacific Region. Asia-Pacific forestry sector outlook study, working paper no. APFSOS/WP/22, Food and Agriculture Organisation (FAO), Bangkok

Hammond D (2001) Development of Maori owned indigenous forests, MAF technical paper no: 2003/4, Ministry of Agriculture and Forestry (MAF), Policy Section, Wellington

Harris D (1978) Adaptation to a tropical rainforest environment: aboriginal subsistence in north-eastern Queensland. In: Reynolds V, Blurton-Jones W (eds) Human behaviour and adaptation. Halstead Press, New York, pp 113–133

Head L (1993) The value of the long-term perspective: environmental history and traditional ecological knowledge. In: Williams N, Bains G (eds) Traditional ecological knowledge: wisdom for sustainable development. Centre for Environmental and Resource Studies, Canberra, pp 66–70

Hill R (2003) Frameworks to support indigenous managers: the key to fire futures. In: Cary G, Lindenmayer D, Dovers S (eds) Australia burning: fire ecology, policy and management issues. Commonwealth Scientific and Industrial Research Organisation (CSIRO), Canberra, pp 175–186

Holzknecht H, James R (2009) A review of the use of portable sawmills in Papua New Guinea and the Solomon Islands: identifying the factors for success. ACIAR project no. FST/2003/049, Australian Centre for International Agricultural Research (ACIAR), Canberra

Hooper A (ed) (2005) Culture and sustainable development in the Pacific. ANU E Press and Asia Pacific Press, Canberra

Horowitz L (2002) Stranger in one's own home: Kanak people's engagements with a multinational nickel mining project in New Caledonia. RMAP working paper no. 30, Australia National University, Canberra

Howells W (1973) The Pacific Islanders. Weidenfield and Nicolson, London

Howitt R, Connell J, Hirsch P (1996) Resources, nations and indigenous peoples. Oxford University Press, Melbourne

Hunter-Anderson R (2009) Savanna anthropogenesis in the Mariana Islands, Micronesia: re-interpreting the paleoenvironmental data. Archaeol Ocean 44:125–141

Huth J (2001) The bunya pine—the romantic *Araucaria* of Queensland. Paper presented at international Araucariaceae symposium, Auckland, 14–17 Mar 2002

Hynes R, Chase A (1982) Plants, sites and domiculture: aboriginal influence upon plant communities in Cape York Peninsula. Archaeol Ocean 17:38–50

Institute for Aboriginal Development (1985) Punu, Yankunytjatjara plant use. Angus and Robertson, Sydney

Isaacs J (1987) Bush food: aboriginal food and herbal medicine. Weldon Owen Pty Ltd., Sydney

Jansen T, Tutua J (2001) Indigenous knowledge of forest food plants: a component of food security in the Solomon Islands. In: Bourke R, Allen M, Salisbury J (eds) Food security for Papua New Guinea. ACIAR proceedings no. 99, Canberra, pp 112–123

Jones R (1969) Fire-stick farming. Aust Nat Hist 16:224–228

Kamminga J (2002) Australian aboriginal timber quick search. Australian Institute of Canberra: Aboriginal and Torres Strait Islander Studies (AIATSIS), Canberra

Kanowski P, Sinclair D, Freeman B (1999) International approaches to forest management certification and labelling of forest products: a review. Agriculture, Fisheries and Forestry Australia, Canberra

Kanowski P, Holzknecht H, Perley C (2005) Oceania—islands of contrast. In: Mery G, Alfaro R, Kanninen M, Lobovikov M (eds) Forests in the global balance—changing paradigms, vol 17, IUFRO world series. International Union of Forest Research Organizations (IUFRO), Helsinki, pp 281–302

Kennedy J, Clarke W (2004) Cultivated landscapes of the southwest Pacific. RMAP working papers no. 50, Australian National University, Resource Management in Asia-Pacific [RMAP], Canberra

Kilham E (2004) Aboriginal communities and government agencies: partners in natural resource management. Thesis, Australia National University, School of Resources Environment and Society, Canberra

Kingsley J, Aldous D, Townsend M, Phillips R, Henderson-Wilson C (2009) Investigating health, economic and socio-political factors that need consideration when establishing Victorian Aboriginal land management projects. Aust J Environ Manage 16:34–44

Kirch P (1979) Subsistence and ecology. In: Jennings J (ed) The prehistory of Polynesia. Australia National University Press, Canberra, pp 286–307

Knudtson P, Suzuki D (1992) Wisdom of the elders. Allen and Unwin, Toronto

Lane M (1997) Regional forest agreements: resolving resource conflicts or managing resource politics? Aust Geogr Stud 37(2):142–153

Langton M (1998) Burning questions. Northern Territory University, Centre for Indigenous Natural and Cultural Resource Management, Darwin

Lindberg K, Furze B, Staff M, Black R (1997) Ecotourism and other services derived from forests in the Asia-Pacific region: outlook to 2010. Asia-Pacific forestry sector outlook study working paper series, no. APFSOS/WESTERN PACIFIC/24, Food and Agriculture Organisation (FAO), Rome

Lloyd D, Van Nimwegen P, Boyd W (2005) Letting indigenous people talk about their country: a case study of cross-cultural (mis)communication in an environmental management planning process. Geogr Res 43(4):406–416

Macpherson J (1939) The eucalyptus in the daily life and medical practice of the Australian aborigines. Mankind 2(6):175–180

Maiden J (1889) The useful native plants of Australia. Trubner, London

Manner H (2005) Traditional agroecosystems. In: Mueller-Dombois D, Bridges K, Daehler C (eds) Biodiversity assessment of tropical forest ecosystems: PABITRA manual for interactive ecology and management. Asia-Pacific Network for Global Change Research (APN), Honolulu, pp 89–104

Manning M, Hughes P (2008) Acquiring land for public purposes in Papua New Guinea and Vanuatu. In: Making land work, volume two, case studies on customary land and development in the Pacific. AusAID, Canberra, pp 241–264

Manuel G, Polsuns M (1974) The fourth world: an Indian reality. The Free Press, New York

McDermott C, Maynard B, Gale F, Prasetyo F, Bewang I, Bun Y, Muhtaman DR, Shahwahid M, Wairiu M (2006) Forest certification in the Asia-Pacific region: regional overview. In: Cashore B, Gale F, Meidinger E, Newsome D (eds) Confronting sustainability: forest certification in developing and transitioning countries. Yale School of Forestry and Environmental Studies, report no. 8, Yale University, New Haven

McIntosh A, Zygadlo F, Matunga H (2004) Rethinking Maori tourism. Asia Pacific J Tour Res 9(4):331–352

Merlin M (2000) A history of ethnobotany in remote Oceania. Pac Sci 54(3):275–287

Merlin M, Raynor W (2005) Kava cultivation, native species conservation and integrated watershed resource management on Pohnpei Island. Pac Sci 59(2):241–260

Metcalf L (2006) Know your New Zealand trees. New Holland Publishers, Auckland

Miller R, Dickinson Y, Reid A (2007) Maori connections to forestry in New Zealand. In: Feary SA (ed) Forestry for indigenous peoples: learning from experiences with forest industries. Papers from technical session 130, XXII World Congress 2005, IUFRO, 8–13 Aug 2005, Brisbane, Australia. ANU Fenner School of Environment and Society Occ. Paper 1, Australia National University, Canberra, pp 13–22

Molloy B (ed) (1995) Riccarton Bush: Putaringamotu. Riccarton Bush Trust, Christchurch

Montreal Process Implementation Group (2008) Australia's state of the forests report 2008. Bureau of Rural Sciences, Canberra

Moore A (1995) Art in the religions of the Pacific: symbols of life. Cassell, London

Morgan K (2007) Traditional approaches to forest management and the Mauri model. Paper presented at sharing indigenous wisdom: an international dialogue on sustainable development conference, Oneida, 11–15 June 2007

Mosko M (2006) Self-Scaling the earth: relations of land, society and body among North Mekeo, Papua New Guinea. In: Rueter T (ed) Sharing the earth, dividing the land. ANU E Press, Canberra, pp 279–297

Mountain M (1993) Bones, hunting and predation in the Pleistocene of Northern Sahul. In: Smith M, Spriggs M, Fankhouser B (eds) Sahul in review. Occasional papers in prehistory, no. 24, Australian National University, Department of Prehistory, Research School of Pacific Studies, Canberra, pp 123–136

Mueller-Dombois D (2008) Pacific Island forests: successionally impoverished and now threatened to be overgrown by aliens. Pac Sci 62(3):303–308

Mulvaney J, Kamminga J (1999) Prehistory of Australia. Allen and Unwin, Sydney

Nalail E (1996) Integration of environmental, social and economic sustainability for Vanuatu—sustainable development: an information package for Vanuatu. Report to South Pacific Regional Environmental Program, Apia

Nuttall R (1981) The impact of exotic forestry on Maori land in Northland: a point of view. N Z J For 26(1):112–117

New Zealand Department of Conservation (2006) Nga Whenua Rahui. http://www.doc.govt.nz/templates/summary.aspx?id=43139. Cited 25 Aug 2006

Olsson M (1994) Okari ecoenterprises: a snapshot of participatory rural development. In: Stevens M, Bourke R, Evans B (eds) South Pacific indigenous nuts. Proceedings of a workshop, 31 Oct–4 Nov 1994. Le Lagon Resort, Port Vila, Vanuatu. ACIAR proceedings no. 69, Australian Centre for International Agricultural Research (ACIAR), Canberra, pp 94–99

Oy I (2003) Benchmarking the Australian forestry standard. Forest and Wood Products Research and Development Corporation, Melbourne

Paijams K (ed) (1976) New Guinea vegetation. CSIRO Press and ANU Press, Canberra

Patterson M (2002) Moving histories: an analysis of the dynamics of place in North Ambrym, Vanuatu. Aust J Anthropol 13(2):200–218

Petty A, Bowman D, Johnston F (2006) Sustainable northern landscapes and the connection to aboriginal health. In: People, practice and policy: a review of social and institutional research. Land and Water Australia, Canberra, pp 64–67

Pollack D (2001) Indigenous land in Australia: a quantitative assessment of indigenous landholdings in 2001. Discussion paper 221, Australia National University, Centre for Aboriginal Economic Policy Research, Canberra

Powell J (1976) Ethnobotany. In: Paijams K (ed) New Guinea vegetation. CSIRO Press and ANU Press, Canberra, pp 106–183

Rangan H, Lane M (2001) Indigenous peoples and forest management; comparative analysis of institutional processes in Australia and India. Soc Nat Resour 14:145–160

Reid A (1995) Continuity and change in the Austronesian transition to Islam and Christianity. In: Bellwood P, Fox J, Tyron D (eds) The Austronesians: historical and comparative perspectives. ANU E Press, Canberra, pp 333–350

Republic of Vanuatu (2000) Vanuatu country report. Asia-Pacific Forestry Commission 18th Session 15–19 May 2000. Food and Agriculture Organisation (FAO), Rome

Reuter T (2006a) Land and territory in the Austronesian world. In: Reuter T (ed) Sharing the earth, dividing the land. ANU E Press, Canberra, pp 11–38

Reuter T (ed) (2006b) Sharing the earth, dividing the land: land and territory in the Austronesian world. ANU E Press, Canberra

Riley M (2005) Know your Maori weaving. Viking Seven Seas, Paraparaumu

Roche M (1990) History of New Zealand forestry. New Zealand Forestry Corporation Limited, Wellington

Rolls E (2005) A land changed forever. In: Keeney J (ed) In the living forest. ETN Communications Pty Ltd., Sydney, pp 16–18

Roposi N (1994) Research and development on edible tree crops in Solomon Islands. In: Stevens M, Bourke R, Evans B (eds) South Pacific indigenous nuts: proceedings of a workshop 31 Oct–4 Nov 1994. Le Lagon Resort, Port Vila, Vanuatu. ACIAR proceedings no. 69, Australian Centre for International Agricultural Research (ACIAR), Canberra, pp 113–115

Rose D (1996) Nourishing terrains: Australian aboriginal views of landscape and wilderness. Australian Heritage Commission, Canberra

Ruis B (2001) No forest convention but ten tree treaties. Unasylva 52(3):8–17

Sahlins M (1972) Stone age economics. Tavistock, London

Sahlins M (2005) On the anthropology of modernity, or, some triumphs of culture over despondency theory. In: Hooper A (ed) Culture and sustainable development in the Pacific, 2nd edn. Asia Pacific Press and Australian National University, Canberra, pp 44–61

Schieffelin E (1975) Felling the trees on top of the crop: European contact and the subsistence ecology of the Great Papuan Plateau. Oceania XLVI(1):25–39

Sculthorpe G (2005) Recognising difference: contested issues in native title and cultural heritage. Anthropol Forum 15(2):171–193

Sinclair K (2000) A history of New Zealand. Penguin, Auckland

Singh G, Kershaw AP, Clark R (1981) Quaternary vegetation and fire history in Australia. In: Gill A, Groves R, Noble I (eds) Fire and Australian biota. Australia Academy of Science, Canberra, pp 23–54

Siwatibau S (1984) Traditional environmental practices in the South Pacific—a case study of Fiji. Ambio 13(5–6):365–368

Siwatibau S (2003) Forests, trees and human needs in Pacific countries. Invited paper submitted to the XII World Forestry Congress, Quebec City

Siwatibau S (2009) Emerging issues in Pacific Island countries and their implications for sustainable forest management. In: Leslie R (ed) The future of forests in Asia and the Pacific: outlook for 2020. RAP publication 2009/03. Food and Agriculture Organization (FAO), Asia-Pacific Forestry Commission, Rome

Smyth D (2001) Joint management of national parks. In: Baker R, Davies J, Young E (eds) Working on country: contemporary indigenous management of Australia's land and coastal regions. Oxford University Press, Melbourne, pp 75–91

Smyth D, Sutherland J (1996) Indigenous protected areas: conservation partnerships with indigenous landholders. Environment Australia, Canberra

SPC/GTZ (2005) Integrated land use planning in the Drawa Block: Pacific German regional forestry project updates. Pac Isl For Trees 1:12–16

Specht A, Specht R (2005) Historical biogeography of Australian forests. In: Dargavel J (ed) Australia and New Zealand forest histories: short overviews. Australian Forest History Society, Canberra, pp 1–8

Stevens M, Bourke R, Evans B (1996) South Pacific indigenous nuts: proceedings of a workshop 31 Oct–4 Nov 1994, Le Lagon Resort, Port Vila, Vanuatu. ACIAR proceedings no. 69, Australian Centre for International Agricultural Research (ACIAR), Canberra

Suchet S (1996) Nurturing culture through country: resource management strategies and aspirations of local landowning families. Aust Geogr Stud 34(2):200–215

Sullivan S (1978) Aboriginal diet and food gathering methods in the Richmond and Tweed River valleys, as seen in early settler records. In: McBryde I (ed) Records of times past. Australian Institute of Aboriginal and Torres Strait Islander Studies, Canberra, pp 101–114

Sveiby K-L, Skuthorpe T (2006) Treading lightly: the hidden wisdom of the world's oldest people. Allen and Unwin, Sydney

Tamla H (2002) A community role in conflict management: lessons from managing the Vanuatu forestry sector. Dev Bull 60:24–27

Te Ara–the Encyclopedia of New Zealand (2009) R wiri taonui: te ngahere–forest lore—M ori relationship with the forest. Available via. http://www.TeAra.govt.nz/en/te-ngahere-forest-lore/1. Cited 17 Jan 2007

Thaman R (1989) Agrodeforestation and the neglect of trees: threat to the wellbeing of Pacific societies. SPREP occasional paper no.5, Secretariat of the Pacific Regional Environment Programme (SPREP), New Caledonia

Thaman R (2002) Trees outside forests. Int For Rev 4(4):268–276

Thaman R, Elevitch C, Wilkinson K (nd) Traditional Pacific island agroforestry systems. The Overstory 49:1–5. Available via. http://agroforestry.net/overstory/overstory49.html. Cited 9 Jan 2009

Thancoupie GF (2007) Thanakupi's guide to language and culture: a Thaynakwith dictionary. Jennifer Isaacs Arts and Publishing, Sydney

Thomson D (1939) The seasonal factor in human culture. Proc Prehist Soc 5:209–221

Thorp G (2003) Ngati Tuwharetoa—from landlords to forest owners. N Z J For 48(1):13–14

Thorp G (2006) Forestry on Maori land. N Z J For 51(1):28–33

Trigger D (1997) Land rights and the reproduction of aboriginal culture in Australia's gulf country. Soc Anal 41(3):84–106

United Nations (2003) Ecotourism development in the Pacific Islands. ESCAP tourism review no. 23. United Nations, New York

Vegter A (2005) Forsaking the forests for the trees: forestry law in Papua New Guinea inhibits indigenous customary ownership. Pacific Rim Law Policy J 14(2):545–574

Venn T (2004) Socio-economic evaluation of forestry development opportunities for Wik people on Cape York Peninsula. Dissertation, University of Queensland, St Lucia

von Strokirch K (2008) The region in review: international issues and events, 2007. Contemp Pac 20(2):424–448

Wairiu M (2006) Forest certification in the Solomon Islands. In: Cashore B, Gale F, Meidinger E, Newsome D (eds) Confronting sustainability: forest certification in developing and transitioning countries. School of forestry and environmental studies report no. 8, Yale University, New Haven, pp 137–162

Walker R (2004) Ka whawhai tonu matou: struggle without end. Penguin, Auckland

Wheen N (2002) Foul play? Government and the SILNA forests. N Z J Environ Law 6:279–296

White JP, O'Connell J (1982) A prehistory of New Guinea and Sahul. Academic, Sydney

Wilkie M, Eckelmann C, Laverdiere M, Mathais A (2002) Forests and forestry in small island developing states. Int For Rev 4(4):257–267

Woinarski J, Mackay B, Nix H, Traill B (2007) The nature of Northern Australia: natural values, ecological processes and future prospects. ANU E Press, Canberra

Wolfe P (1994) Nation and miscegeNation: discursive continuity in the post-Mabo era. Soc Anal 36(October):93–151

Chapter 12
Globalization, Local Communities, and Traditional Forest-Related Knowledge*

Jesús García Latorre and Juan García Latorre

Abstract Traditional forest-related knowledge (TFRK) has allowed human communities to adapt to specific locales. However, this local context is being dramatically affected by changes introduced through globalization. This chapter explores the different paths through which globalization is affecting local and indigenous communities and their traditional forest-related knowledge, and the potential for these communities to adapt to, or counteract, the impacts of globalization. We start with a reflexion on globalization and its links with local communities. Current globalization can be regarded as the most recent phase of a long-term process initiated by European expansion 500 years ago. Following a brief discussion of the positive effects and potential benefits of globalization on local communities, the remainder of the chapter considers the more disquieting aspects of this topic. European countries provide examples of how globalization has affected local communities in capitalist industrial economies as well as under communism. We then address the long-lasting influence of European colonialism, and explore how local communities are still being affected by political ideas developed in Europe during the eighteenth and nineteenth centuries and introduced to the colonies, including centralized control of forests and their management. We continue with a focus on developing countries and the influence of environmental policies and the market economy as important facets of globalization's impact on local communities. This discussion includes an

*This paper represents the opinions of the authors and does not reflect the position of their institutions.

J. García Latorre (✉)
Federal Ministry of Agriculture, Forestry, Environment and Water Management, Vienna, Austria
e-mail: jesus.garcia-latorre@lebensministerium.at

J. García Latorre
Association for Landscape Research in Arid Zones, Almería, Spain
e-mail: jglatorre2@gmail.com

J.A. Parrotta and R.L. Trosper (eds.), *Traditional Forest-Related Knowledge: Sustaining Communities, Ecosystems and Biocultural Diversity,* World Forests 12, DOI 10.1007/978-94-007-2144-9_12,
© Springer Science+Business Media B.V. (outside the USA) 2012

examination of the application of violence in the framework of a market economy. Finally, we discuss how local communities deal with globalization, as well as the importance of participation and consultation processes to support these communities.

Keywords Colonialism • Globalization • Environmental policy • Forest conflict • Forest governance • Forest history • Indigenous peoples • Local communities • Market economy • Traditional knowledge

12.1 Introduction

Traditional ecological knowledge, which includes traditional forest-related knowledge (TFRK), is characterized by its strong local character: embedded in the local cultural milieu (Berkes 1999) and bounded in space and time (Banuri and Apffel Marglin 1993, cited in Berkes 1999). Knowledge of local species and ecosystems has allowed human communities to adapt to specific places and environmental conditions (Berkes 1999). However, the local context within which traditional knowledge exists is being dramatically affected as a result of the changes introduced by the world globalization. This chapter explores the impacts of globalization on rural communities and their traditional forest-related knowledge.

The term globalization represents one of the most fashionable buzzwords of contemporary political and academic discussions. In a popular sense it is perceived as a synonym for such phenomena as free market, economic liberalization, or internet revolution (Scheuermann 2006). A more precise definition stresses the fundamental changes in the spatial and temporal contours of social existence causing a compression of distance or space, which leads in turn to a profound alteration in the organization of human affairs (Scheuermann 2006). So, globalization can be defined as the 'worldwide interconnectedness of places and people' (Lambin et al. 2001), linking 'distant localities in such a way that local happenings are shaped by events occurring many miles away and vice versa' (Giddens 1990). Furthermore, globalization acts by 'removing regional barriers and strengthening global at the expense of national connections' (Lambin et al. 2001).

Being involved in weblike and netlike connections is an intrinsic feature of human groups, and history shows many examples of regional linkages well before AD 1400 (Wolf 1982). However, the European expansion initiated in the fifteenth century 'brought the regional networks into worldwide orchestration, and subjected them to a rhythm of global scope' (Wolf 1982). 'From then on,' continues anthropologist Eric R. Wolf, 'events in one part of the globe would have repercussions in other parts.' Against this background we can regard globalization as the most recent phase of a long-term process that took off 500 years ago (Tainter 2007; Pelizaeus 2008) and was dramatically accelerated with the innovations introduced by the industrial revolution in the fields of transportation and communication (Giddens 1990; Sheuermann 2006).

The different paths through which people located at distant places have woven their webs of connections have increased through time. So, in prior globalization waves, trade played an important role (Wolf 1982), while currently the worldwide interconnectedness occurs 'through, for example, global markets, information and capital flows, and international conventions' (Lambin et al. 2001). This diversity of paths assures the far-reaching implications of the current globalization affecting every facet of human life, including culture, politics, and economics (Friedman 2005; Stiglitz 2006; Scheuermann 2006).

Regarding cultural globalization, a frequent message of critics refers to its destructive implications. Martin Heidegger, a pioneer of the globalization discussion, warned early about the global homogenizing effect of television, a medium that has definitively contributed to the 'abolition of distance,' rendering human experience monotonous (Heidegger 1971, cited in Scheuermann 2006). Recently Thomas L. Friedman recognized that globalization has homogenizing tendencies but, in contrast to the previous view, pointed also to its particularizing ones. Globalization represents a broad, deep, and complex phenomenon that involves new forms of communication and innovation. Particularly the internet is equalizing people, i.e., allowing them to interact with more equal power and opportunities and promoting the globalization of the local, thus counteracting homogeneity (Friedman 2005).

In the field of political life, new information technologies support stakeholders and activists to disseminate their concerns and requests across frontiers and to have a stake in political discussions. Furthermore, political globalization expresses itself also in the tendency towards supranational forms of socioeconomic lawmaking and regulation (Sheuermann 2006), such as the shift to global environmental politics since the late 1980s (Porter and Borwn 1996).

For decades, international policies have also shaped the globalization of the economy, but in the course of time the international economic system has become stronger and better organized than the international political system (Stiglitz 2006). Globalization is frequently identified with the spread of capitalism, markets, and trade. Karl Marx and Friedrich Engels (1848) already acknowledged capitalism as a basic globalization driver: 'The need of a constantly expanding market for its products chases the bourgeoisie over the entire surface of the globe. It must nestle everywhere, settle everywhere, establish connexions everywhere,' they wrote. Compared with earlier networks of economic connections, capitalism introduced for the first time a genuinely global order rested upon economic power (Wallerstein 1974). Sheuermann (2006) points to the disagreements regarding the precise causal forces behind globalization, since some authors question the strong economic focus of the Marxist approach. Nonetheless, the Nobel laureate Joseph Stiglitz (2006) insists that the private economy plays a leading role in driving current globalization. Because economic globalization has been so strong, democratic structures are not well-developed at the international level, and individual countries have become weak in their ability to control economic actors, which has led to a deficit in democracy (Stiglitz 2006; Sheuermann 2006). In fact, observers believe that the economic development of recent decades has led the world into a qualitatively new phase. World trade volume has increased by 20 times since 1950, and foreign investment flow almost doubled in merely a couple of years (1997–1999) (Levin Institute 2011). This wave of globalization has been driven, since the Second World War, by policies that include free-market, reductions in barriers to commerce and international trade, and opening economies. International institutions such as the World Bank and the International Monetary Fund have played an important role, thereby promoting a global frame benefitting particularly rich countries (Stiglitz 2006). Thus, a frenetic self-reliant economy has provided globalization with an international industrial and financial business structure (Levin Institute 2011).

Such a market-centric approach is typical of neo-liberalism (Harvey 2005; Mudge 2008) and has serious consequences when it incorporates local traditional communities into its sphere (Martínez Alier 2002; Harvey 2005; Tainter 2007). Focussing firstly only on the market, economic time (guided by capital flow and profit rate) is much faster than the ecological time of traditionally managed systems. This antagonism damages both traditional cultures and their natural resources (Martínez Alier 2002). Moreover, David Harvey (2003, 2005) brings out how neo-liberalism forced open non-capitalist territories—not only to trade but also to privatization of natural resources, monetization of exchange, forceful expulsion of peasant populations, and suppression of alternative forms of production and consumption—thereby making available cheaper raw materials, labour power, and land. The process, denominated by Harvey as 'accumulation by dispossession,' assures growing profit rates and constitutes an inherent part of neo-liberalism. This is facilitated by the disjuncture in scaling between the flow of materials and the flow of information at local and regional levels. Peripheries incorporated into global systems continue maintaining 'a local scale of information, even as the factors that affect them expand to the national and international arenas' (Tainter 2007). Local people lose autonomy, being even unaware that they have done so (Tainter 2007), while dependency and environmental deterioration is promoted (Martínez Alier 2002; Harvey 2005; Tainter 2007). Joseph Stiglitz (2006) concedes a point to critics who argue that globalization has caused too many losers, an involuntary group joined by many members of indigenous and local communities.

Summing up, although a certain amount of trade-driven reciprocity among distant groups existed even in preindustrial societies, this interconnectedness was by far not so extensive as the worldwide integration achieved since the nineteenth century. By representing distances typically measured in time rather than in kilometres, modern innovations in transport, communication, and information technology that accompanied industrialization have resulted in an effective compression of distance or space (Scheuermann 2006). This, in turn, has led to the blurring of geographical and territorial barriers, accelerating the interactions of culturally differentiated groups. Consequently, traditional societies accustomed to managing their local affairs have become more and more confronted with beliefs, habits, interests, and practices originated in other parts of the world. This is globalization. The new communication advances do allow local communities to make their voices heard even at the international level. Nevertheless, the economic-driven character of such advancement is constraining subsistence households and undermining their traditional knowledge systems. Some introductory examples are shown in Box 12.1.

Box 12.1 Convergent Trajectories

Local communities incorporated into global systems might loss the control over their affairs. Tainter (2007) emphasizes the convergent trajectories observed in different places. Additionally, similar events can be identified at different moments within the long-term process of globalization.

Rural communities in Madhya Pradesh and Maharashtra (India) and Sierra de Segura (Spain) have faced similar misfortunes. Máximo Fernández Cruz (1916–1986) lived as shepherd in Sierra de Segura (Gómez Mena 1987); shepherds have been keepers of traditional forest-related knowledge, such as managing trees for fodder. Fernández Cruz was adversely affected by a centralized forest administration, which had aimed to expand the area of national forests, in part through acquisition and afforestation of pastures, since the beginning of the twentieth century. Máximo refused to sell his land and the administration tried to overcome his resistance. 'They beat me and took me to the police barracks, the gendarmerie looked for me at any time during the night; if a forest fire initiated anywhere they named me as suspect, took me to prison and bashed me; and all this because they want to displace me from my property' (Máximo Fernández Cruz, quoted by Gómez Mena 1987).

In 1974 7,000 complaints were filed against shepherds in the court of justice (De la Cruz Aguilar 1994). One shepherd, depressed by debts, hung himself. Máximo Fernández Cruz, sentenced several times for disrespect to authorities, was sent to prison in 1982. According to the forest administration he had stolen 100 m of wire fencing. At the time he was 66 years old, lame, and half blind (Varillas 1987). Consequently, forest policies have generated resentment among rural communities and conflicts with the administration, whose goal was to displace mountain populations (De la Cruz Aguilar 1994).

In India, the economic situation in some remote hamlets in the tribal district of Satna and Nandurbar leads to infantile malnutrition (Jamwal 2008a; Chibber 2008). The use of medicinal plants is widespread in poor countries (Shand 1997, cited in Ribeiro 2005). The Tharus (Nepal), for instance, are able of living near dense malaria-infested forests thanks to their traditional healthcare system (Ghimire and Bastakoti 2008). But the forest department deprives communities in Satna and Nandurbar of access to the forests (Jamwal 2008a; Chibber 2008; Nidhi Jamwal, personal communication 2009).

These examples illustrate how state interventionism with different purposes (revenue generation, outside trade) might lead to ecological poverty (Brieger and Sauer 2000). Interventionism in rural affairs by imposing official expertise had occurred in Europe since the sixteenth century and supported the access of states to local resources. These political ideas were disseminated as technical facts during colonial times and still shape environmental policies.

(continued)

Box 12.1 (continued)

However, the market, new customs, or the aspirations of poor people for alternative opportunities also play a role. Traditional nutrition in India based on forest resources loses out to the market-dependent practices of urban areas, adversely affecting the health and economies of tribal populations. 'Our children,' says an elderly in Chichati, 'do not even know that forests can give food' (Pallavi 2008). Modern healthcare systems act negatively on the Tharus. The Nepalese government does not support their traditional practices and the youth prefer modern medicine (Ghimire and Bastakoti 2008). Thus, local and indigenous communities lose their traditions when they are subjected to external forces—a situation that also happened in the past in industrial countries.

The exposure of Western European mountains to broader markets provoked their depopulation, the specialized use of more productive parts, and the abandonment of most of the territory (Lasanta Martínez and Ruiz Flaño 1990). Many had left their villages by the twentieth century and emigrated to urban centres, resulting in great loss of traditional knowledge.

The goal of this chapter is to explore the different paths through which globalization is affecting local and indigenous communities and their traditional forest-related knowledge as well as the possibilities for these communities to adapt to or counteract the new circumstances emerging in this globalized world. Certainly, as Joseph Stiglitz (2006) concludes, for most of the world globalization is like a pact with the devil; a few have got richer but many have lost. Nevertheless, we will start our contribution optimistically by dealing with some positive achievements of globalization from which local communities can benefit. The rest of the chapter addresses more disquieting aspects of this topic.

First, we discuss how European countries provide examples of how globalization has affected local communities in industrial economies. After a short digression on local communities within communist regimes, we address colonialism. An essential role fell to Europe in bringing together the Old World and the New (Wolf 1982), with long-lasting influences reaching even the current institutional organization in post-colonial countries (Pelizaeus 2008). We show how political ideas developed in Europe during the eighteenth and nineteenth centuries and introduced in the colonies as technical facts are still affecting local communities as well as the opinion of government officials regarding traditional forest-related knowledge. Then we continue our focus on developing countries and deal with the influence of environmental policies and the market economy as important facets of globalization, complemented with information regarding the application of violence in the framework of a market economy. How local communities deal with globalization is the topic of the penultimate section, while the last one emphasizes the importance of participation and consultation processes to support these communities.

12.2   The Globalization of the Local

Globalization is a complex phenomenon with varied outcomes; while some places and people have flourished, for others it has represented a badly arranged programme (Stiglitz 2006). Correspondingly, its appreciation requires caution. 'The iron law of globalization,' writes Thomas L. Friedman (2005), 'is very simple: If you think it is all good or all bad, you don't get it.' In this section we focus on the good things.

Local communities might use the opportunities presented in the framework of globalization to expose their concerns and requests at the international level. It is an interaction from the local to the global sphere. This book is an example; it has resulted from the efforts made by people of different countries within the IUFRO Task Force on Traditional Forest Knowledge and contains information about local communities from all around the world. Thus, globalization currently allows broadly distinguishing the value of traditional forest-related knowledge and pointing to its importance. Another path communicating the local to the global is provided by side events in international negotiations and conventions, where representatives of indigenous and local communities expose their people's requests and concerns (see Fig. 12.1). Moreover, the United Nations has indeed passed a resolution on indigenous peoples, and in decisions of different conventions these groups are on occasion taken into consideration.

Interactions also occur from the global to the local level. Some internationally accepted concepts, such as integrated pest management, find application to the traditional practices of local communities, so contributing to their legitimization (Martínez Alier 2002). International organizations such as Oil Watch support tropical

Fig. 12.1 Representatives of indigenous peoples stress their concerns and requests in the framework of side events and demonstrations during international conventions. Demonstrators on the photo demanded more attention for their rights with respect to REDD during the climate change conference in Cancún (December 2010). 'No rights no REDD' was their slogan (Photo: Jesús García Latorre)

countries threatened by the oil industry (Martínez Alier 1999). The interaction between the local and the global might also be reciprocal. The environmental efficiency of traditional management systems has been appreciated by international agroecological movements, such as CLADES (Latin American Consortium on Agroecology and Sustainable Development), which, in turn, support the communities applying such practices (Martínez Alier 2002).

Furthermore, technological tools like geographic information systems, remote sensing, or accurate weather forecasts might benefit local communities managing natural resources in traditional ways. But of all the innovations of recent decades, the internet doubtlessly represents one of the most powerful for the issue in question. In discussing globalization, Friedman (2005) highlights the flattening of the world. Flatness describes 'how more people can plug, play, compete, connect, and collaborate with more equal power than even before.' It means equalizing power and opportunity. The internet has contributed considerably to this; it has created a global platform for sharing efforts and knowledge (Friedman 2005). Andean indigenous communities, for instance, used open-source software for developing a web-based multimedia community biocultural register (Swiderska 2007).

The internet also allows the worldwide diffusion of publications and experiences. One homepage offers a database containing publications that describe conflicts regarding local communities and natural resources (http://conflictbiography.recoftc.org/html/index.htm), while another (http://www.redd-monitor.org/) documents the spread of REDD projects (Reducing Emissions from Deforestation and Forest Degradation). Different organizations, such as Survival International[1] inform the world about the demands of indigenous peoples and abuses against their communities. Without leaving home, sitting comfortably in front of your computer, one can learn from Interpol about the risks to local communities from the negotiations addressing emissions from deforestation (Vidal 2009). One can also learn that these risks have become a reality for specific poor communities in Colombia (XLSemanal 2010). One can also access reports on the efficiency of shifting cultivation systems and other traditional management practices (Fox 2000), on human rights and environmental abuses, on the tricks behind concepts such as sustainable forest management (Global Witness 2009), and much more. A vast amount of information concerning traditional local communities is available with just a couple clicks of a mouse.

In Friedman's view, 'this flattening of the playing field is the most important thing happening in the world today.' It is supporting the globalization of the local (Friedman 2005), with a crucial consequence. If the parties to an international convention agree on resolutions, or if individual countries support policies that, if implemented, involve negative impacts for local communities, no one is now able to say 'we didn't know.'

Thus, globalization is not all bad, but, likewise, not all good. When balancing the pros and cons of globalization for local communities, its negative effects most probably would tip the scale. The following sections address how particular beliefs,

---

[1] http://www.survivalinternational.org/

different policies, and the economy act at the local level to dismantle traditional communities and their knowledge systems.

12.3   Impacts of Globalization on Local Communities and Their Traditional Forest-Related Knowledge

While the disintegration of traditional natural resource management systems is currently ongoing in developing countries, the cessation of old practices has already happened and become generalized in industrial economies. Here we summarize the impacts of globalization on local communities and their traditional forest-related knowledge separately for developed and developing countries. Addressing industrial countries first and focussing particularly on the European case is not eurocentrism but rather a necessity, because what occurred earlier in Europe set the stage for many of the scenarios that subsequently played out in developing countries. The similarities of current globalization processes in widely separated places, highlighted by Joseph A. Tainter (2007), are impressive, but no less so are similarities between current processes and those that occurred in Europe over the past three centuries.

12.3.1   Traditional Forest-Related Knowledge in Europe: An Almost Lost Heritage

Ancient pollarded trees in the middle of high forests and unmanaged coppiced woodlands containing old stools, nowadays merely used as game enclosures, are some expressions of traditional forest-related knowledge that can be observed in rural areas of industrial countries, as, for instance, in Western European countries (see Fig. 12.2a, b). These and other objects have an archaeological character constituting evidences of practices carried out in the long past. To mention two concrete examples, the tradition of pollarding, i.e., cutting the upper branches of a tree for gathering fodder or wood, goes back to more than 6,000 years in the Alps (Nicolas Haas 2002), while silvopastoral systems were already present in southwestern Spain in the Copper Age (Stevenson and Harrison 1992). Ethnobotanists' work has helped to save part of the experience of the rural population in managing woodlands and trees, as well as their knowledge in using plant species for many different purposes. Since traditional management activities have today almost completely disappeared, researchers apply a historical approach. For example, interviews with older farmers have revealed some knowledge they received from their elders. Historical documents constitute another important source of information (for examples in Europe see Rackham 1978, 1980; Peterken 1981; González Bernáldez 1989; Kirby and Watkins 1998; Pott and Hüppe 1991; Mesa and Delgado 1995; Stiven and Holl 2004; Jansen and van Benthem 2005; García Latorre and García Latorre 2007).

Fig. 12.2 Two examples of abandoned pasture woodlands with pollarded trees. (**a**) In the Sierra Nevada, Almería, southeastern Spain, spaces between the oaks (*Quercus rotundifolia*) have been colonized by shrubs. (**b**) Old beech trees (*Fagus sylvatica*) in an area close to the city of Bonn (Germany) have thick, straight branches, indicating that these trees have not been pollarded since long ago. Other stands in this forest are densely stocked with young beech trees that have regenerated since abandonment of traditional management. Both woodlands, formerly used for grazing and provision of fuelwood and charcoal, contain considerable amounts of dead wood (Photos: Jesús García Latorre)

The ecological effects of these management practices are of particular interest. They were frequently applied in commons—community-owned and collectively used land—according to the rules of the community. Communal ownership is characterized by its contribution to the conservation of natural resources (Berkes 1989). In northwestern Spain, for example, communally managed woodlands in the past were an important refuge for brown bears (*Ursus arctos*) and even now constitute a valuable habitat (Grande del Brío et al. 2002). Finally, considerable amounts of dead wood and many veteran trees, both essential components of forest ecosystems that contribute to the maintenance of biodiversity (such as, for example, saproxylic insects and cavity-nesting birds), can be found in wood pastures with pollarded trees (Casas 1991; Alexander 1998; Méndez et al. 2010).

A sort of globalization at the European scale occurred during the past 300 years, with late global consequences for the rest of the world, by introducing radical socioeconomic changes in rural populations and causing the abandonment of traditional management practices. Two definitive elements of this process were a long tradition of state interventionism in the use of forest resources and the expansion of a market economy. Enlightenment intellectuals, among other actors, contributed to the dissemination of ideas across Europe supporting both 'scientific forestry' applied in the frame of centralized states, and the privatization and marketization of natural resources.

12.3.1.1 State Interventionism: Top-Down Policies of Former Times

European monarchies and the rise of centralized states in the seventeenth and eighteenth centuries limited the access of rural populations to forest resources during those times. Authorities typically justified restrictive regulations and by-laws as necessary measures to protect resources from the farmers who were considered the enemies of the forest. A large body of legal documents concerning this matter, together with frequent disputes caused by the annexations and interventionism on the part of the ruling class, have helped disseminate the persistent idea that traditional practices are detrimental to the forest. Actually, disputes arose among different social groups over different forms of forest resource use (Warde 2006). Thus, historical legal documents contain only the very subjective view of the group that decreed them (the ruling class). Indeed, David Humphreys (2006) remind us that 'who is within the law and who is an outlaw depends on who writes the law.'

In preindustrial times forests represented one of the largest sectors of the state economy in many European countries. Wood was the primary source of energy for important proto-industrial activities such as mines, iron, and salt works. In response to the increasing wood demand and rising prices beginning in the sixteenth century, government officials aimed at expanding the wood supply (Lowood 1990; Ernst 1998). In this context scientific forest management emerged in the eighteenth century in Germany as an aspect of state administration (Lowood 1990) to serve the elite's economic interests (Pulhin et al. 2010). Its principal objective was the quantification of forest resources and the 'rationalization' of their use focusing basically on wood. From the beginning the new forest management clashed with local communities, whose practices were characterized by a communal and diversified use of forest resources in the frame of a sustenance economy (Rösener 1991). 'This diversity of utilization … endangered the forest stand in the eyes of the foresters because—according to the new standards—it was unorganized, unscientific and unsustainable' (Ernst 1998). Hegemonic foresters' discourses have excluded local knowledge and practices and misread traditional forest landscapes (Williams 2000). Furthermore, forest management 'led first to the abandonment and then the disappearance of traditional, empirical silviculture' (Ciancio and Nocentini 2000).

Early German forestry, recasting forests as a revenue source and promoting state interventionism, was the starting point for every other national effort in forestry science and management (Lowood 1990; Scott 1998). As an example, in the middle of the nineteenth century the study and practice of forestry was introduced in Spain. The 'importers' of this discipline, Agustín Pascual (1818–1884) and Esteban Boutelou (1823–1883), had studied in Tharand, Germany, with Heinrich Cotta (1763–1844), who is considered together with Georg Ludwig Hartig (1764–1837) as the founders of forestry. Both brought the ideological background of forestry with them, that is, Cotta's conviction that the state represents the most appropriate holder of forests and foresters are the only capable forest managers (Casals Costa 1996). To be sure, state involvement in protecting forest land was essential as nineteenth century enclosures ('desamortizaciones', the Spanish version of enclosures) gained prominence. But this involvement went far beyond mere state resistance to privatizing trends; German influence in fact promoted a centralized approach to forest management that still prevailed in the 1970s (De la Cruz Aguilar 1994) and that oppressed farmers and shepherds in rural communities. Such was the case with Máximo Fernández Cruz, in Sierra de Segura, almost 2,000 km away from Tharand (Box 12.1, above).

The history of conflicts and clashing interests behind the emerging discipline of forestry is still overlooked by some scholars who continue to uncritically glorify the origin of scientific forest management. 'Ein Kind der Not' (a child of necessity), reads Ernst Röhring and Hans Achim Gussone's (1990) interpretation of Central European forestry—i.e., forestry as a mere technical response to wood scarcity not linked to political interests. Another example is provided by Burschel and Huss' (1997) silviculture handbook. They argue that 'in spite of forest prescriptions and regulations … careless forest management [by peasants] led to an almost extensive forest devastation. As a consequence forestry emerged in the eighteenth century, first as an orderly activity and afterwards as a science.' These attitudes are similar to those of official historians, as emphasized by the distinguished historian Josep Fontana (1999), 'from the very first … history always had a social function, generally to legitimate the established order.' As a hegemonic discourse, forestry history has frequently been diffused by active practitioners (Williams 2000), who may not be aware of the social origins of the discipline as they treat it as only a technical field.

12.3.1.2 Expanding Market Economy

In international environmental and forest policy discussions, market-based measures are frequently considered as instruments expected to promote rational use of natural resources. However, opinions on the appropriateness of markets for such a task are far from uniform. How the market economy emerged and expanded during the past centuries and its effects on Western European rural communities represent an accumulated experience of great interest for current international policy processes.

The traditional skills documented by researchers' interviews of elderly mountain peasants probably represent only a last heritage from the enormous body of knowledge that was available 150 years ago, when rural communities still managed (at least partly) their resources collectively. Two important elements of traditional peasant societies were their autarkic character and, consequently, their dependence on many different natural resources and a strong mutual cooperation within their communities (Rösener 1991). In addition to agricultural activities, rural households extracted fruits, mushrooms, herbs, fodder, fuel wood, timber, and other products from the woodlands (Pretty 1990) and used forests as extensive pasture lands (Rösener 1991). The collective use and management of these resources, their commons, was an expression of the reciprocal interdependence in peasants' everyday life. The old institution framing those communities is known as Ancient Régime or feudalism. Regardless of how inefficient and oppressive this organization form was, emphasizes Eric J. Hobsbawm (1962), it provided communities with notable economic and social security. This was a system anchored in customary law that allowed peasants, for instance, in times of meagre crops to fall back on landlord's support. Furthermore, they also had the right of acquiring firewood free of cost or at a low price in the forests of their landlords. This system had worked since immemorial times, but at the end of the eighteenth century it was about to change.

The emergence of the social group called bourgeoisie within the Ancient Régime and the successful implementation of their interests brought radical socioeconomic and political changes that dismantled the prior form of social organization, first in Europe and later in most of the world. Mercantilization of natural resources, substitution of the peasants' subsistence logic with the maximization of individual benefit and pre-eminence of private ownership (Hobsbawm 1962; González de Molina 1993) were among the leading ideas of the new social group. These all are constituents of capitalism and the market as forms of social relations and resource allocation respectively. Indeed, by bourgeoisie is meant the class of modern capitalists, the owners of the means of social production (Marx and Engels 1848); England, where this socioeconomic rearrangement began, represents the first capitalist country.

Extensive commons, along with landlords' and churches' forests, constituted important reserves, particularly during difficult times, for subsistence peasants. But these same fields, meadows, and woodlands were viewed as under-utilized by the capitalists. Furthermore, these resources offered a laboratory to test the new liberal model. And just as landlords for centuries had justified their interventionism in rural community affairs related to collectively owned lands, so now did the bourgeoisie. An argument frequently used by the latter was the 'bad' state of the commons, by which they meant that only private initiative and their inclusion into the market could generate the highest benefit.

But before the test could be carried out, an important institutional change was required. The removal of the Ancient Régime had been already initiated in seventeenth century England. However, its definite abolition in Europe was a direct or indirect consequence of the French revolution (1789–1799), which disseminated bourgeoisie liberal ideas everywhere within the period 1789–1848 (Hobsbawm 1962). This was a globalization at the European scale that deeply affected rural communities and their traditional management practices.

The winners of the liberal revolutions, having abolished the old rules, aimed at the privatization of communal resources (enclosures) in order to incorporate them into the market and, in doing so, to promote their efficient use. From the peasants' perspective their common lands, which provided them with indispensable resources and particularly important for the poorest of the poor (Hobsbawm 1962), were being usurped by the bourgeoisie. Consequently, conflicts between both groups became very frequent. In this regard, it should not be overlooked that how a social group acts on natural resources rests on its particular perception of those resources (Godelier 1981), and that a range of perceptions and values can be identified among different groups within a culture (Worster 1988). Thus, disputes between the bourgeoisie and peasants (as had existed for centuries between peasants and landlords) were based on differing views of ownership rights as well as on opposing forms of social organization and management of natural resources (Fontana 1999).

The consequences of the liberal revolutions for rural communities were catastrophic. In England, for instance, in 1760 more than six millions ha of commons were enclosed by 5,000 fences. By the middle of the nineteenth century, 57% of the cultivated land was owned by only 4,000 holders. Dispossessed poor peasants had to move to the cities and industrial centres (Hobsbawm 1962). And if such urban sinks for rural populations were not available, such as in Prussia, then peasants became emigrants. The redistribution of the land carried out all over Europe did not always generate the sort of enterprising people predicted by the bourgeoisie. Indeed, in some cases the old ruling class was reinforced (Hobsbawm 1962). Poor peasants depending on the commons were particularly affected by the enclosures. They had to experience the extraordinary situation of being penalized just for extracting fuelwood from forests that had been common land until a short time before. An eyewitness to this situation was Karl Marx (1818–1883), who reported how traditional customary practices that had allowed dead wood gathering by the poor were outlawed and replaced by new laws of private interests (Marx 1842).

The generalized incorporation of natural resources into the market also had an ecological effect. In a locally self-sufficient economy, diversified land use systems promoted landscape heterogeneity (Lepart and Debussche 1992; Lenz 1994). But replacing peasants' subsistence logic with the maximization of individual benefit and the incorporation of local communities into world markets led to the specialization and intensification of agroecosystems in order to response to the market's need. This meant dismantling complex agro-silvo-pastoral systems (González de Molina 1993; Naredo 2004); such systems still occurred in southeastern Spain at the beginning of the twentieth century, where a reduction of 17% of the forest and pasture surface has been detected in some areas while agricultural fields expanded by 26%. This was a landscape 'agro-colonization' consequence of the liberal revolutions (Cobo Romero et al. 1992).

Another eminent observer of the enclosures was Alfred Russel Wallace (1823–1913) who, before becoming an eminent naturalist, worked among other things as a surveyor as part of the land enclosure programme in mid Wales. At the time he didn't fully appreciate the political implications but later he railed about this 'land-robbery' (Alfred R. Wallace, quoted by Gribbin 2002). But many other scientists did not.

Fig. 12.3 Traditionally managed trees may live for hundreds of years. This ancient pollarded oak (*Quercus rotundifolia*) in Sierra de Filabres (Almería, southeastern Spain) has known at least two different societies: Muslim communities living in the area until the sixteenth century and Spanish peasants since that time. Local peasants refer to this veteran tree by a name (Carrascón de la Peana) that appears already in a document of the seventeenth century. Traditional tree management practices were abandoned long ago in this mountain range (Photo: Jesús García Latorre)

The concerns of the bourgeoisie were supported by scholars of the Scottish Enlightenment School, such as John Locke (1632–1704), David Hume (1711–1776), and Adam Smith (1723–1790), whose essays reflected capitalist ideology by criticizing among other issues property owned in common (Ostrom 1990; Fontana 1999). During the eighteenth and nineteenth centuries, liberal revolutions spread over Europe, legitimized by scholars whose ideas were rooted in the Scottish movement. So, for instance, in Spain the intellectuals of the Enlightenment also complained about the degraded state of forests. Some of them proposed more governmental interventionism while others supported common land privatization (Casals Costa 1996). The result was the same. Local communities lost their control on natural resources while the bourgeoisie increased their ownership (Cobo Romero et al. 1992).

Contemporary authors have for the most part adopted the classical criticism of rural communities. For example, Burschel and Huss (1997) maintain that traditional tree and woodland management practices such as pollarding, shredding, and wood pastures have caused considerable damage to the trees and forests—a statement inconsistent with observations that traditionally managed trees are known to be able to achieve considerable ages (several hundred years, according to Rackham 1978; see Fig. 12.3). Berkes (1999) reviews this topic and shows how widespread these convictions are among contemporary scientists—an unsurprising outcome considering the centuries-old litany of the negative effects of traditional practices on resources, first promulgated by landlords and then by the bourgeoisie. The ideology of the Scottish Enlightenment School, underpinning the bourgeoisie liberal economic system, constitutes a paradigm of unparalleled success in human history that provided the foundation of modern social sciences and part of popular thinking (Fontana 1999) that still influences discussions on resource use.

From the twentieth century onward the market economy has continued to exert influence on rural communities. Since the end of the Second World War, farmers have reacted to long-lasting low prices of agricultural products, intensifying production through mechanization and the use of chemical fertilizers and pesticides to ensure ever-higher yields (industrialization of agriculture). This has promoted further homogenization of agroecosystems (Velvé 1992; González de Molina 1993; Naredo 2004).

With regard to mountainous regions, at the beginning of the twentieth century mountain ranges in Europe were still well-populated and included traditional management systems, constituting important reserves of traditional forest-related knowledge. But most of this cultural reservoir has already vanished. The incorporation of mountains into a broader and more dynamic socioeconomic frame has led to the intensification and specialization of the most fertile and accessible areas. Unlike in former times, when peasants aimed at managing all available resources in the whole territory, the new system strives for the maximal production per farmer in order to equate the earnings of the mountain populations to those of the lowlands (Lasanta Martínez and Ruiz Flaño 1990). Consequently, less productive marginal areas (steep slopes and remote areas located at high elevations) embracing millions of hectares that were formerly traditionally managed have been abandoned, and rural communities have become depopulated. Aside from regional particularities, this process is generalizable for Western European mountains (Lasanta Martínez and Ruiz Flaño 1990).

Thus, becoming embedded in larger systems has meant a transformation from self-sufficiency to dependency on commercial economy and the government, and a trend toward environmental deterioration (Tainter 2007). And the very internal mechanisms of capitalism, particularly market-oriented agricultural production, have played an important role (González de Molina 1993).

12.3.2   Local Communities and Communism

The perspective described in the previous section does not imply that communist regimes have done better. Communism was an idea of European origin that also experienced a globalization process. The same ambitions that elsewhere inspired state-supported industrialization are also characteristic of communism: state control and economic growth (McNeill 2001). Furthermore, communists believed in the absolute control of society over nature through the application of science (resembling in this regard the capitalist model) and, particularly in the former Soviet Union and in Eastern Europe, had a clear preference for big projects such as collective farms (McNeill 2001). But centralized power and gigantism are not compatible with local subsistence communities.

Focusing on Russia, indigenous people currently live under lower standards than the rest of the population; their level of subsistence is low and their culture is endangered. About ten indigenous nations are on the border of extinction (Diatchkova 2001) as a consequence of the Soviet period. Within the totalitarian system imposed by the communist party, collectivization constituted a major factor of socioeconomic activities; any alternatives outside collective farms were forbidden.

Traditional livelihoods were integrated into Soviet society by force. Most indigenous groups had led a nomadic life, but in order to ensure its control the Soviet government made them settle down (Diatchkova 2001). Moreover, the Russification of schools during the latter half of the twentieth century exacerbated the difficulties encountered by families who tried to preserve and translate their traditions. Diatchkova (2001) concludes that 'during the Soviet period the policy of collectivization destroyed the very foundations of the traditional subsistence system.'

Democratization of former communist countries since the late 1980s has brought some improvements for local communities. In the Skole district of the Ukrainian Carpathian, for example, although the cultural heritage of local villages still is threatened by economic globalization (among other factors), the situation is at least better than it was under Soviet large-scale economic production. When the Soviet era ended, traditional villages re-appeared as a way of subsistence for local people; those villages today play a key role in the maintenance of culture and society in the region (Angelstam and Elbakidze 2006). And in Russia, democratization has helped indigenous peoples to appreciate their past. Interestingly, the mass media, which in other countries promote the introduction of Western values into traditional societies, have become in Russia an important mean of ethnic mobilization. Native enterprises have been organized in relation to traditional economies and are improving their skills in dealing with a market economy (Diatchkova 2001). Still recovering from the sequels of communism, these traditional local communities deal now with the constraints of capitalism.

12.3.3   Colonialism and Globalization

Globalization has deep roots that go back to the European expansion initiated in the fifteenth century. Five hundred years are more than time enough for ideas and beliefs affecting local communities to expand worldwide. Such diffusion has been the case with respect to scientific forestry and other forms of natural resource management of European origin.

Certainly, Europe was not impermeable to the influence from the colonies and received, among many other good products, tomatoes, potatoes, and sugar (Pelizaeus 2008). However, regarding forest management, it was not the Australian aboriginal people who came to England and applied their experience burning shrublands to manage coppiced oak woodlands. Rather, it was European (particularly

German) foresters who, on their arrival in Australia, concentrated their activities on the scarce tall and medium eucalypt forests, while disregarding the low wood-lands of little commercial value on steep and stony slopes (Dargavel 2000). It is scientific forestry that has been globalized, and not traditional forest-related knowledge.

European colonialism has had an enduring influence on colonized countries that embraces language, institutions, state organization, religion, and techniques and modes of production. In all these different fields, colonial rulers were convinced of their (supposed) cultural supremacy (Pelizaeus 2008), including the management of forest resources. The whole 'forestry package,' including technique and ideology, was transferred. This transfer brought with it: disregard of traditional knowledge and communal management; limited access to forest resources for local communities; rational forest management focusing on wood and forests as a revenue source; state interventionism; and pre-eminence of foresters' skills. These ideas still pervade forest policies in once-colonized countries.

As discussed earlier, local communities have a long tradition of communal natural resource management, but this custom was not acknowledged as legitimate by European colonialists, as noted by Williams (2000):

> The imperial and colonial political overlordship of the past was convinced of the inferiority of native practices which were consequently repressed, and there was a strong conviction that the application of Western development and its science and organization was superior.

In Algeria and India, colonial bourgeoisie showed as little understanding for traditional agricultural societies as they had shown for European peasant communities. The complex network of rights—such as the avoidance of private ownership in Algeria, or the collectively owned land in India that could be neither bought nor leased—was incomprehensible for European liberals, who instead applied their rational and individualistic system, leading to the absolute pauperization of the farmers (Hobsbawm 1962).

In Latin America, it wasn't until 1850 that newly independent states initiated an effective assault on communal ownership (Hobsbawm 1962). And still today millions of hectares of land historically owned and governed at the community level through customary arrangements in developing countries worldwide are threatened by governmental dispossession (Alden Wily 2010; World Bank 2010). Customary rights of land frequently have uncertain official recognition, and states have considered such lands as empty and, correspondingly, the property of the state. This, in turn, facilitates transfers with few safeguards to foreign investors (Alden Wily 2010; World Bank 2010). 'The tendency to neglect existing rights often derives from a legal framework inherited from colonial days,' according to the World Bank (2010), which reflects the extensive European experience in enclosing common land. As we have seen, during the eighteenth and nineteenth centuries, the bourgeoisie also had viewed fallow lands as underused and demanded its privatization, even though it actually represented reserves of land or land extensively used by subsistence communities (see Sect. 12.3.1.2).

The consequences of this legacy can be found all around the developing world. In Zambia, most of the country's land is governed within the framework of customary rights; however, these rights can be neither registered nor surveyed. In Indonesia, 70% of the country is classified as forest estate and administered by the Forest Department; traditional local communities are vulnerable to displacement since their customary rights are kept unrecognized. The same occurs in Liberia, where official governments deny customary land recognition (World Bank 2010). And even in countries that have developed appropriate policy frameworks regarding community consultations during land acquisition by investors, actual implementation may be unsatisfactory. For instance, in Mozambique, laws on management of land include provisions for community consultations and hearings when land uses or users change. However, district authorities may have more incentive to support the interests of investors and, thus, consultations are in practice fairly limited (Cotula et al. 2009).

Beyond disregarding customary rules, colonialists introduced their entire forestry package. Ciancio and Nocentini (2000) relate how 'the concept of the forest as a timber resource had been exported from Europe to many parts of the world during the colonial era, and left assumptions that persist today, very much to the detriment of the world's forested countries and of those who live in them,' still inspiring 'laws and professional values that underlie forestry.' And Geneviève Michon et al. (2007), after referring to the 'exclusion of farmlands, peasants, and local tree management practices' that occurred in European history, comment:

> This perception was transferred through colonial regimes to most tropical countries... [where] forest agencies still consider that forests should be managed exclusively by professionals under a comprehensive legal, administrative, and technical regulatory framework. ... Consequently, forest people are almost never considered as legitimate and knowledgeable forest managers.

With the aim of exemplifying the consequences of the colonialism heritage, we summarize in Table 12.1 the cases regarding seven countries.

Misconceptions abound about traditional management practices. An exemplary case is provided by shifting cultivation, a system in which patches of forests are burned and cultivated for several years and then left to lie fallow, allowing secondary forest to regrow. Shifting cultivation has a long tradition in many parts of the world, such as in mainland Southeast Asia (Fox 2000).[2] It is a common belief that this system contributes to environmental degradation. International organizations such as the FAO and the World Bank have indeed recommended to governments to eliminate and transform traditional management. Correspondingly, governments have implemented policies that include eviction, subsidizing permanent and commercial agriculture, and outright banning of shifting cultivation. Customary rights to swidden fields have not been recognized, and fallows have

---

[2]Different forms of management that include controlled fires were common in Europe in former times; however, fire and foresters do not always fit well together. For example, in Italy, the forest administration disapproves controlled burns in chestnut stands, a traditional management practice of these woodlands (Grove and Rackham 2001).

Table 12.1 Examples of how different aspects of scientific forestry were introduced by Europeans in colonized countries and continue exerting a strong influence on forest management and on local communities

Country	Colonial background	Current influence
India	India's forest management has a European legacy: British colonial rule. Tribal communities' resources were eroded as state agencies and the private sector established greater control. Even after independence the British influence on forest policy and administration persisted.	Most of India's forest land remains under state control. Access of local communities to productive forestland is limited by government control.
Nepal	Forest policy was directly influenced by the British; British experts helped the Rana rulers establish the Department of Forest in 1942; this department started the nationalization of forest land and perpetuated the colonial notion of scientific forestry.	The Department of Forests retains greater control of high-value forests.
Philippines	Forest management represents a legacy of Spanish and American systems. The first forestry bureau was established by the Spanish administration in 1863. A forestry school was established during the American colonial period supported by Gifford Pinchot, the founder of the U.S. Forest Service.	Forest management is based on scientific forestry. Less productive forests are transferred to local communities. The more productive lands are still under private timber concessions or under government-controlled protected areas.
Cameroon	The state's tendency to retain valuable forest lands is rooted in colonial tradition (German, British, and French).	Forest land is classified into non-permanent and permanent forest estates. Local communities are entitled to access and use the (less productive) non-permanent forests, while high-value forests are reserved for commercial logging and protected areas.
Thailand	British timber companies took control of the teak industry and established a symbiotic relationship with the central power in Bangkok by establishing the Royal Forest Department (1896).	At the beginning of the 1990s a powerful alliance of army generals, forestry officials, and pulp and paper companies unsuccessfully tried to implement a non-participatory, ecologically unsound and repressive state forest policy that included eviction and *Eucalyptus* plantations.
Morocco	French foresters interpreted Moroccan landscapes as degraded and promoted forest regulations that favoured imperial interests over indigenous interests, thereby facilitating land dispossession and the destruction of Moroccans' traditional livelihoods.	In the 1980s the centralized and coercive forest administration tried to eliminate the Agdal practices (a system based on complex ownerships, with positive environmental effects) applied by rural communities to manage their forests in Central High Atlas. Between 1964 and 2002 the forest area increased by 3.1% within the Agdal areas, while deforestation took place beyond them.
Indonesia	Dutch colonial state law and trade arrangements supported increasing state interventionism in forest lands. This has pervaded Indonesian forestry policy long after colonial times.	Foresters still consider themselves to be the exclusive forest custodians. A national extractive model assures a continued flow of resources from the forests to the government centre.

Sources: Pulhin et al. (2010) for India, Nepal, Philippines and Cameroon; Pye (2005) for Thailand; Davis (2005) and Aubert et al. (2008) for Morocco; Robbins (2004) for Indonesia

been perceived as unused or abandoned. These policies are, again, a legacy of the colonial era (Fox 2000). The disregard of shifting cultivation is based on misconceptions of how the system works, because forested ecosystems in which swidden agriculture is applied are actually characterized by their complexity, dynamic successional stages, and structural diversity. In general terms these ecosystems have a low deforestation rate, in contrast to the areas with permanent commercial agriculture (Fox 2000).

Table 12.1 shows how 'the states' tendency to retain valuable forestland is rooted in their colonial tradition and perpetuated by modern forest bureaucracies' (Pulhin et al. 2010). Thus, the set of ideas of Western origin that pervades these bureaucracies has a functional component—namely, to promote government control over resources serving the economic interests of elites. In Indonesia policymakers often decide to convert forest to industrial plantations without proper consultation. And in Liberia customary land frequently is assigned to companies without compensation for local communities (World Bank 2010). States in Southeast Asia typically prefer to devote areas under traditional shifting cultivation to commercial agriculture and plantation forestry (Fox 2000). Thus, traditional forms of forest management, with a recognized social and environmental legitimacy (Fox 2000; Michon et al. 2007; Pulhin et al. 2010), are replaced by industrial plantations that, in turn, frequently lead to fatal clashes and conflicts among local communities and firms or forest departments (World Bank 2010).

Many foresters apply their accredited skills to the state's interest, 'even though they rarely view their policies or implementation as political acts' (Pulhin et al. 2010). In fact, such foresters are applying a belief and value system that is several hundred years old, expressing the same convictions that served the interests of both landlords and the bourgeoisie in Europe. These ideas now legitimate interventions in rural land-rights systems in developing countries that serve the interests of governments, with tragic consequences for the poorest people.

12.3.4 Traditional Forest-Related Knowledge in Developing Countries: A Decaying Heritage

Unlike in many industrial countries, traditional forest-related knowledge in developing economies constitutes a daily instrument applied by rural populations (Fig. 12.4). In fact, resources managed within these traditional systems play an essential role for meeting the needs (from firewood and vegetables to game and medicinal plants) of countless millions of poor households (Chaudhary et al. 2008). The economic importance of non-timber forest products, for instance, can be considerable, accounting for more than half the total employment and contributing to half the total annual income of households (Nanjundaiah 2008). Moreover, these subsistence activities are frequently embedded in philosophies reflected

Fig. 12.4 Farmer with a
jackfruit (*Artocarpus
heterophyllus*) in an
agroforest also containing
timber-producing trees in
Lantapan, Mindanao
(Philippines) (Photo:
Manuel Bertomeu)

simultaneously in religion, social institutions, and land-use practices (e.g., sacred forests), something not typically understandable for Western managers (Godelier 1981). Such traditional systems, encompassing all livelihood aspects, feature a high complexity and have been termed 'domestic forests' by Michon et al. (2007).

In spite of this, when facing the influence of modern land-use models, the defenders of traditional knowledge are seldom local people, but rather often foreign scientists and non-governmental organization activists (Feintrenie and Levang 2008). And today not only alternative management practices but completely new ways of life are being offered to local communities, even in remote areas, making changes in their customs and habits unavoidable.

Half a century ago the German philosopher Martin Heidegger prophesied that television would abolish every possibility of remoteness (Scheuermann 2006). What happen when young people gain access to knowledge of alternative livelihoods as shown on television? Hendarti and Youn (2008) relate the case of Kasepuhan people, an ethnic group living in the Halimun Mountains in West Java. Their daily life, including the management of natural resources, is integrated in a philosophy called Tatali Piranti Karuhun. But the youth, influenced by television, have lost respect for this tradition and are more interested in 'working in the city to earn cash income for buying cloth, mobile phones, etc.' In this regard, many other examples could be mentioned. Tainter (2007) comments how peasants in Epirus, a remote mountainous area in northwest Greece, were stimulated to acquire material goods as cash became increasingly important after World War II. Indigenous and local communities in developing countries, such as the Kasepuhan people, are similar to the Greek example; some villagers, influenced by non-Kasepuhan middle-men who promise them cash money, lose respect for their sacred forests and engage in illegal logging. Such changes are not surprising. According to Brian Hayden (1993), an archaeologist who has lived with indigenous communities on different continents, 'we are all materialistic'. 'I can say categorically', he continues, 'that the people of all the cultures I have come in contact with exhibit a strong desire to have the benefits of industrial goods that are available. I am convinced that the "nonmaterialistic culture" is a myth.' However, many cases of the 'environmentalism of the poor' referred to below would relativise Hayden's categorical statement.

Aside from the impact of the media and other external influences, indigenous and local communities in developing countries are subjected to diversified destabilizing factors that adversely affect them and provoke the loss of traditional forest-related knowledge. Global markets, information and capital flows, and international conventions constitute basic paths of globalization (Lambin et al. 2001). In this context we have indentified the integration of natural resources within a market economy and improper environmental policies as major forces exerting influence on local rural communities. On the one side, the neoliberal economic expansion of past decades has promoted the inclusion into the market economy of all aspects of natural resources, even in very remote areas. Furthermore, since the end of the 1980s, a shift to global environmental policies supports the implementation at the local scale of measures recognized at the international level, including international conventions. Many of these measures contain some of the ideas and beliefs disseminated during colonial times that were described in the previous section, such as the disregard of traditional forms of natural resource management.

12.3.4.1 Market Economy

Toledo et al. (2003) emphasize that peasant farmers around the world carry out a dual economy: 'They produce goods for the market and buy goods using cash yet, at the same time, they produce basic commodities for their own consumption.' Correspondingly, they adopt a strategy that addresses both subsistence and market production, aiming at the multiple use of spatio-temporal and natural resources. The goal is to maximize the diversity and number of available options and, simultaneously, to guarantee subsistence and minimize risk. Nevertheless, their integration into society at national and indeed international levels through the market (linking them to infrastructure projects, the media, or educational programmes, for example) exposes them to external threats.

Regarding the effects of the market on the environment, Tacconi (2000) points out that it is not the market but the dynamics of the economic process that might lead to environmental problems. Such dynamics include changes in values, increasing demands, and specialization—which also reflect the effects of the market on traditional local communities.

The response to the market demand for some products can be accompanied by changes in the value and belief frame of a community, as was noted above with the case of the Kasepuhan people in West Java (Hendarti and Youn 2008). Traditional belief systems underpinning the management of natural resources by Malayali tribal people, the inhabitants of the Kolli Hills (eastern Namakkal district, India), also have disintegrated in past decades. Modern developments have contributed to this situation: encroachment of cash crops like tapioca and the introduction of other economically important plants; the dilution of traditional beliefs through formal education; rising economic status; and declining interest in traditional forest-related knowledge among the youth. This interweavement of factors is promoting the degradation of the Sami Sholai, the sacred forests traditionally managed by the Malayali people (Israel and King 2008).

A comparative study (Ruiz-Pérez et al. 2004) has shown that commercial trade drives a process of intensified production and specialization among forest peoples, leading to higher incomes. How production strategies are influenced by the transition from subsistence to cash economy depends on the degree of integration into the latter. In particular, increasing market demand for a product is identified as an important destabilizing factor that restricts the productive amplitude of traditional households (Martínez Alier 1993; Toledo et al. 2003).

Increasing demands promote the replacement of locally produced agricultural and forest plants by high yielding varieties 'since they provide a short-run economic advantage (in the chrematistic sense)' (Martínez Alier 1993). Additionally, traditional management practices, applied in the framework of a subsistence economy, are replaced by standardized systems (such as plantation forestry) (Fox 2000). The result is the abandonment and loss of traditional forest-related knowledge.

An example of this is offered by the current expansion of oil palm plantations in response to the increasing worldwide demand for biofuels. Poor farmers in developing countries, stimulated by the expectation of earning some income, use their small plots of land for growing oil palms. Although under appropriate conditions biofuels could offer opportunities for poverty alleviation—as, for instance, in the form of a mixture of plantations and agroforests (Feintrenie and Levang 2008)—the dominant agro-industrial model imposes dramatic constraints on local communities. Smallholders become bonded to oil palm companies by the credit provided by these firms, accumulated debt, etc. One of the most important oil palm producers is Indonesia. According to Bailey (2007) 'about 30% of Indonesian palm oil is produced by smallholders, supporting up to 4.5 million people. Most of these are drawn from local communities and indigenous peoples that lost their land to the advancing plantations.' The production of palm oil has not excluded the displacement of indigenous and local communities in Colombia (XLSemanal 2010).

12.3.4.2   Environmental Policies and Traditional Forest-Related Knowledge

Globally agreed-upon environmental policy measures can also potentially lead to the marginalization of traditional groups and their practices. On the one side, the rights and values of local communities and their traditional management systems have been recognized and disseminated by international environmental and forest policy processes. However, constraints on local communities and disregard for their traditional knowledge are being imposed, too, in the name of nature conservation and rational resource use.

Geneviève Michon and her colleagues (2007) relate how in the frame of the current conservationist movement, 'a new kind of self-appointed legitimate forest manager has recently emerged: environmentalists and conservationists, active at the local level through the creation of parks and reserves, and at the global level through the development of international conventions.' These 'legitimate managers' frequently ignore and even despise traditional practices and promote their replacement with Western techniques, exacerbating the problems that were originally intended to be solved. Even worse, they might invent environmental problems or present their severity in a dramatically exaggerated way.

Such is said to already have been the case in savannah ecosystems 'compromised by inappropriate outside interventions, which in the environmentally conscious 1990s cynically see this degrading environment as the way to gain money, justify budgets and keep environmental institutions solvent by attracting major international funding for rehabilitation' (Williams 2000).

Thomas Bassett and Koli Bi Zuéli (2000) have studied the savannah forest mosaics in the Côte d'Ivoire, where (as in other African countries) the World Bank (WB) has supported the development of National Environmental Action Plans (NEAP). These plans place 'the identification of environmental problems and their underlying causes as the first step,' and assume 'that most environmental issues are easy to identify and can be classified along a simple colour scheme.' But the participatory planning of the World Bank 'did not involve consultations with ordinary men and women living in rural areas about what they considered to be the most important environmental issues.'

The NEAP for Côte d'Ivoire presented a grim scene of environmental degradation, blaming peasants for the deforestation of the savannas. It is stated that as a consequence of shifting cultivation, bush fires, and overgrazing, the tree savannah is being replaced by the grass savannah: 'we are increasingly witnessing a decline in vegetation cover' (République de la Côte d'Ivoire, quoted by Bassett and Bi Zuéli 2000). But actually, the contrary is true. By interviewing peasants, comparing old and recent aerial photographs, and doing field work, Thomas Bassett and Koli Bi Zuéli demonstrate that the landscape is becoming more wooded and simultaneously the herbaceous component of these ecosystems is decreasing, as a consequence of changing controlled fire regimes and grazing pressure.

One of the grave outcomes of the hegemonic ecological managers' discourses of degradation is their detrimental effect on local communities. Rural populations might become impoverished through the imposition of fines on the use of forest resources (Williams 2000). Regarding the Côte d'Ivoire example, the NEAP provides a highly regulatory and intrusive cookie-cutter planning model that includes: outlawing logging, encouraging tree planting, controlling bush fires, and creating a forestry police - measures that are at heart reformist and technocentrist. 'Such policy recommendations,' conclude Bassett and Bi Zuéli (2000), 'can be seen as misconceived and a waste of limited resources.'

Bassett and Bi Zuéli (2000) emphasize that 'identifying environmental problems and their causes is one of the most difficult and time-consuming stages in environmental planning and policy making. ... One of the challenges ... is that so little data exist with any meaningful time depth.' A consequence of this information gap is that well-grounded decision-making is still lacking in many environmental projects, affecting local and indigenous communities. Thomas Griffiths (2005) has documented numerous cases of projects carried out by the Global Environmental Facility (GEF, the main institution for international funding for the Convention on Biological Diversity) that have caused quite grave circumstances for poor rural communities. As the World Bank 'experts' did in the Côte d'Ivoire, the planners of GEF-funded projects have frequently presumed the unsuitability of rural people for managing their own resources and have introduced radical changes negatively affecting those small stakeholders. For instance, almost half a million villagers living within and around several protected areas in India were affected by an eco-development project (1996–2003) financed by a GEF grant ($20 million) and a World Bank loan ($28 million) that aimed at biodiversity conservation. The conservation strategy included such measures as voluntary relocation and provision of alternative non-forest based livelihoods. Some reports suggest that in Periyar Tiger Reserve (Kerala), attention was paid to gender, land rights, and forest access, so they consider that this initiative succeeded (Griffiths 2005). But overall, most affected communities consider the project as detrimental to their rights and interests. The effects of existing land use practices on biodiversity were not assessed by baseline field studies, drastically limiting the scientific legitimacy of the project. In the Nagarhole National Park (Karnataka), the forest department was strengthened by the project and local communities were forcibly relocated. Other villagers complained that alternative livelihood plans were developed exclusively by the forest department, which disregarded villagers' proposals based on their own traditional knowledge. Indigenous people have questioned the scientific basis for this interventionism: 'where is the science to show that a specific local resource use practice is damaging to biodiversity conservation? In many cases, restrictions on resource use and new livelihoods are introduced without first establishing what the most genuine threats to biodiversity are' (Griffiths 2005).

The disregard for the human component of ecosystems is detected also in initiatives developed by governmental institutions. Developing countries' governments might try meeting commitments achieved in the framework of international environmental policy processes, increasing their efforts for nature protection. This is the case of India, 1 of the 12 "megadiverse" countries in the world and a member of the Convention on Biological Diversity. The Indian State has enforced protected areas and supported ex-situ conservation projects (Israel and King 2008). However, 'in situ on-farm conservation by rural and tribal women and men largely remains unrecognized and unrewarded' (Swaminathan 2000). Recently a new law was passed (Recognition of Forest Rights, December 2006) that aims at giving tribals and other traditional forest dwellers 'responsibility and authority for sustainable use, conservation of biodiversity and maintenance of ecological balance...' (Forest Rights Act 2006, cited in Misra 2008a). However, the implementation of this act is so far plagued by complex procedures leading to chaos, confusion, and officers overburdened with paperwork (Misra 2008a; Jamwal 2008b). Not surprisingly, less than 2% of the state's 2001 tribal population had submitted the form by mid 2008 (Misra 2008b).

Cases like the Indian one described above are not the exception. In fact, a generalized acute asymmetry between the attention drawn to environmental and social aspects of conservation is the norm; 'biological concerns have gained policy backing and financial resources toward their practical implementation (park establishment), while social approaches remain under-designed and under resourced (Cernea and Schmidt-Soltau 2006). It should be considered that top-down one-sided conservation approaches adversely affect both biodiversity and people. Kapp (2008) and van Vliet et al. (2008) have shown how rural communities in Africa, confronted with the strengthening of restrictive environmental policies (indirectly supported by international NGOs and development agencies), fall back on more intensive agricultural use, with deleterious effects for soils and forest vegetation. In extreme situations thousands of persons belonging to local communities can be displaced and impoverished in the frame of nature conservation policies (referred to as 'eviction'; see for instance Cernea and Schmidt-Soltau 2006; Brockington and Igoe 2006).

Finally, measures adopted within climate policy negotiations might potentially affect indigenous and local communities in the near future[3] (Vidal 2009). This is the case with REDD (Reduction of Emissions from Deforestation and Forest Degradation) in the context of the United Nations Framework Convention on Climate Change (UNFCCC). One of the criticisms made regarding that process is the absence of indigenous and local community representatives in the negotiations; being addressed so far exclusively by governmental officials, REDD constitutes a classic top-down policy approach (Griffiths 2007). This is of great concern to many since potential substantial rewards for forest conservation could represent perverse incentives to enforce unjust forest laws as well as promote eviction and expropriation, as has occurred with some of the environmental measures addressed in this section.

12.3.4.3 Environmental Policies and the Market: Not So Far Away from Each Other

The distinction made regarding the respective impacts of the market and environmental policies on traditional forest-related knowledge facilitates an understanding of some of the paths through which local communities and their traditional practices are being affected by globalization. Nevertheless, this procedure is rather a

---

[3]Examples of negative impacts of REDD initiatives on indigenous communities in the frame of voluntary carbon trading have been recently documented. A tribe's leader in Papua New Guinea was forced at gunpoint by "carbon cowboys", aiming to develop a voluntary carbon trade project, to sign away the carbon rights to his people's forest (Astill 2010; Greenpeace 2010): "They came and got me in the night... police came with a gun. They threatened me. They told me, You sign. Otherwise, if you don't sign, I'll get a police and lock you up," said the leader (Greenpeace 2010).

simplification. In reality such impacts frequently result from a closely interweavement between global environmental policy processes and the economy.

In some cases this relationship is quiet evident. The expansion of oil palm plantations, encouraged by the increasing demand, offers an example. Bailey (2007) relates how some policies aiming at reducing emissions through the enhancement of the proportion of biofuel in the transportation sector in countries like United States and the European Union may be disastrous for poor people. Regarding particularly the EU, the biofuel share is expected to move to a share of 10% by 2020. Bailey (2007) emphasizes that 'a scramble to supply the European market is taking place in the South, and poor people are getting trampled.' Indonesia, for instance, plans to expand the land area devoted to oil palm production by 14 million ha by 2020 (Bailey 2007). In 2008 there were more than reported 500 conflicts related to oil palm plantations in that country (Steni 2010). Recently Persson and Azar (2010) suggest that EU and U.S. biofuels policies are contributing to deforestation in the tropics.

The market may appear later, interweaving with environmental policies in unanticipated ways. An example is provided by 'eviction for conservation,' which is the 'displacement resulting from the establishment and enforcement of protected areas' (Brockington and Igoe 2006). Cernea and Schmidt-Soltau (2006) report how 19% of the hunters resettled in 2000 from Korup National Park (Cameroon) 'have increased their hunting due to better access to markets' depending 'nearly entirely on the old hunting grounds in the national park.'

But the interweavement between global environmental policy processes and the market goes beyond this more or less direct relationship. According to Martin O'Connor's (1993) reflexion on the assimilation of natural resources by the economy, conventional environmental policy would be just a component of the capitalization of nature, an aspect of the so-called 'ecological phase of capital,' with the term capital understood in its rigorous definition—a form of social relations (Baudrillard 1980, cited in O'Connor 1993). Thus, global environmental policy might be, in fact, facilitating the access to natural resources. Some details will be helpful in order to explore Martin O'Connor's perspective with regard to local communities.

The ecological phase constitutes capital's strategic response to decreasing natural resources, increasing competition for them, and social discontentedness. This phase is undergoing a semiotic expansion. Terms such as rational use, sustainable development, stock designation, property rights, and consideration of externalities are characteristic of current environmental policy rhetoric (Martínez Alier 2002), and an inherent constituent of capital's discourse in its ecological phase. The strategy is to institute harmony by means of those signs. Moreover, the process of semiotic expansion is 'aided by the co-option of individuals and social movements in the "conservation game"' (O'Connor 1993). But the real goals are quite different:

> Such self-interest in profits does not equate to an authentic interest in these sources as life-forms or social ends in themselves! The dominant responses of capitalism to environmental crisis and to demands for respect of cultural difference continue to be premised on an instrumental, if not downright cynical, treatment of nature and human nature. There continues to

be direct appropriation of supposedly "free" natural domains, which in general means exclusion of other human groups ... through this process of capitalization of all domains of raw materials and services, through the internalization via the extension of the price system considered as susceptible to giving account of everything and to directing all processes (O'Connor 1993).

Thus the final achievement is neither harmony nor conservation, but a struggle to have particular interests and capitals valorised at the expense of others, 'the fact and imminence of annihilation for the less favoured interests and beings who are the "used" and "abused"' (O'Connor 1993).

Against this background we assess now how emerging global environmental regimes in the fields of forest policies and access and benefit sharing interweave with the economy and might negatively affect local and indigenous communities.

Capital's control on international environmental policy has been highlighted by David Humphreys (2006). Tropical forests represent sources of valuable timber exploited by concessionaries supervised by governmental administrations (Michon et al. 2007). Developed governments push for promoting private investment in sustainable forestry in the tropics, while developing countries are prepared to grant use rights to private companies in the form of concessions (Humphreys 2006). In this regard, the use of the term sustainable and the belief that market forces injected into new domains provide for public goods are constituents of capital's ecological phase. In fact, to assure their access to resources, corporate and political interests promoting trade 'wish any international instruments on the environment and human rights to be kept soft and outside the purview of the WTO [World Trade Organization].' Such instruments 'are kept entirely separate from international trade law, thus ensuring that they neither compromise neoliberal objectives nor are subject to WTO compliance mechanisms.' Furthermore, 'the influence of neoliberalism,' Humprheys writes, 'informs all aspects of global forest discourse.'

Consequently, there is no agreement to negotiate a convention to avoid deforestation. Within the United Nations Forum on Forests, which aims at promoting '... the management, conservation and sustainable development of all types of forests and to strengthen long-term political commitment to this end'[4], only non-legally binding instruments are approached (Humphreys 2006). These instruments are of uncertain value but rest 'comfortably with neoliberal logic' since they contain only soft commitments and are adopted by states on a voluntary basis (Humphreys 2006).

Nevertheless, negotiations in the framework of climate policies (such as Reducing Emissions from Deforestation and Degradation, REDD) might introduce stronger forest policies addressing deforestation. 'Tropical forests have been recently invested,' emphasize Geneviève Michon and her colleagues, 'with a key role in the protection of the global environment against global warming.' This approach has an underlying economic background, since those carbon sinks would be 'regulated and managed through global economic and financial instruments' (Michon et al. 2007). Joan Martínez Alier (1993) advised against the limitations that such international regimes might introduce, depriving 'not only local communities but even independent

---

[4]http://www.un.org/esa/forests/about.html

states of control over the forests, which would be vested in international ecological managers.' In fact, governments might benefit.

REDD negotiations began at the end of 2005 and still have an uncertain way ahead. Market-based approaches are among the instruments being discussed for deforestation reductions in developing countries. Observers predict substantial rewards for forest conservation, whereas 'payments would in large part be made to government ministries or treasuries' (Griffiths 2007). Consequently, avoiding deforestation also features a strong economic component. Thus, forests would become lucrative carbon reservoirs 'deemed by the government and the courts to be "state" lands,' according to Thomas Griffiths (2007), who summarized the potential negative effects of such a regime on rural livelihoods. He mentions among others: increased state and 'expert' control over forests, perverse incentives for government and business to expropriate indigenous land, evictions, reinforce of unjust forest laws, etc. Moreover, Interpol recently warned about a speculated threat that a REDD regime might initiate: 'the chances are very high that criminal gangs will seek to take advantage of REDD schemes in African and Asian countries,' (Vidal 2009). In Colombia, even in remote areas, pamphlets offer local communities incredible fortunes for the carbon saved in their forests (XLSemanal 2010; see also footnote 3, earlier).

A second example regarding the infiltration of economic interests into environmental policy processes comes from the international regime addressing the access to genetic resources—Access and Benefit Sharing (ABS)—which is being negotiated within the Convention on Biological Diversity (CBD). This convention emphasizes states' sovereignty over their natural resources as well as their endeavour to create conditions to facilitate access to these resources. Furthermore, parties are to promote the wider application of traditional knowledge of indigenous and local communities with their approval and encourage the equitable sharing of benefits (Secretariat of the Convention on Biological Diversity 2005). This approach has received scholars' criticisms regarding different issues (Sharma 2005 Shiva 2005; Ribeiro 2005; Swiderska 2007; Frein and Meyer 2008). Here, we focus some of its economic aspects.

According to Ribeiro (2005) and Swiderska (2007), parties to the convention prefer mechanisms for sharing benefits from the commercial use of traditional knowledge that are consistent with intellectual property rights (IPRs). In fact, IPR regimes are becoming increasingly strong, supported by the agreements within the WTO and bilateral free trade agreements. This framework is accelerating the commercial use and privatization of indigenous knowledge and resources (Swiderska 2007). Moreover, since CBD emphasizes states' sovereignty over natural resources, few ABS laws require the previous informed consent of communities (Swiderska 2007). Against this background, Krystyna Swiderska concludes that 'the ABS framework effectively facilitates access by outsiders to community resources.'

A further critique refers to the sharing of benefits arising from the commercial use of genetic resources in a fair and equitable way. Martínez Alier's (1993) reflexion on the distributional obstacles to international environmental policy casts serious doubts on the real benefits. For example, pharmaceutical businesses realize profits of millions of dollars (Frein and Meyer 2008); even accepting these firms' willingness to share, indigenous people's 'piece of the pie' would be very small compared to the firms' returns. The reason for this is that 'peasants and indigenous peoples are likely to set a low price to their hypothetical Farmer's Rights, not because they themselves attribute a low social value to their labour and agronomic knowledge ... but also because they are poor' (Martínez Alier 1993); and, as Martínez Alier emphasizes, 'the Poor sell cheap.' An example of this is provided by the drug jeevani, which has anti-fatigue and anti-stress properties and was developed based on the knowledge of Kani tribes in Kerala. Its estimated commercial value is in the range of at least US \$50 million to 1 billion; the Kani tribes so far have received about US \$12,000 (Sharma 2005). In general terms, the experience gathered in the field of 'bioprospecting' since the 1990s shows that the sums paid by pharmaceutical companies for their access to genetic material have been in the end smaller than anticipated (Pagiola et al. 2002). Thus, in spite of the recognition of community ownership of environmental capital, the final effect is 'a mobilization of the resources—their entry into the sphere of exchange value—in the larger interests of capitalism as a dominant social form' (O'Connor 1993).

Summing up, global policies targeting environmental issues such as forest conservation or the access to genetic resources might also play to concrete economic interests and negatively affect local and indigenous communities. Regarding these two fields, David Humphreys (2006) concludes that 'the trade liberalisation of forest products and the contentious principle of access and benefit sharing each aim explicitly at the continued exploitation of forests for private gain' (Humphreys 2006). Researchers have pointed out the similitude between this current privatization process and the enclosure of the commons that occurred in Europe during the eighteenth and nineteenth centuries (see Sect. 12.3.1.2) (Ribeiro 2005; Humphreys 2006).

12.3.4.4 The Economy and Violence

The application of violence to access resources is not necessarily a characteristic of a market economy. But Martin O'Connor points out that supplying the market demand through capitalist firms eventually does not exclude violence. Violent events adversely affect local and indigenous communities, so we address them briefly.

> Many individual capitalist enterprises, hellbent on survival in a competitive world, would prefer indeed to continue to treat nature and societies as open-access terrains to be mined and trampled upon at will. This is illustrated not only by military or quasi-military operations of *force majeure*, but also by the great pressures on governments. ... Such attempts at maintaining favourable (to capital) cost and supply conditions for needed raw materials and services (of nature, of labour, of society as socialization force and infrastructure) can involve fairly obvious dispossession, as well as cost-shifting onto local communities (O'Connor 1993).

Dramatic examples of this are provided, for instance, by the exploitation of rainforests, as exemplified by following cases.

During the 1980s, indigenous peoples, small farmers, and rubber tappers were displaced by landholders ('fazendeiros') in the Amazon or murdered by their hired killers. Fazendeiros and 'garimpeiros' (gold miners) promoted deforestation and provoked extensive forest fires. Fazendeiros raised cattle in response to the market demand for hamburger meat, and violently eliminated the resistance offered by rubber tappers—such as Chico Mendes, who was murdered in December 1988—and peasants who were sustainably using forest resources (Moro 2006) (see further discussion in the next section).

With respect to oil palm plantations, Bailey (2007) relates how 'in Colombia paramilitary groups are forcing people from their land at gunpoint, torturing and murdering those that resist, in order to plant oil palms, often for biofuels. Many of these violent acts occur in the traditional territories of indigenous peoples.' Local communities might try to defend themselves by organizing labour associations. However, in parts of Latin America, varied circumstances thwart unionization (obstructive legislation, intimidation, lack of worker rights). In Colombia, again, palm oil trade unionists have been tortured and murdered (Bailey 2007). At the beginning of 2010, the online magazine XLSemanal further confirmed the situation described by Robert Bailey and reported on several cases of human rights abuses (XLSemanal 2010). Worldwide, several millions of indigenous people have been affected by the deforestation of their land to make way for biofuel plantations (Bailey 2007).

12.4 Adaptations and Reactions of Local Communities to Globalization

Previous sections have shown some of the almost insuperable forces that oppress local communities, causing the loss of their traditional forest-related knowledge. However, the following cases demonstrate that rural communities are also able of offering their own solutions to cope with the problems derived from globalization.

Research in the field of traditional forest-related knowledge constitutes an important support for the conservation and even improvement of local and indigenous communities' management systems. So, for instance, the production and quality of almaciga resin in Palawan (Philippines) has been increased by scientific research, in turn positively affecting tappers' incomes (Ella 2008). Similarly, the adoption of new crops, together with valuable timber species and their management, have enhanced the annual revenues generated by the shifting cultivation system applied in Yunnan (China) (Liang et al. 2008). Traditional knowledge has also made a considerable difference in many research projects and management strategies. Nevertheless, its wider application still remains elusive (Huntington 2000).

Research with local and indigenous communities has brought a better understanding of subsistence economies, something substantial in order to design well-grounded decision-making and development projects (Godelier 1981). Regarding access to genetic resources, for example, cooperation between scientists and indigenous peoples has emphasized the deep differences that exist between the dominant Western paradigms (such as access and benefit sharing, intellectual property rights) and customary laws and collective custodianship underlying traditional societies. This is a necessary approach to inform the development of more appropriate policies and mechanisms for the protection of genetic resources and traditional knowledge (Swiderska 2007).

Furthermore, investigations carried out by Geneviève Michon and her colleagues (2007) provide evidence for the inherent sustainability of traditional management systems still applied in developing countries, which, in turn, might represent by itself a guarantee of resilience against external pressures. Traditional agroforests, for example, constitute a better contribution to the maintenance of the forest cover in the Philippines than the unsuccessful afforestations supported by the government during recent decades (Bertomeu et al. 2008). In spite of this, the term forest applied by the Food and Agriculture Organization (FAO) does not include agroforests. By contrast, plantations, such as rubber plantations, are considered by the FAO as forests (FAO 2006). Obviously, traditional forest-related knowledge still requires more recognition.

An important feature of traditional knowledge is its dynamic character and openness to change (Berkes 1999). This allows local communities to adapt actively to the new situations originated in a globalized world. In Indonesia, as mentioned above, these communities frequently opt for the conversion of their traditionally managed agroforests into commercial plantations of rubber or oil palms (Feintrenie and Levang 2008). However, these systems managed by farmers deeply differ from the plantations carried out by large companies:

> Farmers try integrating oil palm into their own cropping practices, especially agroforesty, applying less fertilisers and chemicals. They favour a less intensive cropping system, and associate oil palm with several other perennial crops. … By integrating oil palm into their existing farming systems, by adapting techniques to their constraints, objectives and knowledge, farmers make it their own and reinvent tradition (Feintrenie and Levang 2008).

This adaptable character of traditional systems would merit further research since scholars envisage future models as a mixture of monospecific plantations and agroforests (Feintrenie and Levang 2008).

Apart from adaptation, local communities have sometimes actively reacted, denying the unlimited inclusion of their resources in the market system. A well-known case is that of Chico Mendes and the 'seringueiros' (rubber tappers) in the Brazilian Amazon (Martínez Alier 1993), as mentioned earlier. Mendes organized fellow workers into the National Council of Rubber Tappers to protest the cutting of the trees by cattle ranchers. The union succeeded eventually in negotiating government support for the creation of 'extractive reserves,' which protect small areas of land for sustainable use such as rubber tapping.

Many other examples can be found through the developing world. In northern Peru at the beginning of the 1990s local famers supported by estate engineers opposed the firm INCAFOR, which intended to manage forests in a protected area (Martínez Alier 2002). Also by that time thousands of farmers in Thailand refused to abandon their villages and to comply with authorities who intended to implement the Khor Jor Kor, a state forestry programme backed by a coalition of army generals, the forest department, and pulp companies (Pye 2005). If implemented this programme would have caused eviction while promoting deforestation and eucalyptus plantation. But farmers organized themselves and resisted the project, and Khor Jor Kor was scrapped in July 1992 (Pye 2005). This success supported in turn the emergence of a network of rural activists within the organization Samatcha Khon Chon (Assembly of the Poor), which proposes a democratic form of forest and land management including integrating farming methods and community forests (Pye 2005). Moreover, regarding the issue of access to genetic resources, indigenous organizations in Chiapas (Mexico) successfully opposed to a U.S. bioprospecting project aimed at collecting plants used in traditional Mayan medicine as well as at patenting drugs and knowledge (Sharma 2005). Worldwide many other social initiatives exist that emphasize a 'moral economy' and react 'against the threats coming from the generalized market system against the livelihood of the poor' (Martínez Alier 1993); such initiatives have been labelled by Joan Martínez Alier as 'the environmentalism of the poor' (Martínez Alier 1993, 1999, 2002).

In wealthy countries, much research effort is being devoted to regional development in order to find measures that can counteract the marginalization processes in mountainous and other rural areas. In this regard, the results of projects such as ISDEMA (Innovative Structures for the Sustainable Development of Mountain Areas) or EUROLAN (Strengthening the Multifunctional Use of European Land: Coping with Marginalisation) highlight the importance of the following aspects (Dax and Hovorka 2004):

- institutional local development;
- emphasis on local initiatives and resources;
- reinforcement of local identity;
- a regional (rather than sectoral) approach;
- integration of small and medium enterprises supplying different but complementary products;
- a multi-sectoral approach, e.g. through the enlargement of the relationships among agriculture, forestry, trade, and tourism, including both market mechanisms and public measures; and
- long-term development projects.

Pettenella et al. (2008) describe the application of this framework to mushroom gathering in the forests of Borgotaro (Italy). Six enterprises deal with the commercialisation of mushrooms, and several others sell additional products gathered in coppiced beech, oak, and chestnut woodlands. A considerable number of people are engaged in different activities ranging from silvicultural interventions to hiking trail maintenance. And, particularly important for the scope of this chapter, in order to favour the production of mushrooms, woodlands are actively coppiced, which is a traditional form of woodland management. Finally, research studies such as those already mentioned have also accentuated the enormous importance of enhancing the broad participation of local communities and their initiatives in the framework of regional development (Dax and Hovorka 2004). This issue is of great relevance, too, in developing countries and to which we devote the final section.

12.5 Participation, a Requisite for Adaptation

In 2006, one of the authors of this chapter attended a seminar on biodiversity organized by the Chatham House in London. During the discussions, participants repeatedly mentioned projects carried out by the Global Environmental Facility (GEF) in developing countries. The author raised the issue that projects developed by international organizations such as the GEF and the World Bank sometimes have negative impacts on poor rural communities. The statement was supported by various papers (Bassett and Bi Zuéli 2000; Griffiths 2005). Participants, except for one, remained silent. The only reacting person argued that the GEF and the World Bank were the most important institutions funding such projects in developing countries, and excitedly rejected the critical statement. This reaction illustrates the dilemma of whether the good done by such organizations should prevent discussion of the resulting problems of very poor people. Such discussion, in fact, is important particularly for two reasons (among others): their consequences for local communities and the difficulties these communities encounter in trying to participate in the policy process.

The similitude of the misfortunes affecting rural communities in Southern Spain and India to which we referred in Box 12.1 relates also to their final development. The health of Máximo Fernández Cruz, the mountain shepherd dispossessed of his freedom, began declining while he was in prison. Affected by asphyxia and tremors he was taken to the jail ward. Later his progressive health deterioration continued until he passed away in 1986 (Gómez Mena 1987). And in India, in Nandurbar alone from April 2007 to March 2008 more than 40 children below 5 years of age died of malnutrition (Jamwal 2008a), while in Madhya Pradesh 46 children died in 2 months (Chibber 2008). Between 2003 and 2004, nearly 9,000 children under the age of 6 had died directly or indirectly of malnutrition in tribal areas of Maharashtra (Mahapatra 2004). All these communities are economically depressed, a situation exacerbated by the forest departments that do not allow them access to the forest for food gathering (Mahapatra 2004; Chibber 2008; Nidhi Jamwal, personal communication 2009). And as we showed in Sect. 12.3.3, such strict measures are inherited from colonial times. Thus, these children, as in the case for Máximo, were victims of, among other factors, restrictive forest policies disseminated worldwide in the process of globalization.

But globalization does not necessarily imply changes that always act in a uniform direction. Rather, mutually opposed tendencies are also possible (Giddens 1990). Globalization so far has caused many losers and it is obvious that a different world is inevitable (Stiglitz 2006). But the bidirectionality of the process also offers a chance for better changes. In this regard, reforming globalization is a question of politics (Stiglitz 2006), which implies taking into account a wide range of ideas within the framework of collective decision-making (Latour 2004, cited in Trosper 2007)—that is, participation.

There are, indeed, global tendencies supporting positive changes in the South, such as the shift in forest tenure since the mid 1980s embracing some 200 million ha of forest land legally transferred to local communities and indigenous people (Larson et al. 2010). This trend matches the increasing recognition that reinforcing forest tenure for local communities leads to better welfare of people, supports them against outside claimants, and improves forest resource conservation and management (Pulhin et al. 2010). Nonetheless, forest policies continue to be severe and coercive, counteracting recent forest tenure reforms (Larson and Ribot 2007; Pulhin et al. 2010), while effective participation, despite the widespread discourse, remains very limited (Larson and Ribot 2007; Cotula et al. 2009).

Correspondingly, promoting discussion on the topics addressed by this chapter is an urgent necessity. However, like the participants at the workshop in London mentioned earlier, probably many delegates attending international environmental conventions are unaware of these issues. Indeed many scientists, as shown by Berkes (1999), still have a dismissive attitude to traditional knowledge. Environmental policy processes certainly should be broadened to allow local communities and indigenous peoples the ability to represent their own interests and needs in the decision-making process (Swiderska 2007). Furthermore, such inclusion would help reverse the main North–South direction of globalization and would allow enriching Western forest management science with the experience from other cultures. 'Indigenous knowledge,' insists Berkes (1999), 'holds much promise for insights and applications, provided care is taken not to use it out of context.'

A last example in this chapter illustrates this point. Forests in the Menominee Indian Reservation (Wisconsin) are managed by Menominee tribes. They use harvest rates longer than 200 years, which enables large growing stocks and many non-timber products, application of controlled fires, and adaptation of the mill plan to forest management considerations. An important fact here is that thanks to their self-determination and political power, Menominee people have been able to express their values and beliefs in the management of the forests (Trosper 2007).

This demand for participation is made by non-governmental organizations as well as by scientists, such as David Smorffitt (2008) of James Cook University in Cairns, with whose quotation we conclude:

> There should be no doubt whatsoever that national and international policy development needs to undertake a greater level of consultation with local communities whose intimate knowledge of the area and communities in which they live, can play an important role in 'getting in right' the first time around. The personal and commercial interest of stakeholders and the political bias at various levels ensures this is a long-term process but is required in order to develop policies that are the best for all concerned.

References

Alden Wily L (2010) Whose land are you giving away, Mr. President? Paper presented at the annual bank conference on land policy and administration, World Bank Headquarters, Washington, DC, 26–27 Apr 2010

Alexander KNA (1998) The links between forest history and biodiversity: the invertebrate fauna of ancient pasture-woodlands in Britain and its conservation. In: Kirby KJ, Watkins C (eds) The ecological history of European forests. CAB International, Wallingford, pp 73–80

Angelstam P, Elbakidze M (2006) Sustainable forest management in Europe's East and West: trajectories of development and the role of traditional knowledge. In: Parrotta J, Agnoletti M, Johann E (eds) Cultural heritage and sustainable forest management: the role of traditional knowledge: proceedings of the IUFRO conference held in Florence, Italy, 8–11 June 2006, vol 2. MCPFE Liaison Unit, Warsaw, pp 349–357

Astill J (2010) Seeing the wood. A special report on forests. The Economist, 23 Sept 2010. Available via. http://www.economist.com/node/17062713. Cited 20 Mar 2011

Aubert P-M, Maya L, Lurent A (2008) Moroccan forestry policies and local forestry management in the High Atlas: a cross analysis of forestry administration and local institutions. In: Buttoud G (ed) Small-scale rural forest use and management: global policies versus local knowledge. AgroParisTech-ENGREF, Nancy, pp 19–26

Bailey R (2007) Bio-fuelling poverty: why the EU renewable-fuel target may be disastrous for poor people. Oxfam briefing note. Oxfam International, Oxford

Banuri T, Apffel Marglin F (eds) (1993) Who will save the forests? United Nation University/Zed Books, London

Bassett TJ, Bi ZK (2000) Environmental discourses and the Ivorian savannah. Ann Assoc Am Geogr 90(1):67–95

Baudrillard J (1980) L'Echange symbolique et la mort. Gallimard, Paris

Berkes F (ed) (1989) Common property resources: ecology and community-based sustainable development. Belhaven, London

Berkes F (1999) Sacred ecology: traditional ecological knowledge and resource management. Taylor & Francis, Philadelphia

Bertomeu M, Roshetko JM, Fay C (2008) Domesticating landscapes through farmer-led tree cultivation: an agroforestation approach to reforestation in Southern Asia. In: Buttoud G (ed) Small-scale rural forest use and management: global policies versus local knowledge. AgroParisTech-ENGREF, Nancy, pp 27–33

Brieger T, Sauer A. 2000. Narmada Valley: planting trees, uprooting people. Economic and Political Weekly, Oct 20:3795–3797.

Brockington D, Igoe J (2006) Eviction for conservation—a global overview. Conserv Soc 4(3):424–470

Burschel P, Huss J (1997) Grundriß des Waldbaus. Pareys Studientexte, Berlin

Casals CV (1996) Los ingenieros de montes en la España contemporánea: 1848–1936. Ediciones del Serbal, Barcelona

Casas VM (1991) Árboles viejos, aves y hombres. Quercus 67:40–41

Cernea MM, Schmidt-Soltau K (2006) Poverty risks and national parks: policy issues in conservation and resettlement. World Dev 34(10):1808–1830

Chaudhary P, Aryal KP, Bawa KS (2008) Poverty and biodiversity loss: rhetoric and reality. In: Parrotta JA, Liu J, Sim H-C (eds) Sustainable forest management and poverty alleviation: roles of traditional forest-related knowledge, vol 21, IUFRO world series. International Union of Forest Research Organizations (IUFRO), Vienna, pp 23–26

Chibber N (2008) 24 deaths in 2 months. Down to Earth: August 31, 2008. Available via. http://downtoearth.org.in/node/4943. Cited 20 Mar 2011

Ciancio O, Nocentini S (2000) Forest management from positivism to the culture of complexity. In: Agnoletti M, Anderson S (eds) Methods and approaches in forest history. CAB International, New York, pp 47–58

Cobo Romero F, Cruz S, González M (1992) Privatización del monte y protesta campesina en Andalucía Oriental (1836–1920). Agric Soc 65:253–302

Cotula L, Vermeulen S, Leonard R, Keeley J (2009) Land grab or development opportunity? Agricultural investment and international land deals in Africa. IIED/FAO/IFAD, London

Dargavel J (2000) In the wood of neglect. In: Agnoletti M, Anderson S (eds) Forest history: international studies on socio-economic and forest ecosystem change. CAB International, New York, pp 263–277

Davis D (2005) Potential forests: degradation narratives, science, and environmental policy in protectorate Morocco, 1912–1956. Environ Hist 10(2):211–238

Dax T, Hovorka G (2004) Innovative regionalentwicklung im berggebiet: institutionelle grundbedingungen für wirksame regionale strategien. Ländlicher Raum 3. Available via. http://www.laendlicher-raum.at/article/archive/15031. Cited 20 Mar 2011

De la Cruz AE (1994) La destrucción de los montes: claves histórico-jurídicas. Universidad Complutense de Madrid, Madrid

Diatchkova G (2001) Indigenous peoples of Russia and political history. Can J Native Studies XXI(2):217–233

Ella A (2008) Almaciga resin gathering by indigenous people of Palawan province in the Philippines. In: Parrotta JA, Liu J, Sim H-C (eds) Sustainable forest management and poverty alleviation: roles of traditional forest-related knowledge, vol 21, IUFRO world series. International Union of Forest Research Organizations (IUFRO), Vienna, pp 46–48

Ernst C (1998) An ecological revolution? The 'schlagwaldwirtschaft' in western Germany in the eighteenth and nineteenth centuries. In: Watkins C (ed) European woods and forests: studies in cultural history. CAB International, New York, pp 83–92

Feintrenie L, Levang P (2008) Is traditional knowledge a guarantee for good practices in forest management? The case of rubber agroforests in Jambi, Indonesia. In: Buttoud G (ed) Small-scale rural forest use and management: global policies versus local knowledge. AgroParisTech-ENGREF, Nancy, pp 69–77

Fontana J (1999) Historia: análisis del pasado y proyecto social. Crítica, Barcelona

Food and Agriculture Organisation (FAO) (2006) Global forest resources assessment 2005. FAO, Rome

Fox JM (2000) How blaming 'slash and burn' farmers is deforesting mainland Southeast Asia. East-West Centre, Honolulu

Frein M, Meyer H (2008) Die biopiraten: milliardengeschäfte der pharmaindustrie mit dem bauplan der natur. Econ, Berlin

Friedman TL (2005) The world is flat: the globalized world in the twenty-first century. Penguin, London

García Latorre J, García Latorre J (2007) Almería: hecha a mano—una historia ecológica. Cajamar, Almería

Ghimire K, Bastakoti R (2008) Traditional knowledge of Tharu of Nepal on medicinal plants and healthcare systems. In: Parrotta JA, Liu J, Sim H-C (eds) Sustainable forest management and poverty alleviation: roles of traditional forest-related knowledge, IUFRO world series. International Union of Forest Research Organizations (IUFRO), Vienna, pp 61–63

Giddens A (1990) The consequences of modernity. Blackwell, Oxford

Global Witness (2009) Trick or treat? REDD, development and sustainable forest management. Global Witness, London

Godelier M (1981) Instituciones económicas. Editorial Anagrama, Barcelona

Gómez Mena J (1987) Vida y muerte en Cazorla. Quercus 24:42–43

González BF (1989) Influencia humana en los ecosistemas forestales. Quercus 37:34–38

González de Molina M (1993) Historia y medio ambiente. Eudema, Madrid

Grande del Brío R, Hernando Ayala A, Piñeiro MJ (2002) El oso pardo en el noroeste peninsular. Amarú Ediciones, Salamanca

Greenpeace (2010) Papua New Guinea: not ready for (REDD). Greepeace, Sydney

Gribbin J (2002) Science: a history, 1534–2001. Allen Lane, London

Griffiths T (2005) Indigenous peoples and the Global Environment Facility (GEF)—indigenous peoples' experiences of GEF-funded biodiversity conservation: a critical study. Forest People Programme, Moreton-in-Marsh

Griffiths T (2007) Seeing 'RED'? 'Avoided deforestation' and the rights of indigenous peoples and local communities. Forest People Programme, Moreton-in-Marsh

Grove AT, Rackham O (2001) The nature of Mediterranean Europe: an ecological history. Yale University Press, New Haven

Harvey D (2003) The new imperialism. Oxford University Press, New York

Harvey D (2005) A brief history of neoliberalism. Oxford University Press, New York

Hayden B (1993) Archaeology: the science of once and future things. W.H. Freeman, New York

Heidegger M (1971) The Thing. In: Poetry, language, thought. Harper & Row, New York

Hendarti L, Youn YC (2008) Traditional knowledge on forest resource management by Kasepuhan people in the upland area of west Java. In: Parrotta JA, Liu J, Sim H-C (eds) Sustainable forest management and poverty alleviation: roles of traditional forest-related knowledge, vol 21, IUFRO world series. International Union of Forest Research Organizations (IUFRO), Vienna, pp 73–76

Hobsbawm EJ (1962) The age of revolution: Europe 1789–1848. Weidenfeld and Nicolson, London

Humphreys D (2006) Public goods, neoliberalism and the crisis of deforestation. Paper presented to the annual conference of the British International Studies Association, University College Cork, Cork, 18–20 Dec 2006

Huntington H (2000) Using traditional ecological knowledge in science: methods and applications. Ecol Appl 10(5):1270–1274

Israel ED, King O (2008) Strategies for conservation of sacred forests of Kolli Hills, Tamil Nadu, India: a study on botany, ecology and community interactions. In: Parrotta JA, Liu J, Sim H-C (eds) Sustainable forest management and poverty alleviation: roles of traditional forest-related knowledge, UFRO world series. International Union of Forest Research Organizations, Vienna, pp 133–136

Jamwal N (2008a) Consumed by hunger. Down to Earth Aug 31, 2008. Available via. http://downtoearth.org.in/node/4964. Cited 20 Mar 2011

Jamwal N (2008b) Split approach. Down to Earth, July 15, 2008. http://downtoearth.org.in/node/4739. Cited 20 Mar 2011

Jansen P, van Bethem M (2005) Historische boselementen: geschiedenis, herkenning en beheer. Probos, Geldersch Landschap, Geldersche Kasteelen, Wageningen

Kapp S (2008) Gabon (Parc National de la Lope): local production systems and conservation polices: an impossible dialogue? In: Buttoud G (ed) Small-scale rural forest use and management: global policies versus local knowledge. AgroParisTech-ENGREF, Nancy, pp 256–257

Kirby KJ, Watkins C (eds) (1998) The ecological history of European forests. CAB International, Wallingford

Lambin EF, Turner BL, Geist HJ et al (2001) The causes of land-use and land-cover change: moving beyond the myths. Glob Environ Chang 11:261–269

Larson AM, Ribot JC (2007) The poverty of forestry policy: double standards on an uneven playing field. Sustain Sci 2(2):189–204

Larson AM, Barry D, Ram Dahal G (2010) Tenure change in the global south. In: Larson AM, Barry D, Ram Dahal G, Pierce Colfer CJ (eds) Forests for people: community rights and forest tenure reform. Earthscan, London, pp 3–18

Lasanta Martínez T, Ruiz Flaño P (1990) Especialización productiva y desarticulación espacial en la gestión reciente del territorio en las montañas de Europa occidental. In: García Ruiz JM (ed) Geoecología de las áreas de montaña. Geoforma Ediciones, Logroño, pp 267–295

Latour B (2004) Politics of nature: how to bring the sciences into democracy. Harvard University Press, Cambridge, MA

Lenz RJM (1994) Consequences of a 150 years' history of human disturbances in some forest ecosystems. In: Boyle TJB, Boyle CEB (eds) Biodiversity, temperate ecosystems, and global change. Springer, Berlin, pp 265–289

Lepart J, Debussche M (1992) Human impact on landscape patterning: Mediterranean examples. In: Hansen AJ, de Castri F (eds) Landscape boundaries: consequences for biotic diversity and ecological flows. Springer, New York, pp 76–105

Levin Institute (2011) What is globalization? In: Globalization 101. The Levin Institute, State University of New York. http://www.globalization101.org/What_is_Globalization.html. Cited 4 Mar 2011

Liang L, Shen L, Xinkai Y (2008) Building traditional shifting cultivation for rotational agroforestry: experiences from Yunnan, China. In: Parrotta JA, Liu J, Sim H-C (eds) Sustainable forest management and poverty alleviation: roles of traditional forest-related knowledge, vol 21, IUFRO world series. International Union of Forest Research Organizations (IUFRO), Vienna, pp 98–100

Lowood HE (1990) The calculating forester: quantification, cameral science, and the emergence of scientific forestry management in Germany. In: Frängsmyr T, Heilbron JL, Rider RE (eds) The quantifying spirit in the 18th century. University of California Press, Berkeley, pp 315–342

Mahapatra R (2004) Ecological poverty: the new serial killer in India. 06 Oct 2004, Available via. http://sspconline.org/article_details.asp?artid=art12. Cited 1 Nov 2010

Martínez AJ (1993) Distributional obstacles to international environmental policy: the failures at Rio and prospects after Rio. Environ Value 2:97–124

Martínez AJ (1999) Introducción a la economía ecológica. Rubes, Barcelona

Martínez Alier J (2002) The environmentalism of the poor: a study of ecological conflicts and valuation. Elgar, Cheltenham

Marx K (1842) Debatten über das Holzdiebstahlsgesetz. Rheinische Zeitung 298, 300, 303, 305, 307. Available via. http://www.mlwerke.de/me/me01/me01_109.htm. Cited 20 Aug 2011

Marx K, Engels F (1848) Manifesto of the Communist Party. Marxists Internet Archive. Available via. http://www.marxists.org/. Marx/Engels Internet Archive (marxists.org) 1987, 2000. Cited 20 Mar 2011

McNeill J (2001) Something new under the sun: an environmental history of the twentieth-century world. Norton, New York

Méndez M, Micó E, Nieto A (2010) Tesoros vivos en la madera muerta. Quercus 297:82

Mesa S, Delgado A (1995) Los cornicabrales y la cultura pastoril. Quercus 112:8–10

Michon G, de Foresta H, Levang P, Verdeaux F (2007) Domestic forests: a new paradigm for integrating local communities' forestry into tropical forest science. Ecol Soc 12 (2):1. Available via. http://www.ecologyandsociety.org/vol12/iss2/art1/. Cited 20 Mar 2011

Misra SS (2008a) State inaction. Down to Earth July 15 2008. http://www.downtoearth.org.in/node/4733. Cited 20 Mar 2011

Misra SS (2008b) Out of form. Down to Earth July 15, 2008. Available via. http://www.downtoearth.org.in/node/4736. Cited 20 Mar 2011

Moro J (2006) Senderos de libertad. La lucha por la defensa de la selva. Seix Barral, Barcelona

Mudge SL (2008) What is neo-liberalism? Socio-Econ Rev 6:703–731

Nanjundaiah C (2008) The economic value of indigenous environmental knowledge in ensuring sustainable livelihood needs and protecting local ecological services: a case study of Nagarhole National Park, India. In: Parrotta JA, Liu J, Sim H-C (eds) Sustainable forest management and poverty alleviation: roles of traditional forest-related knowledge, vol 21, IUFRO world series. International Union of Forest Research Organizations (IUFRO), Vienna, pp 124–126

Naredo JM (2004) La evolución de la agricultura en España (1940–2000). Universidad de Granada, Granada

Nicolas HJ (2002) 6000 years of tree pollarding and leaf-hay foddering of livestock in the Alpine area. Centralblatt für das Gesamte Forstwesen 119(3/4):231–240

O'Connor M (1993) On the misadventures of capitalist nature. Capital Nat Social 4(3):7–40

Ostrom E (1990) Governing the commons: the evolution of institutions for collective action. Cambridge University Press, Cambridge

Pagiola S, Landell-Mills N, Bishop J (2002) Market-based mechanisms for forest conservation and development. In: Pagiola S, Bishop J, Landell-Mills N (eds) Selling forest environmental services: market-based mechanisms for conservation and development. Earthscan, London, pp 1–13

Pallavi A (2008) Against the grain. Market-dependent urban food practices are swallowing up indigenous food habits. Down to Earth, October 31, 2008. Available via. http://indiaenvironmentportal.org.in/node/266958. Cited 20 Mar 2011

Pelizaeus R (2008) Der kolonialismus: geschichte der Europäischen expansion. Marixverlag, Wiesbaden

Persson M, Azar C (2010) Preserving the world's tropical forests. A price on carbon may not do. Environ Sci Technol 44:210–215

Peterken G (1981) Woodland conservation and management. Chapman & Hall, London

Pettenella D, Maso D, Secco L (2008) The "Net-system model" in NWFP marketing: the case of mushrooms. In: Buttoud G (ed) Small-scale rural forest use and management: global policies versus local knowledge. AgroParisTech-ENGREF, Nancy, pp 167–175

Porter G, Borwn JW (1996) Global environmental politics. Westview Press, Oxford

Pott R, Hüppe J (1991) Die Hudelandschaften Nordwestdeutschlands. Westfälisches Museum für Naturkunde, Münster

Pretty J (1990) Sustainable agriculture in the Middle Ages: the English manor. Agric Hist Rev 38(1):1–19

Pulhin JM, Larson AM, Pacheco P (2010) Regulations as barriers to community benefits in tenure reform. In: Larson AM, Barry D, Ram Dahal G, Pierce Colfer CJ (eds) Forests for people: community rights and forest tenure reform. Earthscan, London, pp 139–159

Pye O (2005) Khor Jor Kor: forest politics in Thailand. White Lotus, Bangkok

Rackham O (1978) Trees and woodland in the British landscape. J.M. Dent, London

Rackham O (1980) Ancient woodland: its history, vegetation and uses in England. Edward Arnold, London

Ribeiro S (2005) The traps of "benefit sharing". In: Burrows B (ed) The catch: perspectives in benefit sharing. The Edmonds Institute, Edmonds, pp 37–80

Robbins P (2004) Political ecology: a critical introduction. Blackwell, Oxford

Röhring E, Gussone HA (1990) Waldbau auf ökologischer Grundlage. Paul Parey, Hamburg

Rösener W (1991) Bauern im mittelalter. C.H. Beck, München

Ruiz-Pérez M, [and 29 others] (2004) Markets drive the specialization strategies of forest peoples. Ecol Soc 9(2):4. Available via. http://www.ecologyandsociety.org/vol9/iss2/art4. Cited 20 Mar 2011

Scheuermann W (2006) Globalization. In: Zalta EN (ed) The Stanford encyclopedia of philosophy. Available via. http://plato.stanford.edu/archives/fall2008/entries/globalization/. Cited 20 Mar 2011

Scott JC (1998) Seeing like a state: how certain schemes to improve the human condition have failed. Yale University Press, New Haven

Secretariat of the Convention of Biological Diversity (2005) Handbook of the convention on biological diversity including its Cartagena protocol on biosafety, 3rd edn. Secretariat of the Convention of Biological Diversity, Montreal

Shand H (1997) Human nature: agricultural biodiversity and farm-based food security. Food and Agriculture Organisation (FAO), Rome

Sharma D (2005) Selling biodiversity: benefit sharing is a dead concept. In: Burrows B (ed) The catch: perspectives in benefit sharing. The Edmonds Institute, Edmonds, pp 1–13

Shiva V (2005) The traps of benefit sharing. In: Burrows B (ed) The catch: perspectives in benefit sharing. The Edmonds Institute, Edmonds, pp 37–80

Smorfitt D (2008) The role of government policy in facilitating investment in forestry: local and national issues. In: Buttoud G (ed) Small-scale rural forest use and management: global policies versus local knowledge. AgroParisTech-ENGREF, Nancy, pp 205–213

Steni B (2010) Indonesia. Better governance in the forestry sector, or business as usual? In: Fenton E (ed) Realising rights, protecting forests: an alternative vision for reducing deforestation. Accra Caucus, pp 11–12. Available via. http://www.careclimatechange.org/files/reports/Accra_Report_English.pdf. Cited 22 Mar 2011

Stevenson AC, Harrison RJ (1992) Ancient forest in Spain: a model for land-use and dry forest management in South-west Spain from 4000 BC to 1900 AD. Proc Prehist Soc 58:227–247

Stiglitz J (2006) Making globalization work. Norton, New York

Stiven R, Holl K (2004) Wood pasture. Scottish Natural Heritage, Redgorton

Swaminathan MS (2000) Government–industry–civil society partnerships in integrated gene management. Curr Sci 78(5):555–562

Swiderska K (2007) Protecting traditional knowledge: a framework based on customary laws and bio-cultural heritage. In: Haverkort B, Rist S (eds) Endogenous development and bio-cultural diversity: the interplay of worldviews, globalization and locality, vol 6, Compas series on worldviews and sciences. COMPAS, Centre for Development and Environment, Leusden, pp 358–365

Tacconi L (2000) Biodiversity and ecological economics: participation, values and resource management. Earthscan, London

Tainter JA (2007) Scale and dependency in world-systems: local societies in convergent evolution. In: Hornborg A, McNeill J, Martínez Alier J (eds) Rethinking environmental history: world-system history and global environmental change. Altamira Press, Lanham, pp 361–377

Toledo VM, Ortiz-Espejel L, Moguel CP, Ordoñez MDJ (2003) The multiple use of tropical forests by indigenous peoples in Mexico: a case of adaptive management. Conserv Ecol 7(3):9 (online) Available via. http://www.ecologyandsociety.org/vol7/iss3/art9/inline.html. Cited 20 Mar 2011

Trosper RL (2007) Indigenous influence on forest management on the Menominee Indian Reservation. For Ecol Manag 249:134–139

van Vliet N, Sunderland T, Slayback D, Assembe S, Asaha S (2008) Deforestation for cocoa plantations to secure land tenure and income around a proposed national park in south west Cameroon: how to reduce leakages in conservation policies? In: Buttoud G (ed) Small-scale rural forest use and management: global policies versus local knowledge. AgroParisTech-ENGREF, Nancy, pp 231–237

Varillas B (1987) El cojo de la fresnedilla. Quercus 24:43

Velvé R (1992) Saving the seed. Genetic diversity and European agriculture. Earthscan, London

Vidal J (2009, Oct 5). UN's forest protection scheme at risk from organized crime, experts warn. Available via. http://www.guardian.co.uk/environment/2009/oct/05/un-forest-protection. Cited 20 Mar 2011

Wallerstein I (1974) The modern world system. Academic, New York

Warde P (2006) Fear of wood shortage and the reality of the woodland in Europe, c.1450–1850. Hist Work J 62:28–57

Williams M (2000) Putting 'flesh on the carbon-based bones' of forest history. In: Agnoletti M, Anderson S (eds) Methods and approaches in forest history. CAB International, New York, pp 35–46

Wolf E (1982) Europe and the people without history. University of California Press, Berkeley

World Bank (2010) Rising global interest in farmland: can it yield sustainable and equitable benefits? World Bank, Washington, DC

Worster D (1988) Doing environmental history. In: Worster D (ed) The ends of the earth: perspectives on modern environmental history. Cambridge University Press, Cambridge, pp 289–307

XLSemanal (2010) Colombia: la guerra del CO2. XLSemanal 1163:7–13. Available via. http://xlsemanal.finanzas.com/web/articulo.php?id=48218&id_edicion=4927. Cited 20 Mar 2011

Chapter 13
Traditional Forest-Related Knowledge and Climate Change

John A. Parrotta and Mauro Agnoletti

Abstract The holders and users of traditional forest-related knowledge are on the front lines of global efforts to deal with climate change and its impacts. Because of their close connection with, and high dependence on, forest ecosystems and landscapes, indigenous and local communities are among the first to witness, understand, and experience the impacts of climate change on forests and woodlands as well as on their livelihoods and cultures. The history of forest and agricultural landscape management practices of indigenous and local communities based on their traditional knowledge offer insights into principles and approaches that may be effective in coping with, and adapting to, climate change in the years ahead. Global, regional, national and local efforts to mitigate and adapt to climate change, however, have not yet given adequate attention either to the forest-related knowledge and practices of traditional communities, or to the interests, needs and rights of local and indigenous communities in the formulation of policies and programmes to combat climate change. Due consideration of, and a more prominent role for, traditional forest-related knowledge and its practitioners could lead to the development of more effective and equitable approaches for facing the challenges posed by climate change while enhancing prospects for sustainable management of forest resources.

Keywords Adaptation • Agroforestry • Biofuels • Carbon markets • Climate change • Environmental policy • Forest management • Traditional communities • Mitigation • Traditional agriculture • Traditional knowledge

J.A. Parrotta (✉)
U.S. Forest Service, Research and Development, Arlington, VA, USA
e-mail: jparrotta@fs.fed.us

M. Agnoletti
Dipartimento di Scienze e Teconolgie Ambientali Forestali, Facoltà di Agraria,
Università di Firenze, Florence, Italy
e-mail: mauro.agnoletti@unifi.it

J.A. Parrotta and R.L. Trosper (eds.), *Traditional Forest-Related Knowledge:*
Sustaining Communities, Ecosystems and Biocultural Diversity,
World Forests 12, DOI 10.1007/978-94-007-2144-9_13,
© Springer Science+Business Media B.V. (outside the USA) 2012

13.1 Introduction

Traditional forest-related knowledge and associated management practices have sustained local and indigenous communities throughout the world under changing environmental, social and economic conditions, long before the advent of formal forest science and 'scientific' forest management. Through long experience of inter-annual and longer-term variability in climate, many such communities have developed significant bodies of knowledge, transmitted through the generations, on how to cope with local climatic shifts and impacts of extreme weather events and other natural disasters. This includes knowledge related to weather prediction, of wild plant and animal species and their management, and agricultural techniques for managing and conserving water and soil resources.

Traditional knowledge and management practices have historically enabled these communities to adjust their agricultural calendars, crop selection and management practices, and to cope with changing abundances of preferred forest and woodland species used for food, medicine and other purposes. Management practices based on traditional knowledge have been applied over extensive closed forest areas, small woodlands, agroforestry and shifting cultivation systems, and even to the management of single trees occurring in agricultural fields and pastures.

In this chapter, we discuss how traditional forest-related knowledge and practices have helped forest-dependent and other rural communities anticipate and adapt to past and present impacts of climate change on forests and forested landscapes. We examine how this knowledge, through its application in a variety of management practices, can contribute to climate change mitigation and adaptation—by providing models for increasing resilience of forest and related agricultural resource management. These models reduce local dependency on fossil fuels and maintain or enhance terrestrial carbon stocks, while contributing to livelihoods, food security, and maintenance of ecological, social and cultural values associated with forested landscapes in the face of environmental uncertainty. Finally, we explore opportunities and risks for traditional forest-related knowledge and the holders and users of this knowledge (local and indigenous communities) associated with ongoing international climate and forest policy debates and emerging climate change mitigation measures. Examples include the expansion of biofuels production, carbon markets, and other mechanisms to enhance terrestrial carbon stocks and reduce greenhouse gas emissions from forested landscapes.

13.2 Current and Projected Impacts of Climate Change on Forest Ecosystems and Forest-Dependent Communities

The historical and projected effects of climate change on forests are influenced by ecosystem-specific factors that include human activities, natural processes, and different dimensions of climate, i.e., temperature, rainfall patterns, and wind. Climate change over the past half century has already affected forest ecosystems in a number of ways, including changing patterns of tree growth and mortality, outbreaks of damaging insects and diseases, and shifts in species distributions and seasonality of ecosystem processes. Because of the complexity of both natural and human systems, and their interactions, scientific quantification of ecological and socio-economic impacts and prediction of future impacts is fraught with difficulties and uncertainties (Lucier et al. 2009).

Nonetheless, under most scenarios outlined in the Fourth Assessment Report of the Intergovernmental Panel on Climate Change (IPCC4: Parry et al. 2007a), climate change is projected to change forest productivity (increasing in some regions, decreasing in others) and the distribution of forest types and their species composition in all biomes. Under most IPCC4 scenarios, boreal forests will be particularly affected by climate change and are expected to shift towards the poles, temperate forests are expected to be less effected than other forest types, and the effects on subtropical and tropical forests are expected to be highly variable depending on location; forest decline is considered likely in many drier tropical forest regions, and biodiversity declines are expected in wetter tropical forest regions (Seppälä et al. 2009). In regions where temperature and rainfall patterns become more favourable for agriculture and forest management purposes, traditional communities engaged in these activities could realize benefits from climate change. Many others will not, particularly where climate change effects on local ecosystems adversely affect traditional livelihoods.

The future IPPC4 scenarios suggest a number of broad regional trends that are likely to affect forests, and therefore forest-dependent communities:

- In Africa, mean annual temperature increases are likely to be larger than global increases throughout the continent and in all seasons, with drier subtropical regions warming more than moister tropical (equatorial) regions. Projections suggest decreasing rainfall in much of the African Mediterranean, northern Sahara, and southern Africa, but increasing rainfall in East Africa.
- In Asia, warming is likely to be above the global mean in central Asia, the Tibetan plateau, eastern Asia and South Asia. Winter precipitation is likely to increase in northern Asia, the Tibetan plateau, eastern Asia and in southern areas of Southeast Asia, while summer precipitation is likely to increase in northern Asia, East Asia, South Asia and most of Southeast Asia, but decrease in central Asia. Extreme rainfall events and severe winds associated with tropical cyclones are likely to increase in East Asia, Southeast Asia and South Asia.
- In Latin America, annual precipitation is likely to decrease in most of Central America and in the southern Andes, and summer rainfall is likely to decrease in the Caribbean and increase in southeastern South America.
- In the boreal regions of North America, Scandinavia and Russia, which are currently experiencing rapid and severe climate change, future temperatures are likely to rise faster than the global mean, particularly during the winter months.
- Elsewhere in North America, mean annual warming is predicted to be greater than the global mean, with notably higher winter temperatures in northern

North America and higher summer temperatures in the southwest. Mean annual precipitation is likely to increase in Canada and northeastern United States, and decrease in the Southwest, with significant changes in the seasonal distribution of precipitation throughout most of Canada and the United States.

• In the Pacific region, all North and South Pacific islands, as well as Australia and New Zealand, are considered very likely to warm. Annual precipitation is likely to increase in equatorial Pacific regions and in the west of New Zealand's South Island, and to decrease in southern Australia in winter and spring with increased risk of drought.

Indigenous and local communities in many parts of the world are already feeling the effects of current climate change on their lands in ways that jeopardize their cultures, livelihoods, food security and, in some cases, their very survival (UNPFII 2007; Macchi et al. 2008; Nilsson 2008; Collings 2009; Osman-Elasha et al. 2009). Despite our limited ability to accurately quantify and predict the magnitude of future climate change impacts of forests in specific localities, the expected or potential impacts of climate change on forest ecosystem structure and function are likely to have far-reaching consequences for the well-being of people in affected areas. At risk are such goods and services such as: timber, fuelwood and essential non-wood forest products, sufficient clean water for consumptive use, and other services required to meet people's basic nutritional, health and cultural needs (Easterling et al. 2007; Colfer et al. 2006; Osman-Elasha et al. 2009).

The observations and experiences to date of indigenous communities throughout the world have been compiled in an overview of more than 400 research studies, case studies and projects published by the United Nations University's Institute of Advanced Studies' Traditional Knowledge Initiative (Galloway-McLean 2010). Some of the major trends and examples are summarized below.

In drought-prone regions of Africa, increasing desertification threatens food security for communities who rely on rain-fed agriculture, as well as for pastoralists who struggle to sustain the fodder needs of their cattle and goats (Simel 2008). Shrinking water resources in such regions has direct implications for human health. Limited options for migration and mobility, degradation of forest and woodland resources, and increasing conflicts over natural resources limit the potential for adaptation within these communities. In tropical Asia, changing weather patterns, including increasingly severe typhoons and cyclones, are also affecting food and water security and increasing vulnerability to water- and vector-borne diseases. Indigenous peoples in some parts of Asia also report increased hardship that are the result of climate change mitigation projects that result in deforestation or loss of access to traditionally used forest resources. These projects include: biofuel planta-tion expansion; building of dams under the Clean Development Mechanism (estab-lished under the Kyoto Protocol); uranium mining; and inclusion of tropical forests under schemes to reduce emissions from deforestation and degradation. In the Andes, alpine warming and the shrinking of glaciers threatens water supplies, and thus food and health security, by causing shifts in populations of wild plants used for food, traditional medicine, as well as resources needed to sustain grazing animal populations and animals hunted for food.

In the Arctic boreal forest regions of North America and Eurasia, vegetation cover is changing, as are the populations and distributions of many fish, game and wild plant species that people in local and indigenous communities have traditionally relied upon for their food security needs and maintenance of their social and cultural identities. Changes in reindeer migration and foraging patterns resulting from shorter winters and increasingly unpredictable weather patterns are a particular problem for the estimated 100,000 or more reindeer herders of over 20 indigenous communities who manage an estimated 2.5 million semi-domesticated reindeer over four million km^2 of land in Alaska, Canada, Greenland, Norway, Sweden, Finland, Russia and Mongolia. Significant changes are also occurring in animal and fish populations, and in the behaviour and migration of bird, mammal and fish species that are important to indigenous (and other) communities in temperate regions of North America.

In low-lying coastal regions worldwide, but notably in many small islands in the Pacific, sea level rise caused by melting glacier ice, expansion of warmer seawater, and melting of sea ice in the Arctic and Antarctic is leading to accelerated erosion of coastal zones, increasing exposure to natural disasters such as floods and extreme weather events, and salt water intrusion into groundwater, with direct negative consequences for food security (UNPFII 2007). In the Pacific, sea level rise is already forcing migration of local and indigenous communities from low-lying islands and relocation to other communities' traditional territories, with a number of adverse social, spiritual, cultural and economic implications for affected communities (Collings 2009)

The current and potential future impacts of climate change on traditional communities who depend on forest ecosystems and landscapes are therefore cause for deep concern. Despite having contributed least to anthropogenic climate change due to their 'low carbon' lifestyles, traditional communities are disproportionately vulnerable to the effects of climate change since they generally live in ecosystems that are particularly prone to climate change impacts—polar and montane regions, small islands, tropical and subtropical forests, and deserts (Crump 2008; Macchi et al. 2008; Nilsson 2008; Rai 2008; Smallacombe 2008). Further, they are usually heavily dependent on low-input (often rain-fed) agriculture to meet their food security needs and on forest resources for other basic needs including wild foods, medicines, fuel and other non-timber forest products (Galloway-McLean 2010)

The particular vulnerabilities of many such communities to these impacts are due not only to their geographical location and greater reliance on local ecosystems to meet their livelihood and food security needs, but also to low incomes and limited institutional capacities. In many parts of Africa, for example, where communities have developed local adaptation strategies for coping with extreme weather events such as prolonged droughts, the impact of decreasing rainfall on water availability, agricultural production and human health add to existing strains on the coping capacity of communities in affected regions (Simel 2008; Galloway-McLean 2010).

13.3 The Value of Traditional Forest-Related Knowledge for Anticipating Climate Change and Its Ecological Effects

People in many indigenous and local communities are well known for their knowledge of their natural environment, and often possess keen insights into meteorological phenomena, animal behaviour, tree phenology, and changes in ecosystem structure and dynamics. Generations of experience, particularly within indigenous communities that have been able to maintain their cultural identities and traditional lifestyles, retain rich bases of knowledge that include historical climatic data (on temperature and rainfall patterns, frequency of climatic events, etc.) and current fine-scale information relating to water and forest resources, biodiversity, agriculture, food security and human health (Salick and Byg 2007; Galloway-McLean 2010). There is a growing body of scientific literature and reports documenting the observations of local and indigenous communities regarding climate and environmental trends in many forested regions of the world (c.f. Parry et al. 2007b; Crump 2008; Galloway-McLean 2010). Vlassova (2006), for example, summarizes the results of interviews with indigenous communities on climate and environmental trends in the Russian boreal forest region.

Traditional knowledge is very commonly used in traditional communities to monitor and forecast weather and plan agricultural activities. For example, local farmers in various parts of the African Sahel have developed complex systems for gathering, predicting, interpreting observations of weather and climate that they use to make critical decisions on cropping patterns and planting dates (Ajibade and Shokemi 2003). A study in Burkina Faso found that the indicators most used for traditional weather forecasting include early dry season temperature patterns, the timing of fruiting in particular local tree species, water levels in streams and ponds, and nesting behaviour of certain bird species (Roncoli et al. 2001). Many other case studies have documented indigenous weather forecasting skills for predicting rainfall and droughts, in Gujarat, India; predicting weather through observations of sea birds and clouds, in Russia; and use of meteorological indicators and animal behaviour for weather prediction in Sri Lanka (Galloway-McLean 2010).

In a recent study conducted in the Philippines, Galacgac and Balisacan (2009) documented and critically evaluated the observations of natural phenomena, and their interpretation, of 204 traditional weather forecasters in numerous locations in Ilocos Norte Province. Traditional weather forecasters in this region use a variety of meteorological observations (e.g., temperature and wind patterns, cloud formations, sky colour, rainbow appearance, and visibility of stars and the moon), sea conditions; phenology of wild and cultivated plant species; and the behaviour of many species of birds, mammals, amphibians, and insects as the basis for prediction of rainfall patterns and specific events, including the onset of the monsoon and tropical storms (i.e., typhoons). This study found that the short- and long-term (up to 1 year) predictions made by traditional forecasters were, in general, highly reliable and often more accurate and extensive than those of national weather bureau. People in these rural communities rely heavily on this knowledge of these traditional knowledge holders (mostly elderly farmers, housewives and fisher folk) to plan their agricultural and agroforestry activities and to prepare for severe weather events. Traditional weather forecasting has been recognized by the Philippine Atmospheric, Geophysical and Astronomical Services Administration (PAGASA), which considers such traditional forecasting techniques to be supplemental to scientific weather forecasting.

Several countries have indigenous knowledge-sharing programmes for weather forecasting. In Australia, for example, the Indigenous Weather Knowledge Website Project, funded by the government, displays the seasonal weather calendars of Aboriginal Australians (Bureau of Meteorology 2011).

13.4 The Value of Traditional Agricultural and Forest Management Practices for Climate Change Mitigation and Adaptation

It is increasingly recognized that agricultural and forest management strategies to deal with climate change should consider all relevant knowledge and historical experiences that have helped societies adapt to changing environmental conditions. Traditional forest-related knowledge and practices developed and utilized throughout the world for centuries have an important role to play in dealing with climate change and its social and economic impacts. In response to ongoing and anticipated climate-induced changes in forest structure and species composition, traditional knowledge of forest-dependent people about the uses of a wide range of forest plant and animal species may enhance prospects for meeting their needs for wild foods, medicinal plants and other non-timber forest products by substitution of increasingly rare species with those that may become more abundant. Land management practices that have historically helped local and indigenous communities survive in the face of extreme climatic conditions merit study as models for production systems that contribute to both climate change mitigation and adaptation (Kronik and Verner 2010).

For cultivated forest species, as for agricultural crop species and varieties, traditional knowledge and associated tree and forest management practices can play a key role in maintaining forest productivity, biodiversity, and food security, in the face of climate change. Case studies from around the world reveal a diversity of traditional ecosystem management approaches under conditions of environmental uncertainty that may increase communities' capacity to adapt to the effects of climate change (Adger et al. 2007; Roberts et al. 2009). In Italy, for example, centuries-old practices for managing chestnut orchards have enabled local communities to cultivate this important woodland food source on sites well beyond the natural climatic limits of the species (Agnoletti 2007).

In the context of climate change mitigation, it is instructive to consider the variety of traditional forest, woodland, and agricultural management practices that have been used to meet the livelihood needs of people on a sustainable, long-term basis in the face of environmental uncertainty. Of particular relevance are practices that, unlike modern energy-intensive industrial agriculture (and many forestry practices), neither depend on significant fossil fuel inputs (in the forms of inorganic fertilizers, pesticides, and mechanization) to maintain their productivity, nor result in deforestation and CO_2 emissions from both forest conversion and agricultural management activities.

In dry and semiarid regions of the world, indigenous and local societies have developed a wide variety of strategies based on traditional forest-related knowledge to cope with recurrent droughts that periodically threaten their food security. These include traditional technologies developed to manage wild and cultivated trees that provide food, fodder, and other non-timber forest products; and practices to harvest and conserve scarce water resources in traditional silvopastoral and agroforestry systems (Laureano 2005; Osman-Elasha et al. 2009). Traditional societies worldwide have devised a range of agricultural techniques designed to conserve soil fertility (and soil carbon) and maintain productivity of a range of food crops even during extended periods of drought. In the Sahel and elsewhere on the African continent where the magnitude and intensity of droughts, and their impacts on rural communities have increased over the past century (Benson and Clay 1998; Somorin 2010), local communities have survived by relying on traditional soil and water conservation practices, tree crop management innovations, use of locally developed early-maturing and drought-hardy crop varieties, and soil management techniques such as zero-tilling and mulching based on traditional knowledge (Schafer 1989; Osunade 1994; Dea and Scoones 2003; Gana 2003; Nyong et al. 2007). A number of recent projects involving indigenous communities in various parts of the world have helped to revitalize traditional practices for soil- and water-conservation methods and harvesting that have yielded important livelihood benefits within these communities (Galloway-McLean 2010).

At the other extreme, in extremely high rainfall areas such as in northeastern India, the soil and vegetation management practices developed over generations by local communities who practice shifting agriculture have been shown to enhance soil fertility and minimize erosion losses in cultivated areas (Ramakrishnan 2007). Such techniques, developed over generations by people in indigenous and local communities, may be seen as methods to 'activate' environmental resources without the use of external energy inputs derived from fossil fuels, such as chemical fertilizers and mechanization (Bertolotto and Cevasco 2000).

In the following sections we consider selected technologies, innovations, and practices developed by traditional societies that have sustained a variety of agricultural, agroforestry, and forest management systems over generations, long before dependence on fossil fuels resulted in their displacement in many parts of the world. These traditional practices, having evolved under conditions of environmental uncertainty and change, illustrate the adaptive capacity (and in some cases the vulnerabilities) of local and indigenous communities and their knowledge systems.

13.4.1   Energy Use Efficiencies of Traditional Agriculture and Forest Management Systems

While the need to reduce consumption of fossil fuels and develop renewable energy resources in industrialized and urban environments is widely recognized, rather less attention has been paid to rural landscapes—arable lands, pastures, and forests. These managed landscapes cover an increasing proportion of the earth's surface, supply the overwhelming majority of the world's food, and contribute very significantly to both carbon sequestration and carbon emissions. The development of strategies to mitigate climate change could benefit from a re-evaluation of rural development models, including those based on traditional knowledge and practices. In such evaluations, consideration should be given to energy budgets and energy use efficiencies (the ratio of energy produced to energy consumed), as well as to landscape-level impacts and trade-offs with respect to the range of environmental, economic, social, and cultural values of forests and associated agricultural landscapes (Haberl et al. 2004; Marull et al. 2010).

Several recent studies have assessed the role played by traditional knowledge and practices in sustainable forest management and conservation of cultural landscapes in (Farina 2000; Agnoletti 2006). Historical changes affecting land use were usually found to be the result of changing management of agroforestry and pastoral systems that were linked to changes in the production and consumption of energy. Landscape change can be seen as the reflection of changing metabolic relationships between a given society and the natural systems sustaining it (Tello et al. 2006). The historical process of industrialization may be viewed as a transition from a rural socio-metabolic regime, with a land-based solar-energy-controlled system, towards an industrial regime with a fossil-fuel-based energy system (Kuskova et al. 2008). Landscapes characterized by traditional activities with low mechanization tend to use energy in a more efficient way, and above all, are not highly dependent on fossil fuels, as is the case for landscapes characterized by agricultural intensification and specialization (Sieferle 2001; Krausmann et al. 2008).

Input-output analysis of the production and consumption of energy in agricultural, forestry, and pastoral systems can be used to assess the relative energy efficiencies of socio-ecological systems, and therefore of the landscape. Such analyses evaluate all variables related to the production and consumption of energy, such as agricultural and animal products, use of fuel and of mechanical, fertilizers, taking into account all the energy flows related to production activities, including inputs (animal and human labour, fertilizer, pesticides, fuels, etc.); outputs (animal and vegetable products, residues, etc.); and socio-economic stocks (humans, domesticated animals). Studies of this type can be used to quantify energy inputs from fossil fuels and from renewable sources, and to estimate the production of greenhouse gases that contribute to climate change.

Studies using input-output analysis have shown that traditional agroforestry systems, despite lower overall productivity, can make more efficient use of solar energy so that the energy produced is greater than the energy consumed. In contrast, modern agricultural landscapes consume more energy than they produce, and require higher energy inputs (particularly those based on fossil fuels, fertilizers, and pesticides) to achieve their high productivity. For example, Cussó et al. (2006) compared the rural landscape of Catalonia in 1860 with that of 1999 and found that the ratio of energy produced to energy consumed declined from 1.67 in 1860 to 0.21 in 1999. Studies of other socio-ecological systems in Europe have also shown low energy efficiencies of landscapes dominated by intensive industrialized practices. Most of the energy used today in industrialized agriculture and forestry is derived from non-renewable energy sources, and in particular from fossil fuels, making agricultural production dependent on the availability and price of oil. As a result, today's intensive agricultural landscapes are both energy inefficient and highly polluting.

Traditional fallow management practices used by shifting cultivators throughout the tropics and subtropics represent another means by which traditional knowledge has contributed, and in many regions continues to contribute, to climate change mitigation. While it usually is argued that these systems are less productive than more energy-intensive agricultural practices, traditional shifting agricultural and forest management systems are typically more energy-efficient and, in fact, may actually be more productive in terms of agricultural yields than alternative, less sustainable, higher energy-input systems—as discussed in the case of traditional agroforest landscape management in northeastern India, in Chap. 9 of this volume.

13.4.2 Agroforestry

Agroforestry—land use systems and practices in which woody perennials are deliberately integrated either in spatial mixtures or temporal sequences with crops and/or animals on the same land unit—have been developed and used by traditional communities worldwide for centuries. Agroforestry systems, both traditional and introduced, involve careful selection of species and careful management of trees and crops to optimize the production and positive effects within the system and to minimize negative competitive effects and the need for chemical fertilizers and other external energy inputs to maintain high productivity. A wide diversity of traditional and recently modified or introduced agroforestry systems are used in densely populated areas of South Asia and Southeast Asia, as well as in less densely populated tropical, subtropical and temperate regions of the world (Fig. 13.1).

While traditional agroforestry systems have evolved with changing agro-ecological, socioeconomic, and demographic conditions, many of these have been weakened, or abandoned, by communities for a variety of reasons, including population growth, land-use conflicts, deforestation, and degradation of forests, watersheds, and agricultural soils. In regions where shifting cultivation is practiced, these pressures have often led to shortened fallow periods and extension of cultivation into areas unsuited

Fig. 13.1 Traditional temperate agroforestry system. Such systems, formerly common in Italy and in other Mediterranean countries, involve terracing and mixed cultivation, and accomodate multiple land uses within a limited land area

to these practices. Many communities that practice permanent agriculture face similar problems connected to their over-exploitation of natural resources to meet increasing demands for food, fodder, fuel, and water resources. This has resulted in decreased agricultural productivity, reduced incomes, poverty, overgrazing of range lands, deforestation, and forest degradation. In such situations, redevelopment of agroforestry practices may offer not only a means to improve the sustainability of agricultural production and reduce degradation of forests, soil erosion, and desertification (Tejwani and Lai 1992; Carmenza et al. 2005), but also a means to better adapt to the negative impacts of climate change and variability.

Scientific interest in tropical agroforestry can be traced to the early to mid nineteenth century, when British colonial foresters in Burma began adapting local (traditional shifting agriculture) fallow management practices for the purposes of establishing teak (*Tectona grandis*) plantations. This practice, useful for establishing plantations of a variety of high-value timber species, soon spread to nearby hill regions of northeast India and present-day Bangladesh, central and southern India, and many parts of Africa from the mid-nineteenth to early twentieth centuries (King 1987). Building on these experiences, the successes of countless generations of traditional agroforestry practitioners, and the failures of mid-twenty first century agricultural development policies and practices that were promoted to replace traditional tropical agroforest management systems, the scientific discipline of agroforestry emerged in the late 1960s and1970s with the establishment of the International Institute of Tropical Agriculture (based in Nigeria) and the International Council for Research in Agroforestry (currently the World Agroforestry Centre, based in Kenya). Over the past 40 years, agroforestry research, development, and training activities

Box 13.1 Advantages of Agroforestry Systems in Dryland Agriculture

- *Diversification of production.* Agroforestry systems yield a wide range of products, including food, fodder, fuel, fruits, green manure, etc.
- *Risk management.* By diversifying production, agroforestry systems reduce economic risk to farmers should in the case of crop failures.
- *Management flexibility.* The presence of trees in agroforestry systems expands management options for farmers. For example, tree crops can be harvested for food (in the case of fruit-yielding species), fuel, fodder, poles, or small timber according to farmers' requirements and market demands. During the fallow season, time can be effectively used for harvesting, pruning, and marketing of the tree components.
- *Utilization of off-season precipitation.* While most of the drylands are cropped during a limited season, perennial tree components can capture and use rainfall throughout the year.
- *Soil and water conservation.* The perennial component of agroforestry systems help to reduce soil and water loss.
- *Moderation of soil microclimate.* Tree components of agroforestry systems help to stabilize soil temperatures and moisture particularly during summer months, protecting soil micro-organisms. Trees also serve as shelter for cattle.
- *Soil fertility enhancement.* The deep-rooted nature of most trees results in tapping nutrients from deeper soil horizons and returning them to the plough layer through leaf litterfall and decomposition.
- *Carbon sequestration.* The presence of permanent (tree and shrub) system components enhances carbon storage above- and below-ground in biomass and soil organic matter.

have been promoted by government agencies, universities, training institutions, NGOs, and local organizations, though not with a view to their potential to address climate change issues until very recently.

Traditional and recently modified agroforestry systems can offer advantages over conventional agricultural and forest production methods. These systems both increase productivity of food crops, fodder, fruits, and other non-timber forest products, as well as provide economic and social benefits; and ecological goods and services, including biodiversity conservation and increased sequestration of carbon in biomass and soils (Box 13.1). In addition to their capacity to sequester and store carbon, sustainable agroforestry systems offer other climate-change-related benefits, through: their potential to reduce deforestation that would result should they be converted to permanent 'modern' agricultural systems; reduced degradation pressures on forest and woodlands by providing fodder and fuelwood; and increased diversity of on-farm tree crops and tree cover that may help to buffer farmers' livelihoods and food security against the effects of climate change. In regions susceptible to drought—as in most of Africa, the Mediterranean, and drier regions of the Americas, Asia, and Australia—the provision of fodder in the form of foliage from trees cultivated in agroforestry systems may be particularly important, providing a measure of insurance during seasons (or years) when fodder resources from grazing lands are scarce.

Nitrogen-fixing species cultivated in agroforestry systems that produce nutrient-rich leaf litter—such as *Sesbania sesban, Tephrosia vogelii, Gliricidia sepium,* and *Faidherbia albida*—have been shown to improve maize yields in sub-Saharan Africa. For example, a 10-year experiment in Malawi demonstrated that using trees such as *Tephrosia vogelii* and *Gliricidia sepium* increased maize yields to an average of 3.7 tonnes per hectare, compared to 1 tonne per ha in plots without these trees or application of mineral fertilizers (World Agroforestry Centre 2011a). Research with *F. albida* in Zambia showed that mature trees can sustain maize yields of 4.1 tonnes per ha over several years compared to 1.3 tonnes per ha beyond the canopy of the tree. Unlike other trees, *Faidherbia* sheds its nitrogen-rich leaves during the rainy crop growing season, precluding its competition with food crops for light, nutrients, and water. The regrowth of foliage during the dry season provides land cover and shade for crops when it is most needed (World Agroforestry Centre 2011b).

Agroforestry research in India (Bharara 2003) has documented a number of tree and shrub species that are common components of traditional agroforestry systems in the villages of arid zones, especially khejri (*Prosopis cineraria,* a tree) and ber (*Ziziphus mauritiana,* a gregarious shrub). These species, typically planted on field borders, or retained within fields, play an important role in providing high-quality fodder, fuel, and other locally used products as well as providing shade to adjacent crops. Within fields, khejri trees are lopped for fodder and fuel each year following the harvest of agricultural crops, while ber is harvested before planting of crops, allowed to re-grow with the associated crops (without competition for soil moisture and light), and again cut after the harvest of field crops. Khejri and ber grown in shelterbelts are not harvested, except for collection of twigs and small branches for fuel and thatching. In these traditional systems, fodder from both species is available every year regardless of rainfall due their deep-rooted habit and drought-tolerance, and are known as desert farmers' 'natural insurance.' During good rain years, yields of both grain crops and fodder around *P. cineraria* and *Z. mauritiana* are often higher than in other parts of the same field, attributed to their beneficial effects on soil microclimate and accumulation of humus.

One of the best examples of European agroforestry systems, known as 'mixed cultivation,' is a traditional practice found in the Mediterranean region dating back to at least 200 BC. In Italy, mixed cultivation was the most dominant land use in past centuries (Braudel 1966), representing approximately 45% of the total arable land of the country at the beginning of the twentieth century. Mixed cultivation practices allowed peasant farmers to exploit the cultivable areas of their lands and provide for their subsistence needs for both food and domestic energy. These practices also contributed significantly to the maintenance of tree cover in agricultural landscapes and meeting the energy requirements of the rural population. At the beginning of the twentieth century, the density of trees in mixed cultivation areas in northern Italy was about 160 trees ha^{-1}, higher than that of many forests, such as chestnut groves (which averaged 90–120 trees ha^{-1}). At that time, the total production of fuelwood from the Italian countryside was twice that from its forests. Those woodlands helped the country to be self-sufficient for fuelwood despite its limited forest area, which was 10% of the total land area (Agnoletti 2005, 2010a). Although mixed cultivation in Italy declined precipitously over the past 100 years with the industrialization of the agriculture sector, it is still practiced in some localities, an expression of the values expressed by these practices, which have survived despite attempts to replace them with technical and productive models that have yet to prove their historical sustainability.

13.4.3   The Oasis System

Local knowledge and practices associated with the creation and management of oases and other agricultural and forest management systems in arid regions of the world offer dramatic examples of the importance and contribution of traditional forest-related knowledge for human adaptation to environmental extremes (Laureano 2005). Such knowledge and the complex agricultural and forest resource management systems that sustain communities under conditions of water scarcity, are likely to become increasingly important in the future for regions where climate change is likely to exacerbate desertification.

Oases, usually relatively small areas within large expanses of sandy deserts as in the Sahara, are not naturally occurring habitats but rather the result of human action through the application of traditional knowledge, skills, and management techniques suited to the harsh environment in which they are found and passed on from generation to generation. The date palm (*Phoenix dactylifera*), an emblematic oasis plant probably native to North Africa and southwestern Asia, is the product of centuries (or longer) of domestication and cultivation. Palm groves established in oases have been planted, tended, and watered using special techniques to capture and efficiently utilize very scarce water resources.

In other arid landscapes that occur in very diverse environments, such as the mountain regions of North Africa, people have also developed highly productive agricultural systems through careful management of scarce rainfall and other available water sources. An example of this may be found in the Telouet valley on the eastern slopes of the Atlas range in Morocco, a locale characterized by dry climate and arid soils. As shown in Fig. 13.2, the land next to the 'oued' (river, in Arabic) is intensively cultivated, while the surrounding mountain slopes are totally denuded. Farmed land in the valley includes a large number of palm and other tree species, supported by a complex system of irrigation channels that are continuously managed by farmers. In this community and others in the region, wheat, fruit, and fodder production ensure self-sufficiency and food security for the local population,

Fig. 13.2 Telouet Valley, Morocco. Traditional techniques have made it possible to live in desert areas of the world without the use of modern technologies, by developing agricultural systems based on the use of trees and characterized by low energy inputs (Photo: Mauro Agnoletti)

while maintaining the beautiful, fine-grained, landscape that preserves the cultural identity of the area.

13.4.4   Traditional Fodder Production

Adaptation to climatic change, as in the case of coping with drought, has often focused on maintaining fodder availability during even the driest years. A number of traditional techniques for this have been developed in different parts of the world. Through their long experience with recurrent drought, farmers in Ethiopia, for example, have developed their own criteria for selection and evaluation of indigenous fodder trees and shrubs, and have also perfected protocols for the propagation and effective conservation of such species (Balehegn and Eniang 2009).

Technical interventions generated by research organizations that aim to improve the quality of life of people in traditional communities, or the nutritional status of their livestock, have generally failed to realize the value of traditional knowledge and innovations such as those of these farmers in Ethiopia. It is now increasingly recognized that unlike research-generated technologies, farmers typically aim to meet a variety of sometimes conflicting objectives through the practices they use to provide feed for their animals and adapt to existing agro-climatic changes in their localities (Thorne et al. 1999), and use their traditional knowledge and experience to develop practices to balance these objectives and optimize outcomes.

While work is underway in many regions to develop varieties of staple food crops that are adapted to existing climatic changes, identification and selection of adaptable forage options remains a task for farmers. Several studies of indigenous and local communities in Africa and Asia have shown that there is a diverse array of indigenous criteria and strategies for evaluation, utilization, and conservation of such trees that are linked to the adaptation needs of each agro-ecological zone (Jimenez-Ferrer et al. 2007). In a number of instances, laboratory analyses of the nutritive value of fodder from tree species have been found to correlate positively with their value according to indigenous evaluation criteria (Roothaert and Franzel 2001). For some species, however, laboratory analyses of fodder nutritive value does not correlate or even correlates negatively with the values according to indigenous criteria, as for example *Ficus thonningii* in Ethiopia (Mekoya et al. 2008). This suggests that these farmers, in their effort to adapt their fodder production practices to severe drought, give less emphasis to conventional criteria (e.g., nutritive value), than to survival of the fodder-yielding trees. Consequently, species that are not conventionally considered (by animal husbandry professionals) to be suitable fodder species are sometimes preferred by local communities for good reason.

Given the existence of site-specific traditional knowledge-based practices used by local people for selecting adaptable fodder species, such knowledge should be considered and integrated with scientific and other conventional evaluation criteria for the selection and promotion of tree fodder species to better assist farmers in different parts of the world adapt to climate change impacts in their localities, as was done in the study of local farmers' knowledge of fodder trees in Ethiopia cited earlier (Balehegn and Eniang 2009). Research scientists concerned with the development of adaptation strategies for forest and on-farm tree management would benefit from greater exposure to, and understanding, of traditional knowledge and practices to deal with the impacts of drought and extreme weather events that have served rural communities well for generations.

An example of sustainable agroforestry practices combining fodder production and controlled use of fire is the 'alnocolture' system, used by farmers in the Italian Apennine region until the end of the nineteenth century (Bertolotto and Cevasco 2000). This system, requiring no chemical fertilizers (i.e., fossil fuel-based inputs) involved a several-stage cycle combining cereal production, fodder production, and timber production using the nitrogen-fixing species *Alnus incana*, which increased the efficiency of the system through its fertilizing effects on soils under its influence of its root system as well as its periodic burning (Daniere et al. 1986). Fragmentary descriptions of this traditional system indicate variability in the time span of the different practices involved in the cycle (Fig. 13.3), which may have been due to differences in land morphology, soil conditions, livestock carrying capacity, and individual farmer preferences. This system was used in a variety of woodland types—in alder woods, and in mixed woods with alder, beech (*Fagus sylvatica*), and Turkey oak (*Quercus cerris*). The felling of alder (stage 1 of the cycle) occurred at intervals from 3 to 12 years, the cycle length related to the time required for production of fuelwood and fodder. Alderwood is specific for the practice of 'ronco'; its wood is also used to make 'fornaci' (confined fire, during stages 2 and 3), after which the ashes are spread to fertilize soils for subsequent crop production. Grazing of sheep, goats, and, rarely, cattle was allowed after harvest of the last agricultural crop, by which time grazing would not jeopardize the growth of the young trees. Controlled fire in the alnoculture cycle helped to clear the undergrowth around hawthorns and particularly brambles, which formed temporary hedges used to protect the cultivated grain crops from the surrounding grazing area. It is probable that the 'ronco coperto' system, in which certain trees were retained, helped to save useful fruit trees such as cherry, and fodder

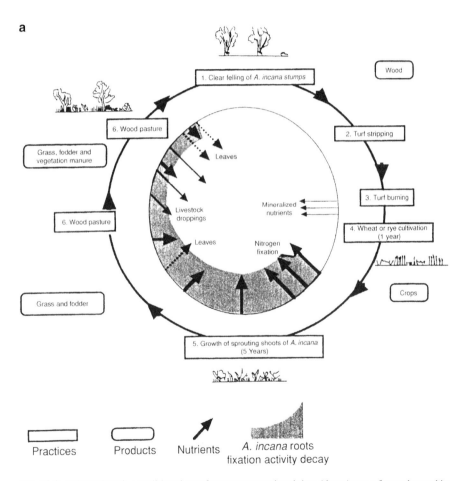

Fig. 13.3 'Alnocolture', a traditional agroforestry system involving *Alnus incana* formerly used in the Italian Apennines. (**a**) schematic of the alnocolture cycle and associated nutrient inputs to soils; (**b**) Selected stages of the alnocolture cycle: (i) turf burning; (ii) cultivation of rye and oats; (iii) wood pasture

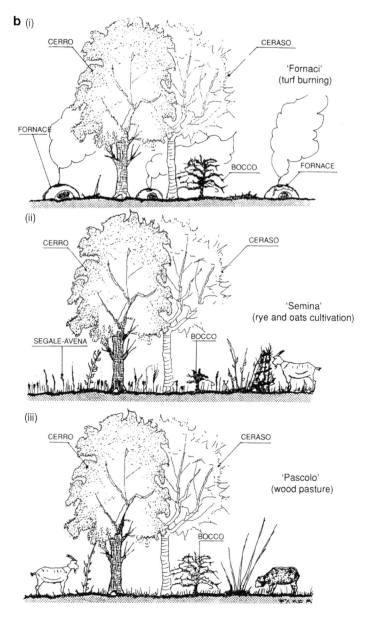

Fig. 13.3 (continued)

trees such as Turkey oak and ash (*Fraxinus* spp.). Very large, old individuals of these species survive in historical alnoculture sites to this day. What is notable about this traditional system is the fact that it did not reduce or degrade available resources over successive cycles.

13.4.5   The Value of Terraces and Dry-Stone Structures for Climate Change Adaptation in Agriculture and Forestry

One of the most serious threats posed by climate change is sea level rise, which is expected to cause flooding and increase salt water intrusion into groundwater in many coastal areas around the world, reducing the availability of cultivated land and freshwater resources for agriculture and other uses. Throughout history, people have adapted their agricultural practices in the vicinity of low-elevation wetlands and have developed management systems to sustain agricultural production in marginal hill regions to meet increasing food needs of growing populations.

Terracing, practiced since Neolithic times, is one of the most widespread techniques used to protect soils from erosion and to capture and channel water. These are self-regulating systems characterized by high aesthetic quality that model the landscape while integrating with its natural features. In the Mediterranean region, terraces have been used since ancient times for the cultivation of a variety of agricultural, horticultural, and other tree crops (Ron 1966; Rackham and Moody 1992; Grove and Rackham 2001). Never an isolated landscape element, terraces are rather part of a system of land-shaping works that are the product of traditional knowledge of construction techniques. They demonstrate an exquisite understanding of local hydrogeology and climate that allow local populations to make the most of their environmental resources (Fig. 13.4).

Terracing is used throughout the world today in agriculture as well as forestry. These systems are characteristic of rice civilization in Southeast Asia, exemplified

Fig. 13.4 Terracing in the Gargano massif, southern Italy. Terracing is commonly used to cultivate trees on mountain slopes in many parts of the world, and are an effective means to manage scarce water resources, particularly in drier regions. Sea level rise resulting in the loss of low-elevation agricultural lands may require restoration or extension of traditional terracing practices in upland areas to cope with such losses (Photo: Mauro Agnoletti)

in the Philippines' Banaue rice terraces of the Ifugao, a UNESCO World Heritage Site in Ifugao Province in Luzon (see Chap. 10, Box 10.3), or those in the Sapa region of northern Vietnam.

In Italy there is a prevalence of terraces that originated in the Middle Ages, when the abandonment of water control structures after the fall of the Roman Empire created vast flooded areas in the lowlands that forced the population to develop agriculture on hill slopes (Sereni 1997). Terracing, in the form of grass-covered contour terraces and terraces supported by dry-stone walls, is thus one of the main forms of adaptation to difficult environmental conditions. Terraces are used to produce not only agricultural crops but also trees of economic importance. Chestnut groves, in particular, are often grown on terraces of various designs in southern Europe (Pitte 1986), including structures held by dry-stone walls to 'lunettes'—semi-circular walls surrounding each individual tree. Terracing was also widely used in the reforestation and watershed restoration in mountain regions, as in most of the Alpine countries of Europe during the latter half of the nineteenth century.

Terraces provide a number of important services, notably land stabilization, by reducing soil erosion and loss (Llorens et al. 1992). In many areas, terrace construction required enormous human effort—cutting into the rocky matrix with sledge-hammers and pickaxes, and carrying soil up from valleys on people's shoulders or mules' backs to fill the terraces. They are sometimes Cyclopean works of monumental character, both for their size and work forces required to build them. Many terraces, especially in farmed areas, have been replaced over the years with 'rittochino' arrangements, with planted crop rows following the maximum acclivity lines (i.e., directly upslope), since modern machines are powerful enough to work on steep slopes. Mechanization has also favoured larger agricultural fields, which has led to the total or partial removal of many terrace structures to increase field sizes, without replacement of equally effective structures to prevent soil loss. As a result, erosion on such farms is often extremely high, as documented in a study conducted in Tuscany which found that erosion rates increased 900% between 1954 and 1976 (Zanchi and Zanchi 2008). Chemical fertilizers are increasingly required to compensate for soil fertility losses.

Today, to ensure sustainability and landscape protection, it is necessary to minimize or reverse soil erosion and hydrogeological deterioration both by using conservative agronomic techniques that are adapted to specific local situations, and by restoring and maintaining traditional structures with high landscape value such as terraces (Zanchi and Zanchi 2008). Dry-stone walls used in traditional terrace construction in the Mediterranean as in other regions offer additional advantages to crops by absorbing heat during the hot hours of the day and releasing it during the cooler hours. Although the maintenance costs of terraces are higher, they guarantee the conservation of agriculture's most important resource—soil. In Italy, the Ministry of Food Agriculture and Forestry supported a survey of terracing as part of its research for a National Catalogue of Historical Rural Landscapes (Agnoletti 2010b), and made the conservation of terraces a pre-requisite for farmers requesting subsides in the framework of the EU Common Agricultural Policy.

Fig. 13.5 A 'pantesco garden,' on the Mediterranean island of Pantelleria (Ben Ryé) between Sicily and Tunisia. The traditional practice of constructing dry-stone walls, which protect trees within them from strong winds, allows the cultivation of trees without irrigation under semiarid conditions (Photo courtesy of Giuseppe Barbera)

Dry stone is used for purposes other than terraces in the traditional management of trees in different parts of the world. In regions where water shortages and high winds have created difficult conditions for cultivation of forest and horticultural tree species, a number of traditional techniques have been developed using locally available materials to overcome these limitations. Examples can be found in many Mediterranean islands as well as in the Canary Islands.

On the volcanic island of Pantelleria (also known as Ben Ryé, from the Arabic *Bent El Riah*), located between Sicily and Tunisia, the *orti panteschi* (pantesco gardens) are currently under study by Italy's National Council of Research as an adaptation to dry climate. Rounded dry-stone structures, up to 4 m tall with walls about 1 m thick, are used to shelter fruit (usually citrus) trees that could not otherwise be grown due to their vulnerability to high winds and high water requirements (Fig. 13.5). The shade, wind protection, and condenzation of dew on the surface of the lava stones used to construct these shelters, create a microclimate that allows the trees to grow and bear fruit even without supplemental irrigation. Local farmers here have also traditionally cultivated drought-tolerant, wind-resistant species using techniques to maximize soil water retention and reduce surface runoff by improving soil drainage. Specific tree management practices—for example, the pruning of olive trees to create collapsed canopies with their branches resting on the ground—are also used to maximize the protection from wind and sea air afforded by dry-stone walls.

On the volcanic island of Lanzarote (in the Canary Islands), high winds present a major challenge for management of tree crops on young soils. Farmers there have developed ingenious methods for cultivating citrus and fig trees, as well as grapes, under these difficult conditions. These include creation of depressions in which these crops are planted, and construction dry-stone walls on their windward edges to provide additional protection from strong winds. These traditional techniques both reduce wind exposure and take advantage of the heat storage capacity of these soils to create a microclimate highly favourable for the growth of planted trees. These farmers have created a unique landscape based on traditional knowledge, with clear social and economic benefits. Wine-making is an important industry on the island, challenging a common opinion that traditional knowledge and practices are not compatible with current agricultural production standards (Fig. 13.6).

13.4.6 Coppice Woods Management

Of the traditional forest management techniques that are relevant to climate change mitigation and adaptation, one of the more useful is coppice management. This short-rotation system is particularly suited to CO_2 sequestration because of its focus on repeated harvesting of wood on rotations of 5–25 years. Coppice systems typically provide higher, and more frequent, yields of fuelwood compared to high forest stands (Agnoletti 2005), and allow grazing within stands when pollarding methods are used. Coppice woods have been documented in Europe during the Roman period (750 BC–450 AD) or even earlier (as discussed in Sect. 6.3), and have been widely distributed in the Mediterranean landscape for the past 150 years (Agnoletti 2005, 2010a). They are usually low woods, managed with variable cutting regimes, found on mountains, hills, and along the coast. The spread of coppice management practices is due to their great capacity for integration with agricultural activities, and their high yield of fuelwood (one of their main products), which was usually converted into charcoal on the spot.

A broad array of coniferous and broadleaved tree and shrub species have traditionally been managed in coppice systems in Europe as well as in many other parts of the world (Ciancio and Nocentini 2004). These highly developed traditional systems involve a variety of management practices depending on the species used, local soil and climatic conditions, the products desired, and the integration of coppice management with other production activities (such as grazing or crop production).

Among the many existing traditional forms of coppice management in Europe is one applied to scrub woods (such as heathland and maquis formations), which were once managed throughout the Italian peninsula to Sicily. Heaths are evergreen species that thrive on dry, acidic, and siliceous soils that are highly inflammable (pyrophytic), and whose reproduction by seed and vegetative regrowth are favoured by fire. These species are often found on very degraded and arid soils, and can rapidly

Fig. 13.6 Horticultural applications of traditional dry-stone structures under extreme environmental conditions in Lanzarote, Canary Islands (Photos: Mauro Agnoletti)

colonize areas otherwise unsuitable for tall vegetation, including bare hill slopes that are protected from landslides and erosion by heath growth .

Traditionally managed scrub woods, typically associated with grazing and agriculture, involve the heath species *Erica scoparia* and *E. arborea,* which occur in many scrublands along with a variety of other shrub species and in the understory of oak (e.g., cork oak—*Quercus suber,* holm oak—*Q. ilex*) and pine forests. Both above- and below-ground parts of these species are used for a variety of purposes. Traditional managed *Erica* coppice systems in Italy yield vineyard poles at cutting intervals of about 8–10 years; foliage for silkworm raising and hut roofs; firewood for bread ovens, brick furnaces, charcoal, and other purposes; branches used for brooms; and high-quality wood from roots for carving pipe bowls. The extent of heath formations has been greatly reduced during the past century because of their declining utility to farmers, abandonment of the practice of land burning, the shrinking of pasture areas, and reforestation of scrublands to 'improve' them (since they were regarded as a degradation of high forest stands, and formal forestry regards *Erica* and associated shrub species as weeds because they compete with trees). Heath management was never dealt with systematically in official silviculture texts, and the traditional knowledge regarding their management is unfortunately being progressively lost with the decline of traditional rural livelihoods (La Mantia et al. 2006, 2007)

13.5   Traditional Forest-Related Knowledge in International Climate Change Assessments, Policy Forums and Programmes

13.5.1   *Traditional Knowledge and Scientific Assessments of Climate Change*

The forest-related knowledge, experience, and practices of local and indigenous communities have a critical role to play in the assessment of climate change trends and impacts, as well as in the development of viable and effective strategies for climate change mitigation and adaptation (Salick and Byg 2007). Although the translation of traditional forest-related knowledge into the language of formal forest science can be a helpful step towards adapting and applying traditional forest-related knowledge, often such translation fails to include the full depth of traditional knowledge. Collaborative efforts are often need to apply both traditional knowledge and formal science to new or changing environmental, social and economic contexts. Traditional forest-related knowledge may find important applications in forest rehabilitation, restoration, and adaptive management in the face of climate change (Berkes et al. 2000; Parrotta and Agnoletti 2007).

Understanding traditional knowledge and associated forest landscape management practices may also further the development of policies that help to close the existing gap between global environmental paradigms and related strategies with the local level, where they often fail. The need to incorporate these views in sustainable forest management has also been recognized by important political bodies, such as the Ministerial Conference on the Protection on Forest in Europe, through specific resolutions and scientific guidelines for implementing them (IUFRO 2007). The Arctic Climate Impact Assessment (ACIA),[1] which includes assessments of climate change and its impacts on boreal forests, is held in high regard by both the scientific community and indigenous peoples and is a good model for integration of traditional and scientific knowledge in climate change assessments.

Of particular relevance to climate change policy and mitigation and adaptation programmes is the work of the Intergovernmental Panel on Climate Change (IPPC), established by the United Nations Environmental Programme (UNEP) and the World Meteorological Organization (WMO) in 1988 to assess the scientific, technical, and socioeconomic information relevant for the understanding of human induced climate change, and its potential impacts and options for mitigation and adaptation. The importance of traditional knowledge has been recognized in the work of the IPPC (specifically within Working Group II, which deals with climate change impacts, adaptation, and vulnerabilities). The IPCC recognizes that traditional and local knowledge has been an important missing element in its previous assessments, and intends to remedy this gap to focus on the knowledge and experience of local and indigenous communities in its 5th Assessment Report (scheduled for completion in 2013–2014), which will place greater emphasis on assessing the socioeconomic aspects of climate change and implications for sustainable development, risk management, and the framing of a response through both adaptation and mitigation.[2]

13.5.2 Traditional Knowledge in Global Climate Change Mitigation and Adaptation Policy

Governments, international organizations, and the private sector have undertaken or are developing a number of measures aimed at reducing emissions of greenhouse gases arising from fossil fuel use. These include: increased production and use of (theoretically less polluting) biomass-based energy sources; mitigating schemes aimed at offsetting fossil fuel-related emissions by increasing CO_2 capture and storage (sequestration) in forest biomass and soils; and development of mechanisms for reducing emissions from terrestrial sources (i.e., agricultural and forest lands). Although mitigation measures are needed to minimize future impacts on forest and agricultural landscapes, and the people who most directly depend on them for their livelihoods, these mitigation approaches have implications (both opportunities and

---

[1]The ACIA is a project of the Arctic Council and the International Arctic Science Committee: http://www.acia.uaf.edu/

[2]http://www.ipcc.ch/activities/activities.shtml

threats) for the holders and users of traditional forest-related knowledge, i.e., the local and indigenous communities who are at the receiving end of the impacts of problems that were not their creation.

As discussed earlier, these communities, and the knowledge possessed by their elders, could make significant contributions towards the development of policies as well as the design and implementation of sustainable climate change mitigation measures that involve carbon sequestration, forest conservation, renewable energy production, low-energy agricultural systems and agroforestry systems, and rehabilitation of degraded agricultural and forested landscapes. Traditional lifestyles and livelihoods present approaches and specific practices and technological innovations that serve as examples for development of more widely applicable models for reducing dependency on fossil fuels (and other environmentally damaging energy sources such as uranium). These include, among others, production practices involving low (or no) consumption of fossil fuels, efficient and sustainable utilization of forest resources (biomass for energy and animal fodder), agricultural practices that conserve water and soils (including maintenance or enhancement of soil carbon storage), and local technological innovations for energy production and conservation from solar energy, biomass, and small-scale hydro-power (Galloway-McLean 2010).

Philips and Titilola (1995) note that it is widely recognized that the key to sustainable social and economic development is knowledge, not financial capital. Building on proven traditional knowledge and land management practices, the local institutions in which they have developed and evolved can thus facilitate the design and implementation of climate change mitigation and adaptation strategies that serve the broader social and economic development needs of communities who must contend with the impacts of climate change (Nyong et al. 2007).

Unfortunately, in the climate change policy arena, the rights, interests, and knowledge of local and indigenous communities are, on the whole, greatly undervalued (Macchi et al. 2008). As discussed in Chap. 1 of this book, traditional societies face an uphill struggle to make their voices heard within the countries in which they live, particularly in the international environmental policy forums including the climate change arena. Despite their knowledge, experience, and wisdom, such communities (with a few exceptions) are typically at a distinct economic and political disadvantage in influencing sectoral policies, strategies, laws, programmes, institutional arrangements, and investments related to climate change mitigation and adaptation and questions related to forests and land use (Garcia-Alix 2008; Gerrard 2008; Nilsson 2008).

13.5.3 Global Forest-Related Climate Change Mitigation Strategies: Carbon Markets and REDD

Carbon emissions trading, a system designed to offset carbon emissions from one activity (such as burning fossil fuels to generate electricity) with another more efficient or less polluting activity remains a much-debated topic in the international community. It is also a contentious issue among organizations representing the rights and interests of indigenous peoples in forest and climate change policy arenas.

Market-based systems for climate change mitigation measures such as carbon trading and agro-fuels (or biofuels) production have generated considerable mistrust among traditional forest communities given the number of cases in which the implementation of projects has violated the rights of indigenous communities (Bailey 2007; CBD 2007; UNPFII 2007; Collings 2009). For example, certain Clean Development Mechanism[3] projects initiated over the past decade have created serious conflicts involving communities who refused to give up territories on which such projects were planned. One example can be seen with the Benet people, an indigenous community living on Mt. Elgon in Uganda, who have struggled to regain their rights to live in their ancestral forest lands following implementation of a tree-planting carbon offset project initiated in the early 1990s (UNPFII 2007). Other communities have been seriously affected by the construction of hydro-power facilities that flooded their lands (as in Burma and Thailand), geothermal plants displacing sacred sites (in the Philippines), and nuclear power plants that have jeopardized human health in Native American communities in the United States (Galloway-McLean 2010).

Other local and indigenous communities in various parts of the world, however, have seen or are anticipating benefits from carbon trading projects (Galloway-McLean 2010; Collings 2009). An example of traditional knowledge holders' collaboration with the private sector is the West Arnhem Land Fire Abatement Project in Australia.[4] In this climate change mitigation programme, Aboriginal communities are receiving significant funds from a private energy firm (Darwin Liquefied Natural Gas) for carrying out their traditional fire management practices, which have been shown to reduce greenhouse gas emissions as compared to naturally recurring wildfires. Also in Australia, the oil company ConocoPhillips agreed in 2007 to pay a group of indigenous communities (practicing traditional fire management) in northern Australia AUD $1million per year for a 17-year period to offset their refinery's greenhouse gas emissions (Collings 2009).

In many countries, local and indigenous communities are preparing for engagement in carbon trading markets in hopes of benefiting financially from their traditional forest management practices that help to maintain forest carbon stores. This often involves collaboration and assistance from NGOs and international research organizations, such as the World Agroforestry Center (ICRAF) through its RUPES programme.[5] In the Philippines, for example, the Ikalahan community, which manages their traditional forest areas for both conservation and various utilization

---

[3]The Clean Development Mechanism is one of three mechanisms (the others being Emissions Trading and Joint Implementation, created under the Kyoto Protocol; the Protocol was established under the United Nations Framework Convention on Climate Change (UNFCC) in 1997 and entered into force in 2005 http://unfccc.int/kyoto_protocol/items/2830.php).

[4]http://www.savanna.org.au/al/fire_abatement.html

[5]RUPES: Rewards for, Use of and shared investment in Pro-poor Environmental Services http://rupes.worldagroforestry.org/overview

purposes, have received support and technical assistance from ICRAF to prepare for possible Clean Development Mechanism funding, and have reportedly engaged with Mitsubishi UFJ Securities for this purpose.

REDD (Reduced Emissions from Deforestation and Forest Degradation) is among the more controversial approaches for climate change mitigation involving forests and forest-dependent communities. REDD is a strategy that seeks to address the perceived shortcomings of the Kyoto Protocol, in which mitigation policies related to deforestation and forest degradation were excluded due to the complexity of measurements and monitoring for such diverse ecosystems and land use changes. REDD expands the scope of carbon offset activities for which carbon credits could be earned to include forests. It is an evolving approach that aims to use market and financial incentives to reduce emissions of greenhouse gases from deforestation and forest degradation—estimated to contribute between 15% and 25% of total CO_2 emissions worldwide (Kindermann et al. 2008; Van der Werf et al. 2009)—while also delivering 'co-benefits' such as biodiversity conservation and poverty alleviation.

In 2005, at the 11th Conference of the Parties (COP-11), the Coalition of Rainforest Nations[6] proposed inclusion of greenhouse gas emissions from deforestation in climate negotiations. Discussions on this evolved to REDD+, which was included in the 'Bali Action Plan' that emerged from the UNFCCC COP-13 meeting in 2007. REDD+ includes "policy approaches and positive incentives on issues relating to reducing emissions from deforestation and forest degradation in developing countries; and the role of conservation, sustainable management of forests and enhancement of forest carbon stocks in developing countries" (UNFCCC Decision 1/CP.13, Bali Action Plan).

A deadline for agreement on the specifics of an international REDD mechanism was set for 15th Conference of the Parties to the UNFCCC (COP-15), held in Copenhagen in December 2009. The Copenhagen Accord of 18 December 2009 included recognition of the crucial role of REDD and REDD+ and the need to provide positive incentives for such actions by enabling the mobilization of financial resources from developed countries. While REDD and REDD+ mechanisms are still under discussion, it is likely that future climate agreements will allow for offsetting carbon emissions through 'savings' from avoided deforestation.

REDD activities are undertaken by national or local governments, large NGOs, the private sector, or a combination of these. The World Bank, other International Financial Institutions and the UN support REDD by setting up the bases for the carbon market and the legal and governance frameworks of countries receiving REDD support. A number of governments of industrialized countries,[7] NGOs

---

[6] http://www.rainforestcoalition.org/

[7] At the 2007 UNFCCC Conference in Bali in 2007, for example, the Norwegian government announced their International Climate and Forests Initiative, which provided $500 million towards the creation and implementation of national-based, REDD activities in Tanzania. Since then a number of other industrialized countries' governments including Australia, Denmark and Spain have followed with major financial commitments to UN-REDD and other international as well as national REDD/REDD+ programmes.

(particularly some larger conservation organizations), development agencies, research institutes and international organizations provide support to countries that wish to engage in REDD activities. These include the World Bank's Forest Carbon Partnership Facility, the UN-REDD Programme, and Norway's International Climate and Forest Initiative.

The World Bank presently plays a leading role in the facilitation of REDD activities. As one of the financial contributors for the REDD programme, the World Bank has created a $300 million fund, the Forest Carbon Partnership Facility (FCPF), for initiation of REDD activities in developing countries. The UN-REDD Programme[8] was established in 2008 by UNDP, UNEP, and FAO to help developing countries address measures needed to effectively participate in the REDD mechanism, including development, governance, engagement of indigenous peoples, and technical support. The Programme also supports global activities on measuring, verifying, and reporting systems, engagement of indigenous peoples and other forest-dependent communities, and governance. In 2010, eight countries—Bolivia, Democratic Republic of Congo, Zambia, Cambodia, Papua New Guinea, Paraguay, the Philippines, and the Solomon Islands—received funding totaling US$51 million from the Programme for development of full or initial national programmes for REDD+ readiness and implementation (UN-REDD 2011).

The success of REDD activities will largely depend on the effective engagement of people in local and indigenous communities whose livelihoods are derived from forests. In the climate change policy arena, many groups are actively promoting 'safeguards' and other means to ensure this. The Accra Caucus on Forests and Climate Change, for example, works to place the rights of indigenous and forest communities at the centre of REDD negotiations and to ensure that efforts to reduce deforestation promote good governance and are not a substitute for emission reductions in industrialized countries.[9] Many grassroots organizations are working to develop REDD activities with communities that include benefit-sharing mechanisms to help ensure that REDD funds reach rural communities as well as governments.

Critics of REDD, such as the International Indigenous Peoples Forum on Climate Change (IIPFCC), argue that REDD mechanisms could facilitate logging operations in primary forests, increase violations of indigenous peoples' rights, displace local populations in the interest of 'conservation,' threaten indigenous agricultural practices, destroy biodiversity and cultural diversity, cause social conflicts, and increase expansion of tree plantations at the expense of native forests. Deficiencies in current major REDD programmes cited by critics include: weak consultation processes with local communities, the lack of criteria to determine when a country is ready to implement REDD projects, the potential negative impacts on forest ecosystems and their biodiversity due to insufficient planning; the lack of safeguards to protect indigenous peoples' rights, and the lack of regional policies to stop

---

[8]http://www.un-redd.org/

[9]The Accra Caucus is a network of NGOs representing approximately 100 civil society and indigenous peoples' organizations from 38 countries. It was formed at the UNFCCC meeting in Accra, Ghana, in 2008 (http://www.rainforestfoundationuk.org/Accra_Caucus).

deforestation; and current operational definitions of 'forests' that do not distinguish between natural forests and plantations, which could result in expansion of plantation forests at the expense of natural forest ecosystems and the communities who depend upon them (Butler 2009). The World Bank's declared commitment to fight against climate change, and its role in promoting REDD, is also viewed with caution by many civil society organizations and grassroots movements who consider the processes being developed under the FCP to be problematic (Dooley et al. 2011).

A number of technical, scientific, economic, political, and social questions likely to affect the success or failure of REDD/REDD+ programmes and activities remain unresolved. Optimizing carbon sequestration and environmental and socioeconomic 'co-benefits' such as biodiversity conservation envisioned under REDD+ schemes will be a challenge given the current state of knowledge regarding the relationships between forest carbon storage, biodiversity and ecosystem services, and how these relationships are affected by forest management activities, including forest restoration (White et al. 2010; Nasi et al. 2011; Sasaki et al. 2011). These topics are receiving increasing attention within the scientific community, as well in policy forums including the Convention on Biological Diversity (SCBD 2006, 2009).

For local and indigenous communities who manage forested landscapes using practices based on their traditional knowledge, these uncertainties revolve around several key issues (Peskett et al. 2008):

- Accountability, particularly under conditions of poor governance, and the fair and equitable distribution of REDD benefits to forest communities that minimizes capture of these benefits by national governments or local elites;
- Full and effective participation of indigenous peoples and forest-dependent communities in the design, implementation, and monitoring of REDD activities, and respect for their human rights;
- Prevention of 'carbon leakage,' caused by the displacement of deforestation to other areas;
- Achieving and balancing multiple benefits (conservation of biodiversity, ecosystem services, and social benefits, e.g., income and improved forest governance);
- Concern that by putting a monetary value on forests, the spiritual value they hold for indigenous peoples and local communities will be neglected; and
- The need in many countries for prior reform in forest governance and more secure tenure systems, without which fair distribution of REDD benefits will not be achieved.

Pirard and Treyer (2010) argue that in order for REDD+ mechanisms to be effective, supportive agricultural policies will be required. These policies should (1) foster changes in agricultural technologies that may contribute to forest conservation objectives; (2) harmonize sectoral public policies that have a direct or indirect impact on forest cover (e.g., agriculture, transportation, etc.); and (3) adopt the principle of payments for environmental services (PES) to promote measures that condition support for the adoption of sound agricultural technologies on the absence of excessive forest clearing on nearby lands.

Proponents of UN-REDD and other REDD schemes are optimistic that such arrangements could present huge financial opportunities for both biodiversity conservation and for local and indigenous forest communities who could be paid to preserve these forests as carbon sinks. Supporters suggest that the funding promised through REDD programmes to governments and communities will help to overcome the obstacles that have severely hampered efforts to reduce forest loss and degradation over the past century.

Others are more sceptical, pointing out that the underlying causes of tropical deforestation and degradation are complex and multi-faceted, related to a number of political, social, economic, and environmental problems—such as poverty; population growth and increasing scarcity of land and other natural resources; human rights; governance issues; national policies in finance, energy, transportation, agricultural, and other non-forest sector; and the like. Large infusions of money into these contexts is unlikely to produce intended forest conservation results without creating social conflict and other unintended negative outcomes until these pre-existing, often recalcitrant, issues are resolved. Critics further suggest that REDD schemes could also directly threaten these same communities if currently intractable issues related to land rights are not properly addressed to recognize their collective and individual rights and customary laws regarding these forests and the livelihoods they support (UNPFII 2007; Gerrard 2008; Collings 2009; Lyster 2010). Unless these and other governance issues are resolved satisfactorily, these communities may not receive adequate (or any) financial benefits, and could face limitations on their traditional uses of forests (or eviction), which would have serious consequences for their livelihoods, cultures, and the future of their traditional forest-related knowledge.

An analysis of past research on tropical deforestation by Kanninen et al. (2007) suggests that the design and implementation of effective REDD policies will be highly problematic, given the complexity of social, economic, environmental, and political dimensions of deforestation and forest degradation. This study reiterates the findings of many others over the past 20 years that underlying causes of deforestation are generated outside the forestry sector, that alternative land uses tend to be more profitable than conserving forests, and that institutions (local, national, and global) for harmonizing the interests and actions of diverse economic actors with the public interest are generally weak.

The uneven playing field on which local and indigenous communities engage with local elites, national governments, investors, and international organizations, discussed above and in Chap. 12 of this book (Sect. 12.3.4), suggests that there is reason for concern that REDD programmes, and carbon markets generally, may not produce positive outcomes from the point of view of traditional communities or their forests, and could exacerbate ongoing conflicts between local forest communities and the state.

13.5.4   Global Forest-Related Climate Change Mitigation Strategies: Biofuels

Biofuels, which include a wide range of fuels derived from biomass, are gaining increased importance and attention from the public, policy-makers, and the scientific community as an alternative to fossil fuels and their potential to help mitigate climate change. Although traditional biomass resources (fuelwood, charcoal, and agricultural residues) are the primary source of domestic energy for cooking and heating used by an estimated 500 million households worldwide (Bringezu et al. 2009), the demand for energy derived directly or indirectly (i.e., liquid biofuels) from biomass is rising steeply. Transport biofuels include ethanol produced from maize and sugar cane, and biodiesel from rapeseed, palm oil, and soya, which collectively accounted for 1.8% of the world's transport fuel in 2008. The world's largest producers at present are the United States, Brazil, and the European Union, although biodiesel production is expanding rapidly in Southeast Asia (principally in Malaysia, Indonesia, China, and Singapore); Latin America (Argentina and Brazil); and in southeast Europe (in Romania and Serbia). Global investment (both private and public) into biofuel production capacity is increasing, and exceeded US$4 billion in 2007 (Bringezu et al. 2009).

Biofuel production and use raise a variety of technical, environmental, economic, and social issues that have been the subject of considerable study and debate within the scientific community, international policy forms, and the media. Many of these issues have, or could have, important implications for forest-dependent indigenous and local communities. These include effects on: global fossil fuel prices; energy balances and efficiencies; greenhouse gas emissions (and thus climate change); land use change and its implications for food security, agricultural productivity, forest loss, and biodiversity; poverty reduction potential; the sustainability of biofuel production systems; and water resources (Laurance 2007, Persson and Azar 2010). A global assessment of these issues associated with biomass energy is provided by Bringezu et al. (2009). This study covers trends and drivers of biofuels production, life-cycle-wide environmental impacts, impacts associated with increased demand for biofuels and land use change, options for more efficient and sustainable production and use of biomass, strategies and measures to enhance resource productivity, and research and development needs.

The production of biofuels requires energy for growing biomass crops, including agricultural inputs (fertilizers and biocides) required for their production, as well as for transport and conversion into usable fuel products—all of which have environmental consequences, including production of greenhouse gases. The energy balance of biofuels—i.e., the ratio of the energy released from the fuel when it is burned to the energy required for its production—is highly variable, depending in part on the feedstock used, the final product, and the energy inputs involved in the processes of production, transportation, and conversion. According to a study by Pimentel and Patzek (2005), energy outputs from ethanol produced by using corn, switchgrass, and wood biomass were each less than the respective fossil energy inputs. Similar results were obtained for biodiesel using soybeans and sunflower, although the energy cost for producing soybean biodiesel was only slightly negative compared with ethanol production.

The scientific literature on 'life cycle analyses' that calculate production of CO_2 and other greenhouse gas costs through the full cycle of biofuel production from planting to final use of the fuel have yielded highly variable results, depending on the feedstock (crop used), conversion technologies, and other factors. Such studies suggest that under the right circumstances, substitution of fossil fuels by biofuels can result in net savings of greenhouse gas emissions (Bringezu et al. 2009). However, such studies do not take into consideration increases in greenhouse gas emissions that may result from indirect land use changes (as in cases of conversion of forests). Consideration of the broader land use dynamics in the context of biofuel production is therefore very important for the assessment of the climate change mitigation potential of biofuels. Life cycle assessments that have taken this broader landscape view provide support for the assertion that biofuel production under many, perhaps most, circumstances, will result in net increases in greenhouse gas emissions and thus exacerbate global climate change (Bringezu et al. 2009; Searchinger et al. 2009).

More immediate concerns for local and indigenous communities living in forested regions where large-scale production of biofuel crops are either underway or planned are that land and livelihood security will decrease and poverty worsen. Depending on location, large-scale expansion of biofuel feedstock production can involve major diversion of land and water resources that are currently used for production of food crops or for animal husbandry; conversion (i.e., clearing) of non-agricultural land (forests and woodlands, for example); and/or the displacement of farmers (practicing permanent or shifting cultivation) or pastoralists from their traditionally used lands (Degawan 2008). While these issues are important to traditional communities in many parts of the world, they are of particular concern to rural communities in Africa, where food security is a paramount interest (Molony and Smith 2010). In Brazil, rapid expansion of soybean production (used increasingly for biodiesel) is expected to focus on conversion of pasture land and savannah (cerrado) ecosystems. In Southeast Asia, palm oil plantation expansion (for both food and biofuel purposes) is a leading cause deforestation. In Indonesia, for example, where approximately 30% of total palm oil is produced by smallholders, two-thirds of current production expansion is based on conversion of rainforests,[10] the remainder being produced on previously cultivated or fallow lands. Palm oil production in Indonesia, which currently uses approximately six million ha of land, is planned to increase to 20 million ha in the not-distant future (Bringezu et al. 2009). This growing industry supports as many as 4.5 million households, most drawn

---

[10] Approximately 25% of this converted forest land is on peat soils, which when drained for oil palm production result in particularly high greenhouse gas emissions. By 2030, oil palm plantations on peat soils are expected to constitute 50% of the total, by which time the total rainforest area of Indonesia is expected to be reduced by 29% (compared to 2005) or 49% of the county's rainforest area in 1990 (Bringezu et al. 2009).

from local and indigenous communities who lost their lands to expansion of palm oil plantations (Bailey 2007).

According to a study undertaken by the Overseas Development Institute [UK], biofuels could help relieve poverty in some developing countries through increased employment, economic growth multipliers, and lower prices for imported fossil fuels used for transportation, agriculture, and domestic needs (Leturque 2009). However, this study also points out that the impacts with respect to poverty reduction may be quite different in situations where biofuel feedstock production is undertaken on a large scale; or creates pressure on limited agricultural resources such as capital, land, and water, or ultimately increases the price of food. The authors note that production of biofuels is subject to the same policy, regulatory, or investment shortcomings that influence the potential of agriculture to improve rural incomes. Since these limitations require *national* policy solutions, a country-by-country analysis of the potential poverty impacts of biofuels is more appropriate than global analyses.

As discussed earlier in the context of REDD programmes, large-scale biofuel production has a high potential to threaten the livelihoods of local and indigenous communities if issues related to land rights are not properly addressed to recognize their collective and individual rights and customary laws regarding agricultural lands, forests, and water resources (Tauli-Corpuz and Tamang 2007; Degawan 2008). As noted above, these problems are already arising in countries where rising demand for palm oil (and other biofuel crops) has resulted in large-scale plantation establishment on sites recently occupied by forests, as discussed also in Sect. 12.3.4. In Indonesia, for example, an estimated 38,000 km² have been converted to palm oil plantations since 1996 (Bailey 2007). Worldwide, several millions of indigenous people have been affected by the deforestation of their lands to make way for biofuel plantations (Bailey 2007).

Rising concerns about the negative environmental and social side effects of biofuels' production have prompted some governments and international organizations to promote the development of criteria for sustainable bioenergy production. However, as pointed out by Bringezu et al. (2009) and many others, these standards and certification schemes rely on life-cycle analysis methods that often fail to take into account impacts (specifically greenhouse gas emissions) along the full biofuel production chain, including landscape-level impacts related to land-use changes (Dale et al. 2010). A recent review by van Dam et al. (2010) evaluated 67 ongoing certification initiatives to safeguard the sustainability of bioenergy, most focusing on sustainability of liquid biofuels. Their analysis confirmed concerns that while environmental impacts are generally considered in standards developed by these initiatives, indirect land-use change and socio-economic impacts of bioenergy production generally are not. An exception to this is an initiative in Brazil known as the 'social fuel label,' which aims to protect small-scale farmers in areas of large-scale biofuel production. They suggest that certification schemes have the potential to influence direct, local impacts related to environmental and social effects of direct bioenergy production provided that unwanted land-use changes and governance issues are properly addressed.

13.6   Conclusions and Recommendations

Traditional forest-related knowledge, innovations, practices, and associated social institutions have supported the livelihoods and cultures of local and indigenous communities for generations. Such communities have often been very successful in adapting to environmental change. Their traditional knowledge—an important component of their cultural capital—represents an important community asset that can enhance the adaptive capacity of indigenous and local communities and reduce their vulnerability to climate change (Salick and Byg 2007; Osman-Elasha et al. 2009).

Although traditional forest-related knowledge represents an important element of adaptive capacity for local and indigenous communities to cope with climate change, it is important to recognize their limitations for adaptation to changing environmental conditions (Eastaugh 2008; FAO 2005, 2009; Collings 2009). While traditional knowledge, practices, and associated social institutions have historically been dynamic, their capacity to adapt quickly enough to the more dramatic climate-change impacts on forests envisaged for some regions cannot be assumed (Adger 1999; Adger et al. 2003, Macchi et al. 2008). This may be particularly problematic in cases where these impacts exceed the limits of applicability of this knowledge, as for example in traditional hunting and fishing communities when climate change impacts create major changes in wildlife ranges and habitat structure. Exacerbating this is the fact that traditional knowledge is already disappearing from local and indigenous communities in many parts of the world for a number of reasons (c.f., Newing et al. 2005; Galloway-McLean 2010), as discussed in earlier chapters of this volume. Because such knowledge is rarely codified and often restricted to a few members (usually the elders) of local, particularly indigenous, communities, its vulnerability to inter-generational loss is increased.

For traditional societies that have maintained their systems of reciprocity (sharing knowledge among communities), coping with change may be easier and may not create conflict within and among communities. As Nuttall (2008) pointed out with reference to indigenous peoples, the adaptive capacity and resilience of communities in the face of significant environmental change depends on several factors: the strength of relationships between people and the environment; community cohesiveness and cultural identity; strong social relations; and the existence of enabling policies and governance that support traditional livelihood practices, food security, and other economic, social and cultural needs.

Opportunities exist to preserve and/or revitalize traditional forest-related knowledge and to develop appropriate strategies for adaptation of forest resource management to climate change that combine traditional and scientific forest-related knowledge (Kruger 2005). Traditional knowledge can complement formal science in various ways, for example in monitoring the effects of climate change (Vlassova 2002). The scientific community has an important role to play in this through collaborative and participatory research with the holders and users of traditional forest-related knowledge to build on existing knowledge (both traditional and formal scientific) to better understand and deal with uncertainties that climate change presents. Such research collaboration could encompass a wide range of topics and objectives. Examples include: monitoring and modelling of climate change impacts on forest ecosystem structure and processes; development of improved practices for increasing resilience in managed forest ecosystems; and addressing threats to productivity in traditional agricultural systems such as shifting cultivation and agroforestry affected by climate change (e.g., shifting rainfall and temperature patterns, inadequate surface and groundwater resources, and salt water intrusion in coastal groundwater supplies due to rising sea levels).

Studies carried out in a manner consistent with ethical principles using research methodologies appropriate for the study of traditional forest-related knowledge (discussed in Chap. 14) can help elucidate underlying ecological principles that could enable wider application and further development of traditional knowledge and practices. Knowledge gained from such research could be translated into innovative forest and agricultural management practices to help a broader spectrum of rural communities adapt to the impacts of climate change (Roberts et al. 2009).

In recent years, indigenous communities have been among the first to react to the impacts of climate change and measures taken at the international level towards its mitigation. Traditional communities have raised the awareness of the global public and policy-makers by recording and publicizing their observations of changes in climate and its effects on the natural environments in which they live. They are also reacting to climate change mitigation approaches that involve significant changes in land use and forest management regimes affecting their livelihoods and cultures. By doing so they are helping to inform the ecological, socio-economic, legal, and human rights dimensions of climate change and policies being developed to combat and/or adapt to its impacts.

Policy makers and other decision makers should give greater attention to the experience and insights that have developed effective practices and related social institutions for coping with environmental change to ensure their food- and forest-based livelihood security (Nilsson 2008; Roberts et al. 2009). Rural communities are more likely to adopt mitigation and adaptation strategies based on principles derived from traditional knowledge that are cost-effective, participatory, and compatible with existing social and governance structures (Robinson and Herbert 2001; Hunn 1993). A study in Zimbabwe, for example, found a greater willingness of farmers to use seasonal weather forecasts based on scientific (meteorological) predictions when such information was presented and compared with local indigenous climate forecasts (Patt and Gwata 2002).

The importance of traditional forest-related knowledge to local and indigenous communities for climate change adaption needs to be recognized, even though TFRK has declined in most regions of the world. Supportive policies and actions should be implemented to facilitate its preservation and further development. (Roberts et al. 2009). Both this knowledge, and the rights, interests, and traditional practices and social institutions of those communities who are responsible for its development and preservation need to be respected in the development of policies and emerging global mechanisms for climate change mitigation and adaptation.

Recent experiences of local and, particularly, indigenous communities with global policy-making bodies, international financial institutions, and others involved in the creation and promotion of climate change mitigation mechanisms (such as carbon markets, Clean Development Mechanism projects, and biofuels development in particular) have been mixed but generally unfavourable to the interests of these local communities (Degawan 2008; Garcia-Alix 2008). It remains to be seen whether emerging mitigation mechanisms such as REDD+ will be an improvement from the point of view of traditional forest-dependent communities. Much will depend on the extent to which the playing field in national and international climate change and forest policy debates can be levelled. At present, the cultural capital of traditional societies is presently no match for the financial capital of the interests that dominate these policy debates worldwide (i.e., investors, international financial institutions, and large international conservation organizations, among others). Greater appreciation of alternative approaches to mitigating and adapting to climate change by policy makers, the scientific community, and the general public could change this equation, however, with potential benefits for all societies.

References

Adger WN (1999) Social vulnerability to climate change and extremes in coastal Vietnam. World Devt 27(2):249–269
Adger WN, Brown K, Fairbrass J, Jordan A, Paavola J, Rosendo S, Seyfang G (2003) Governance for sustainability: towards a 'thick' analysis of environmental decisionmaking. Environ Plan 35:1095–1110
Adger WN, Agrawala S, Mirza MMQ, Conde C, O'Brien K, Pulhin J, Pulwarty R, Smit B, Takahashi K (2007) Assessment of adaptation practices, options, constraints and capacity. In: Parry ML, Canziani OF, Palutikof JP, van der Linden PJ, Hanson CE (eds) Contribution of working group II to the fourth assessment report of the intergovernmental panel on climate change. Cambridge University Press, Cambridge/New York, pp 717–743
Agnoletti M (2005) Osservazioni sulle dinamiche dei boschi e del paesaggio forestale italiano fra il 1862 e la fine del secolo XX. Società Storia 108:377–396
Agnoletti M (ed) (2006) The conservation of cultural landscapes. CAB International, Wallingford/Cambridge
Agnoletti M (2007) The degradation of traditional landscapes in a mountain area of Tuscany during the 19th and 20th centuries: implications for biodiversity and sustainable management. For Ecol Manag 249(1–2):5–17
Agnoletti M (2010a) Paesaggio Rurale. Edagricole, Milano
Agnoletti M (ed) (2010b) Paesaggi rurali storici—per un catologo nazionale [Historical rural land-scapes—for a national register]. Laterza, Bari
Ajibade LT, Shokemi O (2003) Indigenous approaches to weather forecasting in Asa L.G.A., Kwara State, Nigeria. Indilinga Afr J Indig Knowl Syst 2:37–44
Bailey R (2007) Bio-fuelling poverty: why the EU renewable-fuel target may be disastrous for poor people. Oxfam briefing note. Oxfam International, Oxford
Balehegn M, Eniang EA (2009) Assessing indigenous knowledge for evaluation, propagation and conservation of indigenous multipurpose fodder trees towards enhancing climate change Adaptation in Northern Ethiopia. In: Parrotta JA, Oteng-Yeboah A, Cobbinah J (eds) Traditional forest-related knowledge and sustainable forest management in Africa,

vol 23, IUFRO World Series. International Union of Forest Research Organizations, Vienna, pp 39–46

Benson C, Clay EJ (1998) The impact of drought on sub-Saharan economies. The World Bank technical paper no. 401. World Bank, Washington, DC

Berkes F, Colding J, Folke C (2000) Rediscovery of traditional ecological knowledge as adaptive management. Ecol Appl 10:1251–1262

Bertolotto S, Cevasco R (2000) The "alnocolture" system in the Ligurian Eastern Apennines: archived evidence. In: Agnoletti M, Anderson S (eds) Methods and approaches in forest history. CAB International, Wallingford/New York, pp 189–202

Bharara LP (2003) Agroforestry development and management through human resource development at the grassroot level. In: Seeland K, Schmithuesen F (eds) Indigenous knowledge, forest management and forest policy in South Asia. D.K. Printworld (P) Ltd, New Delhi, pp 195–209

Braudel F (1966) Le Mediterranee et le monde Mediterraneen a l'epoque de Philippe II, 2nd edn, 2 vols. Librairie Armand Colin, Paris

Bringezu S, Schütz H, O'Brien M, Kauppi L, Howarth RW, McNeely J (eds) (2009) Towards sustainable production and use of resources: assessing biofuels. United Nations Environment Programme, Washington, DC. Available via http://www.unep.fr/scp/rpanel/pdf/Assessing_Biofuels_Full_Report.pdf. Cited 25 Mar 2011

Bureau of Meteorology [Australian Government] (2011) Indigenous weather knowledge. Available via http://www.bom.gov.au/iwk/. Cited 22 Mar 2011

Butler R (2009) Are we on the brink of saving rainforests? Available via http://news.mongabay.com/2009/0722-redd.html. Cited 23 Mar 2009

Carmenza R, Kanninen M, Pedroni L (2005) Tropical forests and adaptation to climate change: in search of synergies. Center for International Forestry Research, Bogor

CBD (2007): See Convention on Biological Diversity, Working Group on Article 8(j) [CBD]

Ciancio O, Nocentini S (2004) Il bosco ceduo: Selvicoltura, assestamento, gestione. Accademia Italiana di Scienze Forestali, Florence

Colfer CJP, Sheil D, Kaimowitz D, Kishi M (2006) Forests and human health: assessing the evidence. CIFOR occasional paper 45. Center for International Forestry Research, Bogor

Collings N (2009) Environment. In: United Nations (ed) The state of the world's indigenous peoples, Report no. ST/ESA/328. Department of Economic and Social Affairs, Division for Social Policy and Development, Secretariat of the Permanent Forum on Indigenous Issues, New York, pp 84–127

Convention on Biological Diversity, Working Group on Article 8(j) [CBD] (2007) Revision of the second phase of the composite report on the status and trends regarding the knowledge, innovation and practices of indigenous peoples and local communities: Africa. UN Doc. UNEP/CBD/WG8J/AG/2/2/Add.2 April 2007. Available via http://www.cbd.int/doc/meetings/tk/acpow8j-02/official/acpow8j-02-02-add1-en.pdf. Cited 22 Mar 2011

Crump J (2008) Many strong voices: climate change and equity in the Arctic and Small Island Developing States. Indig Aff 1–2(2008):24–33. Available via http://www.iwgia.org/. Cited 25 Mar 2011

Cussò X, Garrabou R, Olarieta JR, Tello E (2006) Balances energéticos y usos del suelo en la agricultura catalana: Una comparación entre mediados del siglo XIX y finales del siglo XX. Hist Agrar 40:471–500

Dale VH, Kline KL, Wiens J, Fargione J (2010) Biofuels: implications for land use and biodiversity. Biofuels and sustainability reports. Ecological Society of America, Washington, DC. Available via www.esa.org/biofuelsreports. Cited 25 Mar 2011

Daniere C, Capellano A, Moiroud A (1986) Dynamique de l'azote dans un peuplement naturel d'*Alnus incana* (L.) Moench. Acta Oecol 7:165–175

Dea D, Scoones I (2003) Networks of knowledge: how farmers and scientists understand soils and their fertility: a case study from Ethiopia. Oxf Dev Stud 31:461–478

Degawan M (2008) Mitigating the impacts of climate change: solutions or additional threats. Indig Aff 1–2(2008):52–59. Available via http://www.iwgia.org/. Cited 25 Mar 2011

Dooley K, Griffiths T, Martone F, Ozinga S (2011) Smoke and mirrors: a critical assessment of the Forest Carbon Partnership Facility. Brussels and Moreton-in-Marsh (UK): FERN and Forest Peoples Programme. Available via http://www.fern.org/smokeandmirrors. Cited 15 Apr 2011

Eastaugh C (2008) Adaptations of forests to climate change: a multidisciplinary review. IUFRO occasional paper 21. International Union of Forest Research Organizations, Vienna

Easterling WE et al (2007) Food, fibre, and forest products. In: Parry ML, Canziani OF, Palutikof JP, van der Linden PJ, Hanson CE (eds) Contribution of working group II to the fourth assessment report of the intergovernmental panel on climate change. Cambridge University Press, Cambridge/New York, pp 273–314

FAO (2005, 2009): See Food and Agricultural Organization of the United Nations

Farina A (2000) The cultural landscape as a model for the integration of ecology and economics. BioScience 50:313–320

Food and Agricultural Organization of the United Nations [FAO] (2005) State of the world's forests 2005. FAO, Rome

Food and Agricultural Organization of the United Nations [FAO] (2009) State of the world's forests 2009. Food and Agriculture Organization of the United Nations, Rome

Galacgac ES, Balicacan CM (2009) Traditional weather forecasting for sustainable agroforestry practices in Ilocos Norte Province, Philippines. For Ecol Manag 257:2044–2053

Galloway-McLean K (2010) Advance guard: climate change impacts, adaptation, mitigation and indigenous peoples—a compendium of case studies. United Nations University-Traditional Knowledge Initiative, Darwin. Available via http://www.unutki.org/news.php?doc_id=101&news_id=92. Cited 20 Mar 2011

Gana FS (2003) The usage of indigenous plantmaterials among small-scale farmers in Niger state agricultural development project: Nigeria. Indilinga Afr J Indig Knowl Syst 2:53–60

Garcia-Alix L (2008) The United Nations permanent forum on indigenous issues discusses climate change. Indig Aff 1–2(2008):16–23. Available via http://www.iwgia.org/. Cited 25 Mar 2011

Gerrard E (2008) Climate change and human rights: issues and opportunities for indigenous peoples. Univ N S W [UNSW] Law J 31(3):941–952

Grove AT, Rackham O (2001) The nature of Mediterranean Europe—an ecological history. Yale University Press, New Haven/London

Haberl H, Fischer-Kowalski M, Krausmann F, Weisz H, Winiwarter V (2004) Progress towards sustainability? What the conceptual framework of material and energy flow accounting (MEFA) can offer. Land Use Policy 21(3):199–213. doi:10.1016/j.landusepol.2003.10.013. Cited 20 March 2011

Hunn E (1993) What is traditional ecological knowledge? In: Williams N, Baines G (eds) Traditional ecological knowledge: wisdom for sustainable development. Centre for Resource and Environmental Studies, Australian National University, Canberra, pp 13–15

International Union of Forest Research Organizations [IUFRO] (2007) Guidelines for the implementation of social and cultural values in sustainable forest management, a scientific contribution to the implementation of MCPFE—Vienna resolution 3. IUFRO occasional paper no.19. International Union of Forest Research Organizations, Vienna

Jimenez-Ferrer G, Perez-Lopez H, Soto-Pinto L, Nahed-toral J (2007) Livestock, nutritive value and local knowledge of fodder trees in fragmented landscapes in Chiapas, Mexico. Interciencia 32(4):274–280

Kanninen M, Murdiyarso D, Seymour F, Angelsen A, Wunder S, German L (2007) Do trees grown on money? The implications of deforestation research for policies to promote REDD. Forest perspectives 4. Center for International Forestry Research, Bogor

Kindermann G, Obersteiner M, Sohngen B, Sathaye J, Andrasko K, Rametsteiner E, Schlamadinger B, Wunder S, Beach R (2008) Global cost estimates of reducing carbon emissions through avoided deforestation. Proc Nat Acad Sci USA 105:10302–10307

King KFS (1987) The history of agroforestry. In: Steppler HA, Nair PKR (eds) Agroforestry—a decade of development. International Council for Research in Agroforestry, Nairobi, pp 1–11. Available via http://www.worldagroforestrycentre.org/downloads/publications/PDFs/07_Agroforestry_a_decade_of_development.pdf. Cited 20 Mar 2011

Krausmann F, Erb K-H, Gingrich S, Lauk C, Haberl H (2008) Global patterns of socioeconomic biomass flows in the year 2000: a comprehensive assessment of supply, consumption and constraints. Ecol Econ 65:471–487

Kronik J, Verner D (2010) The role of indigenous knowledge in crafting adaptation and mitigation strategies for climate change in Latin America. In: Mearns R, Norton A (eds) Social dimensions of climate change: equity and vulnerability in a warming world. The World Bank, Washington, DC, pp 145–169. doi:10.1596/978-0-8213-7887-8, Cited 26 Mar 2011

Kruger LE (2005) Community and landscape change in southeastern Alaska. Landsc Urban Plan 72:235–249

Kuskova P, Gingrich S, Krausmann F (2008) Long term changes in social metabolism and land use in Czechoslovakia, 1830–2000: an energy transition under changing political regimes. Ecol Econ 68:394–407

La Mantia T, Giaimi G, La Mela Veca DS, Tomeo A (2006) The cessation of utilization of Erica arborea: an example of cancellation of a cultural landscape and its naturalistic values. In: Parrotta JA, Agnoletti M, Johann E (eds) Cultural heritage and sustainable forest management: the role of traditional knowledge, vol 1. Ministerial conference on the protection of forests in Europe. Liaison Unit Warsaw, Warsaw, pp 58–65. Available via http://www.iufro.org/science/task-forces/traditional-forest-knowledge/publications/

La Mantia T, Giaimi G, La Mela Veca DS, Pasta S (2007) The role of traditional Erica arborea L. management practices in maintaining northeastern Sicily's cultural landscape. For Ecol Manag 249:63–70

Laurance W (2007) Switch to corn promotes Amazon deforestation. Science 318(5857):1721. doi:10.1126/science.318.5857.1721b

Laureano P (2005) The water atlas—traditional knowledge to combat desertification. Laia Libros, Barcelona, 437 p

Leturque H (2009) Biofuels: could the south benefit? Briefing paper 48. Overseas Development Institute, London

Llorens P, Latron J, Gallart F (1992) Analysis of the role of agricultural abandoned terraces on the hydrology and sediment dynamics in a small mountainous basin (High Llobregat, Eastern Pyrenees). Pirineos 139:27–46

Lucier A, Ayres M, Karnosky D, Thompson I, Loehle C, Percy K, Sohngen B (2009) Forest responses and vulnerabilities to recent climate change. In: Seppälä R, Buck A, Katila P (eds) Adaptation of forests and people to climate change—a global assessment report, vol 22, IUFRO World Series. International Union of Forest Research Organizations, Helsinki, pp 29–52

Lyster R (2010) REDD+, transparency, participation and resource rights: the role of law. Sydney law school legal studies research paper no. 10/56. University of Sydney, Sydney. Available via http://ssrn.com/abstract=1628387. Cited 20 Mar 2011

Macchi M, Oviedo G, Gotheil S, Cross K, Boedhihartono A, Wolfangel C, Howell M (2008) Indigenous and traditional peoples and climate change. IUCN issues paper. International Union for Conservation of Nature, Gland. Available via http://cmsdata.iucn.org/downloads/indigenous_peoples_climate_change.pdf. Cited 26 Mar 2011

Marull J, Pino J, Tello E, Cordobilla MJ (2010) Social metabolism, landscape change and land-use planning in the Barcelona Metropolitan Region. Land Use Policy 27(2):497–510. doi:10.1016/j.landusepol.2009.07.004

Mekoya A, Oosting SJ, Fernandez-Rivera S, Van der Zijpp A (2008) Farmers' perceptions about exotic multipurpose trees and constraints to their adoption. Agrofor Syst 73:141–153

Molony T, Smith J (2010) Biofuels, food security, and Africa. African Affairs (Lond) 109 (436):489–498. doi:10.1093/afraf/adq019

Nasi R, Putz FE, Pacheco P, Wunder S, Anta S (2011) Sustainable forest management and carbon in tropical Latin America—The case for REDD+. Forests 2:200–217. doi:10.3390/f2010200. Available at http://www.mdpi.com/1999-4907/2/1/200/pdf. Cited 26 Mar 2011

Newing H, Pinker A, Leake H (eds) (2005) Our knowledge for our survival. International Alliance of the Indigenous and Tribal Peoples of the Tropical Forest (IAITPTF) and Centre for International Forestry Research (CIFOR), Chiang Mai, 2 vols. Available via http://www.international-alliance.org/documents/overview-finaledit.pdf. Cited 5 Mar 2011

Nilsson C (2008) Climate change from an indigenous perspective: key issues and challenges. Indig Aff 1–2(2008):8–15. Available via http://www.iwgia.org/. Cited 25 Mar 2011

Nuttall M (2008) Climate change and the warming politics of autonomy in Greenland. Indig Aff Clim Change Indig Peoples 1–2(8):8–15. International Working Group on Indigenous Affairs, Copenhagen

Nyong A, Adesina F, Osman EB (2007) The value of indigenous knowledge in climate change mitigation and adaptation strategies in the African Sahel. Mitig Adapt Strateg Glob Chang 12:787–797

Osman-Elasha B, Parrotta J, Adger N, Brockhaus M, Colfer CJP, Sohngen B, Dafalla T, Joyce LA, Nkem J, Robledo C (2009) Future socio-economic impacts and vulnerabilities. In: Seppälä R, Buck A, Katila P (eds) Adaptation of forests and people to climate change—a global assessment report, vol 22, IUFRO World Series. International Union of Forest Research Organizations, Helsinki, pp 101–122

Osunade MA (1994) Indigenous climate knowledge and agricultural practices in Southwestern Nigeria. Malays J Trop Geogr 1:21–28

Parrotta JA, Agnoletti M (2007) Traditional forest knowledge: challenges and opportunities. For Ecol Manag 249:1–4

Parry ML, Canziani OF, Palutikof JP, van der Linden PJ, Hanson CE (eds) (2007a) Contribution of working group II to the fourth assessment report of the intergovernmental panel on climate change. Cambridge University Press, Cambridge/New York

Parry ML, Canziani OF, Palutikof JP, van der Linden PJ, Hanson CE (2007b) Cross-chapter case study. In: Parry ML, Canziani OF, Palutikof JP, van der Linden PJ, Hanson CE (eds) Contribution of working group II to the fourth assessment report of the intergovernmental panel on climate change. Cambridge University Press, Cambridge/New York, pp 843–868

Patt A, Gwata C (2002) Effective seasonal climate forecast applications: examining constraints for subsistence farmers in Zimbabwe. Glob Environ Chang 12:185–195

Persson UM, Azar C (2010) Preserving the world's tropical forest: a price on carbon may not do. Environ Sci Technol 44:210–215. Available via http://pubs.acs.org/doi/abs/10.1021/es902629x. Cited 15 Apr 2011

Peskett L, Huberman D, Bowen-Jones E, Edwards G, Brown J (2008) Making REDD work for the poor. Poverty Environment Partnership, London. Available via http://www.unep-wcmc.org/climate/pdf/Making%20REDD%20work%20for%20the%20poor%20FINAL%20DRAFT%200110.pdf. Cited 26 Mar 2011

Philips AO, Titiola T (1995) Indigenous climate change knowledge and practices: case studies from Nigeria. NISER, Ibadan

Pimentel D, Patzek TW (2005) Ethanol production using corn, switchgrass, and wood; biodiesel production using soybean and sunflower. Nat Resour Res 14(1):65–76. doi:10.1007/s11053-005-4679-8. Cited 23 March 2011

Pirard R, Treyer S (2010) REDD: Agriculture and deforestation: What role should REDD+ and public support policies play? Idées pour le débat N°10/2010. Institute for Sustainable Development and International Relations, Paris. Available via http://www.iddri.org/Publications/Collections/Idees-pour-le-debat/. Cited 23 Mar 2011

Pitte JR (1986) Terres de Castanide. Fayard, Paris

Rackham O, Moody A (1992) Terraces. In: Berit Wells B (ed) Agriculture in ancient Greece, Acta Instituti Atheniensis regni Sueciae, Series IN 4, XLII. Paul Astroms Forlag, Stockholm. pp 123–130

Rai KKS (2008) Climate change and its impact on indigenous peoples in Nepal Himalaya. Indig Aff 1–2(2008):60–65. Available via http://www.iwgia.org/. Cited 25 March 2011.

Ramakrishnan PS (2007) Traditional forest knowledge and sustainable forestry: a north-east India perspective. For Ecol Manag 249:91–99

Roberts G, Parrotta J, Wreford A (2009) Current adaptation measures and policies. In: Seppälä R, Buck A, Katila P (eds) Adaptation of forests and people to climate change—a global assessment report, vol 22, IUFRO World Series. International Union of Forest Research Organizations, Helsinki, pp 123–133

Robinson J, Herbert D (2001) Integrating climate change and sustainable development. Int J Global Environ Issues 1:130–148

Ron Z (1966) Agricultural terraces in the Judean Mountains. Israel Explor J 16(2):33–49

Roncoli C, Ingram K, Kirshen P (2001) The costs and risks of coping with drought: livelihood impacts and farmers' responses in Burkina Faso. Clim Res 19:119–132

Roothaert RL, Franzel S (2001) Farmers' preferences and use of local fodder trees and shrubs in Kenya. Agrofor Syst 52:239–252

Salick J, Byg A (eds) (2007) Indigenous peoples and climate change. Tyndall Centre for Climate Change Research, Oxford. Available via http://www.bvsde.paho.org/bvsacd/cd68/Indigenous peoples.pdf. Cited 22 Mar 2011

Sasaki N. Asner, GP, Knorr W, Durst PB, Priyadi H, Putz FE (2011) Approaches to classifying and restoring degraded tropical forests for the anticipated REDD+ climate change mitigation mechanism. iFor Biogeosci For. doi:10.3832/ifor0556-004. Available via http://www.sisef.it/iforest/show.php?id=556. Cited 26 Mar 2011

SCBD (2006) Guidance for promoting synergy among activities addressing biological diversity, desertifi cation, land degradation and climate change. CBD technical series no. 25. Secretariat of the Convention on Biological Diversity, Montreal. Available via http://www.cbd.int/doc/publications/cbd-ts-25.pdf. Cited 26 Mar 2011

SCBD (2009) Connecting biodiversity and climate change mitigation and adaptation: report of the Second Ad Hoc Technical Expert Group on Biodiversity and Climate Change. CBD technical series no. 41. Secretariat of the Convention on Biological Diversity, Montreal. Available via http://www.cbd.int/doc/publications/cbd-ts-41-en.pdf. Cited 26 Mar 2011

Schafer J (1989) Utilizing indigenous agricultural knowledge in the planning of agricultural research projects designed to aid small-scale farmers. In: Warren DM, Slikkerveer LJ, Titilola SO (eds) Indigenous knowledge systems: implications for agriculture and international development, vol 11, Studies in technology and social change. Technology and Social Change Program, Iowa State University Research Foundations, Ames, pp 116–120

Searchinger TD, Hamburg SP, Melillo J et al (2009) Fixing a critical climate accounting error. Science 326(5952):527–528. doi:10.1126/science.1178797

Seppälä R, Buck A, Katila P (eds) (2009) Adaptation of forests and people to climate change—a global assessment report, vol 22, IUFRO World Series. International Union of Forest Research Organizations, Helsinki

Sereni E (1997) History of Italian agricultural landscape. Princeton University Press, Princeton

Sieferle RP (2001) The subterranean forest: energy systems and the industrial revolution. The White Horse Press, Cambridge

Simel JO (2008) The threat posed by climate change to pastoralists in Africa. Indig Aff 1–2(2008):34–43. Available via http://www.iwgia.org/. Cited 25 Mar 2011

Smallacombe S (2008) Climate change in the Pacific: a matter of survival. Indig Aff 1–2(2008):72–78. Available via http://www.iwgia.org/. Cited 25 Mar 2011

Somorin OA (2010) Climate impacts, forest-dependent rural livelihoods and adaptation strategies in Africa: a review. Afr J Environ Sci Technol 4(13):903–912. Available online at http://www.academicjournals.org/AJEST. Cited 26 Mar 2011

Tauli-Corpuz V, Tamang P (2007) Oil palm and other commercial tree plantations, monocropping: impacts on indigenous peoples' land tenure and resource management systems and livelihoods. Paper prepared for the sixth session of the UN permanent forum on indigenous issues, New York, 14–25 May 2007. Doc. E/C.19/2007/CRP.6. Available via www.un.org/esa/socdev/unpfii/documents/6session_crp6.doc. Cited 26 Mar 2011

Tejwani KG, Lai CK (1992) Asia-Pacific agroforestry profiles. APAN field document no.1, FAO/Asia-Pacific Agroforestry Network, Bogor

Tello E, Garrabou R, Cussò X (2006) Energy balance and land use: the making of an agrarian landscape from the vantage point of social metabolism (the Catalan Valles County in 1860/1870). In: Agnoletti M (ed) The conservation of cultural landscapes. CAB International, Wallingford/Cambridge, pp 42–56

Thorne PJ, Subba DB, Walker DH, Thapa B, Wood CD, Sinclair FL (1999) The basis of indigenous knowledge of tree fodder quality and its implications for improving the use of tree fodder in developing countries. Agrofor Forum 8(2):45–49

UNPFII (2007) Climate change—an overview. Paper prepared by the Secretariat of the United Nations Permanent Forum on Indigenous Issues, 29 p. Available online at: http://www.un.org/esa/socdev/unpfii/en/climate_change.html. Cited 26 Mar 2011

UN-REDD (2011) 2010 year in review. UN-REDD Programme Secretariat, Geneva. Available via www.un-redd.org. Cited 23 Mar 2011

van Dam J, Junginger M, Faaij APC (2010) From the global efforts on certification of bioenergy towards an integrated approach based on sustainable land use planning. Renew Sustain Energy Rev 14(9):2445–2472. doi 10.1016/j.rser.2010.07.010

Van der Werf GR, Morton DC, DeFries RS, Oliver JGJ, Kasibhatla PS, Jackson RB, Collatz GJ, Randerson JT (2009) CO_2 emissions from forest loss. Nat Geosci 2:737–738. doi:10.1038/ngeo671.Cited 20 March 2011

Vlassova TK (2002) Human impacts on the tundra-taiga zone dynamics: the case of the Russian lesotundra. Ambio spec No. 12:30–36

Vlassova TK (2006) Arctic residents' observations and human impact assessments in understanding environmental changes in boreal forests: Russian experience and circumpolar perspectives. Mitig Adapt Strateg Glob Chang 11:897–909

White A, Hatcher J, Khare A, Liddle M, Molnar A, Sunderlin WD (2010) Seeing people through the trees and the carbon: mitigating and adapting to climate change without undermining rights and livelihoods. In: Mearns R, Norton A (eds) Social dimensions of climate change: equity and vulnerability in a warming world. The World Bank, Washington, DC, pp 277–301. doi:10.1596/978-0-8213-7887-8, Cited 26 Mar 2011

World Agroforestry Centre (2011a) Evergreen agriculture. Available via http://www.worldagroforestrycentre.org/evergreen_agriculture. Cited 20 Mar 2011

World Agroforestry Centre (2011b) Turning the tide on farm productivity in Africa: an agroforestry solution. Available via http://www.worldagroforestry.org/newsroom/highlights/turning-tide-farm-productivity-africa-agroforestry-solution. Cited 20 Mar 2011

Zanchi C, Zanchi B (2008) Le sistemazioni idraulico agrarie collinari quale fondamento della sostenibilita' produttiva e della tutela paesaggistica ed ambientale. In: Marinai V (ed) Paesaggio e sostenibilità Studi e progetti. Edizioni ETS, Pisa, pp 195–212

Chapter 14
Ethics and Research Methodologies for the Study of Traditional Forest-Related Knowledge

Christian Gamborg, Reg Parsons, Rajindra K. Puri, and Peter Sandøe

Abstract This chapter examines some of the main research methodologies for studying traditional forest-related knowledge (TFRK). Initially, we address ethical issues, asking, for example, what constitutes proper handling of research results. The relationship between TFRK and modern science is then discussed from a methodological perspective, after which an account of some of the main methods used for studying such knowledge—including participant observation, interviews, cultural domain analysis, questionnaires, and workshops—is provided. Ethnographic approaches are recommended for documenting both verbal and tacit knowledge embedded in skills and practices, while the tools of cultural domain analysis allow for both quantitative and qualitative analysis of individual variation in knowledge. Finally, recurring elements of best practice are presented. If ethical and methodological questions are not addressed in a consistent and systematic manner from the outset of the research, the rights of TFRK owners may well be

C. Gamborg (✉)
Danish Centre for Forest, Landscape and Planning, University of Copenhagen,
Copenhagen, Denmark
e-mail: chg@life.ku.dk

R. Parsons
Canadian Forest Service, Atlantic Forestry Centre, Natural Resources Canada,
Fredericton, Canada
e-mail: Reg.Parsons@NRCan-RNCan.gc.ca

R.K. Puri
School of Anthropology and Conservation, University of Kent at Canterbury, Canterbury, UK
e-mail: R.K.Puri@kent.ac.uk

P. Sandøe
Danish Centre for Bioethics and Risk Assessment, University of Copenhagen,
Copenhagen, Denmark
e-mail: pes@life.ku.dk

J.A. Parrotta and R.L. Trosper (eds.), *Traditional Forest-Related Knowledge:*
Sustaining Communities, Ecosystems and Biocultural Diversity,
World Forests 12, DOI 10.1007/978-94-007-2144-9_14,
© Springer Science+Business Media B.V. (outside the USA) 2012

infringed, meaning that benefits will not accrue to the owners and that access to resources (such as genetic resources) may be suddenly curtailed. Thus, all parties must address the challenges raised by the maintenance, use, and protection of traditional forest-related knowledge when there is interaction between the holders and users of such knowledge.

Keywords Access and benefit-sharing • Best practices • Intellectual property rights • Participatory research • Research methods • Science ethics • Traditional knowledge

14.1   Introduction

In the IUFRO publication, *Forests in the Global Balance—Changing Paradigms*, Colfer et al. (2005) reviewed the state of traditional knowledge and its relationship to forestry research, environmental management, and human well-being. Its authors identified five important issues for future consideration by researchers: intellectual property rights, internal community variation in knowledge, differing epistemologies, potential exchange between multiple-use forestry and traditional knowledge systems, and links among knowledge, livelihoods and land. They concluded:

> Traditional knowledge... represents a vastly under-recognized and underutilized global good. If addressed respectfully, its increased recognition by the forestry community (and others) has the potential to improve conservation and development efforts, to protect and strengthen traditional ways of life (including livelihoods and rights to land), and to increase the prestige and feelings of self-worth among forest peoples. Such feelings can in turn stimulate greater creativity and further knowledge generation among them. We urge readers to engage with forest peoples; they are often the *legitimate* managers of the forests we find ourselves mandated by law or regulations to manage. The marriage of traditional and scientific knowledge is potentially the most potent combination for both environmental and human well-being (Colfer et al. 2005, p. 180).

In this chapter, we develop the arguments of those authors. We also take a further step: we attempt to facilitate greater engagement with forest peoples by describing a research methodology, considering so-called etic and emic approaches for the study of traditional forest-related knowledge (TFRK). We describe ethnographic methods, both qualitative and quantitative, that account for varying epistemologies and individual variation in knowledge, and in particular, we emphasize the ethical issues and practices that must accompany research design from its very inception. The chapter begins by pinpointing the main ethical issues associated with scientific study of traditional knowledge and proceeds to rehearse the relationship between science and traditional knowledge. Thirdly, main challenges and constraints of methods for studying traditional forest-related knowledge are described before ending with an account of best practices for exchange of information between holders and users of such knowledge.

14.2   Ethical Issues Associated with the Scientific Study of Traditional Forest-Related Knowledge

It is accepted today that the use of traditional forest-related knowledge (TFRK) in decision-making processes needs to be based on a partnership in which TFRK owners, scientists, and policy makers create and share knowledge (Berkes 2004; Colfer et al. 2005). However, there is considerable potential for misappropriation of such knowledge and its eventual loss by the original holders (ICSU 2002). Moreover, there is increasing evidence that there is a serious lack of recognition by users of the rights of indigenous peoples and other holders of such knowledge (Ramakrishnan et al. 2000; Laird 2002). At the same time, there are problems with the enforcement of national and international commitments, such as international policies (Forest Peoples Programme et al. 2005; Sandøe et al. 2008a). Many scientists are involved in forest management-related work in which the study of traditional knowledge is an integral component. Such scientists are responsible for the way in which TFRK is acquired and used during their research. According to Article 31.1 of the United Nations Declaration on the Rights of Indigenous Peoples from 2007:

> Indigenous peoples have the right to maintain, control, protect, and develop their cultural heritage, traditional knowledge and traditional cultural expressions, as well as the manifestations of the sciences, technologies and cultures, including human and genetic resources, seeds, medicines, knowledge of the properties of fauna and flora, oral traditions, literatures, designs, sports and traditional games and visual and performing arts. They also have the right to maintain, control, protect and develop their intellectual property over such cultural heritage, traditional knowledge and traditional cultural expressions (UN 2007).

Consequently, scientists have an obligation to ensure that Indigenous peoples, as well as other local peoples who hold traditional forest-related knowledge, freely consent to research involving their knowledge; and this may require a level of participation in research design and implementation that scientists may not have experienced before. Gaining consent may require a long process of negotiation ending in signed agreements by a variety of implicated parties. It is worth noting that a signed agreement could be meaningless to a holder of traditional knowledge, for example, if they do not subscribe to the written word; in such cases, oral agreement might be what is needed. These agreements (oral or signed) may cover the scope of research, methods, and tools that can be used; rules for meeting and talking with participants, sharing research results back with those who shared, and paying for their assistance and services; possible benefit-sharing; and acknowledgement in publications and other forms of dissemination. The questions remain: how much of the responsibility here resides with the researchers and how much with the community holding the traditional forest-related knowledge? What does the responsibility consist of and which methods should be applied to study such knowledge in a mutually fruitful, yet ethically acceptable, manner? These issues are addressed in Sects. 14.2 and 14.3.

Another important issue concerns the protection of traditional forest-related knowledge. Several attempts have been made to address this issue (McManis 2007;

Battiste and Youngblood Henderson 2000). Major intergovernmental forums dealing with traditional knowledge, access to genetic resources, and benefit-sharing, include: the Convention on Biological Diversity (CBD), the World Intellectual Property Organization (WIPO), the Food and Agriculture Organization (FAO), the World Trade Organization, the United Nations Conference on Trade and Development (UNCTAD), the World Health Organization (WHO), and the United Nations Educational, Scientific and Cultural Organization (UNESCO).

However, fewer initiatives have been made in the field of acquisition and use of traditional forest-related knowledge. Most recently, in 2010, at the Conference of the Parties to the Convention on Biological Diversity, COP-10 in Nagoya, Japan, adopted the 'Nagoya Protocol on Access to Genetic Resources and the Fair and Equitable Sharing of Benefits Arising from their Utilization,' which includes traditional knowledge associated with genetic resources that is held by indigenous and local communities. The non-legally binding 2002 Bonn Guidelines on Access to Genetic Resources and Fair and Equitable Sharing of the Benefits Arising out of their Utilization are relevant, one of the main objectives being '… to contribute to the development by Parties of mechanisms and access and benefit-sharing regimes that recognize the protection of traditional knowledge, innovations and practices of indigenous and local communities, in accordance with domestic laws and relevant international instruments' (Secretariat of the Convention on Biological Diversity 2002, p. 3). According to the World Intellectual Property Organization (WIPO) Intergovernmental Committee on Intellectual Property and Genetic Resources, Traditional Knowledge and Folklore, in its fourth session, in 2002, some of the main objectives of protecting traditional forest-related knowledge (TFRK) are: to safeguard against third-party claims of intellectual property rights (IPR) over TFRK issues, to protect TFRK subject matter against unauthorized disclosure or use, and to protect unique TFRK-related commercial products. These issues are dealt with in Sect. 14.4.

14.3   Relationship Between Traditional Forest-Related Knowledge and Science

In several parts of the world efforts are being made to supplement science with traditional knowledge and at the same time help local communities to manage their natural resources. This raises the question: What is the nature of traditional forest-related knowledge—what are its main differences from, and similarities to, modern science? How do these differences bear on the study of traditional forest-related knowledge?

Traditional knowledge and modern science are often framed as incommensurable. It is stated that there are some fundamental epistemological differences between traditional knowledge and science—or that a loosely drawn set of properties distinguish one from the other (Duerden 2005; Colfer et al. 2005). However, an exclusive focus on such properties may lead to unfounded generalizations and the stereotyping of both frameworks of knowledge (Agrawal 1995; Dickson 2003). With this proviso, however, some key characteristics can certainly be discerned (Ellen and Harris 2000).

Traditional forest-related knowledge includes all aspects of the forest (biological, socioeconomic, cultural, and spiritual) and is local, and therefore context-dependent (Smylie et al. 2004). It has been described as a 'lived' knowledge (Parsons and Prest 2003), a way of life and arising from the day-to-day activities of those who hold it (McGregor 2008). Many scientific theories, in contrast, are to a much greater degree context-*in*dependent and ideally universal. In other words, traditional forest-related knowledge is often very specific to particular plant species or soil types or habitats found in the territories where local people live; knowledge associated with plant management—including harvesting techniques—and processing may be developed with particular cultural uses in mind and shaped, historically, by social, demographic, and ecological characteristics that are particular to a locality. Such knowledge may be less useful to others with different needs and access to different natural resources. Often traditional knowledge is embedded in myths, stories, and songs, which are filled with symbolic references particular to a local language and culture; often it is held within a community by a limited number of people. These expressions of local cosmology might be perceived as universal by their owners, despite the fact that their knowledge and experience of the world outside their territories is limited.

Traditional forest-related knowledge is holistic, whereas modern science is fragmented and compartmentalized into disciplines; interdisciplinarity is more the exception than the rule in modern science. In traditional knowledge, humans are seen as an integral part of nature (biocentric or ecocentric worldview); this contrasts to the classical scientific ideal in which humans are seen as external observers or manipulators (anthropocentric worldview). In traditional knowledge, knowledge transfer also takes place through direct oral communication, imitation, and demonstration, instead of being written down and taught via textbooks—although obviously some aspects of scientific laboratory work and fieldwork are also taught through imitation and demonstration.

Another major difference between traditional forest-related knowledge and modern science turns on the way knowledge is obtained and validated. Often, in the past, traditional knowledge has been dismissed because of its anecdotal nature and its reliance on intuition, for being non-systematic and non-empirical, as well for lacking ways of validating knowledge (Becker and Ghimire 2003). However, direct observation and reliance on experience as well as experimentation and interpretation are also tenets of traditional forest-related knowledge, which can be described as follows:

> ...the consequence of practical engagement in everyday life and is constantly reinforced by experience, trial and error, and deliberate experiment. This experience is characteristically the product of many generations of intelligent reasoning and since its failure has immediate consequences for the lives of its practitioners its success is very often a good measure of Darwinian fitness (Ellen and Harris 2000, p. 4).

Modern science, by comparison, is based on a systematic way of acquiring new knowledge involving a principle of transparency with regard to hypotheses, explanations, predictions, and observations. Controlled experiments are used to refute theories or provide the best-founded explanations. Science seeks to work from principles such as testability, publicity, novelty, explanatory power, consistency, and objectivity through inter-subjectivity; in theory it exposes results and theories to refutation and criticism by peers (Stehr and Grundmann 2005).

Methods of assuring validity in the transmission process may not be evident initially, but they do exist. For example, in the Pacific Northwest of the United States and Canada, oral knowledge is not accepted as valid until it has been witnessed and approved by people known for their expertise (Wa and Uukw 1992). When stories are told in public, the audience can comment on whether or not the story has been told well. Moreover, in modern science, scientific facts or assertions accepted by scientists have been subjected to peer review; but as Trosper (2009, p. 96) argues in the case of the Nisga'a, Gitksan, Wetsuwet'en, and Nuu-chah-nulth in the Pacific Northwest, '[o]ne might say that review by Chiefs is a kind of peer review.'

A point often stressed as a notable difference (Moller et al. 2004) is that conventional scientific practices often require technology and specialized skills. As such, these practices are not always practicable in remote places of resource use—places where no tradition for, or trust in, science exists. Moreover, as ordinary users have no time to engage in complicated monitoring methods, there is a call for the traditional, indigenous monitoring methods, which are often rapid and less costly (Berkes 2008).

However, there are also some notable similarities. Empirical observations play an important part in traditional forest-related knowledge as well as in modern science. As in science, traditional knowledge may be shared, but it is also fragmentary (never held entirely by one person) and segmented (knowledge may vary by age, sex, occupation, wealth, and education). More importantly, traditional knowledge is as dynamic and changing as any body of scientific knowledge; innovation, experimentation, and borrowing are all important processes that propel both of these knowledge systems forward. Traditional knowledge and science are both used to inform decision-making processes and actions. Within traditional forest-related knowledge and the scientific disciplines related to ecology, there is awareness that components in the environment are interconnected (Fernandez-Gimenez et al. 2006). Moreover, in many cases, sustainability is viewed as a fundamental value in relation to the management of forest ecosystems.

As with any other system of knowledge, traditional knowledge and modern science are both embedded within specific worldviews. Take, for example, the view one might have of man's relation to nature. This is more instrumentalist in character in science than it is in traditional forest-related knowledge, where the symbiotic relationship between the natural world and humans is stressed (International Council for Science (ICSU) 2002).

This issue may have practical ramifications. For example, practices that appear, initially, to be an expression of superstition to an outside observer (e.g., the scientific researcher) will, as additional knowledge is gained about worldviews on culture and the environment, appear appropriate and relevant. A reluctance to take into account differences in worldviews may lead to the misinterpretation of both the elements and the whole, potentially leading to premature dismissal of traditional forest-related knowledge (see Colfer et al. 2005). Moreover, as has been pointed out by Moller et al. (2004) in a comparative study of monitoring of wildlife for management (in which traditional ecological management and science came to basically the same conclusions and recommendations), simple and inexpensive yet robust monitoring methods are wanted for management purposes. Although science is expensive, and harvesting takes place in developing countries or in poor regions, traditional forest-related knowledge may provide such methods. For example, local communities in Alaska who hunt caribou (*Rangifer tarandus*) have a simple monitoring system based on the body condition (e.g., the fat content) of the caribou (Berkes 2008). In New Zealand, the Rakiura Maori manage their harvesting of the chicks of the bird known in Maori as tītī, the sooty shearwater (*Puffinus griseus*), by tracking the rate at which they are catching the chicks (Kitson 2004). The harvest rate, also known as the catch per unit of effort (CPUE) measurement, is one of the most practical ways of monitoring populations for such users—provided the total population is known, and provided the behaviour of populations as their numbers decline is understood (Moller et al. 2004).

So what should scientists be aware of when using traditional forest-related knowledge? As stated previously, much of this knowledge is both culturally and ecologically context-dependent, so its relevance will often be limited to similar contexts. Similarly, when science makes generalizations, such as those involved in theoretical models, that appear to have wide applicability, particular circumstances in certain locations may negate the applicability of that general knowledge. Some authors propose that by combining traditional forest-related knowledge and science to monitor forest resources, for example, management for sustainable use can be greatly enhanced (Donovan and Puri 2004; Folke 2004). This new paradigm is a so-called 'co-existence model;' it recognizes the value of scientifically obtained knowledge as well as traditional knowledge and uses relevant information from each type of knowledge domain (Becker and Ghimire 2003). Here, from a scientific perspective, adding traditional forest-related knowledge might help to provide: (i) a better understanding of forest ecosystems; (ii) knowledge of appropriate management techniques, including harvesting (e.g., Puri 2001a); and (iii) qualitative information on trees and other plants as well as forest-related animals and other natural phenomena. The addition of traditional knowledge might also: (iv) raise awareness of (subtle) changes and patterns in ecosystems, and (v) offer assistance in sampling and surveying efforts as well as monitoring, with both holders and users of traditional knowledge learning (e.g., Puri 2001b; Sheil et al. 2006). As Fortmann (2008, p. 7) suggests, both 'conventional science' and 'civil science' are needed, and the question that applies to both is: How can we learn what we need to know and understand in order to create, sustain, and enhance healthy ecosystems and human communities? The question then arises of how to add such knowledge to the knowledge pool in a mutually beneficial way—that is, in a manner allowing the respectful and appropriate use of traditional forest-related knowledge and discouraging extraction of traditional knowledge without benefitting those who developed it?

14.4   Methods for Studying Traditional Forest-Related Knowledge—Challenges and Constraints

One clear problem in using traditional forest-related knowledge alongside scientific research, as well as in resource assessments and resource management, is that traditional knowledge is rarely written down. In the Arctic there is a saying: 'When an elder dies, a library burns.' Thus, one of the main characteristics of traditional forest-related knowledge is that it exists as a (sometimes shared, interpersonal) oral or visual pool of knowledge that is not necessarily easily accessed, and in a declarative, behavioural, or performance-related form (Puri 2005). It is often stated (e.g., Alaska Native Science Commission 2004) that traditional knowledge is living knowledge, living with the elders, typically. The implication is that great care should be taken when accessing and recording such knowledge, thereby transforming it into what might be termed non-living knowledge for which no one has specific responsibility to pass on.

Sources of traditional forest-related knowledge include interviews, oral histories, archived transcripts and recordings, public meeting minutes, written ethnographies, and inventories. One may also learn traditional knowledge, as do many holders themselves, through observation, imitation and practice, and even apprenticeship. Traditional forest-related knowledge is traditionally described as long-term wisdom, the result of intuitive and experiential learning; it is often holistic in nature; its teaching takes the form of storytelling or learning by doing and experiencing (Mazzocchi 2006). However, if traditional knowledge is to be used in an immediate form in a project setting, then there is a need to document and describe it through the exchange of information between local experts and external scientists, that is, through an exchange between holders and users of traditional and scientific forest knowledge related to forest management. However, a question also arises from this context: Should this knowledge be made available for all to use outside of the context from which it came? Does the knowledge of the holders have the same meaning once written?

Traditional forest-related knowledge is often locally and culturally specific, combining facts and values. It does not come out 'looking the standard Western notion' (Berkes 2008, p. 267), but rather blends 'knowledge, practice, authority, spiritual values, and local and social cultural organization; a knowledge space' (Turnbull 1997, p. 560). Moreover, as noted earlier, explanations in TFRK are based, not on hypotheses, theories, and models, but on examples, parables, and anecdotes. For a scientist accustomed to working in an analytical manner, driven by models or hypotheses with hierarchical differentiation that exclude the supernatural, it can be quite arduous trying to make sense of what she or he hears.

Social scientists describe this problem in terms of the notions 'etic' and 'emic'.[1] Etic (not to be confused with ethic) implies an outside perspective on the

---

[1] The terms etic and emic were originally introduced by the American linguist Kenneth Pike (1967)—the words being derived from phonetic and phonemic—and further developed by the anthropologists Marvin Harris (see Headland et al. 1990) and Ward Goodenough (1970).

culture or system, using prior (and in this context, scientific) knowledge and one's own (here: research) vocabulary to describe and understand in a neutral or objective way. Emic means that the researcher looks at a culture or system from the inside in the sense that she or he assembles a description that can be recognized and verified by those within the system who hold the knowledge. Focal terms and concepts are the ones that give meaning in the local culture. If the goal is to understand a culture, an emic approach is important. On the other hand, if the aim is to compare cultures, an etic approach is unavoidable. Box 14.1 provides an example of the distinction between emic and etic in relation to resource mapping. The main message is that careful consideration of the cultural context is of utmost importance when studying traditional forest-related knowledge, and when choosing methods to do so, because insiders and outsiders ask different questions (Bishop 2005).

Although much traditional forest-related knowledge (TFRK) is encompassed in what is known as the (natural) scientific domain, the methods for studying and documenting TFRK derive from the social sciences. However, many forest scientists are not familiar with methods and approaches used in the social sciences. As a consequence, they may fail to acquire information from the holders of TFRK. The reverse can happen as well: a social scientist, not being expert in a specific natural or physical science, may not be respected by, for example, an expert fisherman (Johannes 1989). Moreover, documenting TFRK can be a long process—sometimes unexpectedly so—and it may sometimes be hard to judge whether the effort is worthwhile (Huntington et al. 2004, cf. Vayda et al. 2003). This is the reason why, for example, forest ecologists and geneticists team up with environmental anthropologists or sociologists when they undertake research (e.g., Sheil et al. 2003, 2006). In addition, in many cases, traditional knowledge is not only about (site-specific) knowledge and resource management practices, but also a question of a different ethical outlook, alternative environmental ethics, and contrasting conceptions of nature; for example, witness many First Nations, such as the Cree (Roslin 2005). Often, the outlook or worldview embraced by TFRK holders emphasizes the more symbiotic nature of the relationship between humans and the natural world. Thus, there is a question of not just improving the knowledge pool but also of shifting value orientation, or grasping or respecting other environmental ethical outlooks, to gain a greater understand of the TFRK. Hence, using community knowledge is more than gathering 'soft data' and (maybe) transforming it into 'hard evidence' through quantification.

In the remainder of this section, five, predominantly qualitative methods for studying traditional forest-related knowledge are described, some in more detail than others. The methods are: participant observation, interviews (individual and focus group), cultural domain analysis, questionnaires and community workshops (based on Newing et al. 2010; Bernard 2000, 2005; Huntington 2000; Weller and Romney 1988). The methods are discussed in relation to challenges of reciprocal exchange between traditional forest-related knowledge holders and researchers.

Qualitative approaches emphasize the importance of the ways in which people view the world, and how they act in it (Interagency Advisory Panel on Research

Box 14.1 Documenting Traditional Forest-Related Knowledge and Resource Use in Malinau, Indonesia

The Center for International Forestry Research [CIFOR]'s Multidisciplinary Landscape Assessment (MLA) project in rainforest-dependent communities in East Kalimantan, Indonesian Borneo (1999–2001), set out to discover what matters, in terms of forest resources, to local communities and why. Following a process of free and prior informed consent among village members, a team of local, national, and international researchers set out to map the area's resources from both an etic and emic point of view; local community members participated in the research design process during the pilot phase; collected and identified plant specimens, soil samples, and animals; led surveys to establish plots; and provided their opinions and knowledge on all aspects of their relationship to their forest and riverine environment.

Researchers used a variety of interviews and community mapping workshops to identify local land cover and land use categories that were then assessed using standard biodiversity inventory techniques. The maps also contained lists of what each group (differentiated by age, gender, and ethnicity) considered to be their most important plant and animal species.

Once a base map had been constructed, community members were invited to an evening meeting in a large community centre. They undertook a mapping exercise, which usually lasted around 2 h, although some groups would take considerably longer to perfect their drawings. The mapping exercise usually started with the community members finding their orientation with respect to the map, naming and drawing in numerous tributaries, and identifying their direction of flow. The groups were then asked to draw additional reference sites (such as old village locations and hill tops) and to start locating positions associated with specific land cover types, resources, features, or activities, including special or unusual sites. A key of specific symbols or colorings was developed.

The resulting maps were generally pinned up on a wall where they could be viewed by community and team members and updated as needed. They then served as a basis for further discussions between the community and the village-based research team, and for selecting sites for biodiversity assessment and inventory by the field team. In all, 200 plots were inventoried for plants, animals and human uses; 600 soil samples were taken; and 15,000 plant voucher specimens were collected. Community members continually revised the maps over the course of the following weeks. These further refinements required the combined efforts of both the field- and village-based research teams, because discussions or field observations during the day often led to minor changes or additions. Before the research teams left each village, a master map—combining data from all the initial maps, revisions

(continued)

Box 14.1 (continued)

and additions—was neatly drawn and copies were left with the village leaders. CIFOR organized a workshop for local government officials where the results of all seven of the community mapping exercises were presented by community leaders. In addition, posters, or open books, were produced for each village that displayed results using pictures, graphics, and text. A video documentary was also made, in Indonesian and English, and follow-up surveys, additional research and community assistance by NGOs have all followed in the past decade. Based on these results, CIFOR was able to dissuade local government from proceeding with an oil palm plantation in the area, arguing that the forest was more valuable for the area and its people if left standing (Sandker et al. 2007).

Source: Sheil et al. (2003, 2006). (See also the MLA website at: www.cifor. cgiar.org/mla/.)

Ethics 2010). The task is to try to understand how people ascribe meaning and how they interpret their surroundings. A wide variety of approaches, methodologies, and techniques are used in qualitative research in order to engage with research participants' particular environments. Ethnographic methods to study traditional forest-related knowledge, such as participant observation, often centre on direct interaction with participants and on the gaining of insights into their perception and knowledge of the world and their behaviour. Another feature of such studies, often highlighted, is the dynamism of the research situation—requiring flexibility and reflexivity in a diverse and often evolving social setting (see Puri 2010c).

In terms of data collection, and given that research is time-limited, most qualitative approaches emphasize depth rather than breadth: they pursue understanding through rich sources of information rather than being concerned with representativeness and statistical significance. However, it is simple enough to analyze qualitatively derived data using quantitative techniques if there has been some attempt to select or sample respondents according to some criteria of interest, such as analysing perceptions of forest change among respondents of differing age and gender (see Bernard 2005; Newing et al. 2010). In general, the more structured the method, the more control the researcher has over the data, and the more amenable it is to quantitative analyses (Newing et al. 2010). Nevertheless, data collection mostly relies on a variety of sources of information and data-gathering strategies to strengthen the quality of the information and to aid validation, a process often referred to as triangulation. Such data collection strategies often entail extended contact between the researcher and the participants. This enables the researcher to collect trustworthy data through, for example, participant observation and interviews. It may easily result in the research being a highly collaborative (and sometimes a somewhat open-ended) process.

14.4.1   Participant Observation

Participant observation is a relatively unstructured interactive method for studying people as they go about doing their daily routines and activities, such as harvesting non-timber forest products. The researcher accompanies one or more people both to observe what they do and say and to participate, to varying degrees, in the activities being studied. The method is unstructured in the sense that the researcher has to follow the schedule and activities of her/his participants, rather than imposing a framework that interferes with their normal routine. However, the researcher does maintain some control over which activities to participate in, how frequently to participate, and the kinds of question that are driving the study (see Puri 2010c). This allows the researcher to document the use of traditional forest-related knowledge in action, so to speak, and to observe more tacit or nonverbal types of traditional knowledge being used. Puri (2005) spent several years learning to hunt with the Penan of Borneo in order to be able to understand the kinds of knowledge required to hunt successfully, including knowledge of animals, plants, the seasons, and weather (Puri 2007) and the various tools and strategies for finding, tracking, and killing prey. He demonstrated the importance of 'performance knowledge' possessed by experts, which allows them to adapt to contingencies and manipulate a variety of variables to ensure success during the course of a hunt. It is unlikely that one would ever uncover the extent of traditional forest-related knowledge through interviews alone. That said, participant observation is quite easily combined with informal and casual interviewing. An often-heard weakness of such an approach is that it requires a long period of immersion and close personal contact with people—for instance, to learn local languages or be allowed to participate in certain activities. There have also been questions raised about the dangers of culture shock and the maintenance of researcher objectivity in such contexts; however, with time and practice most researchers can develop self-awareness of their own biases, at least with regard to the research topic being studied (Bernard 2005).

14.4.2   Interviewing

Interviewing is perhaps the key approach used for studying traditional forest-related knowledge. It may be conducted with one or more participants, with varying degrees of formality, ranging from a conversation to a fully structured and directed interview (e.g., a questionnaire). Cultural domain analyses, discussed below, are often conducted during structured interviews with individuals. The kind of approach used depends on the purpose and time available. Interviews may be conducted on a one-to-one basis, which allows for a more in-depth coverage and the establishing of rapport, essential for building good working relationships and trust. The ethnobotanist Nancy Turner (2005) has pointed out that discussions with, for example, a plant in hand greatly improve communication between a scientist and an elder. Interviews may be useful in an exploratory phase (e.g., casual conversations over a drink about the research topic), or for working with key experts or participants (such as a healer). Interviews can and should be conducted more than once, allowing for the checking of facts, previous statements, and new opinions, as well as the continued development of personal relationships, which allows one to gauge the reliability of participants and the accuracy of the information being given.

In some cases, focus-group interviews involving a small number of participants may offer a more culturally appropriate setting than a one-to-one interview. Participants are free to talk with each other in the group and can discuss issues in an interactive setting. Focus-group interviews allow for insight into the knowledge and attitudes of various cultural and social groups. The drawback, of course, is that focus groups can replicate social divisions, hierarchies, and biases present in the group. In the CIFOR-MLA project described in the box above, four focus groups were used to prevent gender and age differences from inhibiting participation, and to simultaneously uncover any internal community variation in traditional forest-related knowledge.

Roughly speaking, less structured and less directed interviews tend to take up more time but may encompass more topics and deliver more detail than a structured interview. On the other hand, some structuring ensures important information is not missed. Often, the chosen path is a semi-structured interview, where the researcher may have a list of questions or issues to discuss, but where she or he is also content for unexpected associations and lines of thought to be brought up by the participants in the course of the conversation. The semi-structured interview is also helpful in that it becomes less of a question-and-answer session and more of a conversation, which is helpful if participants are not entirely comfortable with the situation or direct questions, or, as often happens, the question is understood differently from the way intended (see Bernard 2005; Newing et al. 2010).

14.4.3 Cultural Domain Analysis

Various tools and methods originating in cognitive anthropology aim at understanding cultural domains, that is cognitive categories used by people to organize the objects and experiences making up their world, such as 'plants', 'animals' and 'forest products'. Since the aim is to understand how participants see these domains—rather than how outside researchers perceive them—this is clearly an emic approach. Cultural domain analysis is not a single method. It includes structured interviews intended to define relevant cultural domains, a number of methods to investigate the content and structure of these domains, and finally a number of unstructured interviews to gain a better understanding of their significance. Among the methods most widely used to examine the domains are 'freelisting' (defining all of the items in a cognitive domain), and 'pile sorts' and 'triads' (examining the internal structure of the domain through a basic perceptual relationship: similarity). Freelisting entails asking participants to list, for example, all the trees they know about, or the kinds of medicinal plants in a certain area. Pile sorts elicit judged similarities. Respondents are asked to sort cards with names of domain items (e.g., trees) or values (e.g., autonomy) into piles, according to how similar the respondents think they are, albeit only on a non-quantitative basis. An alternative to pile sorts for measuring similarity is the triad test, which basically includes sets of three items chosen to represent a certain cultural domain. Participants are asked to look at each set and choose which of the three items they judge to be most dissimilar to the other two (and sometimes to explain why). A triad test can be used to isolate some of the components judged most salient to a cultural domain or system, such as certain forest management schemes. It can also be used to examine internal community variation in, for example, ethnobotanical beliefs (Martin 2004; Puri 2010a).

14.4.4 Questionnaire

Using questionnaires to study traditional forest-related knowledge is relevant when the respondents are fluent in the researcher's language, when the researcher knows what she or he is looking for, and when key goals include being able to quantify in some way certain relationships (e.g., '60% of respondents found tree roots to be…') or being able to compare respondents by having categories of standardized answers. Generally speaking, questionnaires are cheaper and less time consuming than interviews and require less direct effort by the researcher. Questionnaires can be designed in many different forms (e.g., with a greater or fewer number of open-ended questions), giving respondents an opportunity to provide more detail or to allow for interpretations of a particular question. Another reason for using a questionnaire to study traditional forest-related knowledge is that some participants may be more comfortable filling out a questionnaire than participating in an interview (possibly with an interpreter), depending on cultural context. However, questionnaires are not well-suited to research aims where the exploratory element is dominant. Thus, interviewing may be a relevant study method on its own, or in combination with questionnaires.

14.4.5 Workshop

A workshop approach usually involves a smaller number of participants working over a specific period of time jointly on specified tasks, interacting and exchanging information (see the description of an example of a community mapping workshop in Box 14.1). Formal or ad hoc workshops among scientists and the holders of traditional forest-related knowledge in the community can take many forms. But instead of gathering new information and compiling data, one aim of community workshops may be to achieve a better understanding of what is already known—in other words, collaborative analysis and interpretation (Huntington et al. 2002). Such workshops may also help achieve a better understanding of various perspectives among users and holders of traditional knowledge and may eventually result in some shared understanding.

14.4.6 Other Methods

In addition to the above-described methods several other, mainly participatory, methods are frequently used to study traditional forest-related knowledge. The idea of 'collaborative field work' is that relationships may be formed in the field over an extended period of time and that the mutual interests of scientists and the holders of traditional forest-related knowledge—and especially local researchers— may lead to unanticipated results, for example, in terms of the interpretation of field results (Davis and Reid 1999). Training local researchers is a form of building local capacity, which may greatly facilitate collaborative forest management in the future. Local researchers may be trained in 'participatory resource mapping' and inventory (for ground truthing), which enables calibration of other information such as remote-sensing data (Puri 2010b; see Box 14.2). In Canada, traditional land use and occupancy mapping using map biographies is a widely-used technique for chronicling traditional forest-related knowledge (Tobias 2000). Another approach is 'participatory action research,' which essentially involves participants—local residents and outside researchers—examining, in an active way, issues that are experienced as problematic, with the aim of improving the problem situation. Participants may draw on their traditional forest-related knowledge to offer solutions to such questions as what trees should be used to reforest a logged area, or what plants should be propagated in a nursery (see IIRR 1996; Reason and Bradbury 2001).

In all qualitative and quantitative methods for studying traditional forest-related knowledge, an important consideration is the selection of participants. It is necessary to ask: Who are the holders of knowledge? Often, active selection of key participants may fare better than randomly sampling the community; however, as discussed above, because of the fragmentary and segmented nature of knowledge, it may be important to understand variation in knowledge held by a community—for instance, whether the study is of the transmission of traditional forest-related knowledge or an attempt to understand its loss. Sometimes 'chain referral' or 'snowball sampling'—in which participants suggest other local holders of knowledge—may prove useful because communities can themselves identify the individuals they respect as knowledgeable (Bernard 2005). Many insights follow from the assumption that a community has ways of singling out its own experts. Clearly, there may well be cases of contradictory information, but applying other sources of feedback and group reviews may help resolve such conflicts. As Fortmann (2008) points out, credibility is another important and crosscutting issue for any method of adding to, or studying, traditional knowledge. Here, it is addressed in terms of credible ways

---

Box 14.2 Mapping for Community Use Zones in Crocker Range Park, Sabah, Malaysia

---

The Global Diversity Foundation (GDF) conducted a 3 year (2004–2007) ethnobiological assessment of key resources and anthropogenic landscapes that are important for the indigenous Dusun communities living inside and adjacent to the Crocker Range Park in Sabah, Malaysian Borneo. The Crocker Range Park Management Plan allows for 'community use zones' that are specifically set aside for continued local community use. GDF, PACOS (a local NGO), and Sabah Parks worked with a team of eight community field assistants plus community leaders, key informants, and local researchers to obtain baseline data and develop methodologies for the future monitoring of natural resource use in and around the community use zone.

Participatory mapping exercises, along with other ethnobiological methods, demonstrated the extent and intensity of non-timber forest product gathering, swidden agriculture, hunting, and freshwater fishing by the local communities. In follow-up projects, the GDF teams have pioneered the use of participatory GIS and participatory three-dimensional modelling (P3DM) to increase the ability of communities to present their research results in a technologically sophisticated, powerful manner. The results will guide the formulation of rules and regulations for the community use zone management agreement that will govern the joint management of the community use zone by the local communities and Sabah Parks.

Source: Global Diversity Foundation (GDF). (www.globaldiversity.org.uk)

---

of doing science and who will be seen as credible 'knowers.' Some groups or individuals may readily be excluded from the category of credible knowers because they do not have the right so-called social markers (e.g., an academic degree, majority ethnicity, or gender); or they may only be incorporated if that incorporation is mediated by those with the 'right' social markers. According to Fricker (2007), wronging a person in her or his capacity as a knower is a case of 'epistemic injustice.' Scientists should not commit such injustice by refusing credibility to the knowledge of indigenous peoples, or by listening only to, for example, the men.

Lastly, before engaging in the selection of methods for the actual data gathering, two of the most important things to consider for any study of traditional forest-related knowledge are the cultural context in which the research takes place and adherence to appropriate ethical standards of traditional forest-related knowledge research conduct. There is often a need to use more than one method in an investigation involving traditional forest-related knowledge, and it should also be recognized that there is no one 'cookie cutter' approach to conducting research where such knowledge is concerned.

14.5 Best Practices for Exchange of Information Between Holders and Users of Traditional Forest-Related Knowledge

The increasing use of traditional knowledge has prompted the need to establish guidelines to ensure that people and the information acquired are treated appropriately. As stated in many research guidelines in connection with study of traditional knowledge, great care needs to be taken to deal with the aforementioned ethical issues raised by access to such knowledge, the gathering of data, the use of data, and publication of outcomes. Some of the main issues to consider include: free and prior informed consent, confidentiality and the relationship between researcher and the participants/respondents/informants, and fair sharing of the benefits of research. Some of these issues can be identified during the initial research planning and design phases, whereas others may arise during the research process. It is important to have policies or guidelines to deal with the issues here. In some cases, existing norms for handling scientific dishonesty can be used. In others, and many relating to the use of traditional forest-related knowledge, new norms and requirements will have to be introduced to the scientific community. Thus, ethical guidelines for working with indigenous or other local peoples are an expansion of already accepted ethical standards. Dealing with ethical conduct in research requires discretion, sound judgment, and flexibility in relation to risks and benefits arising from the research. As there are more and more indigenous communities developing and implementing their own research protocols, such as the Whitefeather Forest Research Cooperative (see Box 14.3), the first thing researchers should check is whether the indigenous community in question has their own guidelines which—at a minimum—should be followed.

Research has historically been viewed as something that, by definition, is pursued for the good of humankind. It seems to be generally assumed that science aims at the discovery of the truth, which is a close relative to the good. However, as pointed out in the ethical studies undertaken under the auspices of the Consultative Group on International Agricultural Research (CGIAR) (Sandøe and Jensen 2008; Sandøe et al. 2008b), a number of developments have undermined this common view of science: (i) personal interests (e.g., regarding careers and success in science) may create incentives for scientists to engage in various forms of dishonesty, such as fabricating data or stealing ideas; (ii) increased dependence on external funding has led to growing competition among different research groups, and this might tempt scientists, not only into various forms of scientific dishonesty, but also into delivering results that are wanted by those who pay for the research, saying what is convenient rather than what is true; (iii) in the process of creating scientific results, human subjects may be used, for example, in medical experiments, and the results may be achieved at a cost to the health, well-being, or integrity of those experimented on; and (iv) natural (e.g., genetic) resources may be used as raw material in scientific inventions that benefit those who do the research and those who fund it without benefiting those who live in the areas where the resources were

Box 14.3 The Whitefeather Forest Research Cooperative

The Pikangikum First Nation in Ontario, Canada, has established the Whitefeather Forest Research Cooperative (WFRC), bringing together several research partners—the University of Manitoba, Sault College, Lakehead University, and the University of Winnipeg—to develop 'the Whitefeather Forest Initiative in the form of a knowledge network where Pikangikum people are in the driver's seat regarding the research programme' (WFRC 2004).

What is special about this cooperative? The First Nation has taken the lead in meeting its own goals with respect to the use of traditional forest-related knowledge (referred to as forest-based indigenous knowledge by the WFRC) in forest management planning. All research initiatives undertaken by this cooperative must first be reviewed by a research advisory committee, consisting of members from the community and from all the cooperative member organizations. The committee then submits recommendations to the Pikangikum First Nation Band Council for approval, and researchers must follow the protocols established by the community in the cooperative agreement. Community members have a say in any proposed research through meetings with the researchers.

All research findings are communicated back to the community and, as much as possible, integrated into the school curriculum for the Pikangikum Education Authority.

The WFRC has also developed a strategy for the establishment of the Whitefeather Forest Indigenous Knowledge Teaching Centre, thereby providing a means of implementing the elders' vision of First Nation-led research partnerships while enabling the community to conduct experiential learning and raise cultural heritage awareness.

Research results are grounded in the traditional forest-related knowledge of the elders in the community. For example, a study on caribou occurrence in the Whitefeather Forest area (O'Flaherty et al. 2007) was carried out using traditional forest-related knowledge to determine how to increase the number of caribou within the Whitefeather Forest while still allowing commercial forest management activities to occur in a sustainable fashion and providing caribou with all the necessities of life (food, shelter, ability to move about the landscape, and mating opportunities). By using the traditional knowledge found within the community, forest management decisions were made relevant to the community and within their context. The research looked at site-specific and landscape-level needs using traditional knowledge, natural science, and social science. This research now contributes toward the planning and management of the Whitefeather forest for the Pikangikum First Nation.

Pikangikum First Nation is very much in control of the research undertaken within their community.

Sources: O'Flaherty et al. 2007; Whitefeather Forest Research Cooperative 2004.

found or the knowledge was obtained. More generally, those who pursue scientific discoveries may be rewarded through patents and other forms of intellectual property rights, and this goes against the ideal of sharing knowledge, which is central in traditional research.

In general, ethical conduct in research refers to all or some of the following principles or key values: equality, respect, responsibility, autonomy, justice, informed consent, collaboration, and benefit-sharing. Respect is one of the main values. It entails regard for the welfare, beliefs, customs, and cultural heritage of all involved in the study of traditional forest-related knowledge, especially the holders of that knowledge. As suggested by the Desert Knowledge Cooperative Research Centre (CRC) in Australia, 'A key aspect of working ethically is for researchers to understand and respect that aboriginal peoples also have their own protocols for behavior— which are effectively local ethical guidelines' (CRC 2007, p. 1). Thus, the first obligation of a researcher is to follow locally defined protocols. Agreed standards of research conduct typically entail clarification of expectations and outcomes for involved parties, fostering a mutual exchange of information and learning, and making a foundation for effective research relationships (e.g., see Clayoquot Alliance for Research, Education and Training 2003).

The purpose of best practice is multifaceted, but it includes: ensuring that appropriate weight is given to traditional and scientific knowledge, facilitating a positive working relationship that builds on trust and collaborative learning, being aware of how traditional forest-related knowledge should be put to appropriate use alongside other bodies of knowledge, and recognizing that traditional knowledge is not just information but an integral part of decision-making processes.

Initiatives to develop best practices for the exchange of information between the holders and users of traditional forest-related knowledge (TFRK) (i.e., ethical guidelines for the use of TFRK in research, and science that is relevant to TFRK) are somewhat scattered. Some relate specifically to forests, but most relate to other types of ecosystems, and to various approaches to environmental management as well. Moreover, many guidelines were created in relation to a specific geographical or cultural region or community/nation. However, these specific guidelines have usually been adapted from a more generic set of traditional knowledge research guidelines, and as such they tend to contain the common elements summarized and discussed below (Alaska Native Science Commission 2004). In 2010, at the Conference of the Parties to the Convention on Biological Diversity, COP-10 in Nagoya, Japan, a decision (COP/10/L.38) was adopted, which includes an ethical code of conduct inviting parties and other governments to make use of this code to develop their own models on ethical conduct for research, access to, and use of information concerning traditional knowledge.

Strong protocols for knowledge exchange create conditions for excellent joint research, whereas weak protocols generate lopsided research programmes and results (Davis and Reid 1999). In Canada, for example, there is a move toward a research approach in which research agreements are negotiated so as to share knowledge and results and ensure benefits to the communities. The major national research funding agencies in Canada, under the umbrella of the Tri-Council, have included guidance on research with human subjects, with special attention given to aboriginal peoples. One of these agencies, the Canadian Institutes for Health Research (CIHR), has responded to shifts in research with aboriginal peoples by developing their own guidelines (CIHR 2007). Canada's Tri-Council subsequently developed a new draft policy statement that included clearer directions for research with aboriginal peoples. In general, numerous protocols have been developed both directly in forest research and for research generally (Davidson-Hunt 2003; McDonald 2004; Medeek 2004). Various techniques have been evaluated, such as workshops pointing out where and when they are applicable, and how they could be used (Huntington et al. 2002, 2006; Davidson-Hunt and O'Flaherty 2007).

Of most direct relevance and applicability to researchers is the work of the International Society for Ethnobiology (ISE), whose many members include social and natural scientists, indigenous peoples, NGO staff, community activists, and policy makers. Upon its foundation in 1988, the ISE initiated discussions on the ethics of research with indigenous peoples about their knowledge, and it agreed and published its Code of Ethics in 2006, with additions in 2008. Among the most comprehensive of its kind so far conceived, the ISE Code of Ethics is based on the concept of traditional resource rights, and acknowledges that biological and cultural harms have resulted from research undertaken without the consent of indigenous peoples. It affirms the commitment of the ISE to work collaboratively, in ways that: support community driven development of indigenous peoples' cultures and languages; acknowledge indigenous cultural and intellectual property rights; protect the inextricable linkages between cultural, linguistic, and biological diversity; and contribute to positive, beneficial, and harmonious relationships in the field of ethnobiology. (International Society for Ethnobiology 2006, p. 2). The key value underlying the ISE Code is the notion of mindfulness, understood as a continual willingness to evaluate one's own understandings, actions, and responsibilities to others. The ISE Code comprises 17 principles (including many that are found in other initiatives):

- Prior rights and responsibilities
- Self-determination
- Inalienability
- Traditional guardianship
- Active participation
- Full disclosure
- Educated prior informed consent
- Confidentiality
- Respect
- Active protection
- Precaution
- Reciprocity, mutual benefit and equitable sharing
- Supporting indigenous research
- The 'dynamic interactive cycle'
- Remedial action
- Acknowledgement and due credit
- Diligence.

Another recent initiative of relevance for traditional forest-related knowledge comes from the CGIAR, following a 10 year effort to identify ethical issues relating to research and advice (Sandøe et al. 2008a; CGIAR 2009). These guidelines deal with such issues as general research ethics (including ways of dealing with scientific dishonesty, publication, and authorship); the use of human subjects in research; the use of genetic resources in research and later applications; intellectual property rights over scientific inventions; consultation; and debate relating to controversial technologies. Although this set of guidelines was designed in connection with agriculture, it has general relevance to forest-related activities. The guidelines, which aim to provide practical recommendations to the CGIAR community, set out five core principles to ensure appropriate, ethical, and consistent use of traditional knowledge:

1. Free and prior informed consent or sometimes just prior informed consent,
2. Publication of traditional knowledge,
3. Giving back research results,
4. Access and benefit-sharing, and
5. Active traditional knowledge protection efforts.

Other often-cited elements or principles (International Institute for Sustainable Development (IISD) 2005) are: mutually agreed terms, community and indigenous participation, and information and transparency. Clearly, each of these principles can be defined and characterized in different ways. Here, a few examples of how to understand and interpret the principles will be given. Besides these rather general principles, many more specific codes of conduct have been developed for specific regions and communities which, effectively, are translated into guidelines applying to general phases of research (before, during, and after research).

14.5.1　Informed Consent

It is a key point that permission from, or the informed consent of, the holders of traditional forest-related knowledge should always be obtained by researchers from the appropriate governing body. Here, it is important to note that researchers often work within an individualistic model in which all that is required is the consent of the individual. With indigenous communities, not only is it important to consider the risks to and seek the consent of individual research participants, but also to consider the risks to and seek the consent of the collective (e.g., chiefs and councils or other representative organizations). Many guidelines require written consent, but there may well be circumstances in which an indigenous community or research participant is not comfortable giving written consent (or can't write). In these cases a clear process for obtaining consent needs to be outlined. Moreover, holders of traditional forest-related knowledge should be acknowledged as experts in their field. 'Free and prior informed consent' is defined by the United Nations Permanent Forum on Indigenous Issues, which is the main body for addressing issues facing indigenous peoples within the UN system, as a '… process undertaken free of coercion or manipulation, involving self-selected decision-making processes undertaken with sufficient time for effective choices to be understood and made, with all relevant information provided and in an atmosphere of good faith and trust.' It is worth noting that free and prior informed consent is defined here as a process, which implies an ongoing set of discussions, consultations, meetings, and agreements in which relevant issues may be returned to more than once.

14.5.2 Publication of Traditional Forest-Related Knowledge: Ownership, Source, and Origins

The primary ownership and origin of the knowledge should be respected by addressing the needs and sensitivities of traditional forest-related knowledge holders and obtaining their approval and involvement. Moreover, due credit should be given to the holders of such knowledge, and it should be ensured that the study of traditional knowledge does not become a matter of information 'harvesting,' but is rather a collaborative undertaking in which the community also holds a right to review the research process in relation to the mutually agreed terms (see below).

14.5.3 Giving Back Research Results: Feedback Loop

New knowledge and key findings should be reported back to communities in non-technical language. Copies of studies should be given to local libraries, or other ways in which communities can store and access research results should be established. Here language and documentation formats ought to be considered. Any benefits arising from the use of traditional forest-related knowledge should be shared equitably with holders of that knowledge (see below).

14.5.4 Access and Benefit-Sharing—Active Traditional Knowledge Protection Efforts

Research and any work undertaken should be based on issues of common interest, and any benefit or credits accruing from the research should be shared in an equitable manner with the community. Appropriate ways of compensating the community, if relevant, should be worked out at the outset of the research. It should be clear for whom, and for what purpose, the research is intended, and who will benefit from the collection of the knowledge. Permission to access traditional forest-related knowledge should be sought from the appropriate governing body, and in the course of obtaining this permission it will be necessary to discover who is expected to negotiate such agreements (see below). The community should work out what specific research elements will require negotiation. Time should be allowed for addressing questions and concerns, and for amending the research proposal in question. Indeed, if possible, the holders of traditional forest-related knowledge should be involved in the development of the research questions at the earliest possible time—even in the development of the research proposal.

14.5.5   Mutually Agreed Terms—Relationship

A strong and lasting relationship should be established—a relationship that often takes time to create, building on openness, mutual trust, and honesty. The community should be contacted by the researcher(s) before work with elders, or other community members, begins. The contributions of indigenous resource people should be acknowledged, and if the holders of traditional forest-related knowledge decide that only part of the information can be shared or used, this decision must be respected. For example, some parts of traditional knowledge may be considered sacred by the community, and some forms of knowledge (e.g., medicine) are sanctioned only for use by a recognized practitioner from the community. Confidentiality of surveys and sensitive material should always be guaranteed.

14.5.6   Community and Indigenous Participation: Collaboration and Cultural Sensitivity

Partnerships should be established on a collaborative basis with mutual learning as a goal. The nature of traditional forest-related knowledge within its spiritual and cultural context should be acknowledged. Indigenous people could be hired and trained to assist in the study, and the first language should be used whenever English, Spanish, Chinese, or another language is the second language. This principle also implies culturally sensitive methods of approaching (e.g., interviewing, or presenting questionnaires to) traditional knowledge holders. Research projects should be reviewed at an early planning stage for their lifetime impact on other stakeholders or communities and the issues requiring dialogue with other communities or stakeholders.

14.5.7   Information and Transparency: Communication and Advice

Continuous communication among partners should be encouraged. The community that is potentially to be affected by the study should be advised about its purpose, time frame, techniques, and positive and negative implications, so that they are in a position to decide whether or not to participate. Researchers should address the need to train community-based researchers and assess adequate forms of existing legal protection for sensitive information. Likewise, non-community-based researchers (scientists) should be trained in the ways of the people who will be involved.

14.6   Conclusions

Some of the main problems in studying traditional forest-related knowledge are: failure to accord recognition to holders of traditional knowledge, misappropriation of such knowledge, and lack of appropriate benefit-sharing. Modern science and traditional knowledge have often been viewed as incompatible, with fundamental differences. However, increasingly, traditional forest-related knowledge and science are recognized as approaches that can co-exist and, moreover, be mutually beneficial.

The main methods of studying traditional forest-related knowledge come from the social sciences—in particular, sociology and anthropology—and place a heavy emphasis on participatory elements. Both quantitative and qualitative methods are used, but particularly the latter. Some of the most commonly used methods and techniques of analysis are participant observation, interviews (individual, group), cultural domain analysis, questionnaires and community workshops. Any method that is used should be culturally sensitive, piloted and triangulated where possible.

With the increasing use and recognition of traditional forest-related knowledge in decision-making and science, there is a serious need to establish and follow ethical guidelines in its acquisition and use. Several regional initiatives exist. The main principles of most guidelines, however, are: free and prior informed consent, mutually agreed terms, transparency, community and indigenous participation, appropriate dissemination of traditional knowledge in the public domain, returning research results in an appropriate form, access and benefit-sharing, and active protection efforts. Each of these principles can be interpreted in various ways. Researchers, as well as the community, bear a responsibility here, and they should take a strong interest in considering and abiding by these principles. However, the important challenge of clarifying how such guidelines are best enforced remains.

References

Agrawal A (1995) Dismantling the divide between indigenous and scientific knowledge. Dev Change 26:413–439

Alaska Native Science Commission (2004) Ethical guidelines for the use of traditional knowledge in research and science. Adapted from traditional knowledge research guidelines—Council of Nations. Available via http://www.cyfn.ca/index.html. Cited 15 July 2009

Battiste M, Youngblood HJ (2000) Protecting indigenous knowledge and heritage. Purish Publishing Ltd., Saskatoon

Becker CD, Ghimire K (2003) Synergy between traditional ecological knowledge and conservation science supports forest preservation in Ecuador. Conserv Ecol 8(1):1

Berkes F (2004) Rethinking community-based conservation. Conserv Biol 18(3):621–630

Berkes F (2008) Sacred ecology, 2nd edn. Routledge, New York

Bernard HR (2000) Social research methods: qualitative and quantitative approaches. Sage, Thousand Oaks

Bernard HR (2005) Research methods in anthropology, 4th edn. Altamira Press, Walnut Creek

Bishop R (2005) Freeing ourselves from neo-colonial domination in research: a Kaupapa Māori approach to creating knowledge. In: Denzin N, Lincoln Y (eds) The Sage handbook of qualitative research, 3rd edn. Sage, Thousand Oaks, pp 109–138

Canadian Institutes of Health Research [CIHR] (2007) CIHR guidelines for health research involving aboriginal people. CIHR, Ottawa, 46 p. Available via http://www.cihr-irsc.gc.ca/e/documents/ethics_aboriginal_guidelines_e.pdf. Cited 15 July 2009

CGIAR (2009): See Consultative Group on International Agricultural Research

Clayoquot Alliance for Research, Education and Training (2003) Standard of conduct of research in Northern Barkley and Clayoquot Sound communities. Available via http://depts.washington.edu/ccph/pdf_files/standardconduct_jun03-1.pdf. Cited 15 July 2009

Colfer CJP, Colchester M, Joshi L, Puri RK, Nygren A, Lopez C (2005) Traditional knowledge and human well-being in the 21st century. In: Mery G, Alfaro R, Kanninen M, Lobovikov M (eds) Forests in the global balance—changing paradigms, vol 17, IUFRO World Series. International Union of Forest Research Organizations [IUFRO], Helsinki, pp 173–182

Consultative Group on International Agricultural Research [CGIAR] (2009) CGIAR guidelines on the acquisition and use of traditional knowledge. Unpublished manuscript, draft 2009. Available via http://cgiar.org/pdf/grpc_24th_meeting_minutes.pdf. Cited 15 July 2009

Cooperative Research Centre [CRC] (Desert Knowledge) (2007) What are ethics? CRC briefing paper 6. Available via http://www.desertknowledgecrc.com.au/socialscience/downloads/DKCRC-SS-BP6-What-are-ethics.pdf

Davidson-Hunt IJ (2003) Indigenous lands management, cultural landscapes and Anishinaabe people of Shoal Lake, Northwestern Ontario, Canada. Environments 31:21

Davidson-Hunt IJ, O'Flaherty MR (2007) Researchers, indigenous peoples, and place-based learning communities. Soc Nat Resour 20:291–305

Davis SM, Reid R (1999) Practicing participatory research in American Indian communities. Am J Clin Nutr 69(Supplement):755S–759S

Dickson D (2003) Let's not get too romantic about traditional knowledge. Editorial. Science and Development Network. Available via http://www.scidev.net. Cited 15 July 2009

Donovan DG, Puri RK (2004) Learning from traditional knowledge of non-timber forest products: Penan Benalui and the autecology of *Aquilaria* in Indonesian Borneo. Ecol Soc 9(3):3, Available via http://www.ecologyandsociety.org/vol9/iss3/art3. Cited 13 Sept 2009

Duerden F (2005) Relations between traditional knowledge and western science (review). Arctic 58(3):309–310

Ellen RF, Harris H (2000) Introduction. In: Ellen R, Parkes P, Bicker A (eds) Indigenous environmental knowledge and its transformations: critical anthropological perspectives. Harwood Academic Press, Amsterdam, pp 1–33

Fernandez-Gimenez ME, Huntington HP, Frost KJ (2006) Integration or co-optation? Traditional knowledge and science in the Alaska Beluga Whale Committee. Environ Conserv 33(4):306–315

Folke C (2004) Traditional knowledge in social-ecological systems. Ecol Soc 9(3):7

Forest Peoples Programme et al (2005) Broken promises: how World Bank group policies fail to protect forests and forest peoples' rights. Available via http://www.forestpeoples.org/documents/ifi_igo/wb_forests_joint_pub_apr05_eng.pdf. Cited 20 July 2009

Fortmann L (ed) (2008) Participatory research in conservation and rural livelihoods: doing science together. Wiley, Hoboken

Fricker M (2007) Epistemic injustice: power and the ethics of knowing. Oxford University Press, Oxford

Goodenough W (1970) Description and comparison in cultural anthropology. Cambridge University Press, Cambridge

Headland TN, Pike K, Harris M (eds) (1990) Emics and etics: the insider/outsider debate. Sage, Newbury Park

Huntington HP (2000) Using traditional ecological knowledge in science: methods and applications. Ecol Appl 10:1270–1274

Huntington HP, Brown-Schwalenberg PK, Frost KJ, Fernandez-Gimenez ME, Norton DW (2002) Observations on the workshop as a means of improving communication between holders of traditional and scientific knowledge. Environ Manag 30(6):778–792

Huntington HP, Suydam RS, Rosenberg DH (2004) Traditional knowledge and satellite tracking as complementary approaches to ecological understanding. Environ Conserv 31(3):177–180

Huntington HP, Trainor SF, Natcher DC, Huntington OH, DeWilde L, Chaplin SFI (2006) The significance of context in community-based research: understanding discussions about wildfire in Huslia, Alaska. Ecol Soc 11(1):40

Interagency Advisory Panel on Research Ethics (2010) Tri-Council policy statement: ethical conduct for research involving humans (TCPS2), 2nd edn. Interagency Secretariat on Research Ethics, Ottawa

International Council for Science [ICSU] (2002) Traditional knowledge and sustainable development. ICSU series on science for sustainable development 4. International Council for Science, Paris. Available via http://www.icsu.org/. Cited 15 July 2009

International Institute for Sustainable Development [IISD] (2005) A guide to using the working draft ABS management tool—a management tool for implementing genetic resource access and benefit sharing activities. IISD, Winnipeg, 72 p. Available via http://www.iisd.org/pdf/2005/standards_abs_mt_user_guide.pdf. Cited 20 July 2009

International Institute of Rural Reconstruction [IIRR] (1996) Recording and using indigenous knowledge: a manual. Cavite, Silang

International Society of Ethnobiology (2006) International Society of Ethnobiology code of ethics (with 2008 additions). Available via http://ise.arts.ubc.ca/global_coalition/ethics.php. Cited 16 Sept 2009

Johannes RE (1989) Fishing and traditional knowledge. In: Johannes RE (ed) Traditional ecological knowledge: a collection of essays. IUCN, Gland, pp 39–43

Kitson JC (2004) Harvest rate of sooty shearwaters (Puffinus griseus) by Rakiura Maori: a potential tool to monitor population trends? Wildl Res 31(3):319–325

Laird SA (ed) (2002) Biodiversity and traditional knowledge: equitable partnerships in practice. Earthscan, London

Martin GJ (2004) Ethnobotany: a methods manual. Earthscan, London

Mazzocchi F (2006) Western science and traditional knowledge. EMBO Rep 7(5):463–466

McDonald JA (2004) The Tsimshian protocols: locating and empowering community-based research. Can J Nativ Educ 28(1–2):80–91

McGregor D (2008) Linking traditional ecological knowledge and western science: aboriginal perspectives from the 2000 state of the lakes ecosystem conference. Can J Nativ Stud 28(1):139–158

McManis CR (ed) (2007) Biodiversity and the law: intellectual property, biotechnology and traditional knowledge. Earthscan, London

Medeek WW (2004) Forests for the future: the view from Gitkxaala. Can J Nativ Educ 28(1–2):8–14

Moller H, Berkes F, Lyver PO, Kislalioglu M (2004) Combining science and traditional ecological knowledge: monitoring populations for co-management. Ecol Soc 9(3):2

Newing H, Eagle C, Puri RK, Watson CW (eds) (2010) Conducting research in conservation, a social science perspective. Routledge, London

O'Flaherty RM, Davidson-Hunt I, Manseau M (2007) Keeping woodland caribou (Ahtik) in the Whitefeather Forest, vol 27, Sustainable Forest Management Network Research Note Series. Sustainable Forest Management Network, Edmonton, Available via http://www.sfmnetwork.ca/docs/e/E27%20Caribou%20in%20the%20Whitefeather%20forest.pdf. Cited 4 Oct 2009

Parsons R, Prest G (2003) Aboriginal forestry in Canada. For Chron 79(4):779–784

Pike KL (1967) Language in relation to a unified theory of structure of human behaviour, 2nd edn. Mouton, The Hague

Puri RK (2001a) Local knowledge and manipulation of the fruit 'mata kucing' (*Dimocarpus longan*) in East Kalimantan. In: Victor M, Barash A (eds) Cultivating forests: alternative forest management practices and techniques for community forestry. RECOFTC, Bangkok, pp 98–110, Available via: http://www.recoftc.org/pubs_interreports.html#Cultivating. Cited 16 Sept 2009

Puri RK (2001b) The Bulungan ethnobiology handbook. Center for International Forestry Research [CIFOR], Bogor, Available via http://www.cifor.cgiar.org/publications. Cited 16 Sept 2009

Puri RK (2005) Deadly dances in the Bornean rainforest: hunting knowledge of the Penan Benalui. KITLV Press, Leiden

Puri RK (2007) Responses to medium-term stability in climate: El Niño, droughts and coping mechanisms of foragers and farmers in Borneo. In: Ellen R (ed) Modern crises and traditional strategies: local ecological knowledge in Island Southeast Asia. Berghahn, Oxford, pp 46–83

Puri RK (2010a) Documenting local knowledge. In: Newing H, Eagle C, Puri RK, Watson CW (eds) Conducting research in conservation, a social science perspective. Routledge, London, pp 146–169, Chapter 8

Puri RK (2010b) Participatory mapping. In: Newing H, Eagle C, Puri RK, Watson CW (eds) Conducting research in conservation, a social science perspective. Routledge, London, pp 187–198, Chapter 10

Puri RK (2010c) Participant observation. In: Newing H, Eagle C, Puri RK, Watson CW (eds) Conducting research in conservation, a social science perspective. Routledge, London, pp 85–97, Chapter 5

Ramakrishnan PS, Chandrashekara UM, Elouard C, Guilmoto CZ, Maikhuri RK, Rao KS, Sankar S, Saxena KG (eds) (2000) Mountain biodiversity, land use dynamics, and traditional ecological knowledge. UNESCO and Oxford and IBH Publishing, New Delhi

Reason P, Bradbury H (eds) (2001) The SAGE handbook of action research: participative inquiry and practice. Sage, Thousand Oaks

Roslin A (2005) The Ndoho Istchee vision. Waswanipi Cree Model Forest, Waswanipi Cree First Nation, Quebec

Sandker M, Suwarno A, Campbell BM (2007) Will forests remain in the face of oil palm expansion? Simulating change in Malinau, Indonesia. Ecol Soc 12(2):37, Available via http://www.ecologyandsociety.org/vol12/iss2/art37/. Cited 16 Sept 2009

Sandøe P, Jensen KK (2008) How should the CGIAR handle ethical challenges? Issues and proposal for a strategic study. In: Consultative Group on International Agricultural Research [CGIAR] (ed) Ethical challenges for the CGIAR: report of three studies. CGIAR Science Council Secretariat, Rome, pp 9–31. Available via ftp://ftp.fao.org/docrep/fao/011/i0501e/i0501e00.pdf. Cited 15 July 2009

Sandøe P, Ejeta G, Lipton M, Jensen KK, Gardiner P (2008a) Ethics and CGIAR mission: study III. In: Consultative Group on International Agricultural Research [CGIAR] (ed) Ethical challenges for the CGIAR: report of three studies. CGIAR Science Council Secretariat, Rome, pp 83–111. Available via ftp://ftp.fao.org/docrep/fao/011/i0501e/i0501e00.pdf. Cited 15 July 2009

Sandøe P, Adair L, Dias C, Jensen KK, Gardiner P (2008b) Ethics and CGIAR research: study II. In: Consultative Group on International Agricultural Research [CGIAR] (ed). Ethical challenges for the CGIAR: report of three studies. CGIAR Science Council Secretariat, Rome, pp 49–74. Available via ftp://ftp.fao.org/docrep/fao/011/i0501e/i0501e00.pdf. Cited 15 July 2009

Secretariat of the Convention on Biological Diversity (2002) Bonn guidelines on access to genetic resources and fair and equitable sharing of the benefits arising out of their utilization. Secretariat of the Convention on Biological Diversity, Montreal, Available via http://www.cbd.int/decision/cop. Cited 15 July 2009

Sheil D, Puri RK, Basuki I, van Heist M, Wan M, Liswanti N, Rukmiyati, Sardjono MA, Samsoedin I, Sidiyasa K, Chrisandini, Permana E, Angi EM, Gatzweiler F, Johnson B, Wijay A (2003) Exploring biological diversity, environment and local people's perspectives in forest landscapes: methods for a multidisciplinary landscape assessment, 2nd edn. Center for International

Forestry Research [CIFOR], Bogor, Available via http://www.cifor.cgiar.org/mla/download/publication/exploring_bio.pdf. Cited 16 Sept 2009

Sheil D, Puri RK, Wan M, Basuki I, van Heist M, Liswanti N, Rukmiyati I, Rachmatika I, Samsoedin I (2006) Local people's priorities for biodiversity: examples from the forests of Indonesian Borneo. Ambio 15(1):17–24

Smylie J, Martin CM, Kaplan-Myth N, Steele L, Tait C, Hogg W (2004) Knowledge translation and indigenous knowledge. Int J Circumpolar Health 63(suppl 2):139–143

Stehr N, Grundmann R (2005) Knowledge: critical concepts. Taylor and Francis, London

Tobias T (2000) Chief Kerry's moose: a guidebook to land use and occupancy mapping, research design and data collection. Union of BC Indians and Ecotrust Canada, Vancouver

Trosper RL (2009) Resilience, reciprocity and ecological economics: Northwest Coast sustainability. Routledge, London

Turnbull D (1997) Reframing science and other local knowledge traditions. Futures 29(6):551–556

Turner NJ (2005) The Earth's blanket, traditional teachings for sustainable living. Douglas and McIntyre, Vancouver

United Nations [UN] (2007) United Nations declaration on the rights of indigenous peoples 2007. Adopted by the General Assembly 13 September 2007. Available via http://www.un.org/esa/socdev/unpfii/en/declaration.html. Cited 15 July 2009

Vayda AP, Walters B, Setyawati I (2003) Doing and knowing: questions about studies of local knowledge. In: Bicker AJ, Sillitoe P, Pottier J (eds) Investigating local knowledge: new directions, new approaches. Ashgate Publishing, London, pp 35–58

Wa G, Uukw D (1992) The spirit in the land: statements of the Gitksan and Wet'suwet'en hereditary chiefs in the Supreme Court of British Columbia, 1987–1990. Reflections, Gabriola

Weller S, Romney A (1988) Systematic data collection. Sage, London

Whitefeather Forest Research Cooperative [WFRC] (2004) Whitefeather Forest Research Cooperative agreement. Pikangikum First Nation, Ontario. Available via: http://www.whitefeatherforest.com/wp-content/uploads/2008/08/wfrc_agreement.pdf. Cited 4 Oct 2009

Chapter 15
The Unique Character of Traditional Forest-Related Knowledge: Threats and Challenges Ahead*

Ronald L. Trosper, John A. Parrotta, Mauro Agnoletti, Vladimir Bocharnikov, Suzanne A. Feary, Mónica Gabay, Christian Gamborg, Jésus García Latorre, Elisabeth Johann, Andrey Laletin, Lim Hin Fui, Alfred Oteng-Yeboah, Miguel Pinedo-Vasquez, P.S. Ramakrishnan, and Youn Yeo-Chang

Abstract This chapter reflects on the major findings of the lead authors of this book regarding traditional forest-related knowledge (TFRK) using five criteria for distinguishing the unique character of traditional knowledge: (1) its attention to sustainability; (2) relationships to land; (3) identity; (4) reciprocity; and (5) limitations on market involvement. Following an explanation of these criteria, we discuss the definition of "traditional forest-related knowledge," with some remarks

*This chapter is the outcome of a meeting of the lead authors of this book held in at the headquarters of the International Union of Forest Research Organizations in Mariabrunn, Austria, November 8–10, 2010 (Fig. 15.1).

R.L. Trosper (✉)
Faculty of Forestry, University of British Columbia, Vancouver, BC, Canada
e-mail: rltrosper@email.arizona.edu

J.A. Parrotta
U.S. Forest Service, Research and Development, Arlington, VA, USA
e-mail: jparrotta@fs.fed.us

M. Agnoletti
Dipartimento di Scienze e Teconolgie Ambientali Forestali, Facoltà di Agraria,
Università di Firenze, Florence, Italy
e-mail: mauro.agnoletti@unifi.it

V. Bocharnikov
Pacific Institute of Geography, Russian Academy of Science, Vladivostok, Russia
e-mail: vbocharnikov@mail.ru

S.A. Feary
Conservation Management, Vincentia, NSW, Australia
e-mail: suefeary@hotkey.net.au

M. Gabay
Directorate of Forestry, Secretariat of Environment and Sustainable Development,
Buenos Aires, Argentina
e-mail: monagabay@yahoo.com

J.A. Parrotta and R.L. Trosper (eds.), *Traditional Forest-Related Knowledge: Sustaining Communities, Ecosystems and Biocultural Diversity*, World Forests 12, DOI 10.1007/978-94-007-2144-9_15,
© Springer Science+Business Media B.V. (outside the USA) 2012

about its resilience. We then consider threats to the maintenance of TFRK, how other definitions of sustainability differ from that used in TFRK, and how relationships that holders of this knowledge have to their land have been weakened and their identities challenged. We highlight how the key role of reciprocity, or the sharing of the utilization of land, is undermined by individualistic motives which are promoted by the global expansion of modern markets (for commodities, ecosystems services and for knowledge itself), which also challenge the policies of traditional knowledge holders to keep market influences under control. We then focus on two notable, but often ignored, contributions of TFRK (and the holders of this knowledge) to forest management today, specifically the preservation of biodiversity, and traditional knowledge-based shifting cultivation practices and their importance for both sustainable management of forests and food security.

C. Gamborg
Danish Centre for Forest, Landscape and Planning, University of Copenhagen,
Copenhagen, Denmark
e-mail: chg@life.ku.dk

J. García Latorre
Federal Ministry of Agriculture, Forestry, Environment and Water Management,
Vienna, Austria
e-mail: jesus.garcia-latorre@lebensministerium.at

E. Johann
Austrian Forest Association, Vienna, Austria
e-mail: elisabet.johann@aon.at

A. Laletin
Friends of the Siberian Forests, Krasnoyarsk, Russia
e-mail: laletin3@gmail.com

Lim H.F.
Forest Research Institute Malaysia, Kepong, Selangor, Malaysia
e-mail: limhf@frim.gov.my

A. Oteng-Yeboah
Department of Botany, University of Ghana, Legon, Ghana
e-mail: alfred.otengyeboah@gmail.com

M. Pinedo-Vasquez
Center for Environmental Research and Conservation (CERC), Columbia University,
New York, NY, USA

Center for International Forest Research (CIFOR), Bogor, Indonesia
e-mail: map57@columbia.edu

P.S. Ramakrishnan
School of Environmental Sciences, Jawaharlal Nehru University, New Delhi, India
e-mail: psr@mail.jnu.ac.ac.in

Youn Y.-C.
Department of Forest Sciences, Seoul National University, Seoul, Republic of Korea
e-mail: youn@snu.ac.kr

Finally, we consider enabling conditions for the preservation and development of TFRK, and examine the role of the scientific community in relation to TFRK and principles for successful collaboration between traditional knowledge holders and scientists.

Keywords Biodiversity • Cultural diversity • Forest management • Forest science • Local communities • Indigenous peoples • Sustainability • Traditional knowledge

15.1   What Have We Learned from This World Survey About Traditional Forest-Related Knowledge?

In light of the surveys provided in the regional and special issue chapters of this book, the lead authors discussed further the meaning of the term, "traditional forest-related knowledge." While we do not see reasons to change the basic definition as originally adopted,[1] we have found that deeper understanding of the three components are possible.

15.1.1   "Traditional"

The word "traditional" carries connotations of "old" or "static" and unchanging. The authority of traditional ideas is perceived to come from their persistence and their inheritance from previous generations. As a consequence, many might perceive that adoption of new ideas, particularly of the results of science, would change and transform traditional knowledge to the point that it would lose its distinctiveness, and therefore its authority and its credibility as a unique way of knowing. What criteria can be applied to determine whether or not a particular knowledge system has retained its distinctive character even as it adapts to new circumstances and adopts tools and ideas that may have originated from outside of the particular system?

   We suggest that some combination of the following five criteria can be used to distinguish the unique character of "traditional" knowledge:

1. *Sustainability*: the goal of understanding remains to maintain the sustainability of the system.
2. *Relationships*: peoples' connections among themselves and to their territory are not severed by the use of new knowledge, ideas or techniques.

---

[1]"A cumulative body of knowledge, practice and belief, handed down through generations by cultural transmission and evolving by adaptive processes, about the relationship between living beings (including humans) with one another and with their forest environment" (UN 2004, adapted from Berkes et al. 2000).

3. *Identity*: people maintain their distinct identity.
4. *Reciprocity:* people maintain their system of benefit sharing among themselves, and
5. *Limits on exchange*: while people may engage in market exchange with the flow of products from the land, the fundamental productivity of the system itself is not viewed as capital to be exchanged.

We now comment on each of these.

A fundamental characteristic of all of the traditional systems considered in the chapters is that the people practicing them pursue sustainability. While the particular content of the systems vary, a key aspect of their traditional character is their long persistence. Not only have the systems persisted, often in the face of significant challenges from outside, the participants strive to keep the land's productive capacity in place for the benefit of future generations

In the process of maintaining the system, a second key aspect is maintaining the relationship of people to their territory. In some situations, people believe that they are part of the territory, and that their connections cannot be severed. Others maintain that the condition of their territory is the product of their long historical relationships with it, creating what are known as "cultural landscapes". In either circumstance, adoption of new plants or animals to include in the system is constrained by the need to maintain people's relationship to that territory. For instance, when traditional hunters adopt guns to replace older methods, they still retain their attitudes of respect for the species hunted (Nadasdy 2003). This is evident in the case of the Cree and their relationships with beaver, moose and caribou (Tanner 1979). Those practicing shifting cultivation may introduce new trees to forest fallows to improve the system's productivity, while maintaining both a fallow period dominated by perennial plants and trees and a farming period dominated by annual or biannual plants. The overriding concern remains maintenance of the productivity of the system. In the Mexican tropics, for example, the rituals associated with the *milpa* system persist even if the dominant crops change away from maize, since a portion of the landscape is still reserved for maize production. In this traditional system, maize is the defining species, and to remove it totally would remove people's ritual relationship to the land (Alcorn and Toledo 1998; Toledo et al. 2007).

A third characteristic of traditional systems is that people seek to maintain their identity with respect to neighbours. Even when a community is close to losing their language and have intermarried with other groups, they maintain ways to identify themselves as a distinct community. That identity can remain even if the symbols and content of the identity change is particularly evident in Indian tribes in the United States. The people of the Confederated Salish and Kootenai Tribes of the Flathead Indian Reservation in Montana, for instance, are predominantly Roman Catholic in religion, speak English, and live intermixed with their non-tribal neighbours on a reservation on which a majority of the residents are non-Indian. But they have maintained their tribal government, increased the amount of land that they own (held in trust by the federal government), are striving to save their language, and established their own educational institutions.

The fourth characteristic, maintenance of systems of reciprocity and sharing, relates closely to retention of traditional land tenure systems. It is widely recognized that traditional societies have utilized various approaches to sharing the use of land, and recent literature has further explained the systems of reciprocity that were and are still used in such societies (Escobar 1995, 2005; Fiske 1991; Gudeman 2008; Henrich et al. 2004; Kolm and Mercier Ythier 2006; Polanyi 2001; Sahlins 1972, 1996).

The fifth characteristic is placement of limitations on the logic of market capitalism. With the expansion of the world-wide market economy, few groups have retained a position of autonomy in which they do not trade with the broader system. Many sell products that travel far from the local origin. The Menominee Tribe in Wisconsin, for example, maintains a lumber mill and harvests timber from their lands in order to supply worldwide markets with certified lumber. But they have not reduced the rotation period of the trees in their forest (150 years) to the shorter rotations that are typical of those who manage standing timber as capital (80 years). They do not regard the forest as an alternative to a bank, and retain a growing stock that is much larger than corporate forestry would keep. Rather, the forest produces multiple benefits to the current generation and is regarded as belonging to future generations. In fact, they have increased their growing stock to compensate for harvest levels that were too large early in the twentieth century. They maintain species that are not currently very profitable, and they do not let the mill management dictate foresters' decisions. The current generation cannot sell the land because they do not have that authority within their own belief system. When the US government in 1961 forced them to transfer the land to a private corporation owned by the members with individual shares, the Menominee resisted. A bank had become trustee for the children's shares, giving the bank control. As children reached voting age, community members gradually obtained sufficient voting power to control this corporation's board of directors, after which they removed the capitalist-oriented management and returned management of the land to better align with their own long-term views. In 1972, the federal government restored the reservation to their control through the Menominee Restoration Act. Today, the Menominee sell wood fibre without adopting all of the characteristics of timber corporations who manage forests only from the profits earned by growing trees (Trosper 2007).

15.1.2 *"Forest-Related"*

As the authors began work on this book, we agreed that the subject matter was forest-*related* knowledge, not merely forest knowledge. Our surveys for the preceding chapters demonstrated that the extent of "related" was vast indeed. Originally, the idea was that a tree-dominated landscape would have other components that would be important, such as the products of the non-tree understory. Forests provide grazing for domesticated and wild animals, fruit from trees and shrubs, habitat for mammals, birds, reptiles, amphibians, and insects, protection for the supply of water, to make just a short list. The widespread practice of using forest fallows in farming, however, further widened the scope of "related" to include agriculture. Whether it is Mayan Indians raising corn and beans in their *milpas*, Andean Indians raising potatoes and yucca, or people in South and Southeast Asia growing upland rice, agricultural lands were definitely related to forested land. On the Kamchatka peninsula of Siberia, and on the Northwest Coast of North America, the anadromous fish of the river are also a vital part of the forest system. Proper care of the forest provides maintenance of the quality of the rivers, and proper care of the salmon provides nutrients for the forest through the uptake of nutrients brought from the ocean to the land.

15.1.3 *"Knowledge"*

Our work revealed some radical views of knowledge. Among the Gitsxan of the Northwest Coast of North America, for instance, leaders insist that the source of law is the land, not humans (Gisday Wa and Delgam Uukw 1992). Other indigenous peoples insist that they must listen carefully to the land to understand what it is telling them. Animals such as moose and caribou are assumed to be sentient, able to draw conclusions as a result of observing the behaviour of humans. That the creatures of the land will listen and observe human behaviour means that humans must take care in what they do. Aboriginal Australians consistently refer to themselves as belonging to the land, rather than the land belonging to them. Damage to the land is tantamount to damage to a person and can make people sick. That the source of knowledge is the land, not humans, marks a dramatic break with the epistemology normally associated with formal science. The sentience of humans is also part of the land, and the land reflects the activities and decisions of humans. This connection of humans to the land is emphasized with regard to the standards for defining tradition, as discussed above.

For most of the indigenous peoples and local communities discussed in this volume, one can assert that the idea of "nature," as an entity separate from human society and identifiable as a subject to be studied without reference to human activity, does not exist. Often, land without human influence is maintained as "pristine" through conscious choice that recognizes the value of such places that are often identified as "sacred."

Even while recognizing human influence on ecosystems, people have been perceived by scientists as an external modifying force and not as an integral, interacting, component of the biotic and abiotic world. This perspective supports the classical perception of two divorced entities: society and nature. The landscape is typically viewed by most ecologists as a mosaic of communities evolving to a supposed climax stage through well-defined successional paths. Humans' actions interrupt these paths and changes the communities. Scientists in the discipline of phytosociology, which originated in Europe through the work of Josias Braun-Blanquet (1884–1980), have classified the landscape in terms of what the "potential vegetation" would be in the absence of the activities of man. Some argue, however, that very little of the landscape in Europe (and in many other parts of the world) in fact is not a "cultural landscape," which has reached its current condition through the history of interaction with humans. The general public is similarly confused. "Natural" and "untouched" are terms frequently used by hikers to emphasize the beauty of a country scene, even when it is in fact a deeply modified cultural landscape.

Undoubtedly, humans have continuously and substantially modified ecosystems and physical environments throughout the world, as discussed extensively in the regional chapters of this book. This long relationship between nature and human activity has produced distinguishable landscapes which reflect cultural differentiation between societies and reveal the ways that humans have responded to the exigencies of the natural environment. Recently, the Anishinaabe people of Iskatewizaagegan No. 39 Independent First Nation, in north-western Ontario, Canada, were offered a system for classifying the landscape in scientific terms by scientists. Elders insisted that while the categories seemed reasonable, the history of their own interactions with the land needed to be added as a consideration. The elders were interpreted as saying that they are "learning as they journey" on the land (Davidson-Hunt and Berkes 2003). Another kind of learning journey (in time) occurs in northeastern India, where those engaged in shifting cultivation have reduced the length of the fallow period and changed the amount of forest cover in response to many changes resulting from population and income growth (Ramakrishnan 1992). Similarly, studies in Amazonia have discovered a deep history of the presence of humans in the forest. The current character of the forest is a consequence of that history, which must not be forgotten in dealing with the Amazon. In India, distinct tribal groups have generated their own landscapes through their histories. These separate areas can be described as "natural cultural systems" which their inhabitants value highly and protect in various ways (Ramakrishnan 1992; Ramakrishnan et al. 2006).

Another aspect of the traditional forest knowledge system highlighted in our surveys was the degree of integration of multiple uses of the forest, what is often described as a "holistic view" of the landscape. This holistic view accommodated the attention given to so many things that were related to the forest but were not the trees themselves.

In summary, each of the three components of the topic of this volume marks very distinctive ideas with considerable depth. It is perhaps remarkable that these characteristics were found to be shared among traditional peoples on all of the continents that we have surveyed.

15.1.4 Resilience of Traditional Forest-Related Knowledge

A characteristic of the socio-ecological systems that have produced traditional forest-related knowledge is their resilience and adaptability. Since traditional societies are highly dependent on these systems, they often have developed management practices and local institutions that are able to adapt to environmental and other challenges, whether internal or external. The climate change chapter stressed this characteristic of traditional systems. The capacity to adapt can be examined in three dimensions: the ability to rearrange and relocate particular landscapes, the ability to select and modify habitats within landscapes, and the ability to utilize knowledge of species to adjust the mix of species utilized within habitats. All of these capacities are the result of TFRK. The result of the capacity to adapt is the ability to maintain levels of ecosystem services, quantities of products produced, and other system outputs, among which is inspiration.

As a consequence of this adaptive capacity, the societies and the socio-ecological systems in which TFRK developed have good records of survival. In some cases these records can be substantiated through written sources, as in Europe and Asia. Where written records do not exist, survival and persistence can be demonstrated through a combination of archaeological evidence and oral histories.

Some of the noteworthy examples of the prior resilience of these systems were described in the preceding chapters. For instance, shifting cultivation has a long history and has occurred in Africa, the Americas, Europe, Asia and the Pacific in a wide range of tropical, subtropical, and temperate forest ecosystems. For example, the Iroquois corn-bean-squash fields were planted on lands subject to shifting locations in the northern hardwood forests of North America. In Chile and Bolivia indigenous peoples used potato, yucca and corn. In India, over 100 groups have relied for countless generations on shifting cultivation *(jhum)* systems (Ramakrishnan 1992). In the mountainous area of mainland Southeast Asia the long history of shifting cultivation casts doubts on the 'primary' character of forests located in accessible areas. In this region shifting cultivation has promoted fairly stable secondary forest vegetation characterized by complex, dynamic and structurally diverse successional stages (Fox 2000).

15.2   Threats to Traditional Forest-Related Knowledge

The previous section began with a list of five characteristics that maintain the "traditional" component of traditional forest-related knowledge: sustainability, relationships with land, identity, reciprocity, and limits on exchange. Although the international interest in sustainability has probably contributed to the current growing interest in TFRK, some of the characteristics of the "traditional" character of TFRK are also the sources of its vulnerabilities.

Regarding sustainability, perhaps the largest challenge relates to the core concept of sustainability itself. For example, the Menominee Tribe in Wisconsin has long been interested in sustained yield management of its forest, and has created a Sustainable Development Institute as part of the College of the Menominee Nation. The Menominee do not define sustained yield in the manner used by the forestry profession, i.e., based on maximum sustained yield or maximization of soil expectation value. The Menominee maintain a growing stock much larger than would be calculated with maximum sustained yield, and thus reduce the growth rate of their forest. They desire to grow large trees and maintain a diverse, structurally complex, forest. The Menominee definition of sustainability involves the idea that they are borrowing the forest from future generations, and seems to correspond roughly to the idea of "strong sustainability" from ecological economics. It means including all the capacity of land, and not just the production of commodities or ecosystem services. This broad definition is challenged by less strict definitions, which can lead to unsustainable policies (Norgaard 2010).

The idea of sustainability supported by scientific approaches to forest management differs from that based on traditional knowledge. The scientific idea of sustainability originated with the success of one type of scientific forestry—the system developed in Germany starting in the sixteenth century. This led to the export of the ideas of 'scientific forestry' to other parts of the world, with the accompanying idea that local knowledge was not valuable (Michon et al. 2007). Scientific forestry was much liked by national governments, as well as colonial governments (from the mid-nineteenth to mid-twentieth centuries). Because of its appeal, governments sought to rationalize the management of forests by removing uses that seemed to contradict or interfered with the primary goal—growing timber. Because the scientific approach supported the interests of states, colonizing powers and post-colonial governments worldwide, both science and the state have played key roles in reducing the use of TFRK in forest management. Thus, sustainability has evolved from a concept concentrating on the level of wood production, or yield, to the present, more comprehensive concept integrating ecological, economic and social aspects of forestry.

The idea that local knowledge was not valuable and that wood fibre production should dominate under a sustained yield forest management regime also began the process of severing connections between local people and the land, the second of the major issues that relate to the maintenance of traditional knowledge. Other factors, including the spread of centralized monotheistic religions (particularly Christianity and Islam) and the modern idea of "nature" as separate from humans also contributed to separating traditional people from their land.

Throughout human history, religion has been abused to justify active suppression of alternative ideas and fuel conflict. Traditional communities, and particularly indigenous peoples and their conceptions of their place in the world and their relationships to forests, have long been on the receiving end. During the European (and later American) global colonization process from the early sixteenth to twentieth centuries, many religious leaders opposed spiritual ideas that differed from their own. Local religious perspectives, including ideas of the sacred in forests, were frequently suppressed and replaced by those of centralized, "top-down" religions. Missionaries sought to change local religions and practices, labelling beliefs in forest deities as "animistic" and not consistent with the monotheism. In the Americas, for example, Christian religion in its varied forms accompanied colonization from 1492 onwards. While European settlers sought to subdue the wilderness and realize economic gain from cutting down forests, Christianity provided justification for man to conquer and dominate the natural world. While traditional people located the sacred in the landscape, for various reasons, religion located it in human-constructed dwellings and in organizations that were exclusively managed by humans.

An important and highly influential image from the Bible is that of the Garden of Eden, a land of great beauty and productivity given to the first humans for their use. In the twentieth century, the idea of Nature acquired Edenic associations. The idea evolved that "pristine" Nature has values for humans, not just for recreation, but for personal development (Shutkin 2000). Recently, pristine nature is seen to provide benefits for humans by preservation of biodiversity. This idea motivates governments to set aside such areas by excluding all uses except visitation. These places, while special, weren't consciously described as "sacred" in the sense of being like the altar in a church. For people in traditional communities, accustomed to living in their sacred lands and protecting their sacred groves, the idea of protecting a landscape from human influence has often had serious consequences. In many cases it meant that they would be removed (as they often have been removed from designated "protected areas" worldwide), just as they are removed from landscapes being developed for economic purposes.

Thus, both applying a narrow definition of sustainability and separating local people from their land undermine traditional forest-related knowledge. A third way to reduce the effectiveness of TFRK is to undermine identity through efforts that result in deculturalization. For example, campaigns of major proselytizing religious sects to displace local concepts of spirituality also contributed to undermining the identity of traditional peoples by changing their religion. Another major factor undermining identity has been national education systems which interfere with the inter-generational transfer of traditional knowledge and wisdom between elders and youth. While not necessarily a global force, we note that the role of youth as people to carry on TFRK is important. In many regions, the holders of TFRK (particularly elders) are attempting to pass along their knowledge and wisdom to the next generation. Women, as holders of knowledge, are responsible for raising children, which provides them with opportunities to impart their knowledge to the very young. Recently NGO organizations are supporting efforts to pass on traditional knowledge and revitalize traditional cultures and languages. Communities are increasingly demanding that researchers, in sharing benefits from research, support documentation of traditional knowledge as well as development of curricular materials. Communities also develop educational media themselves.

In spite of these efforts, we also find that complex pressures are making the cultural transmission of traditional forest-related knowledge difficult. Children often leave their home villages as they pursue education above the primary level. Consequently they miss out on opportunities to obtain profound knowledge from their elders of traditional socio-ecological systems. Urban centres have other appeals besides education, particularly employment. When educated indigenous people become activists for issues of importance for their communities, they generally remain in the urban centres where resources and media are available. While their activism helps protect their original communities, the activists themselves are less likely to participate in traditional knowledge production systems.

The chapter on globalization stressed that the international spread of market institutions contribute to reduction in the utilization of TFRK. It does so because market institutions tend to undermine the final two of the five characteristics that define tradition, namely reciprocity and limiting markets. The idea of reciprocity emphasizes community obligations, both of individuals to their community, and of the community to individuals. Modern market concepts are connected to individualism rather than community. The exclusion of some activities from being governed by market considerations conflicts with the views that market relations can or should cover all forest values. This issue is relevant to a number of policy discussions, including those related to markets for environmental services such as carbon, discussed in the chapter on TFRK and climate change.

Ideas of individualism undermine reciprocity in several ways. As noted above, colonial governments harvested timber over large forest areas for revenue, displacing the traditional residents. In places where the governments did not assert control over forests, conversion of common land to private land, followed by sale of land as property, contributed to displacement of local people from the use of that land. The idea of private property asserts that the owner has full control of the surplus generated from the land, while ideas of reciprocity mean that the surplus is distributed in complex ways to community members. Even the idea of "common land" becomes distorted under the influence of individualistic ideas. "Common land" is frequently thought of as open access land, in which individuals have are free to remove products without community controls on that access. Of course, excessive harvest is an expected result when controls are not present. But traditional land tenure systems are more complex. Hunters are forced to share their catch with relatives, thus reducing the individual incentive to overharvest, while at the same time revealing the amount harvested to public scrutiny. When needed to encourage investment in land productivity, families and individuals may be given exclusive control of parts of the land, under the assumptions that they will share a portion of the harvest and that if they cease to use the land, it returns to the community to be allocated to others. With flexible community-based systems of allocating access to land, excessive division into small parcels could be avoided through community leadership. This kind of flexibility, for instance, is part of many systems of shifting agriculture in which land is allocated to families when farmed, but allocated to other uses when fallow.

When land is part of a system of reciprocal relationships, identifying a unique owner who can sell the land becomes difficult. The whole community has interests in all of the land, and as a result the whole community needs to participate in decisions as momentous as selling a community's land. Many indigenous communities simply refuse to consider land to be a marketable commodity.

This limitation on the sale of land is a part of the fifth and final characteristic of tradition—the placing of limits on the extent to which market relations are allowed to penetrate and affect land use decisions. When some land uses are not valued in markets, then prices are not available for judgment about how to use the land for those purposes. Profit-maximizing calculations cannot be made easily. Those products that a community allows for sale are marketed. The prices received for such products depend upon broader market considerations, and one reason that communities limit their interaction with markets is that such limitations reduce risk connected to prices dropping unexpectedly. Market logic encourages specialization in the products whose current market price is highest, which exposes a community to risk resulting from declines in such prices.

Thus all five characteristics of traditional knowledge – strong sustainability, human connection to land, community identity, reciprocity and control of market relationships—are threatened in the modern era. In spite of these difficulties, the people and their systems that support TFRK have persisted. As a result, they have made contributions to sustainable forest management. The next section considers two such major contributions.

15.3   Contributions of Traditional Forest-Related Knowledge to Sustainable Forest Management and Forest Science

Two areas in which TFRK has made, and continues to make, major contributions to sustainable forest management are the protection of biodiversity and the development and utilization of shifting cultivation practices. This section reviews the contributions of the holders and users of TFRK in these areas, and concludes with a brief summary of other contributions to sustainable management of forests and associated ecosystems.

15.3.1   Conservation and Sustainable Use of Biodiversity

Traditional knowledge holders have a strong record of protecting and even enhancing the biodiversity of their homelands (Belair et al. 2010). The identification and protection of important species, the protection of important plant communities (as in sacred groves), advice on dealing with catastrophic processes such as fire and flood; these are all important ways to support biodiversity. Very often, culturally important species are also ecological keystone species in addition to cultural keystone species. The protection of important forest areas in general, and of areas that are important during key times (such as calving for mammals or spawning for fish) are another characteristic of TFRK. Sacred groves in Asia and Africa are a prime example of such protection. Sometime sacred areas are important for the delivery of ecosystem services, especially water. In other cases, they serve as vital refuges for a variety of valued plant and animal species, including plants used in traditional (and modern) medicine.

Conservation biology owes much to conservation by indigenous peoples, and to knowledge regarding the life history of significant species. For instance, management and preservation efforts species of giant turtles in the Amazon has required knowledge from local people to identify what was needed for their conservation.

Woodlands and trees traditionally used and managed by local communities have important effects on biodiversity.

The open oak-wooded pastures of Mediterranean countries, called *dehesas* in Spain and *montados* in Portugal, are an excellent example. The ecotone nature of these woodland ecosystems allows the development of their characteristically high biological diversity (Campos et al. 1998; Díaz et al. 1997). These woodlands constitute a principal refuge for several rare or endangered bird species in Europe, including the Spanish imperial eagle (*Aquila adalberti*), the cinereous vulture (*Aegypius monachus*) and the black stork (*Ciconia nigra*). A significant proportion of the population of Western European common cranes (*Grus grus*) overwinters in the Spanish dehesas (Muñoz-Pulido 1989). These wooded pastures also support a variety of cavity-nesting bird species that utilize sites in the branches of old managed oaks (Pulido and Díaz 1997).

It is perhaps no accident that there is a strong correlation between the location of the world's centres of forest biodiversity and the cultural diversity represented by indigenous peoples who live in these areas (Galloway-McLean 2010), as discussed in Chap. 1. A high percentage of "biodiversity hotspots" have high resident indigenous peoples, who have maintained and will continue to maintain biodiversity of these areas provided they are allowed to continue their practices. For example in the Brazilian Amazon where, according to Nepstad et al. (2006): "indigenous lands were often created in response to frontier expansion, and many prevented deforestation ... despite high rates of deforestation along their boundaries. The inhibitory effect of indigenous lands on deforestation was strong after centuries of contact with the national society and was not correlated with indigenous population density. Indigenous lands occupy one fifth of the Brazilian Amazon—five times the area under protection in parks—and are currently the most important barrier to Amazon deforestation".

Management of formally designated protected areas can carried out effectively in collaboration with indigenous knowledge holders. Examples of this can be found in the Americas, Africa, Asia and the Pacific. For instance, Gwaii Haanas, a park on the coast of British Columbia, is jointly managed by the Haida and Parks Canada (Mabee et al. 2010). Compared to the situation in other Canadian parks, the Haida have considerable influence on conservation outcomes in Gwaii Haanas (Timko and Satterfield 2008; Timko and Innes 2009). In Brazil, sustainable development reserves, extractive reserves, and indigenous territories all implicitly acknowledge that traditional knowledge is useful for management.

15.3.2   Shifting Cultivation Systems

In the area of shifting cultivation, traditional forest-related knowledge has been instrumental to the recent advances in the management of fallow periods (between rotations of agricultural crops). The value of properly conducted shifting cultivation, particularly in the tropics, has not yet been generally recognized, and these practices still have a bad image in forest management and forest science circles in most countries. This is in spite of recent scientific research that has demonstrated the ways in which shifting cultivators optimize the productivity of their agro-forest systems. For instance, indigenous communities in Mexico have a strong record of applying shifting cultivation successfully (Alcorn and Toledo 1998; Toledo et al. 2007). Shifting agriculture in Yunnan, China has become modernized into a new rotational agroforestry system that is able to sustain the natural resource base in mountainous areas (Liang et al. 2009).

Shifting cultivation's poor reputation is due to its being unfairly blamed for forest degradation which had other causes. Evidence compiled in India indicates that the real causes of degradation have been forest harvesting for export and increased intensity of agricultural use of formerly forest lands, rather than shifting cultivation. Similarly, a study by the National Academies of Science of the India, China and the US concluded that economic pressures and related changes in land use, not population growth, are the principal underlying causes of forest degradation in these three countries (Indian National Science Academy et al. 2001).

A similar situation exists in Southeast Asia, where national policies have banned shifting cultivation and international development agencies have pressured governments to eliminate or transform indigenous agricultural practices. Fox (2000) reports that areas of swidden (i.e., shifting) agriculture have a richness of animal and plant species compared to other areas, that 77–95% of the landscape in five study areas had secondary forest, and that the system is characterized by stability over the large area even though it is dynamic and structurally diverse. Permanent agriculture, by comparison, creates permanent deforestation.

In the scientific literature, there is growing evidence that shifting cultivation as well as other traditional agro-forest management systems have high productivity for the amount of energy and other inputs that they utilize (Sieferle 2001; Tello et al. 2006). The high productivity is realized because of multiple outputs, not all of which are valued in markets, and not all of which are recognized because they occur during the fallow period. The high productivity of other agricultural systems is often based upon inputs of energy (mechanization, irrigation and fertilizer) which have uncounted costs that raise issues of sustainability (Ramakrishnan 1992), including their contributions to climate change through greenhouse gas emissions. Traditional and improved shifting cultivation and agroforestry systems offer distinct advantages (for sustainable agricultural production and biodiversity conservation) over the more intensive farming systems that frequently replace them, particularly in fragile mountain environments (Cairns 2007; NEPED and IIRR 1999; Palm et al. 2005; Swift et al. 1996).

Although the economics-oriented forest science literature used to focus on the optimization of a single product (wood fibre), another more recent ecologically-oriented part of the forest science literature emphasizes multiple products and services from forests. As a result, forest science has become more open to the recognition of the value of shifting cultivation and agroforestry systems than are the more product-oriented agricultural sciences.

Another set of reasons for non-recognition of the value of shifting cultivation result from the resistance from governments, which usually have a great difficulty with mobile communities. This includes the traditional farmers who practice shifting cultivation, as well as those who practice transhumance[2] which is sometimes associated with shifting cultivation. In addition, burning the forest prior to planting, an important part of most traditional shifting cultivation cycles, is generally difficult for governments and the public to accept. This intolerance is increasing as concerns for climate change increase, and burning forests appears to reduce carbon storage. In some countries, shifting cultivation is illegal.

Shifting cultivators are in many regions modifying their management practices while retaining the principle of using forest during the fallow period between agricultural phases (Palm et al. 2005). Such innovations include the abandonment of the use of fire for forest clearing, and planting of new species (forest trees and other valued plant species) in cropping and fallow management Examples of such changes are documented in the regional chapters for India, China, several countries in both Southeast Asia and Africa, and in Mexico.

Thus, traditional forest-related knowledge has made significant contributions to biodiversity conservation and to the modernization of shifting cultivation and agroforestry systems. Such knowledge has contributed to, and could continue to advance, scientific understanding and forest management applications in a number of other areas. These areas include landscape history, climate change, and geographical information systems to be discussed below.

15.3.3   Landscape History, Climate Change, and Geographical Information Systems

Studies of landscape history can use social science methodologies to reconstruct the stories of what areas used to look like. In southeastern Spain, interviews of elders provided alternative reasons and explanations for historical changes in forested landscapes which differed from those which scientists had derived from their examination of current ecological study data (García Latorre and García Latorre 2007). In this case, scientists attempted to reconstruct past forest ecosystem composition and structure based on present floristic composition, without taking into account the management applied to these woodlands by peasants. Peasants interviewed in several mountain ranges in southeastern Spain have shown how oaks as well as many other tree and shrub species in these mountains have been traditionally pollarded, coppiced or shredded (activities never mentioned by phytosociologists, who focus only the current floristic composition). Peasants described the landscape of their youth as

---

[2]Transhumance refers to the seasonal movement of people with their livestock typically as for example in the Himalayan and Alpine regions when herds are moved to higher pastures in summer and to lower valleys where herders have permanent settlements in the winter.

completely cultivated (woody vegetation having been largely eliminated to supply fuel wood needs). The matorral shrublands studied by phytosociologists, which led to erroneous conclusions about past vegetation structure in the region, did not exist in this region prior to the mid-twentieth century, having developed as a consequence of rural abandonment. By incorporating the oral histories into studies such as this, more complete and accurate reconstructions of landscape histories are possible.

Similarly, within indigenous communities on the coast of British Columbia, the oral histories of lineages are carefully monitored for accuracy at feasts. Although these histories maintain the sequence of events, the timing tends to be lost. Archaeologists can date events precisely, while the oral histories provide details on what occurred. Another contribution of traditional knowledge to understanding landscapes is provision of an appropriate method of classifying habitats, as has occurred in the Amazon. These classification systems include information about how humans have modified the landscape.

Regarding phenology and other ecological indicators of climate change, traditional forest-related knowledge has an obvious role to play in the monitoring and analysis of environmental change and its implications for traditional forest-based livelihoods (Parry et al. 2007; Salick and Byg 2007; Galloway-McLean 2010). For instance, Chief Millaman Reinao of the Mapuche in Chile reports that his people have observed many differences in the timing of seasons (Reinao 2008). In Russia, indigenous peoples traditionally used observations of plant phenology (i.e., flowering and fruiting behaviour) to determine when it was time to start hunting. For example, for the Udihe and Oroch peoples of Siberia, changes in larch indicated that it was time to begin their fur hunting season; the flowering of dog rose blooming was an indicator of the initiation of salmon spawning; and a rich harvest of Siberian pine nuts was a sign of that a good hunting season for fur and hoofed animals was ahead.

The utilization of Geographical Information Systems (GIS) is an area of increasing collaboration between scientists and traditional knowledge holders. Maps are a useful way to depict land uses, and the associated data bases are excellent ways to include information from both science and traditional knowledge. Maps can also be used to demonstrate changes over time, both by reproducing what has occurred in the past, and by projecting potential futures. However, care is needed when some of the information obtained from traditional community members is confidential (as discussed in the chapter on research ethics and methodologies). The collaborators on the Foothills Model Forest (now the Foothills Research Institute) in Canada, for instance, provided that some of the information remain under the control of the indigenous partners. Proposed developments would be provided to those partners, who then would report the areas in which conflicts of use existed, without revealing exactly what the indigenous use was. This is needed both for protection of sacred areas and for protection of cultural information. While GIS data bases present a risk similar to information registers, these risks can be addressed by developing adequate controls over the access and use of information from traditional community sources.

15.4   Enabling Conditions for the Preservation and Further Evolution of Traditional Forest-Related Knowledge

While empowerment of local people can support the retention and utilization of traditional knowledge, such policies are not often possible. Two examples concern the relative autonomy of the Menominee Tribe in Wisconsin, and the self-governing powers of lowland indigenous groups in Mexico. Policies that fall short of full empowerment can still have positive effects. For example, in Europe, farmers receive subsidies when they engage in maintenance of traditional agricultural landscapes by organic farming. Similarly, "green agriculture" states in the mountains of India support utilization of traditional knowledge.

Some policies that appear to be designed for preservation of traditional knowledge are resisted. For example, the establishment of knowledge registers does not always receive support by local communities (Agrawal 2002). An example of this occurred in Russia in connection with the UNDP project, "Biodiversity Conservation in the Russian Portion of the Altai-Sayan Ecoregion" (UNDP Focus Areas 2011; Altai Regional Institute of Ecology 2007.). The FAO's "globally important agricultural heritage systems" program has met similar resistance from local people in some countries; the basis of the problem is that the FAO works with governments, while local people are suspicious of their government's further exploitation of their knowledge without compensation. The Traditional Knowledge Information Portal of the Convention on Biological Diversity[3] avoids this controversy by stating it "does not provide or document traditional knowledge per se."

Those interested in traditional forest-related knowledge are often concerned that once lost, it cannot be recovered. The process of forgetting has several stages, and if the existence of the lost knowledge can be recognized at an early stage, living knowledge-holders can be encouraged to revive it. Drivers of the recovery of fading knowledge are the recognition that a problem might have already been solved and the solution be found (or remembered) among remaining knowledge holders. Market demand for traditional products such as baskets can lead to recovery of knowledge related to techniques for harvesting or cultivating plants used for these purposes (Lake 2007).

In North America, some indigenous peoples concerned about the bad health effects of modern foods are seeking to improve their diets by returning to traditional crops and plants harvested from the wild. This provides an impetus for exploring methods of tending plants in the wild, such as berries, wild rice, and other foods. It also enhances the role of elders needed to train the youth in proper harvesting techniques (MacPherson 2009).

When knowledge has apparently been lost in one area, it may have been retained in others. A case occurred in the early twenty-first century in Austria when a windstorm destroyed a forest that had protected a very busy highway

---

[3]CBD Traditional Knowledge Information Portal: http://www.cbd.int/tk/about.shtml/

down-slope. Because of the topographic situation and the heavy road traffic, the timber could not be removed by cable logging. As there were only a few loggers in Austria able to bring the logs down from the hillside by gravity, the forest administration asked Romanian loggers to carry out this task. Their predecessors (farmers and loggers) had been Austrians who were expelled from their home in the Austrian mountains in the eighteenth century due to their religious beliefs (as Protestants). On the advice of the Austrian Emperor they were resettled in Romania. After more than 200 years members of these communities maintained not only their culture but also the traditional knowledge related to logging and timber transport on steep slopes.

The Indian government's desire to strengthen intellectual property protection and prevent the patenting of plants traditionally used in Indian systems of medicine (i.e., Ayurveda, Unani and Siddha) led to a large project to document the uses of these plant species.[4] At the local level, 'people's biodiversity registers'—consisting of records of individual people's knowledge of biodiversity, its use, trade, and efforts for its conservation and sustainable utilization—have been established. These registers are recognized in the Indian Biological Diversity Bill of 2000. These developments have led to both recovery and preservation of traditional knowledge in India.

15.5   The Role of the Scientific Community

The scientific community has shown increasing interest in traditional knowledge, evidence of which is presented below. Mutually beneficial and ethically acceptable collaboration between holders of TFRK and scientists, however, has to address three main sources of power imbalance between scientists and holders of TFRK: differences in intellectual prestige, differences in access to funding, and differences in access to political power. The second part of this section addresses these power imbalances, followed by a discussion of how collaboration among scientists and holders of traditional forest-related knowledge can proceed by taking into account these power imbalances.

15.5.1   Scientific Interest

There is a long history of interest in traditional knowledge within social science disciplines such as anthropology, and among biological scientists working in the field of ethnobotany. However, research within the biophysical sciences such as ecology,

---

[4]See Traditional Knowledge Digital Library. Available online at: http://www.tkdl.res.in/tkdl/langdefault/common/home.asp?GL=Eng. Cited 10 April 2011.

forestry, and agriculture is relatively recent, mostly dating from the 1970s. Although many scientists over the years have been indifferent to traditional knowledge, or regarded it as somewhat inferior to the knowledge generated by 'modern' science, some increase in recognition has occurred recently. For example, recent publications in ecology, such as the two editions of Berkes's *Sacred Ecology* (2008) and a special issue of the journal *Ecological Applications* in 2000 (Ford and Martinez 2000) show increasing respect for traditional knowledge among a growing number of scientists.

Within the global forest science community, traditional forest-related knowledge has been increasingly addressed by the International Union of Forest Research Organizations (IUFRO). IUFRO World Congresses since 1985 onwards have included TFRK-related sessions, particularly those organized by IUFRO units that focus on social and policy science. The IUFRO Research Group Forest and Woodland history, which has been active for decades, has increasingly focussed on traditional forest-related knowledge. At the 2010 IUFRO World Congress 2010 held in Seoul, two Working Parties on traditional knowledge were established within this Research Group, building on its close collaboration with the IUFRO Task Force on Traditional Forest Knowledge[5] between 2005 and 2010 (Fig. 15.1).

As Berkes points out (2008, 13–15), philosophical and political differences become intertwined, with the political differences at times the more important. The holders of TFRK are usually politically weak and the scientists strong. Thus a scientist can assert that scientific knowledge justifies the scientist speaking "for nature". This contrasts with local people, who claim that their intimate knowledge of their landscape gives them the justification to speak for it.

15.5.2   Differences in Power

The use of traditional forest-related knowledge in decision-making processes should be based on a partnership in which TFRK owners, scientists and policy makers create and share knowledge (Berkes 2004; Colfer et al. 2005). However, there is considerable potential for misappropriation of such knowledge and its eventual loss by the original holders (ICSU 2002). The issue may be framed as the scientists' responsibility, vis-à-vis their acknowledged interests, values and aspirations. Alternatively, it may be seen as a matter of scientists as employees or persons allied with powerful groups which are interested in resources or knowledge controlled by traditional knowledge holders. How much of the responsibility to behave properly resides with the researchers and how much with the community holding the traditional knowledge? What does the responsibility consist of and which methods should be applied to study TFRK in a mutually fruitful, yet ethically acceptable, manner? Agreements (oral or signed) should cover the scope of research, methods and tools that can be used, rules for meeting and talking with

---

[5]http://www.iufro.org/science/task-forces/former-task-forces/traditional-forest-knowledge/

participants, reporting research results back with those who shared, and payment for their assistance and services, possible benefit-sharing, and acknowledgement in publications and other forms of dissemination.

Prominent and recent examples of relevance for the study of TFRK include the Convention on Biological Diversity's "Nagoya Protocol on Access to Genetic Resources and the Fair and Equitable Sharing of Benefits Arising from their Utilization" (SCBD 2011), which includes traditional knowledge associated with genetic resources that is held by indigenous and local communities and the CGIAR-led initiative to identify key research ethics and IPR challenges (CGIAR 2009). Requirements that forest concessionaires and others utilizing indigenous lands undertake significant consultation and accommodation of the interests of the knowledge holders has been a principle put forward and followed to some degree. In Russia, for instance, a special law was passed requiring forestry companies to involve the community and scientists in the planning of their timber harvests destined for export (Forest Club 2011). In Canada, the courts are asking that the Provinces significantly consult with indigenous peoples when their aboriginal rights are threatened, and to accommodate those rights by modifying planned activities (Christie 2006).

While scientists are not necessarily in control of decision-making regarding development of forested lands, they have more control over curriculum and programmes within universities. Cooperation is developing both in research and in course content. At the international level, for example, United Nations University–Institute of Advanced Studies' Traditional Knowledge Initiative supports research activities, policy studies, and capacity development related to traditional forest-related knowledge. In Italy, the Engineering and Forestry Faculties at the University of Florence have created a centre for the innovation of local and indigenous systems of knowledge that explores the usefulness of traditional knowledge. In Canada, the Natural Resources Institute at the University of Manitoba has a successful research relationship with the elders of the Pikangikum First Nation. The Faculty of Forestry at the University of British Columbia is in its second strategic plan for its Aboriginal Initiative, which includes both research and the development of a specialization in Community and Aboriginal Forestry. The Faculty also works with the Haida Gwaii Semester in Natural Resource Studies in providing university credit for the courses that are taught at Haida Gwaii.[6] The University of Northern British Columbia and the Tlazt'en First Nation jointly manage the John Prince Research Forest near St. Johns, British Columbia (Grainger et al. 2006).

15.5.3 Collaboration

Collaboration is not an easy task when differences in relative power and difference in world views need to be accommodated. If ethical and methodological questions are not addressed in a consistent and systematic manner from the outset of the

---

[6] www.haidagwaiisemester.com

research, the rights of TFRK owners may well be infringed, meaning that benefits will not accrue to the owners, and that access to resources (e.g., genetic resources) may be suddenly curtailed. Thus, all parties must address the challenges raised by the maintenance, use, and protection of TFRK when there is interaction between the holders and users of such knowledge.

Our chapter on research ethics pointed out that communities are more and more insisting that research be conducted under conditions controlled by written protocols and the free informed consent of the TFRK holders who are involved. In particular that includes free and prior informed consent (FPIC) as well as access and benefit-sharing and active traditional knowledge protection efforts. It is also worth noting that successful efforts have been made to facilitate greater engagement with forest peoples by describing a research methodology, considering so-called *etic* and *emic* approaches for the study of traditional forest-related knowledge.[7] A key point is that careful consideration of the cultural context is of utmost importance when studying TFRK, and when choosing methods to do so, because insiders and outsiders ask different questions (Bishop 2005).

The principles governing ethical research can be applied also to collaboration in forest management. Factors that support successful collaboration are agreement on the goals of management (especially when everyone is concerned with sustainability), respect for cultural and disciplinary differences, wide participation in determining research questions, and sharing of power in decision-making. As was pointed out in the chapter on research ethics, these principles can be interpreted in various ways, so both researchers and the community bear responsibility and should take a strong interest in considering and abiding by these principles. However, an important challenge remains to clarify how such guidelines are best enforced.

15.6   Conclusion

In this era of increased concern about sustainability and sustainable management of forest ecosystems, traditional forest related knowledge has an increasingly important role. In this book, we have documented how people around the world have maintained their connections to their lands and cultures. Although many challenges exist, as have been described in the chapters, scientists are recognizing that traditional knowledge presents more than information to calibrate their own models; traditional forest-related knowledge offers ways to understand based on world views that have persisted precisely because the knowledge (including its assumptions, facts, and values) is valid. In struggling to maintain their way of life and identity,

---

[7]*Etic* (not to be confused with ethic) implies an outside perspective on the culture or system, using prior (and in this context, scientific) knowledge and one's own (here: research) vocabulary to describe and understand in a neutral or objective way. *Emic* means that the researcher looks at a culture or system from the inside in the sense that she or he assembles a description that can be recognized and verified by those within the system who hold the knowledge.

Fig. 15.1 Participants of the lead authors' meeting held at the headquarters of the International Union of Forest Research Organizations at Mariabrunn, Vienna, Austria in November 2010. *Standing* (*left to right*): Jésus García Latorre, Miguel A. Pinedo-Vasquez, Elisabeth Johann, Ronald Trosper, Vladimir Bocharnikov, Renate Prüller (IUFRO headquarters), P. S. Ramakrishnan, Mónica Gabay. *Front row* (*left to right*): Andrey Laletin, Mauro Agnoletti, John Parrotta, Lim Hin Fui; *Missing from photo:* Suzanne A. Feary, Christian Gamborg, Alfred Oteng-Yeboah, Youn Yeo-Chang

traditional knowledge holders have engaged in political processes at local, national and international levels.

The great diversity of local systems precludes strong generalizations about the contributions of traditional forest-related knowledge. This concluding chapter highlighted biodiversity conservation and shifting cultivation as two areas of significant contributions. We can attempt the generalization that the relationship between TFRK and 'formal' science has been hampered by both the philosophical underpinnings of Western science and the historical alliance between national governments and scientists. Seen as key to national economic growth, forest science in support of forest management to maximize timber production or, more recently, biodiversity conservation, used to ignore or even oppose analysis based on traditional knowledge. But experience has demonstrated the wisdom of much of traditional knowledge, leading to a shift of the relationship between traditional forest-related knowledge and other forms of knowledge.

As a result, scientists are increasingly interested in traditional forest-related knowledge. Models have emerged to guide reasonable collaboration among scientists and traditional knowledge holders. Such collaboration is also valuable for involving all local people. Many of these examples are explained in our chapters. While much remains to be done, this book demonstrates much has been accomplished.

References

Agrawal A (2002) Indigenous knowledge and the politics of classification. Int Soc Sci J 173:287–297

Alcorn JB, Toledo VM (1998) Resilient resource management in Mexico's forest ecosystems: the contribution of property rights. In: Berkes F, Folke C (eds) Linking social and ecological systems: management practices and social mechanisms for building resilience. Cambridge University Press, Cambridge, pp 216–249

Altai Regional Institute of Ecology (2007) Creation of databases of traditional knowledge in environmental management of the indigenous population (Altai-Kizhi), Altai Republic (in Russian). Available via http://ethnography.omskreg.ru/res/page000000001202/Files/2.pdf. Cited 23 Mar 2011

Belair C, Ichikawa K, Wong BYL, Molongoy KJ (eds) (2010) Sustainable use of biological diversity in socio-ecological production landscapes. Background to the 'Satoyama initiative for the benefit of biodiversity and human well-being', vol 52, Technical Series. Secretariat of the Convention on Biological Diversity, Montreal

Berkes F (2004) Rethinking community-based conservation. Conserv Biol 18(3):621–630

Berkes F (2008) Sacred ecology, 2nd edn. Routledge, New York

Berkes F, Colding J, Folke C (2000) Rediscovery of traditional ecological knowledge as adaptive management. Ecol Appl 10:1251–1262

Bishop R (2005) Freeing ourselves from neo-colonial domination in research: a Kaupapa Māori approach to creating knowledge. In: Denzin NK, Lincoln YS (eds) The SAGE handbook of qualitative research, 3rd edn. Sage Publications, Thousand Oaks, pp 109–138

Cairns M (2007) Voices from the forest: integrating indigenous knowledge into sustainable farming. Resources for the Future, Washington, DC

Campos P, Díaz M, Pulido FJ (1998) Las dehesas arboladas: un equilibrio necesario entre explotación y conservación. Quercus 147:31–35

CGIAR, see Consultative Group on International Agricultural Research

Christie G (2006) Developing case law: the future of consultation and accommodation. U B C Law Rev 39(1):139–184

Colfer CJP, Colchester M, Joshi L, Puri RK, Nygren A, Lopez C (2005) Traditional knowledge and human well-being in the 21st century. In: Mery G, Alfaro R, Kanninen M, Lovobikov M (eds) Forests in the global balance—changing paradigms, vol 17, IUFRO World Series. International Union of Forest Research Organizations [IUFRO], Helsinki, pp 173–182

Consultative Group on International Agricultural Research [CGIAR] (2009) CGIAR guidelines on the acquisition and use of traditional knowledge. Unpublished manuscript, draft 2009. Available via http://cgiar.org/pdf/grpc_24th_meeting_minutes.pdf. Cited 15 Jul 2009

Davidson-Hunt I, Berkes F (2003) Learning as you journey: anishinaabe perception of social-ecological environments and adaptive learning. Conserv Ecol 8:5. Available via http://www.consecol.org/vol8/iss1/art5/. Cited 25 Mar 2011

Díaz M, Campos P, Pulido FJ (1997) The Spanish dehesas: a diversity in land-use and wildlife. In: Pain DJ, Pienkowski MW (eds) Farming and birds in Europe. Academic, San Diego, pp 178–209

Escobar A (1995) Encountering development: the making and unmaking of the third world, Princeton studies in culture/power/history. Princeton University Press, Princeton

Escobar A (2005) Economics and the space of modernity: tales of market, production and labour. Cult Stud 19(2):139–175

Fiske AP (1991) Structures of social life: the four elementary forms of human relations. The Free Press, New York

Ford J, Martinez D (2000) Traditional ecological knowledge, ecosystem science, and environmental management. Ecol Appl 10:1249–1250

Forest Club (2011) Forest certification—the tool for sustainable forestry? Available via http://www.forest.ru/eng/sustainable_forestry/certification/. Cited 23 Mar 2011

Fox JM (2000) How blaming 'slash and burn' farmers is deforesting mainland Southeast Asia. Analysis from the East-West Center, vol 47. East-West Center, Honolulu, Available via http://scholarspace.manoa.hawaii.edu/bitstream/handle/10125/3832/api047.pdf?sequence=1. Cited 25 March 2011

Galloway-McLean K (2010) Advance guard: climate change impacts, adaptation, mitigation and indigenous peoples—a compendium of case studies. United Nations University-Traditional Knowledge Initiative, Darwin, Australia. Available via http://www.unutki.org/news.php?doc_id=101&news_id=92. Cited 20 Mar 2011

García Latorre J, García Latorre J (2007) Almería: hecha a mano. Una historia ecológica. Cajamar, Almería

Grainger S, Sherry E, Fondahl G (2006) The John Prince research forest: evolution of a co-management partnership in northern British Columbia. For Chron 82(4):484–495

Gudeman S (2008) Economy's tension: the dialectics of community and market. Berghahn Books, New York

Henrich JP, Boyd R, Bowles S, Camerer C, Fehr E, Gintis H (eds) (2004) Foundations of human sociality: economic experiments and ethnographic evidence from fifteen small-scale societies. Oxford University Press, Oxford

Indian National Science Academy, Zhongguo ke xue yuan, National Academy of Sciences (2001) Growing populations, changing landscapes: studies from India, China, and the United States. National Academy Press, Washington, DC

International Council for Science [ICSU] (2002) Science, traditional knowledge and sustainable development, vol 4, ICSU series on science for sustainable development. ICSU, Paris, http://www.icsu.org/Gestion/img/ICSU_DOC_DOWNLOAD/65_DD_FILE_Vol4.pdf. Cited 11 February 2011

Kolm S, Mercier YJ (2006) Handbook of the economics of giving, altruism and reciprocity, 1st edn, Handbooks in economics. Elsevier, Amsterdam

Lake FK (2007) Traditional ecological knowledge to develop and maintain fire regimes in north-western California, Klamath-Siskiyou bioregion: management and restoration of culturally significant habitats. Ph.D. dissertation, Oregon State University, Corvallis

Liang L, Shen L, Yang W, Yang X, Shang Y (2009) Building on traditional shifting cultivation for rotational agroforestry: experiences from Yunnan, China. For Ecol Manag 257(10):1989–1994

Mabee HS, Tindall D, Hoberg G, Gladu JP (2010) Co-management of forest lands: the cases of Clayoquot sound and Gwaii Haanas. In: Tindall D, Trosper R, Perreault P (eds) First nations and forest lands in British Columbia and Canada. Under review by University of British Columbia Press, Vancouver, pp 245–261, Chapter 16

MacPherson NE (2009) Traditional knowledge for health. Master's thesis, University of British Columbia/Siska Traditions Society, Vancouver

Michon G, de Foresta H, Levang P, Verdeaux F (2007) Domestic forests: a new paradigm for integrating local communities' forestry into tropical forest science. Ecol Soc 12(2). Available via http://www.ecologyandsociety.org/vol12/iss2/art1/. Cited 25 Mar 2011

Muñoz-Pulido R (1989) Ecología invernal de la grulla en España. Quercus 45:10–21

Nadasdy P (2003) Hunters and bureaucrats: power, knowledge, and aboriginal-state relations in the southwest Yukon. University of British Columbia Press, Vancouver

NEPED, IIRR (1999) Building upon traditional agriculture in Nagaland. Nagaland Environmental Protection and Economic Development/International Institute of Rural Reconstruction, Nagaland/Silang

Nepstad D, Schwartzman S, Bamberger B, Santilli M, Ray D, Schlesinger P, Lefebvre P, Alencar A, Prinz E, Fiske G, Rolla A (2006) Inhibition of Amazon deforestation and fire by parks and indigenous lands. Conserv Biol 20(1):65–73. doi:10.1111/j.1523-1739.2006.00351.x

Norgaard RB (2010) Ecosystem services: from eye-opening metaphor to complexity blinder. Ecol Econ 69(6):1219–1227

Palm CA, Sanchez PA, Ericksen PJ, Vosti SA (eds) (2005) Slash-and-burn agriculture: the search for alternatives. Columbia University Press, New York

Parry ML, Canziani OF, Palutikof JP, van der Linden PJ, Hanson CE (eds) (2007) Contribution of working group II to the fourth assessment report of the Intergovernmental Panel on Climate Change. Cambridge University Press, Cambridge

Polanyi K (2001) The great transformation: the political and economic origins of our time, 2nd Beacon Paperback edn. Beacon Press, Boston

Pulido FJ, Díaz M (1997) Linking individual foraging behaviour and population spatial distribution in patchy environments: a field example with Mediterranean blue tits. Oecologia 111(434):442

Ramakrishnan PS (1992) Shifting agriculture and sustainable development: an interdisciplinary study from north-eastern India, vol 10, Man and biosphere book series. UNESCO/Parthenon Publishing, Paris/Caernforth/Lancaster

Ramakrishnan PS, Saxena KG, Rao KS (2006) Shifting agriculture and sustainable development of north-east India: tradition in transition. UNESCO/Oxford & IBH, New Delhi

Reinao RM (2008) The Mapuche and climate change in the Chilean neoliberal economic system. Indig Aff 1–2(08):66–71, http://www.iwgia.org/graphics/Synkron-Library/Documents/publications/Downloadpublications/IndigenousAffairs/IA%201-2_08%20Climate%20Change%20and%20IPs.pdf Accessed 23 March 2011

Sahlins M (1972) Stone age economics. Aldine, New York

Sahlins M (1996) The sadness of sweetness: the native anthropology of western cosmology. Curr Anthropol 37(3):395–428

Salick J, Byg A (eds) (2007) Indigenous peoples and climate change. Tyndall Centre for Climate Change Research, Oxford. Available via http://www.bvsde.paho.org/bvsacd/cd68/Indigenouspeoples.pdf. Cited 22 Mar 2011

Secretariat of the Convention on Biological Diversity [SCBD] (2011) Nagoya protocol on access to genetic resources and the fair and equitable sharing of benefits arising from their utilization to the convention on biological diversity: text and annex. SCBD, Montreal. Available via http://www.cbd.int/abs/doc/protocol/nagoya-protocol-en.pdf. Cited 24 Mar 2011

Shutkin WA (2000) The land that could be: environmentalism and democracy in the twenty-first century, Urban and industrial environments. MIT Press, Cambridge

Sieferle RP (2001) The subterranean forest. Energy systems and the industrial revolution. The White Horse Press, Cambridge

Swift MJ, Vandermeer J, Ramakrishnan PS, Anderson JM, Ong CK, Hawkins B (1996) Biodiversity and agroecosystem function. In: Mooney HA, Cushman JH, Medina E, Sala OE, Schulze E-D (eds) Functional roles of biodiversity: a global perspective. SCOPE Series, Chichester, pp 261–298

Tanner A (1979) Bringing home the animals: religious ideology and mode of production of Mistassini Cree hunters. Hurst, London

Tello E, Garrabou R, Cussò X (2006) Energy balance and land use: the making of an agrarian landscape from the vantage point of social metabolism (the Catalan Valles County in 1860/1870). In: Agnoletti M (ed) The conservation of cultural landscapes. CAB International, Wallingford, pp 42–56

Timko JA, Innes JL (2009) Evaluating ecological integrity in national parks: case studies from Canada and South Africa. Biol Conserv 142(3):676–688. doi:10.1016/j.biocon.2008.11.022

Timko JA, Satterfield T (2008) Seeking social equity in national parks: experiments with evaluation in Canada and South Africa. Conserv Soc 6(3):238–254. doi:10.4103/0972-4923.49216

Toledo VM, Ortiz-Espejel B, Moguel P, Ordoñez MDJ (2007) The multiple use of tropical forests by indigenous peoples in Mexico: a case of adaptive management. Conserv Ecol 7(3):9, Available via http://www.consecol.org/vol7/iss3/art9/. Cited 25 March 2011

Trosper RL (2007) Indigenous influence on forest management on the Menominee Indian Reservation. For Ecol Manag 249:134–139

United Nations [UN] (2004) Traditional forest-related knowledge: report of the Secretary-General, United Nations forum on forests 4th session. Document E/CN.18/2004/7. Available via http://daccess-dds-ny.un.org/doc/UNDOC/GEN/N04/261/74/PDF/N0426174.pdf?OpenElement. Cited 5 Mar 2011

United Nations Development Programme [UNDP] (2011) UNDP Focus Areas, Russian Federation, Biodiversity Conservation in the Russian Portion of the Altai-Sayan Ecoregion. Available via http://www.undp.ru/index.php?iso=RU&lid=1&cmd=programs&id=8. Cited 23 Mar 2011

Wa G, Uukw D (1992) The spirit in the land: statements of the Gitksan and Wet'suwet'en hereditary chiefs in the supreme court of British Columbia, 1987–1990. Reflections, Gabriola

Author Index

Detailed Contents

J.A. Parrotta and R.L. Trosper (eds.), *Traditional Forest-Related Knowledge:*   613
Sustaining Communities, Ecosystems and Biocultural Diversity,
World Forests 12, DOI 10.1007/978-94-007-2144-9,
© Springer Science+Business Media B.V. (outside the USA) 2012

Chapter 2 Africa

Chapter 3 Latin America—Argentina, Bolivia, and Chile

Chapter 9   South Asia

Chapter 10   Southeast Asia

Chapter 11　Western Pacific

Chapter 14   Ethics and Research Methodologies for the Study of Traditional Forest-Related Knowledge

Chapter 15   The Unique Character of Traditional Forest-Related Knowledge: Threats and Challenges Ahead